Lecture Notes in Computer Science   13453

Advanced Research in Computing and Software Science

Subline of Lecture Notes in Computer Science

More information about this series at https://link.springer.com/bookseries/558

Michael A. Bekos · Michael Kaufmann (Eds.)

Graph-Theoretic Concepts in Computer Science

48th International Workshop, WG 2022
Tübingen, Germany, June 22–24, 2022
Revised Selected Papers

 Springer

Editors
Michael A. Bekos (iD)
University of Ioannina
Ioannina, Greece

Michael Kaufmann (iD)
Universität Tübingen
Tübingen, Germany

ISSN 0302-9743   ISSN 1611-3349   (electronic)
Lecture Notes in Computer Science
ISBN 978-3-031-15913-8   ISBN 978-3-031-15914-5   (eBook)
https://doi.org/10.1007/978-3-031-15914-5

This Springer imprint is published by the registered company Springer Nature Switzerland AG
The registered company address is: Gewerbestrasse 11, 6330 Cham, Switzerland

Preface

This volume contains the 32 papers presented at the 48th International Workshop on Graph-Theoretic Concepts in Computer Science (WG 2022). The workshop was held in Tübingen, Germany, from June 22 to 24, 2022 with a reception on the evening of June 21. A total of 42 participants attended the workshop in person, with a further 130 registered participants online, although most talks were attended by 20–30 remote participants on average.

WG has a longstanding tradition. Since 1975, WG has taken place 24 times in Germany, five times in The Netherlands, three times in France, twice in Austria, the Czech Republic, and the UK, and once in Greece, Israel, Italy, Norway, Poland, Slovakia, Spain, Switzerland, and Turkey. This was the 25th time the workshop was held in Germany.

WG aims to merge theory and practice by demonstrating how concepts from graph theory can be applied to various areas in computer science, or by extracting new graph theoretic problems from applications. The goal is to present emerging research results and to identify and explore directions of future research. The conference is well-balanced with respect to established researchers and junior scientists.

We received 103 submissions, seven of which were withdrawn before entering the review process. The Program Committee (PC) provided three independent reviews for each submission. The PC accepted 32 papers – an acceptance ratio of exactly 1/3. As in previous years, due to strong competition and limited space there were papers that were not accepted although they deserved to be.

The prize for the Best Paper at WG 2022 was awarded to Csaba Tóth for his paper "Minimum Weight Euclidean $(1 + \varepsilon)$-Spanners". The prize for the Best Student Paper at WG 2022 was awarded to David J. C. Dekker for his paper "Kernelization for Feedback Vertex Set via Elimination Distance to a Forest", coauthored by Bart M. P. Jansen. The program included two inspiring invited talks, by Bettina Speckmann (TU Eindhoven, The Netherlands) on "Maps, Matrices, and Rugs: Algorithms for Rectangular Visualizations" and by Torsten Ueckerdt (Karlsruher Institut für Technologie, Germany) on "Stack and Queue Layouts of Planar Graphs".

Moreover, many individuals contributed to the success of WG 2020. In particular our thanks go to

- All authors who submitted their newest research results to WG.
- The reviewers whose expertise supported the selection process.
- The members of the PC, who graciously gave their time and energy.
- All the members of the Organizing Committee based at the University of Tübingen: Henry Förster, Renate Hallmayer, Julia Katheder, Axel Kuckuk, Maximilian Pfister, and Lena Schlipf.
- The EasyChair system for hosting the evaluation process.
- Springer for supporting the Best Paper Awards.
- Our sponsors: DFG and yWorks.

- The invited speakers, all presenters, the session chairs, and the participants for their contributions and support to make WG 2022 an inspiring event.

July 2022

Michael A. Bekos
Michael Kaufmann

Organization

Program Committee

Cristina Bazgan	Université Paris Dauphine-PSL, France
Michael Bekos (Chair)	University of Ioannina, Greece
Fedor Fomin	University of Bergen, Norway
Cyril Gavoille	University of Bordeaux, France
Carla Groenland	Utrecht University, The Netherlands
Michael Hoffmann	ETH Zurich, Switzerland
Michael Kaufmann (Chair)	Universität Tübingen, Germany
Philipp Kindermann	Universität Trier, Germany
Linda Kleist	TU Braunschweig, Germany
Tereza Klimošová	Charles University, Czech Republic
Piotr Micek	Jagiellonian University, Poland
Martin Milanič	University of Primorska, Slovenia
Debajyoti Mondal	University of Saskatchewan, Canada
Fabrizio Montecchiani	University of Perugia, Italy
Rolf Niedermeier	TU Berlin, Germany
Yota Otachi	Nagoya University, Japan
Sang-il Oum	Institute for Basic Science (IBS) and KAIST, South Korea
Charis Papadopoulos	University of Ioannina, Greece
Dömötör Pálvölgyi	ELTE, Hungary
Dieter Rautenbach	University of Ulm, Germany
Bernard Ries	Université de Fribourg, Switzerland
Ignasi Sau	Université de Montpellier, CNRS, France
Lena Schlipf	Universität Tübingen, Germany
Melanie Schmidt	Universität zu Köln, Germany
Dimitrios Thilikos	LIRMM, Université de Montpellier, CNRS, France
Meirav Zehavi	Ben-Gurion University, Israel

Additional Reviewers

Abu-Khzam, Faisal	Bhattacharya, Anup
Andres, Stephan Dominique	Blum, Johannes
Araujo, Julio	Bläsius, Thomas
Barbay, Jérémy	Bonnet, Édouard
Belmonte, Rémy	Bonomo, Flavia
Bentert, Matthias	Bose, Prosenjit
Bergougnoux, Benjamin	Bougeret, Marin
Bergé, Pierre	Brand, Cornelius

Brandes, Ulrik
Bressan, Marco
Buchin, Kevin
Bérczi-Kovács, Erika Renáta
Casel, Katrin
Chaudhary, Juhi
Chiarelli, Nina
Coulombe, Michael
Cseh, Ágnes
Curticapean, Radu
Dabrowski, Konrad K.
Dallard, Clément
de Kroon, Jari J. H.
de Lima, Paloma
Deligkas, Argyrios
Di Giacomo, Emilio
Debski, Michał
Eiben, Eduard
Ekim, Tinaz
Fernau, Henning
Fiala, Jiri
Fleszar, Krzysztof
Fluschnik, Till
Froese, Vincent
Fujita, Shinya
Fulek, Radoslav
Fuzy, Eric
Förster, Henry
Geniet, Colin
Giannopoulos, Panos
Golovach, Petr
Grelier, Nicolas
Gronemann, Martin
Gupta, Siddharth
Gurjar, Rohit
Gurski, Frank
Hamm, Thekla
Harutyunyan, Ararat
Hatzel, Meike
Heeger, Klaus
Hegerfeld, Falko
Hoang, Hung
Hocquard, Hervé
Hodor, Jedrzej
Izbicki, Mike
Jacob, Ashwin

Jacob, Hugo
Jaffke, Lars
Jartoux, Bruno
Johnston, Tom
Kaiser, Tomas
Katzmann, Maximilian
Kavitha, Telikepalli
Keldenich, Phillip
Keszegh, Balázs
Klemz, Boris
Klobas, Nina
Knop, Dušan
Kobayashi, Yasuaki
Kobayashi, Yusuke
Konstantinidis, Athanasios
Korhonen, Tuukka
Kosinas, Evangelos
Krauthgamer, Robert
Krnc, Matjaž
Krupke, Dominik
Kuckuk, Axel
Kunz, Pascal
Kuszmaul, William
Kwon, O-Joung
Köhler, Noleen
Lafond, Manuel
Langlois, Hélène
Le, Hung
Lee, Euiwoong
Lendl, Stefan
Liedloff, Mathieu
Macajova, Edita
Maniatis, Spyridon
Manlove, David
Mann, Felix
Masařík, Tomáš
McCarty, Rose
Mertzios, George
Mestre, Julian
Milovanov, Alexey
Miltzow, Till
Misra, Neeldhara
Molter, Hendrik
Munaro, Andrea
Naia, Tassio
Nederlof, Jesper

Nelles, Florian
Neuen, Daniel
Nguyen, Kim Thang
Novick, Beth
Obdrzalek, Jan
Ordyniak, Sebastian
Pandey, Sukanya
Panolan, Fahad
Paulusma, Daniël
Pedrosa, Lehilton L. C.
Pfister, Maximilian
Picouleau, Christophe
Pilipczuk, Marcin
Protopapas, Evangelos
Reddy, Meghana M.
Renault, David
Rieck, Christian
Rollin, Jonathan
Roy, Sanjukta
Roy, Shivesh K.
Rzążewski, Paweł
Sagunov, Danil
Sampaio, Rudini
Sandeep, R. B.
Schindl, David
Schmidt, Daniel
Schulz, André
Simonov, Kirill
Skoviera, Martin
Smid, Michiel
Sommer, Frank
Souza, Uéverton
Staals, Frank
Steiner, Raphael

Stojakovic, Milos
Štorgel, Kenny
Stumpf, Peter
Suchy, Ondrej
Surianarayanan, Vaishali
Suzuki, Akira
Swennenhuis, Céline
Sylvester, John
Szilagyi, Krisztina
T. P., Sandhya
Takaoka, Asahi
Takazawa, Kenjiro
Tale, Prafullkumar
Tan, Jane
Tappini, Alessandra
Telle, Jan Arne
Tewari, Raghunath
Togni, Olivier
Tomescu, Alexandru I.
Tzimas, Spyridon
Uehara, Ryuhei
Vaish, Rohit
van der Zanden, Tom
van Leeuwen, Erik Jan
Walczak, Bartosz
Wasa, Kunihiro
Watrigant, Rémi
Wicke, Kristina
Wlodarczyk, Michal
Xiao, Mingyu
Zeitoun, Marc
Zeman, Peter
Zschoche, Philipp

The Long Tradition of WG

WG 1975	U. Pape – Berlin, Germany
WG 1976	H. Noltemeier – Göttingen, Germany
WG 1977	J. Mühlbacher – Linz, Austria
WG 1978	M. Nagl, H. J. Schneider – Burg Feuerstein, near Erlangen, Germany
WG 1979	U. Pape – Berlin, Germany
WG 1980	H. Noltemeier – Bad Honnef, Germany
WG 1981	J. Mühlbacher – Linz, Austria
WG 1982	H. J. Schneider, H. Göttler – Neuenkirchen, near Erlangen, Germany
WG 1983	M. Nagl, J. Perl – Haus Ohrbeck near Onasbrück, Germany
WG 1984	U. Pape – Berlin, Germany
WG 1985	H. Noltemeier – Schloß Schwanberg near Würzburg, Germany
WG 1986	G. Tinhofer, G. Schmidt – Stift Bernried near Munich, Germany
WG 1987	H. Göttler, H. J. Schneider – Kloster Banz near Bamberg, Germany
WG 1988	J. van Leeuwen – Amsterdam, The Netherlands
WG 1989	M. Nagl – Castle Rolduc, The Netherlands
WG 1990	R. H. Möhring – Johannesstift Berlin, Germany
WG 1991	G. Schmidt, R. Berghammer – Fischbachau near Munich, Germany
WG 1992	E. W Mayr – Wilhelm-Kempf-Haus, Wiesbaden-Naurod, Germany
WG 1993	J. van Leeuwen – Utrecht, The Netherlands
WG 1994	G. Tinhofer, E. W. Mayr, G. Schmidt – Munich, Germany
WG 1995	M. Nagl – Haus Eich, Aachen, Germany
WG 1996	G. Ausiello, A. Marchetti-Spaccamela – Cadenabbia near Como, Italy
WG 1997	R. H. Möhring – Bildungszentrum am Müggelsee, Berlin, Germany
WG 1998	J. Hromkovič, O. Sýkora – Smolenice Castle, Slovakia
WG 1999	P. Widmayer – Monte Verità, Ascona, Switzerland
WG 2000	D. Wagner – Waldhaus Jakob, Konstanz, Germany
WG 2001	A. Brandstädt, Boltenhagen near Rostock, Germany
WG 2002	L. Kučera – Český Krumlov, Czech Republic
WG 2003	H. L. Bodlaender – Elspeet, The Netherlands
WG 2004	J. Hromkovič, M. Nagl – Bad Honnef, Germany
WG 2005	D. Kratsch – Île du Saulcy, Metz, France
WG 2006	F. V. Fomin – Sotra near Bergen, Norway
WG 2007	A. Brandstädt, D. Kratsch, H. Müller – Jena, Germany
WG 2008	H. Broersma, T. Erlebach – Durham, UK
WG 2009	C. Paul, M. Habib – Montpellier, France
WG 2010	D. M. Thilikos – Zarós, Crete, Greece
WG 2011	J. Kratochvíl – Teplá Monastery, West Bohemia, Czech Republic
WG 2012	M. C. Golumbic, G. Morgenstern, M. Stern, A. Levy – Israel
WG 2013	A. Brandstädt, K. Jansen, R. Reischuk – Lübeck, Germany
WG 2014	D. Kratsch, I. Todinca – Le Domaine de Chalès, Orléans, France
WG 2015	E. W. Mayr – Garching near Munich, Germany
WG 2016	P. Heggernes – Rumeli Hisarüstü, Istanbul, Turkey
WG 2017	H. L. Bodlaender, G. J. Woeginger – Eindhoven, The Netherlands
WG 2018	A. Brandstädt, E. Köhler, K. Meer – Cottbus, Germany

WG 2019	I. Sau, D. M. Thilikos – Vall de Núria, Catalunya, Spain
WG 2020	I. Adler, H. Müller – Leeds, UK (virtual)
WG 2021	Ł. Kowalik, M. Pilipczuk, P. Rzążewski – Warsaw, Poland
WG 2022	M. A. Bekos, M. Kaufmann – Tübingen, Germany

Contents

Minimal Roman Dominating Functions: Extensions and Enumeration

Faisal N. Abu-Khzam[1], Henning Fernau[2]([⊠]), and Kevin Mann[2]

[1] Department of Computer Science and Mathematics, Lebanese American University, Beirut, Lebanon
faisal.abukhzam@lau.edu.lb
[2] Fachbereich 4 – Abteilung Informatikwissenschaften, Universität Trier, 54286 Trier, Germany
{fernau,mann}@uni-trier.de

Abstract. Roman domination is one of the many variants of domination that keeps most of the complexity features of the classical domination problem. We prove that Roman domination behaves differently in two aspects: enumeration and extension. We develop non-trivial enumeration algorithms for minimal Roman dominating functions with polynomial delay and polynomial space. Recall that the existence of a similar enumeration result for minimal dominating sets is open for decades. Our result is based on a polynomial-time algorithm for EXTENSION ROMAN DOMINATION: Given a graph $G = (V, E)$ and a function $f : V \to \{0, 1, 2\}$, is there a minimal Roman dominating function $\tilde{f}$ with $f \le \tilde{f}$? Here, $\le$ lifts $0 < 1 < 2$ pointwise; minimality is understood in this order. Our enumeration algorithm is also analyzed from an input-sensitive viewpoint, leading to a run-time estimate of $\mathcal{O}(1.9332^n)$ for graphs of order n; this is complemented by a lower bound example of $\Omega(1.7441^n)$.

Keywords: Roman domination · Extension problems · Enumeration

1 Introduction

We combine four lines of research: (a) variations of *domination problems* in graphs, here Roman domination [17,21,28]; (b) *input-sensitive enumeration* of minimal solutions, a topic that has drawn attention also concerning domination problems [2,18,19,26,27]; (c) related to (and motivated by) enumeration, *extension problems* have been examined in particular in the context of domination problems[1] in [3,9,11,12,32,33,40]: is a given set a subset of any minimal dominating set?; (d) *output-sensitive enumeration*: the HITTING SET TRANSVERSAL PROBLEM is the question if all minimal hitting sets of a hypergraph can be enumerated with polynomial delay only: this question is open for four decades by

---

[1] Historically, a logical extension problem [10] should be mentioned, as it has led to [40, Théorème 2.16], dealing with an extension variant of 3-HITTING SET; also see [40, Proposition 3.39] concerning implications for EXTENSION DOMINATING SET.

© Springer Nature Switzerland AG 2022
M. A. Bekos and M. Kaufmann (Eds.): WG 2022, LNCS 13453, pp. 1–15, 2022.
https://doi.org/10.1007/978-3-031-15914-5_1

now and is equivalent to several enumeration problems in logic, database theory and also to enumerating minimal dominating sets in graphs, see [20,22,25,31]. By way of contrast, we show that all minimal Roman dominating functions can be enumerated with polynomial delay, a surprising result in view of the similarities between the complexities of domination and Roman domination problems.

ROMAN DOMINATION comes with a nice (hi)story: namely, it should reflect the idea of how to secure the Roman Empire by positioning armies on the various parts of the Empire in a way that either (1) a specific region r is also the location of at least one army or (2) one region r' neighboring r has two armies, so that r' can afford sending off one army to the region r (in case of an attack) without diminishing self-defense capabilities. For further details, we refer to [45,49].

ROMAN DOMINATION has received a lot of attention from the algorithmic community in the past 15 years [4,15,21,24,35,36,39,43,44,47]. Relevant to our paper is the development of exact algorithms for ROMAN DOMINATION: combining ideas from [35,46], an $\mathcal{O}(1.5014^n)$ exponential-time and -space algorithm (via a transformation to PARTIAL DOMINATING SET) was presented [48]. In [14,16,23,30,34,37,38,42,50–52], more combinatorial studies can be found. A larger chapter on Roman domination is contained in the monograph [29]. There is also an interesting link to the notion of a *differential* of a graph, introduced in [41], see [7], also adding further algorithmic thoughts, as expressed in [1,5,6]. For instance, in [5] a different exponential-time algorithm was published.

One of the ideas leading to the development of the area of *extension problems* (as described in [12]) was to cut branches of search trees as early as possible, in the following sense: to each node of the search tree, a so-called pre-solution U can be associated, and it is asked if it is possible to extend U to a meaningful solution S. In the case of DOMINATING SET, this means that U is a set of vertices and a 'meaningful solution' is an inclusion-wise minimal dominating set. Notice that such a strategy would work not only for computing smallest dominating sets, but also for computing largest minimal dominating set, or for counting minimal solutions, or for enumerating them. Alas, as it has been shown by many examples, extension problems turn out to be quite hard problems. In such a case, the approach might still be viable, as possibly parameterized algorithms exist with respect to the parameter 'pre-solution size'. This would be interesting, as this parameter is small when a big gain can be expected in terms of an early abort of a search tree branch. In particular for EXTENSION DOMINATING SET, this hope is not fulfilled. To the contrary, with this parameterization $|U|$, EXTENSION DOMINATING SET is one of the few problems known to be complete for the parameterized complexity class W[3], as shown in [8].

With an appropriate definition of the notion of minimality, ROMAN DOMINATION becomes one of the few examples where the hope seeing extension variants being efficiently solvable turns out to be true, as we will show in this paper. This is quite a surprising result, as in nearly any other way, ROMAN DOMINATION behaves most similar to DOMINATING SET. Together with its combinatorial foundations (a characterization of minimal Roman dominating functions), this constitutes the first main result of this paper. The main algorithmic exploit of this result is a non-trivial polynomial-space enumeration algorithm for minimal Roman dominating functions that guarantees polynomial delay only, which is the second main result of the paper. As mentioned above, the corresponding question for enumerating minimal dominating sets is open since decades, and we are not aware of any other modification of the concept of domination that seems to preserve any other of the difficulties of DOMINATING SET but the complexity of extension and enumeration. Our enumeration algorithm is a branching algorithm that we analyze with a simple Measure & Conquer approach, yielding a running time of $\mathcal{O}(1.9332^n)$, which also gives an upper bound on the number of minimal Roman dominating functions of an n-vertex graph. This result is complemented by a simple example that proves a lower bound of $\Omega(1.7441^n)$ for the number of minimal Roman dominating functions on graphs of order n.

Most proofs have been suppressed in this extended abstract; we refer to the long version for all omitted details.

2 Definitions

Let $\mathbb{N} = \{1, 2, 3, \dots\}$ be the set of positive integers. For $n \in \mathbb{N}$, let $[n] = \{m \in \mathbb{N} \mid m \le n\}$. We only consider undirected simple graphs. Let $G = (V, E)$ be a graph. For $U \subseteq V$, $G[U]$ denotes the graph induced by U. For $v \in V$, $N_G(v) := \{u \in V \mid \{u, v\} \in E\}$ denotes the *open neighborhood* of v, while $N_G[v] := N_G(v) \cup \{v\}$ is the *closed neighborhood* of v. We extend such set-valued functions $X : V \to 2^V$ to $X : 2^V \to 2^V$ by setting $X(U) = \bigcup_{u \in U} X(u)$. Subset $D \subseteq V$ is a *dominating set*, or ds for short, if $N_G[D] = V$. For $D \subseteq V$ and $v \in D$, define the *private neighborhood* of $v \in V$ with respect to D as $P_{G,D}(v) := N_G[v] \setminus N_G[D \setminus \{v\}]$. A function $f : V \to \{0, 1, 2\}$ is called a *Roman dominating function*, or rdf for short, if for each $v \in V$ with $f(v) = 0$, there exists a $u \in N_G(v)$ with $f(u) = 2$. To simplify the notation, we define $V_i(f) := \{v \in V \mid f(v) = i\}$ for $i \in \{0, 1, 2\}$. The *weight* w_f of a function $f : V \to \{0, 1, 2\}$ equals $|V_1(f)| + 2|V_2(f)|$. The classical ROMAN DOMINATION problem asks, given G and an integer k, if there exists an rdf for G of weight at most k. Connecting to the original motivation, G models a map of regions, and if the region vertex v belongs to V_i, then we place i armies on v.

For the definition of the problem EXTENSION ROMAN DOMINATION, we need to define the order $\le$ on $\{0, 1, 2\}^V$ first: for $f, g \in \{0, 1, 2\}^V$, let $f \le g$ if and only if $f(v) \le g(v)$ for all $v \in V$. We call a function $f \in \{0, 1, 2\}^V$ a *minimal Roman dominating function* if and only if f is a rdf and there exists no rdf g, $g \ne f$, with $g \le f$.[2] The weights of minimal rdf can vary considerably. Consider for example a star $K_{1,n}$ with center c. Then, $f_1(c) = 2$, $f_1(v) = 0$ otherwise; $f_2(v) = 1$ for all vertices v; $f_3(c) = 0$, $f_3(u) = 2$ for one $u \ne c$, $f_3(v) = 1$ otherwise, define three minimal rdf with weights $w_{f_1} = 2$, and $w_{f_2} = w_{f_3} = n + 1$.

---

[2] According to [29], this notion of minimality for rdf was coined by Cockayne but then dismissed, as it does not give a proper notion of *upper Roman domination*. However, in our context, this definition seems to be the most natural one, as it also perfectly fits the extension framework proposed in [13]; see the discussions in Sect. 7.

Problem name: EXTENSION ROMAN DOMINATION, or EXTRD for short
Given: A graph $G = (V, E)$ and a function $f \in \{0, 1, 2\}^V$.
Question: Is there a minimal rdf $\widetilde{f} \in \{0, 1, 2\}^V$ with $f \leq \widetilde{f}$?

As our first main result, we are going to show that EXTRD can be solved in polynomial time in Sect. 4. To this end, we need some understanding of the combinatorial nature of this problem, which we provide in Sect. 3.

The second problem that we consider is that of enumeration, both from an output-sensitive and from an input-sensitive perspective.

Problem name: ROMAN DOMINATION ENUMERATION, or RDENUM for short
Given: A graph $G = (V, E)$.
Task: Enumerate all minimal rdf $f \in \{0, 1, 2\}^V$ of G!

From an output-sensitive perspective, it is interesting to perform this enumeration without repetitions and with polynomial delay, which means that between the consecutive outputs of any two minimal rdf for $G = (V, E)$ that are enumerated, no more than $p(|V|)$ time elapses for some polynomial p, including the corner-cases at the beginning and at the end of the algorithm. From an input-sensitive perspective, we want to upper-bound the running time of the algorithm, measured against the order of the input graph. The obtained run-time bound should not be too far off from known lower bounds, given by graph families where known to have a certain number of minimal rdf. Our algorithm will be analyzed from both perspectives and achieves both goals; see Sects. 5 and 6.

3   Properties of Minimal Roman Dominating Functions

The combinatorial backbone of our algorithm is the following characterization.

Theorem 1. *Let $G = (V, E)$ be a graph, $f : V \to \{0, 1, 2\}$ and abbreviate $G' := G[V_0(f) \cup V_2(f)]$. Then, f is a minimal rdf if and only if the following conditions hold:*

1. $N_G[V_2(f)] \cap V_1(f) = \emptyset$,
2. $\forall v \in V_2(f) : P_{G', V_2(f)}(v) \not\subseteq \{v\}$, *also called* privacy condition, *and*
3. $V_2(f)$ *is a minimal dominating set of G'.*

Proof. To give a flavor of the proof of this crucial result, we show the (easier) "if-direction" in the following, assuming the "only-if-direction" was already proved.

Let f be a function that fulfills the three conditions. Since $V_2(f)$ is a dominating set in G', for each $u \in V_0(f)$, there exists a $v \in V_2(f) \cap N_G[u]$. Therefore, f is a rdf. Let $\widetilde{f} : V \to \{0, 1, 2\}$ be a minimal rdf with $\widetilde{f} \leq f$. Therefore, $\widetilde{f}$ (also) satisfies the three conditions by the "only-if-direction". Assume that there exists a $v \in V$ with $\widetilde{f}(v) < f(v)$. Hence, $V_2\left(\widetilde{f}\right) \subseteq V_2(f) \setminus \{v\}$.

Case 1: $\widetilde{f}(v) = 0, f(v) = 1$. Therefore, there exists a $u \in N_G(v)$ with $f(u) \geq \widetilde{f}(u) = 2$. This contradicts Condition 1.

Case 2: $\widetilde{f}(v) \in \{0, 1\}, f(v) = 2$. Consider any $u \in N_G(v)$ with $f(u) = 0$. This implies $\widetilde{f}(u) = 0$ and $\emptyset \neq N_G[u] \cap V_2\left(\widetilde{f}\right) \subseteq N_G[u] \cap V_2(f) \setminus \{v\}$. Therefore, u cannot be a private neighbor of v w.r.t. f. This contradicts Condition 2 for f.

Thus, $\widetilde{f} = f$ holds and f is minimal. □

We conclude this section with an upper bound on the size of $V_2(f)$.

Lemma 1. *Let $G = (V, E)$ be a graph and $f : V \to \{0, 1, 2\}$ be a minimal rdf. Then $2\,|V_2(f)| \leq |V|$ holds.*

Algorithm 1. Solving instances of ExtRD

1: **procedure** ExtRD SOLVER(G, f)
    Input: A graph $G = (V, E)$ and a function $f \colon V \to \{0, 1, 2\}$.
    Output: Is there a minimal Roman dominating function $\widetilde{f}$ with $f \leq \widetilde{f}$?
2:     $\widetilde{f} := f$.
3:     $M_2 := V_2(f)$. { Invariant: $M_2 = V_2(\widetilde{f})$ }
4:     $M := M_2$. { All $v \in V_2(\widetilde{f})$ are considered below; invariant: $M \subseteq M_2$. }
5:     **while** $M \neq \emptyset$ **do**
6:         Choose $v \in M$. { Hence, $\widetilde{f}(v) = 2$. }
7:         **for** $u \in N(v)$ **do**
8:             **if** $\widetilde{f}(u) = 1$ **then**
9:                 $\widetilde{f}(u) := 2$.
10:                 Add u to M and to M_2.
11:         Delete v from M.
12:     **for** $v \in M_2$ **do**
13:         **if** $N_G(v) \subseteq N_G[M_2 \setminus \{v\}]$ **then**
14:             **Return No.**
15:     **for** $v \in V \setminus N_G[M_2]$ **do**
16:         $\widetilde{f}(v) := 1$.
17:     **Return Yes.**

---

4   A Polynomial-Time Algorithm for ExtRD

By taking care of the conditions of Theorem 4, we can construct an algorithm that solves the problem EXTENSION ROMAN DOMINATION in polynomial time.

Theorem 2. *Let $G = (V, E)$ be a graph and $f \colon V \to \{0, 1, 2\}$. For the inputs G, f, Algorithm 1 returns yes if and only if (G, f) is a yes-instance of ExtRD. In this case, the function $\widetilde{f}$ computed by Algorithm 1 is a minimal rdf.*

Proposition 1. *Algorithm 1 runs in time cubic in the order of the input graph.*

Algorithm 2. A simple enumeration algorithm for minimal rdf

1: **procedure** RD ENUMERATION(G)
  Input: A graph $G = (V, E)$.
  Output: Enumeration of all minimal rdf $f : V \rightarrow \{0, 1, 2\}$.
2: **for** all functions $f : V \rightarrow \{1, 2\}$ **do**
3:   **for** all $v \in V$ with $f(v) = 1$ **do**
4:    **if** $\exists u \in N_G(v) : f(u) = 2$ **then**
5:     $f(v) := 0$.
6:   Build graph G' induced by $f^{-1}(\{0, 2\}) = V_0(f) \cup V_2(f)$.
7:   private-test $:= 1$.
8:   **for** all $v \in V$ with $f(v) = 2$ **do**
9:    **if** $P_{G', V_2(F)}(v) \subseteq \{v\}$ **then**
10:     private-test $:= 0$.
11:   **if** private-test $= 1$ and if $f^{-1}(2) = V_2(f)$ is a minimal ds of G' **then**
12:    Output the current function $f : V \rightarrow \{0, 1, 2\}$.

---

5 Enumerating Minimal RDF for General Graphs

For general graphs, our general combinatorial observations allow us to strengthen the (trivial) $\mathcal{O}^*(3^n)$-algorithm for enumerating all minimal rdf for graphs of order n down to $\mathcal{O}^*(2^n)$, as displayed in Algorithm 2. To understand the correctness of this enumeration algorithm, the following lemma is crucial.

Lemma 2. *Let $G = (V, E)$ be a graph with $V_2 \subseteq V$ such that $P_{G, V_2}(v) \nsubseteq \{v\}$ for each $v \in V_2$ holds. Then there exists exactly one minimal rdf $f \in \{0, 1, 2\}^V$ with $V_2 = V_2(f)$. Algorithm 1 can calculate f.*

Proposition 2. *Let $G = (V, E)$ be a graph. For minimal rdf $f, g \in \{0, 1, 2\}^V$ with $V_2(f) = V_2(g)$, it holds $f = g$.*

Hence, there is a bijection between the minimal rdf of a graph $G = (V, E)$ and subsets $V_2 \subseteq V$ that satisfy the condition of Lemma 2.

Proposition 3. *All minimal rdf of a graph of order n can be enumerated in time $\mathcal{O}^*(2^n)$.*

The presented algorithm clearly needs polynomial space only, but it is less clear if it has polynomial delay. Below, we will present a branching algorithm that has both of these desirable properties, and moreover, its running time is below 2^n. How good or bad such an enumeration is, clearly also depends on examples that provide a lower bound on the number of objects that are enumerated. The next lemma explains why the upper bounds for enumerating minimal rdf must be bigger than those for enumerating minimal dominating sets.

Lemma 3. *A disjoint collection of c cycles on five vertices yields a graph of order $n = 5c$ that has $(16)^c$ many minimal rdf.*

Corollary 1. *There are graphs of order n that have at least $\sqrt[5]{16}^{\,n} \in \Omega(1.7441^n)$ many minimal rdf.*

We checked with the help of a computer program that there are no other connected graphs of order at most eight that yield (by taking disjoint unions) a bigger lower bound.

6   A Refined Enumeration Algorithm

In this section, we are going to present the following result, which can be seen as the second main result of this paper.

Theorem 3. *There is a polynomial-space algorithm that enumerates all minimal rdf of a given graph of order n with polynomial delay and in time $\mathcal{O}^*(1.9332^n)$.*

Notice that this is in stark contrast to what is known about the enumeration of minimal dominating sets, or, equivalently, of minimal hitting sets in hypergraphs. Here, it is a long-standing open problem if minimal hitting sets in hypergraphs can be enumerated with polynomial delay.

In the remainder of this section, we sketch the proof of Theorem 3.

6.1   A Bird's Eye View on the Algorithm

As all along the search tree, from inner nodes we branch into the two cases if a certain vertex is assigned 2 or not, it is clear that (with some care concerning the final processing in leaf nodes) no minimal rdf is output twice. Hence, there is no need for the branching algorithm to store intermediate results to test (in a final step) if any solution was generated twice. Therefore, our algorithm needs only polynomial space, as one has only to store information along one path of the recursion tree.

Because we have a polynomial-time procedure that can test if a certain given pre-solution can be extended to a minimal rdf, we can build (a slightly modified version of) this test into an enumeration procedure, hence avoiding unnecessary branchings. Therefore, whenever we start with our binary branching, we know that at least one of the search tree branches will return at least one new minimal rdf. Hence, we will not move to more than N nodes in the search tree before outputting a new minimal rdf, where N is upper-bounded by twice the order of the input graph. This is the basic explanation for the claimed polynomial delay.

Let $G = (V, E)$ be a graph. Let us call a partial function $f : V \to \{0, 1, 2, \bar{1}, \bar{2}\}$ a *generalized Roman dominating function*, or grdf for short. Extending previously introduced notation, let $\overline{V_1}(f) = \{x \in V \mid f(x) = \bar{1}\}$, and $\overline{V_2}(f) = \{x \in V \mid f(x) = \bar{2}\}$. A vertex is said to be *active* if it has not been assigned a value (yet) under f; these vertices are collected in the set $A(f)$. Hence, for any grdf f, we have the partition $V = A(f) \cup V_0(f) \cup V_1(f) \cup V_2(f) \cup \overline{V_1}(f) \cup \overline{V_2}(f)$.

After performing a branching step, followed by an exhaustive application of the reduction rules, any grdf f considered in our algorithm always satisfies the following **(grdf) invariants**:

1. $\forall x \in \overline{V_1}(f) \cup V_0(f) \, \exists y \in N_G(x) : y \in V_2(f)$,
2. $\forall x \in V_2(f) : N_G(x) \subseteq \overline{V_1}(f) \cup V_0(f) \cup V_2(f)$,
3. $\forall x \in V_1(f) : N_G(x) \subseteq \overline{V_2}(f) \cup V_0(f) \cup V_1(f)$,
4. if $\overline{V_2}(f) \neq \emptyset$, then $A(f) \cup \overline{V_1}(f) \neq \emptyset$.[3]

For the extension test, we will therefore consider the function $\hat{f} : V \to \{0, 1, 2\}$ that is derived from a grdf f as follows:

$$\hat{f}(v) = \begin{cases} 0, & \text{if } v \in A(f) \cup V_0(f) \cup \overline{V_1}(f) \cup \overline{V_2}(f) \\ f(v), & \text{if } v \in V_1(f) \cup V_2(f) \end{cases}$$

The enumeration algorithm uses a combination of reduction and branching rules, starting with the nowhere defined function $f_\perp$, so that $A(f_\perp) = V$. The schematics of the algorithm is shown in Algorithm 3. To understand the algorithm, call an rdf g as *consistent* with a grdf f if $g(v) = 2$ implies $v \in A(f) \cup V_2(f) \cup \overline{V_1}(f)$ and $g(v) = 1$ implies $v \in A(f) \cup V_1(f) \cup \overline{V_2}(f)$ and $g(v) = 0$ implies $v \in A(f) \cup V_0(f) \cup \overline{V_1}(f) \cup \overline{V_2}(f)$. Below, we start with presenting some reduction rules, which also serve as (automatically applied) actions at each branching step, whenever applicable. The branching itself always considers a most attractive vertex v and either gets assigned 2 or not. The running time analysis will be performed with a measure-and-conquer approach. Our simple measure is defined by $\mu(G, f) = |A(f)| + \omega_1 |\overline{V_1}(f)| + \omega_2 |\overline{V_2}(f)| \leq |V|$ for some constants ω_1 and ω_2 that have to be specified later. The measure never increases when applying a reduction rule.

We are now presenting details of the algorithm and its analysis.

6.2   How to Achieve Polynomial Delay and Polynomial Space

In this section, we need a slight modification of the problem EXTRD in order to cope with pre-solutions. In this version, we add to an instance, usually specified by $G = (V, E)$ and $f : V \to \{0, 1, 2\}$, a set $\overline{V_2} \subseteq V$ with $V_2(f) \cap \overline{V_2} = \emptyset$. The question is if there exists a minimal RDF $\tilde{f}$ with $f \leq \tilde{f}$ and $V_2\left(\tilde{f}\right) \cap \overline{V_2} = \emptyset$. We call this problem a *generalized* rdf extension problem, or GENEXTRD for short. In order to solve this problem, we modify Algorithm 1 to cope with GENEXTRD by adding an if-clause after Line 8 that asks if $u \in \overline{V_2}$. If this is true, then the algorithm returns *no*, because it is prohibited that $\tilde{f}(u)$ is set to 2, while this is necessary for minimal rdf, as there is a vertex v in the neighborhood of u such that $\tilde{f}(v)$ has been set to 1. We call this algorithm GENEXTRD SOLVER.

Lemma 4. *Let $G = (V, E)$ be a graph, $f : V \to \{0, 1, 2\}$ be a function and $\overline{V_2} \subseteq V$ be a set with $V_2(f) \cap \overline{V_2} = \emptyset$. GENEXTRD SOLVER gives the correct answer when given the GENEXTRD instance $(G, f, \overline{V_2})$.*

---

[3] This condition assumes that our graphs have non-empty vertex sets.

Algorithm 3. A refined enumeration algorithm for minimal rdf

1: **procedure** REFINED RD ENUMERATION(G, f)
    Input: A graph $G = (V, E)$, a grdf $f : V \to \{0, 1, 2, \overline{1}, \overline{2}\}$.
    Assumption: There exists at least one minimal rdf consistent with f.
    Output: Enumeration of all minimal rdf consistent with f.
2:     **if** f is everywhere defined and $f(V) \subseteq \{0, 1, 2\}$ **then**
3:         Output f and return.
4:     $\{$ We know that $A(f) \cup \overline{V_1}(f) \neq \emptyset. \}$
5:     Pick a vertex $v \in A(f) \cup \overline{V_1}(f)$ of highest priority for branching.
6:     $f_2 := f; f_2(v) := 2.$
7:     Exhaustively apply reduction rules to f_2. $\{$ Invariants are valid for $f_2. \}$
8:     **if** GENEXTRD SOLVER $\left(G, \widehat{f_2}, \overline{V_2}(f_2)\right)$ **then**
9:         REFINED RD ENUMERATION (G, f_2).
10:    $f_{\overline{2}} := f;$ **if** $v \in A(f)$ **then** $f_{\overline{2}}(v) := \overline{2}$ **else** $f_{\overline{2}}(v) := 0.$
11:    Exhaustively apply reduction rules to $f_{\overline{2}}$. $\{$ Invariants are valid for $f_{\overline{2}}. \}$
12:    **if** GENEXTRD SOLVER $\left(G, \widehat{f_{\overline{2}}}, \overline{V_2}(f_{\overline{2}})\right)$ **then**
13:        REFINED RD ENUMERATION $(G, f_{\overline{2}})$.

Let f be a grdf at any moment of the branching algorithm. We can show that GENEXTRD SOLVER could tell us in polynomial time if there exists a minimal rdf that could be enumerated by the branching algorithm from this point on. This is crucial for showing polynomial delay of Algorithm 3. Moreover, Algorithm 3 does not enumerate any solution twice, which allows us to use polynomial space.

6.3 Details on Reductions and Branchings

For the presentation of the following rules, we assume that $G = (V, E)$ and a grdf f is given. The rules are executed exhaustively in the given order.

Reduction Rule LPN (Last Potential Private Neighbor). If $v \in V_2(f)$ satisfies $|N_G(v) \cap (\overline{V_2}(f) \cup A(f))| = 1$, then set $f(x) = 0$ for $\{x\} = N_G(v) \cap (\overline{V_2}(f) \cup A(f))$.

Reduction Rule V_0. Let $v \in V_0(f)$. Assume there exists a unique $u \in V_2(f) \cap N_G(v)$. Moreover, assume that for all $x \in N_G(u) \cap (V_0(f) \cup \overline{V_1}(f) \cup \overline{V_2}(f))$, $|N_G(x) \cap V_2(f)| \geq 2$ if $x \neq v$. Then, for any $w \in N_G(v) \cap A(f)$, set $f(w) = \overline{2}$ and for any $w \in N_G(v) \cap \overline{V_1}(f)$, set $f(w) = 0$.

Reduction Rule V_1. Let $v \in V_1(f)$. For any $w \in N_G(v) \cap A(f)$, set $f(w) = \overline{2}$. For any $w \in N_G(v) \cap \overline{V_1}(f)$, set $f(w) = 0$.

Reduction Rule V_2. Let $v \in V_2(f)$. For any $w \in N_G(v) \cap A(f)$, set $f(w) = \overline{1}$. For any $w \in N_G(v) \cap \overline{V_2}(f)$, set $f(w) = 0$.

Reduction Rule NPD (No Potential Domination). If $v \in \overline{V_2}(f)$ satisfies $N_G(v) \subseteq \overline{V_2}(f) \cup V_0(f) \cup V_1(f)$, then set $f(v) = 1$ (this also applies to isolated vertices in $\overline{V_2}(f)$).

Reduction Rule NPN (No Private Neighbor). If $v \in A(f)$ satisfies $N_G(v) \subseteq V_0(f) \cup \overline{V_1}(f)$, then set $f(v) = \overline{2}$ (this also applies to isolated vertices in $A(f)$).

Reduction Rule Isolate. If $A(f) = \emptyset$ and if $v \in \overline{V_1}(f)$ satisfies $N_G(v) \cap \overline{V_2}(f) = \emptyset$, then set $f(v) = 0$.

Reduction Rule Edges. If $u, v \in \overline{V_2}(f) \cup V_0(f) \cup V_1(f)$ and $e = uv \in E$, then remove the edge e from G.

In the following, we first take care of the claimed grdf invariants.

Proposition 4. *After exhaustively executing the proposed reduction rules, as indicated in Algorithm 3, the claimed grdf invariants are maintained.*

We have now to show the *soundness* of the proposed reduction rules. In the context of enumerating minimal rdf, this means the following: if f, f' are grdf of $G = (V, E)$ before or after applying any of the reduction rules, then g is a minimal rdf that is consistent with f if and only if it is consistent with f'.

Proposition 5. *All proposed reduction rules are sound.*

In order to fully understand Algorithm 3, we need to describe priorities for branching. We describe these priorities in the following in decreasing order for a vertex $v \in A(f) \cup \overline{V_1}(f)$.

1. $v \in A(f)$ and $|N_G(v) \cap (A(f) \cup \overline{V_2}(f))| \geq 2$;
2. any $v \in A(f)$;
3. any $v \in \overline{V_1}(f)$, preferably if $|N_G(v) \cap \overline{V_2}(f)| \neq 2$.

These priorities also split the run of our algorithm into phases, as whenever the algorithm was once forced to pick a vertex according to some lower priority, there will be never again the chance to pick a vertex of higher priority thereafter. It is useful to collect some **phase properties** that instances must satisfy after leaving Phase i, determined by applying the ith branching priority.

- Before entering any phase, there are no edges between vertices u, v if $u, v \in V_0(f) \cup V_1(f) \cup \overline{V_2}(f)$ or if $u \in V_i(f)$ and $v \in \overline{V_i}(f) \cup A(f)$ ($i \in \{1, 2\}$, as we assume the reduction rules have been exhaustively applied.
- After leaving the first phase, any active vertex with an active neighbor is either pendant or has only further neighbors from $\overline{V_1}(f) \cup V_0(f)$.
- After leaving the second phase, $A(f) = \emptyset$ and $N_G(\overline{V_2}(f)) \subseteq \overline{V_1}(f)$.
- After the third phase, $A(f) = \overline{V_2}(f) = \overline{V_1}(f) = \emptyset$, so f is a minimal rdf.

Proposition 6. *The phase properties hold.*

Table 1. The branching vectors of different branching scenarios of the enumeration algorithm for listing all minimal Roman dominating functions of a given graph; we always branch on $v \in A(f) \cup \overline{V_1}(f)$. The $(*)$ refers to the worst branchings w.r.t. our setting of weights of $\omega_1 = \frac{2}{3}$ and $\omega_2 = 0.38488$.

Phase	Scenario	Branching vector
1	$v \in A(f)$	Subcases apply
1.1	$|N_G(v) \cap A(f)| \geq 2$	$(1 - \omega_2, 3 - 2\omega_1)\ (*)$
1.2	$|N_G(v) \cap \overline{V_2}(f)| \geq 2$	$(1 - \omega_2, 1 + 2\omega_2)$
1.3	$\begin{cases} |N_G(v) \cap A(f)| = 1, \\ |N_G(v) \cap \overline{V_2}(f)| = 1 \end{cases}$	$(1 - \omega_2, 2 + \omega_2 - \omega_1)$
2	$v \in A(f)$	Subcases apply
2.1	$N_G(v) \cap A(f) = \{x\}$	$(1 - \omega_2, 2)$
2.2.a	$\begin{cases} N_G(v) \cap \overline{V_2}(f) = \{x\}, \\ N_G(x) \cap \overline{V_1}(f) \neq \emptyset \end{cases}$	$(1 - \omega_2, 1 + \omega_2 + \omega_1)$
2.2.b	$\begin{cases} N_G(v) \cap \overline{V_2}(f) = \{x\}, \\ |N_G(x) \cap A(f)| \geq 2 \end{cases}$	$(1 - \omega_2, 2)$
2.2.c	$\begin{cases} N_G(v) \cap \overline{V_2}(f) = \{x\}, \\ |N_G(x)| = 1 \end{cases}$	$(1 + \omega_2, 1)$
3	$v \in \overline{V_1}(f)$	Subcases apply
3.1	$|N_G(v) \cap \overline{V_2}(f)| \geq 3$	$(\omega_1, \omega_1 + 3\omega_2)$
3.2.a	$\begin{cases} N_G(v) \cap \overline{V_2}(f) = \{u\}, \\ |N_G(u) \cap \overline{V_1}(f)| \geq 2 \end{cases}$	$(\omega_1, 2\omega_1 + \omega_2)$
3.2.b	$\begin{cases} N_G(v) \cap \overline{V_2}(f) = \{u\}, \\ |N_G(u) \cap \overline{V_1}(f)| = 1 \end{cases}$	$(\omega_1 + \omega_2, \omega_1 + \omega_2)\ (*)$
3.3.a	$\begin{cases} N_G(v) \cap \overline{V_2}(f) = \{u_1, u_2\}, \\ |N_G(u_1) \cap \overline{V_1}(f)| = 1 \end{cases}$	$(\omega_1 + \omega_2, \omega_1 + \omega_2)\ (*)$
3.3.b	$\begin{cases} N_G(v) \cap \overline{V_2}(f) = \{u_1, u_2\}, \\ |N_G(u_1) \cap \overline{V_1}(f)| \geq 2 \end{cases}$	$(2\omega_1 + 2\omega_2, 2\omega_1 + 2\omega_2, 2\omega_1 + 2\omega_2, 2\omega_1 + 2\omega_2)$ $(*)$

6.4   A Measure and Conquer Approach

We now present the branching analysis, classified by the described branching priorities. We summarize a list of all resulting branching vectors in Table 1.

Proposition 7. *On input graphs of order n, Algorithm* REFINED RD ENU-MERATION *runs in time* $\mathcal{O}^*(1.9332^n)$.

Proof. We follow the run-time analysis that led us to the branching vectors listed in Table 1. The claim follows by choosing as weights $\omega_1 = \frac{2}{3}$, $\omega_2 = 0.38488$.   □

7   An Alternative Notion of Minimal RDF

So far, we focused on an ordering of the functions $V \to \{0, 1, 2\}$ that was derived from the linear ordering $0 < 1 < 2$. Due to the different functionalities, it might be not that clear if 2 should be bigger than 1. If we rather choose as a basic partial ordering $0 < 1, 2$, with $1, 2$ being incomparable, this yields another ordering for the functions $V \to \{0, 1, 2\}$, again lifted pointwise. Being reminiscent of partial orderings, let us call the resulting notion of minimality PO-minimal rdf. This variation would also lead to a non-trivial notion of UPPER ROMAN DOMINATION, because the minimal rdf $f : V \to \{0, 1, 2\}$ with biggest sum $\sum_{v \in V} f(v)$ is no longer (necessarily) achieved by the constant function $f = 1$. Also, this can be seen as a natural pointwise lifting of the inclusion ordering, keeping in mind that $f \leq_{PO} g$ iff $V_1(f) \subseteq V_1(g)$ and $V_2(f) \subseteq V_2(g)$. We can obtain results that are similar to the notion of minimality considered before; the simple enumeration algorithm is even provably optimal in this case.

Theorem 4. *Let $G = (V, E)$ be a graph, $f : V \to \{0, 1, 2\}$ and abbreviate $G' := G[V_0(f) \cup V_2(f)]$. Then, f is a PO-minimal rdf iff the following conditions hold: (1) $N_G[V_2(f)] \cap V_1(f) = \emptyset$, (2) $V_2(f)$ is a minimal dominating set of G'.*

Theorem 5. *The extension problem ExtPO-RDF is polynomial-time solvable.*

Theorem 6. *There is a polynomial-space algorithm that enumerates all PO-minimal rdf of a given graph of order n in time $\mathcal{O}^*(2^n)$ with polynomial delay. Moreover, there is a family of graphs G_n, with G_n being of order n, such that G_n has 2^n many PO-minimal rdf.*

8   Conclusions

While the combinatorial concept of Roman domination leads to a number of complexity results that are completely analogous to what is known about the combinatorial concept of domination, the two concepts lead to distinctively different results when it comes to enumeration and extension problems. These are the main messages and results of the present paper.

We are currently working on improved enumeration and also on counting of minimal rdf in special graph classes. Our first results are very promising; for instance, there are good chances to completely close the gap between lower and upper bounds for enumerating minimal rdf for some graph classes.

Another line of research is looking into problems that are similar to Roman domination, in order to better understand the specialties of Roman domination in contrast to the classical domination problem. What makes Roman domination behave different from classical domination when it comes to finding extensions or to enumeration?

Finally, let us mention that our main branching algorithm also gives an input-sensitive enumeration algorithm for minimal Roman dominating functions in the sense of Chellali *et al.* [16]. However, we do not know of a polynomial-delay enumeration algorithm in that case. This is another interesting line of research. Here, the best lower bound we could find was a repetition of a C_4, leading to $\sqrt[4]{8} \geq 1.68179$ as the basis.

References

1. Abu-Khzam, F.N., Bazgan, C., Chopin, M., Fernau, H.: Data reductions and combinatorial bounds for improved approximation algorithms. J. Comput. Syst. Sci. **82**(3), 503–520 (2016)
2. Abu-Khzam, F.N., Heggernes, P.: Enumerating minimal dominating sets in chordal graphs. Inf. Process. Lett. **116**(12), 739–743 (2016)
3. Bazgan, C., et al.: The many facets of upper domination. Theoret. Comput. Sci. **717**, 2–25 (2018)
4. Benecke, S.: Higher order domination of graphs. Master's thesis, Department of Applied Mathematics of the University of Stellenbosch, South Africa (2004). http://dip.sun.ac.za/~vuuren/Theses/Benecke.pdf
5. Bermudo, S., Fernau, H.: Computing the differential of a graph: hardness, approximability and exact algorithms. Discret. Appl. Math. **165**, 69–82 (2014)
6. Bermudo, S., Fernau, H.: Combinatorics for smaller kernels: the differential of a graph. Theoret. Comput. Sci. **562**, 330–345 (2015)
7. Bermudo, S., Fernau, H., Sigarreta, J.M.: The differential and the Roman domination number of a graph. Appl. Anal. Discret. Math. **8**, 155–171 (2014)
8. Bläsius, T., Friedrich, T., Lischeid, J., Meeks, K., Schirneck, M.: Efficiently enumerating hitting sets of hypergraphs arising in data profiling. In: Algorithm Engineering and Experiments (ALENEX), pp. 130–143. SIAM (2019)
9. Bonamy, M., Defrain, O., Heinrich, M., Raymond, J.F.: Enumerating minimal dominating sets in triangle-free graphs. In: Niedermeier, R., Paul, C. (eds.) 36th International Symposium on Theoretical Aspects of Computer Science (STACS 2019). LIPIcs, vol. 126, pp. 16:1–16:12. Schloss Dagstuhl - Leibniz-Zentrum für Informatik (2019)
10. Boros, E., Gurvich, V., Hammer, P.L.: Dual subimplicants of positive Boolean functions. Optim. Methods Softw. **10**(2), 147–156 (1998)
11. Casel, K., Fernau, H., Khosravian Ghadikolaei, M., Monnot, J., Sikora, F.: Extension of some edge graph problems: standard and parameterized complexity. In: Gąsieniec, L.A., Jansson, J., Levcopoulos, C. (eds.) FCT 2019. LNCS, vol. 11651, pp. 185–200. Springer, Cham (2019). https://doi.org/10.1007/978-3-030-25027-0_13
12. Fernau, H., Huber, K.T., Naor, J.S.: Invited talks. In: Calamoneri, T., Corò, F. (eds.) CIAC 2021. LNCS, vol. 12701, pp. 3–19. Springer, Cham (2021). https://doi.org/10.1007/978-3-030-75242-2_1
13. Casel, K., Fernau, H., Ghadikolaei, M.K., Monnot, J., Sikora, F.: On the complexity of solution extension of optimization problems. Theoret. Comput. Sci. **904**, 48–65 (2022). https://doi.org/10.1016/j.tcs.2021.10.017
14. Chambers, E.W., Kinnersley, B., Prince, N., West, D.B.: Extremal problems for Roman domination. SIAM J. Discret. Math. **23**, 1575–1586 (2009)
15. Chapelle, M., Cochefert, M., Couturier, J.-F., Kratsch, D., Liedloff, M., Perez, A.: Exact algorithms for weak Roman domination. In: Lecroq, T., Mouchard, L. (eds.) IWOCA 2013. LNCS, vol. 8288, pp. 81–93. Springer, Heidelberg (2013). https://doi.org/10.1007/978-3-642-45278-9_8

16. Chellali, M., Haynes, T.W., Hedetniemi, S.M., Hedetniemi, S.T., McRae, A.A.: A Roman domination chain. Graphs Comb. **32**(1), 79–92 (2016)
17. Cockayne, E.J., Dreyer, P., Jr., Hedetniemi, S.M., Hedetniemi, S.T.: Roman domination in graphs. Discret. Math. **278**, 11–22 (2004)
18. Couturier, J., Heggernes, P., van 't Hof, P., Kratsch, D.: Minimal dominating sets in graph classes: combinatorial bounds and enumeration. Theoret. Comput. Sci. **487**, 82–94 (2013)
19. Couturier, J., Letourneur, R., Liedloff, M.: On the number of minimal dominating sets on some graph classes. Theoret. Comput. Sci. **562**, 634–642 (2015)
20. Creignou, N., Kröll, M., Pichler, R., Skritek, S., Vollmer, H.: A complexity theory for hard enumeration problems. Discret. Appl. Math. **268**, 191–209 (2019)
21. Dreyer, P.A.: Applications and variations of domination in graphs. Ph.D. thesis, Rutgers University, New Jersey, USA (2000)
22. Eiter, T., Gottlob, G.: Identifying the minimal transversals of a hypergraph and related problems. SIAM J. Comput. **24**(6), 1278–1304 (1995)
23. Favaron, O., Karami, H., Khoeilar, R., Sheikholeslami, S.M.: On the Roman domination number of a graph. Discret. Math. **309**(10), 3447–3451 (2009)
24. Fernau, H.: Roman domination: a parameterized perspective. Int. J. Comput. Math. **85**, 25–38 (2008)
25. Gainer-Dewar, A., Vera-Licona, P.: The minimal hitting set generation problem: algorithms and computation. SIAM J. Discret. Math. **31**(1), 63–100 (2017)
26. Golovach, P.A., Heggernes, P., Kanté, M.M., Kratsch, D., Villanger, Y.: Enumerating minimal dominating sets in chordal bipartite graphs. Discret. Appl. Math. **199**, 30–36 (2016)
27. Golovach, P.A., Heggernes, P., Kratsch, D.: Enumerating minimal connected dominating sets in graphs of bounded chordality. Theoret. Comput. Sci. **630**, 63–75 (2016)
28. Haynes, T.W., Hedetniemi, S.T., Slater, P.J.: Fundamentals of Domination in Graphs. Monographs and Textbooks in Pure and Applied Mathematics, vol. 208. Marcel Dekker (1998)
29. Haynes, T.W., Hedetniemi, S.T., Henning, M.A. (eds.): Topics in Domination in Graphs. Developments in Mathematics, vol. 64. Springer, Cham (2020). https://doi.org/10.1007/978-3-030-51117-3
30. Hedetniemi, S.T., Rubalcaba, R.R., Slater, P.J., Walsh, M.: Few compare to the great Roman empire. Congr. Numer. **217**, 129–136 (2013)
31. Kanté, M.M., Limouzy, V., Mary, A., Nourine, L.: On the enumeration of minimal dominating sets and related notions. SIAM J. Discret. Math. **28**(4), 1916–1929 (2014)
32. Kanté, M.M., Limouzy, V., Mary, A., Nourine, L., Uno, T.: Polynomial delay algorithm for listing minimal edge dominating sets in graphs. In: Dehne, F., Sack, J.-R., Stege, U. (eds.) WADS 2015. LNCS, vol. 9214, pp. 446–457. Springer, Cham (2015). https://doi.org/10.1007/978-3-319-21840-3_37
33. Kanté, M.M., Limouzy, V., Mary, A., Nourine, L., Uno, T.: A polynomial delay algorithm for enumerating minimal dominating sets in chordal graphs. In: Mayr, E.W. (ed.) WG 2015. LNCS, vol. 9224, pp. 138–153. Springer, Heidelberg (2016). https://doi.org/10.1007/978-3-662-53174-7_11
34. Kraner Šumenjak, T., Pavlić, P., Tepeh, A.: On the Roman domination in the lexicographic product of graphs. Discret. Appl. Math. **160**(13–14), 2030–2036 (2012)
35. Liedloff, M.: Algorithmes exacts et exponentiels pour les problèmes NP-difficiles: domination, variantes et généralisations. Ph.D. thesis, Université Paul Verlaine - Metz, France (2007)

36. Liedloff, M., Kloks, T., Liu, J., Peng, S.L.: Efficient algorithms for Roman domination on some classes of graphs. Discret. Appl. Math. **156**(18), 3400–3415 (2008)
37. Liu, C.H., Chang, G.J.: Roman domination on 2-connected graphs. SIAM J. Discret. Math. **26**(1), 193–205 (2012)
38. Liu, C.H., Chang, G.J.: Upper bounds on Roman domination numbers of graphs. Discret. Math. **312**(7), 1386–1391 (2012)
39. Liu, C.H., Chang, G.J.: Roman domination on strongly chordal graphs. J. Comb. Optim. **26**(3), 608–619 (2013)
40. Mary, A.: Énumération des dominants minimaux d'un graphe. Ph.D. thesis, LIMOS, Université Blaise Pascal, Clermont-Ferrand, France, November 2013
41. Mashburn, J.L., Haynes, T.W., Hedetniemi, S.M., Hedetniemi, S.T., Slater, P.J.: Differentials in graphs. Utilitas Math. **69**, 43–54 (2006)
42. Mobaraky, B.P., Sheikholeslami, S.M.: Bounds on Roman domination numbers of graphs. Matematitchki Vesnik **60**, 247–253 (2008)
43. Pagourtzis, A., Penna, P., Schlude, K., Steinhöfel, K., Taylor, D.S., Widmayer, P.: Server placements, Roman domination and other dominating set variants. In: Baeza-Yates, R.A., Montanari, U., Santoro, N. (eds.) Foundations of Information Technology in the Era of Networking and Mobile Computing, IFIP 17th World Computer Congress – TC1 Stream/2nd IFIP International Conference on Theoretical Computer Science IFIP TCS, pp. 280–291. Kluwer (2002). Also available as Technical report 365, ETH Zürich, Institute of Theoretical Computer Science, 10/2001
44. Peng, S.L., Tsai, Y.H.: Roman domination on graphs of bounded treewidth. In: The 24th Workshop on Combinatorial Mathematics and Computation Theory, pp. 128–131 (2007)
45. ReVelle, C.S., Rosing, K.E.: Defendens imperium Romanum: a classical problem in military strategy. Am. Math. Monthly **107**, 585–594 (2000). http://www.jhu.edu/~jhumag/0497web/locate3.html
46. van Rooij, J.M.M.: Exact exponential-time algorithms for domination problems in graphs. Ph.D. thesis, Universiteit Utrecht, The Netherlands (2011)
47. Shang, W., Wang, X., Hu, X.: Roman domination and its variants in unit disk graphs. Discret. Math. Algorithms Appl. **2**(1), 99–106 (2010)
48. Shi, Z., Koh, K.M.: Counting the number of minimum Roman dominating functions of a graph. Technical report, arXiv/CoRR, abs/1403.1019 (2014)
49. Stewart, I.: Defend the Roman empire. Sci. Am. **281**(6), 136–138 (1999)
50. Xing, H.M., Chen, X., Chen, X.G.: A note on Roman domination in graphs. Discret. Math. **306**(24), 3338–3340 (2006)
51. Xueliang, F., Yuansheng, Y., Baoqi, J.: Roman domination in regular graphs. Discret. Math. **309**(6), 1528–1537 (2009)
52. Yero, I.G., Rodríguez-Velázquez, J.A.: Roman domination in Cartesian product graphs and strong product graphs. Appl. Anal. Discret. Math. **7**, 262–274 (2013)

Disjoint Compatibility via Graph Classes

Oswin Aichholzer[1], Julia Obmann[1], Pavel Paták[2], Daniel Perz[1]([⊠]),
Josef Tkadlec[3], and Birgit Vogtenhuber[1]

[1] Graz University of Technology, Graz, Austria
{oaich,daperz,bvogt}@ist.tugraz.at, julia.obmann@student.tugraz.at
[2] Czech Technical University in Prague, Prague, Czech Republic
patak@kam.mff.cuni.cz
[3] Department of Mathematics, Harvard University, Cambridge, MA 02138, USA
tkadlec@math.harvard.edu

Abstract. Two plane drawings of graphs on the same set of points are
called disjoint compatible if their union is plane and they do not have
an edge in common. Let S be a convex point set of $2n \geq 10$ points and
let $\mathcal{H}$ be a family of plane drawings on S. Two plane perfect matchings
M_1 and M_2 on S (which do not need to be disjoint nor compatible) are
disjoint $\mathcal{H}$-compatible if there exists a drawing in $\mathcal{H}$ which is disjoint
compatible to both M_1 and M_2. In this work, we consider the graph
which has all plane perfect matchings as vertices and where two vertices
are connected by an edge if the matchings are disjoint $\mathcal{H}$-compatible.
We study the diameter of this graph when $\mathcal{H}$ is the family of all plane
spanning trees, caterpillars or paths. We show that in the first two cases
the graph is connected with constant and linear diameter, respectively,
while in the third case it is disconnected.

Keywords: Compatibility · Convex Set · Matchings

1 Introduction

In this work we study straight-line drawings of graphs. Two plane drawings
of graphs on the same set S of points are called *compatible* if their union is
plane. The drawings are *disjoint compatible* if they are compatible and do not

Research on this work was initiated at the 6th Austrian-Japanese-Mexican-Spanish
Workshop on Discrete Geometry and continued during the 16th European Geometric
Graph-Week, both held near Strobl, Austria. We are grateful to the participants for
the inspiring atmosphere. We especially thank Alexander Pilz for bringing this class
of problems to our attention. D.P. is partially supported by the FWF grant I 3340-
N35 (Collaborative DACH project *Arrangements and Drawings*). The research stay
of P.P. at IST Austria is funded by the project CZ.02.2.69/0.0/0.0/17_050/0008466
Improvement of internationalization in the field of research and development at Charles
University, through the support of quality projects MSCA-IF.

 This project has received funding from the European Union's Horizon 2020
research and innovation programme under the Marie Skłodowska-Curie
grant agreement No 734922.

M. A. Bekos and M. Kaufmann (Eds.): WG 2022, LNCS 13453, pp. 16–28, 2022.
https://doi.org/10.1007/978-3-031-15914-5_2

have an edge in common. For a fixed class $\mathcal{G}$, e.g. matchings, trees, etc., of plane geometric graphs on S the (disjoint) *compatibility graph* of S has the elements of $\mathcal{G}$ as the set of vertices and an edge between two elements of $\mathcal{G}$ if the two graphs are (disjoint) compatible. For example, it is well known that the (not necessarily disjoint) compatibility graph of plane perfect matchings is connected [4,5]. Moreover, in [2] it is shown that there always exists a sequence of at most $\mathcal{O}(\log n)$ compatible (but not necessarily disjoint) perfect matchings between any two plane perfect matchings of a set of $2n$ points in general position, that is, the graph of perfect matchings is connected with diameter $\mathcal{O}(\log n)$. On the other hand, Razen [11] provides an example of a point set where this diameter is $\Omega(\log n/\log \log n)$.

Disjoint compatible (perfect) matchings have been investigated in [2] for sets of $2n$ points in general position. The authors showed that for odd n there exist perfect matchings which are isolated vertices in the disjoint compatibility graph and posed the following conjecture: For every perfect matching with an even number of edges there exists a disjoint compatible perfect matching. This conjecture was answered in the positive by Ishaque et al. [7] and it was mentioned that for even n it remains an open problem whether the disjoint compatibility graph is always connected. In [1] it was shown that for sets of $2n \geq 6$ points in convex position this disjoint compatibility graph is (always) disconnected.

Both concepts, compatibility and disjointness, are also used in combination with different geometric graphs. For example, in [5] it was shown that the flip-graph of all triangulations that admit a (compatible) perfect matching, is connected[1]. It has also been shown that for every graph with an outerplanar embedding there exists a compatible plane perfect matching [3]. Considering plane trees and simple polygons, the same work provides bounds on the minimum number of edges a compatible plane perfect matching must have in common with the given graph. For simple polygons, it was shown in [10] that it is NP-hard to decide whether there exist a perfect matching which is disjoint compatible to a given simple polygon. See also the survey [6] on the related concept of compatible graph augmentation.

In a similar spirit we can define a bipartite disjoint compatibility graph, where the two sides of the bipartition represent two different graph classes. Let one side be all plane perfect matchings of S while the other side consists of all plane spanning trees of S. Edges represent the pairs of matchings and trees which are disjoint compatible. Considering connectivity of this bipartite graph, there trivially exist isolated vertices on the tree side – consider a spanning star, which can not have any disjoint compatible matching. Thus, the question remains whether there exists a bipartite connected subgraph which contains all vertices representing plane perfect matchings.

This point of view leads us to a new notion of adjacency for perfect matchings. For a given set S of $2n$ points and a family $\mathcal{H}$ of drawings on S, two plane perfect matchings M_1 and M_2 (which do not need to be disjoint nor compatible) are *disjoint $\mathcal{H}$-compatible* if there exists a drawing D in $\mathcal{H}$ which is

---

[1] In the flip-graph, two triangulations are connected if they differ by a single edge.

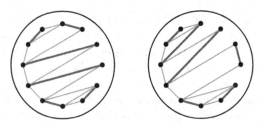

Fig. 1. Two plane perfect matchings (in blue) on the same set of twelve points in convex position which are disjoint $\mathcal{T}$-compatible. The complying disjoint compatible spanning tree is drawn in green. (Color figure online)

disjoint compatible to both M_1 and M_2; see Fig. 1 for an example. The disjoint $\mathcal{H}$-compatibility graph $\mathrm{DCG}_S(\mathcal{H})$ has all plane perfect matchings of S as vertices. We have an edge between the vertices corresponding to M_1 and M_2 if M_1 and M_2 are disjoint $\mathcal{H}$-compatible. In other words, they are two steps apart in the corresponding bipartite disjoint compatibility graph. Rephrasing the above question, we ask whether $\mathrm{DCG}_S(\mathcal{H})$ is connected. Recall that the disjoint compatibility graph for perfect matchings alone is not connected (see [1,2]).

In this work we study the case where S is a set of $2n$ points in convex position and consider the cases where $\mathcal{H}$ is the family $\mathcal{T}$ of all plane spanning trees, the family $\mathcal{C}$ of all plane spanning caterpillars, or the family $\mathcal{P}$ of all plane spanning paths. We show that $\mathrm{DCG}_S(\mathcal{T})$ and $\mathrm{DCG}_S(\mathcal{C})$ are connected if $2n \geq 10$. In that case the diameter of $\mathrm{DCG}_S(\mathcal{T})$ is either 4 or 5, independent of n, and the diameter of $\mathrm{DCG}_S(\mathcal{C})$ is $\mathcal{O}(n)$. On the other hand we show that $\mathrm{DCG}_S(\mathcal{P})$ is disconnected.

From here on, if not said otherwise, all matchings, trees, caterpillars and paths are on point sets in convex position and are plane. Hence, we omit the word 'plane' for these drawings. Further, all matchings considered in this work are perfect matchings. Due to space restrictions most proofs are sketched or omitted. This work is partially based on the master's thesis of the second author [9].

2 Preliminaries

Throughout this article let S be a set of $2n$ points in the plane in convex position. The edges of a drawing on S can be classified in the following way. We call edges, that are spanned by two neighboring points of S, *perimeter edges*; all other edges spanned by S are called *diagonals*. We call matchings without diagonals *perimeter matchings*. Note that there are exactly two perfect perimeter matchings. We label the perimeter edges alternately even and odd. The *even perimeter matching* consists of all even perimeter edges. The *odd perimeter matching* consists of all odd perimeter edges.

Looking at a matching M on S, the edges of M split the convex hull of S into regions, such that no edge of M crosses any region. More formally, we call a set $X \subset M$ of $k \geq 2$ matching edges a *k-semicycle* if no edge of M intersects the

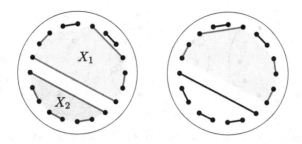

Fig. 2. Left: A matching M and two semicycles X_1 (red edges) and X_2 (blue edges) with their convex hulls. The cycle $\overline{X_1}$ is an inside 4-cycle, since the boundary of the red shaded area contains at least two (in fact three) diagonals. The cycle $\overline{X_2}$ is a 4-ear. Right: The matching resulting from rotating the cycle X_1. (Color figure online)

interior of the convex hull of X. Further, we call the boundary of the convex hull of X a k-cycle, denoted by $\overline{X}$. If $\overline{X}$ contains at least two diagonals of S, then we call X an *inside k-semicycle*. Otherwise, we call X a *k-semiear* (this includes perimeter matchings); see Fig. 2. Analogously, we denote cycles as *inside k-cycles* or *inside k-ears*, respectively.

Consider a perfect matching M and a semicycle X of M. We say that we *rotate* X if we take all edges of M and replace X by $\overline{X} \backslash X$, which gives us a perfect matching M'. So the symmetric difference of M and M' is exactly $\overline{X}$.

3   Disjoint Compatibility via Spanning Trees

In this section we show that for convex point sets S of $2n \geq 10$ points, the disjoint compatibility graph $\mathrm{DCG}_S(\mathcal{T})$ is connected. We further prove that the diameter is upper bounded by 5. The idea is that any matching on S has small distance to one of the two perimeter matchings and those themselves are close to each other in $\mathrm{DCG}_S(\mathcal{T})$. First we show that arbitrarily many inside cycles can be simultaneously rotated in one step.

Lemma 1. *Let M and M' be two matchings whose symmetric difference is a union of disjoint inside cycles. Then M and M' are $\mathcal{T}$-compatible to each other.*

Proof Idea. First we focus on one inside cycle $\overline{X}$. Let $u_1 v_1$ and $u_2 v_2$ be two diagonals of $\overline{X}$, labeled as in Fig. 3. Note that v_1 and u_2 might be the same point if each of M and M' contains one of $u_1 v_1$ and $u_2 v_2$. We take the edges from u_1 to any point between v_1 and u_2 and from u_2 to any point between v_2 and u_1 including u_1. This yields a tree on the points of X except v_1 and v_2.

Now we do this for every inside cycle. The resulting trees and the remaining points can be connected in a greedy way to a spanning tree that is disjoint compatible to both M and M'.   □

We next consider sufficiently large ears. The following lemma states that such ears can be rotated in at most three steps; see Fig. 4 for a sketch of this sequence

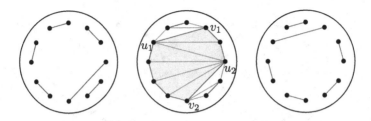

Fig. 3. Two plane matchings (in blue and red) on S which whose symmetric difference is an inside cycle. The complying disjoint compatible spanning tree is drawn in green. (Color figure online)

of rotations, whose proof uses Lemma 1. Note that Lemma 2 also implies that the two perimeter matchings have distance at most 3 in $\mathrm{DCG}_S(\mathcal{T})$.

Lemma 2. *Let M and M' be two matchings whose symmetric difference is a k-ear with $k \geq 6$. Then M and M' have distance at most 3 in $\mathrm{DCG}_S(\mathcal{T})$.*

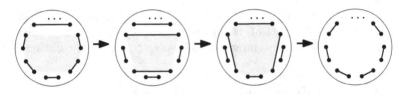

Fig. 4. Rotation of a 6-ear in 3 steps (in each step we rotate the grey inside cycle).

Theorem 1. *For $2n \geq 10$, the graph $\mathrm{DCG}_S(\mathcal{T})$ is connected with diameter* $\mathrm{diam}(\mathrm{DCG}_S(\mathcal{T})) \leq 5$.

Proof Idea. If $2n = 10$, the claim can be checked by constructing the whole graph. For $2n \geq 12$, the idea is to show that all matchings can be quickly transformed either to the odd perimeter matching O or to the even perimeter matching E (or to both – by Lemma 2 we have $\mathrm{dist}(O, E) \leq 3$). In particular, for a fixed matching M we denote by $\mathrm{d_{min}}(M)$ (resp. $\mathrm{d_{max}}(M)$) the distance from M to the closer (resp. further) perimeter matching. Then we prove that the non-perimeter matchings can be split into three classes A_1, A_2, A_3 with the following properties:

1. $\forall M \in A_1$ we have $\mathrm{d_{min}}(M) \leq 1$ (and hence $\mathrm{d_{max}}(M) \leq 1 + 3 = 4$);
2. $\forall M \in A_2$ we have $\mathrm{d_{min}}(M) \leq 2$ and $\mathrm{d_{max}}(M) \leq 3$;
3. $\forall M \in A_3$ we have $\mathrm{d_{max}}(M) \leq 3$ and $\forall M, M' \in A_3$ we have $\mathrm{dist}(M, M') \leq 4$.

See Fig. 5 for a depiction. This guarantees that $\mathrm{diam}(\mathrm{DCG}_S(\mathcal{T})) \leq 5$. □

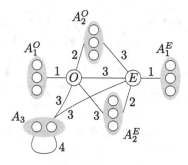

Fig. 5. Depiction of the partitioning of the set of all non-perimeter matchings into subsets $A_1 = A_1^E \cup A_1^O$, $A_2 = A_2^E \cup A_2^O$, and A_3, with bounds on their distances.

3.1 A Lower Bound for the Diameter of $\mathrm{DCG}_S(\mathcal{T})$

Since the diameter of $\mathrm{DCG}_S(\mathcal{T})$ has a constant upper bound, is seems reasonable to also ask for a best possible lower bound. To do so, we first identify structures which prevent that two matchings are $\mathcal{T}$-compatible. Let M and M' be two matchings in S. A *boundary area with k points* is an area within the convex hull of S containing k points of S that is bounded by edges in M and M' such that these edges form at least one crossing and such that all points of S on the boundary of the area form a sequence of consecutive points of S along the boundary of the convex hull of S; see Fig. 6.

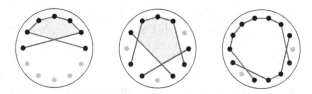

Fig. 6. Boundary areas with five points (left) and three points (middle). The drawing on the right does not show a boundary area; not all points are neighboring on the convex hull of S. (Color figure online)

We next define two special matchings. A *2-semiear matching* is a matching on a set of $4k$ points consisting of exactly k 2-semiears and an inside k-semicycle (with all its edges being diagonals). Similarly, a *near-2-semiear matching* is a matching on a set of $4k + 2$ points consisting of exactly k 2-semiears and an inside $(k + 1)$-semicycle; see Fig. 7.

As for perimeter matchings, we distinguish between *odd* and *even* (near-) 2-semiear matchings. If the perimeter edges of the 2-semiears are labeled 'even' then we call the (near-)2-semiear matching even, otherwise we call it odd.

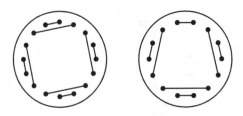

Fig. 7. Left: A 2-semiear matching. Right: A near-2-semiear matching.

Lemma 3. *Let M, M' be two matchings whose symmetric difference is an ear or a boundary area with at least three points. Then M and M' are not $\mathcal{T}$-compatible to each other.*

Proof Idea. Assuming that M and M' are both compatible to a tree T, we show that the points of an ear or boundary area span at most two induced subtrees of T. A counting argument on numbers of edges gives then a contradiction. □

Lemma 4. *Let M be a matching that is $\mathcal{T}$-compatible to an even (odd) 2-semiear-matching. Then M contains no odd (even) perimeter edge.*

Proof Idea. Adding an odd perimeter edge always yields either an ear or a boundary area with at least three points. □

The following lemma can be proven in a similar way.

Lemma 5. *Let M be a matching that is $\mathcal{T}$-compatible to a near-2-semiear-matching M' consisting of k even (odd) and one odd (even) perimeter edge. Then M contains at most one odd (even) perimeter edge, which is the one in M'.*

Lemma 6. *Let M and M' be two $\mathcal{T}$-compatible matchings. Then M and M' have at least two perimeter edges in common.*

Proof Idea. We argue here why M and M' have one perimeter edge in common. The arguments can be extended to show that M and M' have at least two perimeter edges in common.

If M contains a semiear of size at least three, then one of the perimeter edges of this semiear is in M'. Otherwise the union of m and M' contains one of the structures depicted in Fig. 8, which all prevent a disjoint spanning tree.

So we can assume that M only has 2-semiears. If M' contains the perimeter edge of a 2-semiear of M, then we are done. So assume this is not the case. If we have a union of M and M' which looks locally like Fig. 9(a) or Fig. 9(b) then M and M' are not disjoint $\mathcal{T}$-compatible. So the only possibility that M and M' are disjoint $\mathcal{T}$-compatible and do not share a perimeter edge of a 2-semiear is depicted in Fig. 9(c). Out of the 2-semiears of M we choose the one with no further semiear of M on one side of a diagonal d in M'. This is possible since

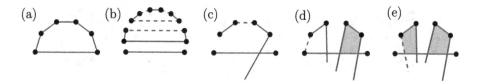

Fig. 8. All possible cases for a semiear of size $k \geq 3$ in a matching M (in red) and a second matching M' (in blue) which does not use any of the perimeter edges in M. The solid edges are the ones defining each case. (Color figure online)

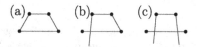

Fig. 9. All possible cases for a 2-ear in a matching M (in red) and a second matching M' (in blue) which does not use the perimeter edges in M. (Color figure online)

the number of semiears is finite and the diagonals in M' cannot intersect each other, therefore there is an ordering of the 2-semiears in M. Since d is a diagonal, there exists a semiear E' on this side of d in M'. Every edge of M on this side of d is a perimeter edge or intersects d, since there does not exist a semiear on this side of d in M. If E' is a 2-semiear and one diagonal in M intersects d, we get another blocking structure. This means that the perimeter edge of E' is also in M.   □

Corollary 1. *Let S be of size $2n \geq 10$. For even n, the distance between an even 2-semiear matching and an odd 2-semiear matching is at least 4.*
For odd n, let M be a near-2-semiear matching with a single even perimeter edge e and let M' be a near-2-semiear matching with a single even perimeter edge e' that shares a vertex with e. Then the distance between M and M' is at least 4.

Proof Idea. We obtain the statement by applying Lemma 4 (for n even) or 5 (for n odd), respectively, and Lemma 6; cf. Fig. 10.   □

4   Disjoint Caterpillar-Compatible Matchings

A natural question is what happens if we do not take the set of all plane spanning trees, but a smaller set.

A *caterpillar* (from p to q) is a tree which consists of a path (from p to q, also called *spine*) and edges with one endpoint on the path. These latter are also called the *legs* of the caterpillar. We denote the set of all plane spanning caterpillars by $\mathcal{C}$. Furthermore, a *one-legged caterpillar* is a caterpillar where every vertex of the spine is incident to at most one leg. We denote the family of all plane spanning one-legged caterpillars in S by $\mathcal{C}_3$. Note that every vertex of a one-legged caterpillar has degree at most 3. Hence, one-legged caterpillars are special instances of trees with maximum degree 3.

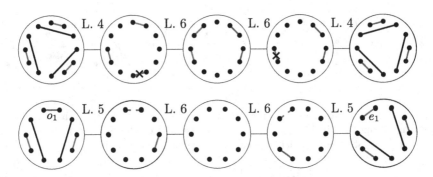

Fig. 10. The distance between two special 2-semiear matchings (top row) and between two special near-2-semiear matchings (bottom row) is at least 4. Even perimeter edges are drawn in red, odd ones are drawn in blue. The numbers next to the edges indicate which Lemma is applied. (Color figure online)

Lemma 7. *For any edge $e = pq$ of a matching M there exists a plane one-legged caterpillar compatible to M from p to q which spans all points between p and q along the boundary of the convex hull of S (on either side of e).*

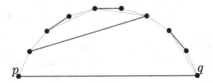

Fig. 11. A matching (in blue) and a compatible caterpillar (in green) constructed in the proof of Lemma 7. (Color figure online)

Proof Idea. We construct the caterpillar C in a greedy way from p to q. Assume we are at a point x and the next two points are y and z. If xy is not an edge of M, we add xy to C and continue from y. Otherwise, if xy is an edge of M, then xz and yz are not edges of M. We add xz and yz to C and continue from z. By construction, every spine vertex has at most one leg. So we constructed a one-legged caterpillar. An example is depicted in Fig. 11.   □

Note that every matching M contains a perimeter edge and by Lemma 7 there also exists a caterpillar which is disjoint compatible to M. Further, by construction p is incident to only one edge.

Lemma 8. *Let M and M' be two matchings whose symmetric difference is an inside cycle. Then M and M' are disjoint C-compatible.*

Proof Idea. For every diagonal of the inside cycle we get a caterpillar by Lemma 7. We merge caterpillars which have a point in common. This yields a set of (in

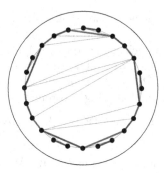

Fig. 12. Constructed caterpillar (in green) for two disjoint $\mathcal{C}$-compatible matchings (in red and blue, common edges in blue). (Color figure online)

general non-spanning) caterpillars on S. One can connect the caterpillars of this set in a zig-zag way and add the remaining points like a fan; see Fig. 12. □

Note that Lemma 8 is a sufficient condition for $\mathcal{C}$-compatibility of matchings similar to Lemma 1 for $\mathcal{T}$-compatibility. Adapting the proof of Theorem 1 to rotate only one cycle (instead of several) per step, and noting that the number of cycles is $O(n)$, we get the following theorem.

Theorem 2. *For $2n \geq 10$, the graph $\mathrm{DCG}_S(\mathcal{C})$ is connected with diameter* $\mathrm{diam}(\mathrm{DCG}_S(\mathcal{C})) = \mathcal{O}(n)$.

Next we consider disjoint $\mathcal{C}_3$-compatible matchings. As before, we first find a sufficient condition for their compatibility.

Lemma 9. *Let M and M' be two matchings whose symmetric difference is an inside 2-cycle. Then M and M' are disjoint $\mathcal{C}_3$-compatible.*

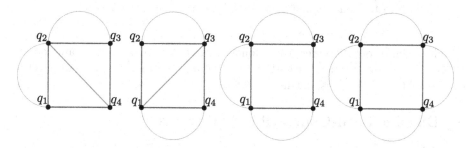

Fig. 13. All possibilities for an inside 2-cycle (in red and blue) with a disjoint compatible caterpillar (in green). The half circles are caterpillars. (Color figure online)

Proof Idea. We have four cases depending on how many edges of the 2-cycle are diagonals and their relative position. The four cases are depicted in Fig. 13.

In the leftmost case, where exactly two diagonals share a point, we take two one-legged caterpillars from q_2 to q_1 and q_3, respectively, constructed as in the proof of Lemma 7. Note that each such caterpillar has degree 1 at its start point. Hence, together with the edge q_2q_4 they form a one-legged caterpillar which is disjoint compatible to both M and M'. The other cases work similarly.   □

With this, we can show the following theorem.

Theorem 3. *For $2n \geq 10$, the graph $\mathrm{DCG}_S(\mathcal{C}_3)$ is connected and has diameter* $\mathrm{diam}(\mathrm{DCG}_S(\mathcal{C}_3)) = \mathcal{O}(n)$.

Proof Idea. We first show that any two matchings M and M' whose symmetric difference is a single inside cycle K are connected in $\mathrm{DCG}_S(\mathcal{C}_3)$. An inside cycle K can always be split into interior-disjoint inside 2-cycles $K_1, \ldots, K_r$; cf. Fig. 14. Note that every edge of K is in exactly one of $K_1, \ldots, K_r$. Further, every edge of K_i, $1 \leq i \leq r$, in the interior of K is in exactly two of $K_1, \ldots, K_r$. Let $M_0, \ldots, M_r$ be matchings such that the symmetric difference of M_{i-1} and M_i is K_i for $i = 1, \ldots r$. Then by Lemma 9, M_{i-1} and M_i are disjoint $\mathcal{C}_3$-compatible. Further, the symmetric difference of M_0 and M_r is K, implying that M and M' are connected in $\mathrm{DCG}_S(\mathcal{C}_3)$.

Combining this result with the proof of Theorem 2, it follows that $\mathrm{DCG}_S(\mathcal{C}_3)$ is connected.

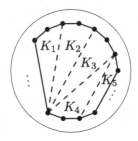

Fig. 14. Subdivision of an inside 6-cycle into five inside 2-cycles.

The bound on diameter then follows from the bound on diameter of $\mathrm{DCG}_S(\mathcal{T})$ in combination with the fact that any set of disjoint inside cycles can be split into $\mathcal{O}(n)$ disjoint inside 2-cycles.   □

5   Disjoint Path-Compatible Matchings

Let $\mathcal{P}$ be the family of all spanning paths on S. Note that paths are special instances of trees and caterpillars. The following proposition states that in contrast to trees and caterpillars, $\mathrm{DCG}_S(\mathcal{P})$ is disconnected.

Proposition 1. *Let M be a plane matching on S with at least three semiears. Then there is no spanning path on S which is disjoint compatible to M, that is, M is an isolated vertex in $\mathrm{DCG}_S(\mathcal{P})$.*

Proof Idea. If there exists a spanning path $P \in \mathcal{P}$ compatible to M, then P has an end in every semiear of M. Since every path has only two ends, P cannot end in more than two semiears. □

From Proposition 1 it follows that $\mathrm{DCG}_S(\mathcal{P})$ contains isolated vertices if S is a set of at least 12 points. Note that there are also matchings with two semiears that are not compatible to any spanning path. On the other hand, one might ask whether all matchings which are disjoint $\mathcal{P}$-compatible to some other matching are in one connected component of $\mathrm{DCG}_S(\mathcal{P})$. The following proposition gives a negative answer to that question.

Proposition 2. *The two perimeter matchings are not connected in* $\mathrm{DCG}_S(\mathcal{P})$.

Proof Idea. We show that every matching that is in the component of $\mathrm{DCG}_S(\mathcal{P})$ containing the even perimeter matching has only even semiears[2]. Assume to the contrary that there exist two disjoint $\mathcal{P}$-compatible matchings M and M' such that M has only even semiears and M' has an odd semiear. Then, roughly speaking, the union of M and M' has either three disjoint semiears (which gives a contradiction by Proposition 1) or at least one ear or one boundary area of size at least 3, which also gives a contradiction by Lemma 3. Since the odd perimeter matching does have an odd ear (is an odd ear), it cannot be in the same component as the even perimeter matching. □

We remark that several more observations on $\mathrm{DCG}_S(\mathcal{P})$ can be found in [9].

6   Conclusion and Discussion

We have shown that the diameter of the disjoint $\mathcal{T}$-compatible graph $\mathrm{DCG}_S(\mathcal{T})$ for point sets S of $2n$ points in convex position is 4 or 5 when $2n \geq 10$.

We conjecture that the diameter of $\mathrm{DCG}_S(\mathcal{T})$ is 4 for all $2n \geq 18$. An open question is the computational complexity of determining whether two given matchings have distance 3 in $\mathrm{DCG}_S(\mathcal{T})$.

For $\mathrm{DCG}_S(\mathcal{C})$ and $\mathrm{DCG}_S(\mathcal{C}_3)$, we showed that their diameters are both in $\mathcal{O}(n)$. Determining whether those two diameters are (asymptotically) the same, and what their precise values are, remains open.

Regarding spanning paths we showed that $\mathrm{DCG}_S(\mathcal{P})$ is disconnected, with no connection between the two perimeter matchings and many isolated vertices.

Further natural open questions include determining whether $\mathrm{DCG}_S(\mathcal{T})$ is connected for general point sets, and whether there exist point sets S such that $\mathrm{DCG}_S(\mathcal{P})$ is connected.

We remark that our main approach for bounding diameters was to rotate inside semicycles. A similar approach has also been used in a different setting of flip graphs of matchings. A difference is that in the flip graph setting, semiears can be flipped, which is not possible in the disjoint $\mathcal{T}$-compatible setting. On the other hand, one can flip only one semicycle, or even only two edges at a time. A recent related work on flip graphs is [8]. There, so-called *centered* flips in

---

[2] Even (odd) semiears have only even (odd) perimeter edges.

matchings on convex point sets are considered. A centered flip is the rotation of an empty quadrilateral that contains the center of the point set. This operation is more restrictive than our rotation of quadrilaterals for $DCG_S(\mathcal{C}_3)$, as can also be seen by the fact that the flip graph of matchings with centered flips is sometimes disconnected.

References

1. Aichholzer, O., Asinowski, A., Miltzow, T.: Disjoint compatibility graph of non-crossing matchings of points in convex position. Electron. J. Comb. **22**, 1–65 (2015). http://www.combinatorics.org/ojs/index.php/eljc/article/view/v22i1p65
2. Aichholzer, O., et al.: Compatible geometric matchings. Comput. Geom. Theor. Appl. **42**(6–7), 617–626 (2009)
3. Aichholzer, O., García, A., Hurtado, F., Tejel, J.: Compatible matchings in geometric graphs. In: Proceeding XIV Encuentros de Geometría Computacional, pp. 145–148. Alcalá, Spain (2011)
4. Hernando, C., Hurtado, F., Noy, M.: Graphs of non-crossing perfect matchings. Graphs Comb. **18**(3), 517–532 (2002). https://doi.org/10.1007/s003730200038
5. Houle, M.E., Hurtado, F., Noy, M., Rivera-Campo, E.: Graphs of triangulations and perfect matchings. Graphs Comb. **21**(3), 325–331 (2005). https://doi.org/10.1007/s00373-005-0615-2
6. Hurtado, F., Tóth, C.D.: Plane geometric graph augmentation: a generic perspective. In: Thirty Essays on Geometric Graph Theory, pp. 327–354. Springer (2013). https://doi.org/10.1007/978-1-4614-0110-0_17
7. Ishaque, M., Souvaine, D.L., Tóth, C.D.: Disjoint compatible geometric matchings. Discrete Comput. Geom. **49**(1), 89–131 (2013)
8. Milich, M., Mütze, T., Pergel, M.: On flips in planar matchings. Discret. Appl. Math. **289**, 427–445 (2021)
9. Obmann, J.: Disjoint Compatibility of Plane Perfect Matchings via other Graph Classes. Master's thesis, Graz University of Technology (2020)
10. Pilz, A., Rollin, J., Schlipf, L., Schulz, A.: Augmenting geometric graphs with matchings. In: GD 2020. LNCS, vol. 12590, pp. 490–504. Springer, Cham (2020). https://doi.org/10.1007/978-3-030-68766-3_38
11. Razen, A.: A lower bound for the transformation of compatible perfect matchings. In: Proceedings of EuroCG, pp. 115–118 (2008)

Testing Isomorphism of Chordal Graphs of Bounded Leafage is Fixed-Parameter Tractable (Extended Abstract)

Vikraman Arvind[1], Roman Nedela[2], Ilia Ponomarenko[3], and Peter Zeman[4(✉)]

[1] The Institute of Mathematical Sciences (HBNI), Chennai, India
`arvind@imsc.res.in`
[2] Faculty of Applied Sciences, University of West Bohemia, Pilsen, Czech Republic
`nedela@savbb.sk`
[3] V. A. Steklov Institue of Mathematics, Russian Academy of Sciences,
St. Petersburg, Russia
`inp@pdmi.ras.ru`
[4] Institut de mathématiques, Université de Neuchâtel, Neuchâtel, Switzerland
`zeman.peter.sk@gmail.com`

Abstract. The computational complexity of the graph isomorphism problem is considered to be a major open problem in theoretical computer science. It is known that testing isomorphism of chordal graphs is polynomial-time equivalent to the general graph isomorphism problem. Every chordal graph can be represented as the intersection graph of some subtrees of a representing tree, and the leafage of a chordal graph is defined to be the minimum number of leaves in a representing tree for it. We prove that chordal graph isomorphism is fixed parameter tractable with leafage as parameter.

Keywords: graph isomorphism · chordal graphs · leafage · fixed parameter tractable problem

1 Introduction

The graph isomorphism problem is one of the few natural problems in NP that is neither known to be NP-complete nor it is known to be polynomial-time solvable. In his breakthrough work, Babai [4] proved that the graph isomorphism problem is solvable in quasipolynomial time, i.e., in time $n^{\mathrm{poly}(\log n)}$, where n is the number of vertices.

The full version of this paper is available on arXiv [2]. Roman Nedela was supported by GAČR 20-15576S and APVV-19-0308. Peter Zeman was supported by the Swiss National Science Foundation project PP00P2-202667. While at Department of Applied Mathematics, Faculty of Mathematics and Physics, Charles University, Peter Zeman was supported by GAČR 20-15576S.

© Springer Nature Switzerland AG 2022
M. A. Bekos and M. Kaufmann (Eds.): WG 2022, LNCS 13453, pp. 29–42, 2022.
https://doi.org/10.1007/978-3-031-15914-5_3

A significant line of research concerns the parameterized complexity of the graph isomorphism problem with respect to some natural graph parameter. These include treewidth [18], degree [14,20], genus [21,23], excluded minors [15,22], etc. It is worth mentioning that in several of these cases, Babai's new techniques have yielded new algorithms with improved running time. For example, Luks's original algorithm with running time $n^{O(k)}$ for degree-k graphs has a modified $n^{\text{poly}(\log k)}$-time algorithm [14,20]. However, in some of these cases a fixed-parameter tractable (FPT) algorithm, i.e., an algorithm with running time $f(k)\,\text{poly}(n)$, have remained elusive. Such an improvement likely cannot be obtained using known techniques and would require some new techniques and ideas.

In our work, we deal with parameterized complexity of the graph isomorphism problem for the class of chordal graphs. An undirected graph is said to be *chordal* if it has no chordless cycle of length at least four. Every chordal graph admits a representation as the intersection graph of subtrees of some tree T [13]. In this case, we say that T is a representing tree for X. The *leafage* $\ell(X)$ of a chordal graph X is the least positive integer such that X has a representing tree with $\ell(X)$ leaves. The notion of leafage was introduced in [17] and is a natural graph parameter for chordal graphs.

It is interesting to note that the well-studied interval graphs are precisely the intersection graphs of paths. It follows that $\ell(X) \leq 2$ if and only if X is an interval graph (and $\ell(X) = 1$ if and only if X is complete). Thus, the leafage of a chordal graph X measures how far it is from being an interval graph, which has interesting algorithmic consequences. For instance, efficient solutions to certain NP-hard problems on interval graphs naturally extend to chordal graphs of bounded leafage; e.g., [25].

Graph Isomorphism restricted to chordal graphs is polynomial-time equivalent to Graph Isomorphism for general graphs [19, Theorem 5]. On the other hand, the problem can be solved in polynomial (even linear time) for interval graphs [19]. The main result of the present paper can be considered as a substantial generalization of the latter.

Theorem 1. *Testing isomorphism of chordal graphs of leafage ℓ is fixed parameter tractable, with ℓ as fixed parameter.*

The leafage of chordal graphs is known to be polynomial-time computable [16]. Denote by $\mathfrak{K}_\ell$ the class of all chordal graphs of leafage at most ℓ. In particular, the graph class $\mathfrak{K}_\ell$ is polynomial-time recognizable.

In order to test if two connected graphs $X, Y \in \mathfrak{K}_\ell$ are isomorphic, it suffices to check if there is a generator of the automorphism group of their disjoint union $X \cup Y$, which swaps X and Y. Since the graph $X \cup Y$ belongs to the class $\mathfrak{K}_{2\ell}$, the graph isomorphism problem for the graphs in $\mathfrak{K}_\ell$ is reduced to the problem of determining the automorphism group of a given graph in $\mathfrak{K}_{2\ell}$. Thus Theorem 1 is an immediate consequence of the following theorem which is proved in the paper.

Theorem 2. *Given an n-vertex graph $X \in \mathfrak{K}_\ell$, a generating set of the group* Aut(X) *can be found in time* $t(\ell)\,\mathrm{poly}(n)$, *where* $t(\cdot)$ *is a function independent of* n.

The function t from Theorem 2 is bounded from above by a polynomial in $(\ell 2^\ell)!$. The running time bound, especially the function t, does not appear to be final and, most likely, it can be significantly improved.

We emphasize that our algorithm does not require that the input X is given by an intersection representation. Indeed, the algorithm works correctly on all chordal graphs and the leafage bound ℓ is required only to bound the running time for inputs from the class $\mathfrak{K}_\ell$.

The proof of Theorem 2 is given in Sect. 7. The main steps involved in the algorithm are: (a) to transform the given graph X efficiently into an order-3 hypergraph $H = H(X)$ (see below), (b) to give an algorithm for computing a generating set for Aut(H), and (c) to recover from it a generating set for Aut(X).

This brings us to the notion of *higher-order hypergraphs*. A usual hypergraph with vertex set V has hyperedge set contained in the power set $\mathcal{E}_1 = 2^V$. The hyperedges of an order-3 hypergraph H will, in general, include order-2 and order-3 hyperedges. These are elements of $\mathcal{E}_2 = 2^{\mathcal{E}_1}$ and $\mathcal{E}_3 = 2^{\mathcal{E}_2}$, respectively. The hyperedge set E of H is contained in $\mathcal{E}_1 \cup \mathcal{E}_2 \cup \mathcal{E}_3$ and can be of triple-exponential size in $|V|$. However, the input size of H is defined to be $|V| \cdot |E|$, for H given as input to an algorithm. The efficient reduction from finding Aut(X) to finding Aut(H) is presented in Sects. 4 and 5. The key point of the reduction is a graph-theoretical analysis of the vertex coloring of the chordal graph X obtained by the 2-dimensional Weisfeiler-Leman algorithm [26]. The reduction takes X as input and computes the colored order-3 hypergraph H such that each vertex color class of H has size at most $b = \ell 2^\ell$, where $\ell = \ell(X)$.

At this point, we deal with the general problem of determining the automorphism group of a colored order-k hypergraph H ($k \geq 1$) by an FPT algorithm with respect to the parameter b which bounds the size of each vertex color class. This problem seems interesting in itself and could find other applications. For ordinary hypergraphs, it was shown to be fixed parameter tractable in [3]. A generalization of that result to order-k hypergraphs is given in Sect. 6. The running time bound we obtain is not FPT in terms of the parameter k.

We complete the introduction with some remarks about H-graphs introduced in [5]. An H-graph X is an intersection graph of connected subgraphs of a subdivision of a fixed graph H. Every graph is an H-graph for a suitable H, which gives a parameterization for all graphs. It is interesting to note that we can get well-known graph classes as H-graphs for suitable choices of H. For instance, interval graphs are K_2-graphs, circular-arc graphs are K_3-graphs, and chordal graphs are the union of all T-graphs, where T is a tree.

Basic algorithmic questions on H-graphs, including their recognition and isomorphism testing, have been studied, e.g., [8,9,12]. It is shown in [1] that isomorphism testing for S_d-graphs, where S_d is a star of degree d, is fixed parameter tractable. Since S_d-graphs are chordal graphs of leafage at most d, our FPT algorithm applied to chordal graphs with bounded leafage significantly extends that result [1].[1] On the other hand, the isomorphism problem for H-graphs is as hard as the general graph isomorphism problem if H is not unicyclic [10]. Thus, it remains open whether isomorphism can be solved in polynomial time for the unicyclic case with fixed number of leaves, which would provide a dichotomy for parameterization by H-graphs. Our work can be also considered a step towards this dichotomy.

2   Preliminaries

General Notation. Throughout the paper, Ω is a finite set. Given a bijection f from Ω to another set and a subset $\Delta \subseteq \Omega$, we denote by f^{Δ} the bijection from Δ to $\Delta^f = \{\delta^f : \delta \in \Delta\}$. For a set S of bijections from Ω to another set, we put $S^{\Delta} = \{f^{\Delta} : f \in S\}$.

Let π be a partition of Ω. The set of all unions of the classes of π is denoted by $\pi^{\cup}$. The partition π is a *refinement* of a partition π' of Ω if each class of π' belongs to $\pi^{\cup}$; in this case, we write $\pi \geq \pi'$, and $\pi > \pi'$ if $\pi \geq \pi'$ and $\pi \neq \pi'$. The partition of $\Delta \subseteq \Omega$ induced by π is denoted by π_{Δ}.

Graphs. Let X be an undirected graph. The vertex and edge sets of X are denoted by $\Omega(X)$ and $E(X)$, respectively. The automorphism group of X is denoted by $\mathrm{Aut}(X)$. The set of all isomorphisms from X to a graph X' is denoted by $\mathrm{Iso}(X, X')$.

The set of all leaves and of all connected components of X are denoted by $L(X)$ and $\mathrm{Conn}(X)$, respectively. For a vertex α, we denote by αX the set of neighbors of α in X. The vertices α and β are called *twins* in X if every vertex other than α and β is adjacent either to both α and β or neither of them. The graph X is said to be *twinless* if no two distinct vertices of X are twins.

Let $\Delta, \Gamma \subseteq \Omega(X)$. We denote by $X_{\Delta,\Gamma}$ the graph with vertex set $\Delta \cup \Gamma$ in which two vertices are adjacent if and only if one of them is in Δ, the other one is in Γ, and they are adjacent in X. Thus, $X_{\Delta} = X_{\Delta,\Delta}$ is the subgraph of X induced by Δ, and $X_{\Delta,\Gamma}$ is bipartite if $\Delta \cap \Gamma = \varnothing$.

Let $\Delta \subseteq \Omega(X)$ and $Y = X_{\Delta}$. The set of all vertices adjacent to at least one vertex of Δ and not belonging to Δ is denoted by ∂Y. The subgraph of X, induced by $\Delta \cup \partial Y$ is denoted by $\overline{Y}$.

For a tree T, let $S(T) = \{\Omega(T') : T' \text{ is a subtree of } T\}$ be the set of all vertex sets of the subtrees of T. A *representation* of a graph $X = (\Omega, E)$ on the tree T (called *tree-representation*) is a function $R \colon \Omega \rightarrow S(T)$ such that for all $u, v \in \Omega$,

$$R(u) \cap R(v) \neq \varnothing \Leftrightarrow \{u, v\} \in E.$$

A graph X is chordal if and only if X has a tree-representation [13]. The leafage $\ell(X)$ of X is defined to be the minimum of $|L(T)|$ over all trees T such that X has a tree-representation on T.

---

[1] After posting our paper on the arXiv [2], we found a paper [7] containing an FPT algorithm testing isomorphism of T-graphs for every fixed tree T. This gives an alternative FPT algorithm for chordal graphs of leafage ℓ.

Colorings. A partition π of Ω is said to be a *coloring* (of Ω) if the classes of π are indexed by elements of some set, called *colors*. In this case, the classes of π are called *color classes* and the color class containing $\alpha \in \Omega$ is denoted by $\pi(\alpha)$. Usually the colors are assumed to be linearly ordered. A bijection f from Ω to another set equipped with coloring π' is said to be *color preserving* if the colors of $\pi(\alpha)$ and $\pi'(f(\alpha))$ are the same for all points $\alpha \in \Omega$.

A graph equipped with a coloring of the vertex set (respectively, edge set) is said to be *vertex colored* (respectively, *edge colored*); a graph that is both vertex and edge colored is said to be *colored*. The isomorphisms of vertex/edge colored graphs are ordinary isomorphisms that are color preserving. To emphasize this, we sometimes write $\mathrm{Aut}(X, \pi)$ for the automorphism group of a graph X with coloring π.

Let X be a colored graph with vertex coloring π. Consider the application of the Weisfeiler-Leman algorithm (2-dim WL) to X [26]. For the purpose of the paper, it suffices to understand that 2-dim WL iteratively colors pairs of vertices of X until the coloring satisfies a specific regularity condition (where the vertex coloring corresponds to the coloring of diagonal pairs (α, α)). The resulting coloring of pairs is just what is called a *coherent configuration*.

The output of 2-dim WL defines a new vertex coloring $\mathrm{WL}(X, \pi) \geq \pi$ of X. We say that π is *stable* if $\mathrm{WL}(X, \pi) = \pi$. In the language of coherent configurations, π is stable precisely when the classes of π are the fibers of a coherent configuration (details can be found in the monograph [11]). In the sequel, we will use some elementary facts from theory of coherent configurations. The following statement summarizes relevant properties of stable colorings.

Lemma 1. *Let X be a graph and π be a stable coloring of X. Then*

(1) for $\Delta, \Gamma \in \pi$, the number $|\delta X \cap \Gamma|$ does not depend on $\delta \in \Delta$,
(2) if $\Delta \in \pi^{\cup}$ or $X_\Delta \in \mathrm{Conn}(X)$, then the coloring π_Δ is stable.

A coloring π of the vertices of a graph X is said to be *invariant* if every class of π is $\mathrm{Aut}(X)$-invariant. In this case, the coloring $\mathrm{WL}(X, \pi)$ is also invariant and stable. Since the coloring of the vertices in one color is invariant and the Weisfeiler-Leman algorithm is polynomial-time, in what follows we deal with invariant stable colorings.

Hypergraphs. Let V be a finite set. The set $\mathcal{E}_k = \mathcal{E}_k(V)$ of the order-k *hyperedges* on V is defined recursively as follows:

$$\mathcal{E}_0 = V, \qquad \mathcal{E}_k = \mathcal{E}_{k-1} \cup 2^{\mathcal{E}_{k-1}} \text{ for } k > 1.$$

So, we consider elements of V as order-0 hyperedges and the order-k hyperedges include all order-$(k-1)$ hyperedges and their subsets.

Let $U \subseteq V$ and $e \in \mathcal{E}_k$ ($k \geq 1$). We recursively define the *projection* of e on U as the multiset

$$e^U = \begin{cases} e \cap U & \text{if } k = 1, \\ \{\{\tilde{e}^U : \tilde{e} \in e\}\} & \text{if } k > 1. \end{cases}$$

We extend this definition to all sets $E \subseteq \mathcal{E}_k$ by putting $E^U = \{e^U : e \in E\}$.

Definition 1 (order-k hypergraph). *An order-k hypergraph ($k \geq 1$) on V is a pair $H = (V, E)$, where $E \subseteq 2^{\mathcal{E}_k}$; the elements of V and E are called vertices and hyperedges of H, respectively.*

Clearly, order-1 hypergraphs are usual hypergraphs. Moreover, higher-order hypergraphs (i.e., order-k hypergraph for some k) are combinatorial objects in the sense of [6]. The concepts of isomorphism and coloring extend to higher-order hypergraphs in a natural way.

Let $k \geq 2$. The $(k-1)$-*skeleton* of an order-k hypergraph $H = (V, E)$ is an order-$(k-1)$ hypergraph $H^{(k-1)}$ on V with the hyperedge set

$$E^{(k-1)} = \{\tilde{e} \in \mathcal{E}_{k-1} : \tilde{e} \text{ is an element of some } e \in \mathcal{E}_k \cap E\}.$$

It is easily seen that for every order-k hypergraph $H' = (V', E')$

$$\mathrm{Iso}(H, H') = \{f \in \mathrm{Iso}(H^{(k-1)}, H'^{(k-1)}) : e \in E^{(k)} \Leftrightarrow e^f \in E'^{(k)}\}. \quad (1)$$

where for each order-k hyperedge $e = \{e_1, \ldots, e_a\}$ we set $e^f = \{e_1^f, \ldots, e_a^f\}$.

Let $H_1 = (V_1, E_1)$ be an order-k hypergraph for some k and $H_2 = (V_2, E_2)$ be a usual hypergraph such that $V_2 = E_1$. Then each hyperedge $e \in E_2$ is a subset of hyperedges of H_1. We define the *hypergraph composition* of H_1 and H_2 to be the order-$(k+1)$ hypergraph

$$H := H_1 \uparrow H_2 = (V, E_1 \cup E_2).$$

When the hypergraphs H_1 and H_2 are colored, the vertex coloring of H is defined in the obvious way. The color $c(e)$ of $e \in E(H)$ is defined as follows: if $e \in E_1 \setminus E_2$ then $c(e)$ is the color $c_1(e)$ of e in H_1. If $e \in E_1 \cap E_2$ then $c(e)$ is defined as the triple $(0, c_1(e), c_2(e))$, where $c_2(e)$ is the color of e in H_2. Finally, if $e \in E_2 \setminus E_1$ then $c(e) = (1, c_1(e'), c_2(e))$, where e' is the set of elements of e.

In computations with high order hypergraphs, every hyperedge is considered as a rooted tree and the size of a high order hypergraph is defined to be the sum of sizes of these trees.

3   Chordal Graphs

3.1   Stable Colorings in Chordal Graphs

We now present some auxiliary statements on the structure of subgraphs of a chordal graph induced by one or two color classes of a stable coloring.

Lemma 2. *Let X be a chordal graph and π a stable coloring of X. Then for every $\Delta, \Gamma \in \pi$, the following statements hold:*

(1) $\mathrm{Conn}(X_\Delta)$ *consists of cliques of the same size,*
(2) *if $|\mathrm{Conn}(X_\Delta)| \leq |\mathrm{Conn}(X_\Gamma)|$, then $\mathrm{Conn}(X_\Delta) = \{Y_\Delta : Y \in \mathrm{Conn}(X_{\Delta \cup \Gamma})\}$,*
(3) *if the graphs X_Δ and X_Γ are complete, then $X_{\Delta, \Gamma}$ is either complete bipartite or empty.*

Lemma 3. *Let X be a connected chordal graph and let π be a stable partition of Ω. There exists $\Delta \in \pi$ such that the graph X_Δ is complete.*

3.2 Estimates Depending on the Leafage

The two lemmas in this subsection show bounds that are crucial for estimating the complexity of the main algorithm.

Lemma 4. *Let X be a chordal graph, Δ a subset of its vertices, $X - \Delta$ is the subgraph of X induced by the complement of Δ, and*

$$S = S(X, \Delta) = \{Y \in \mathrm{Conn}(X - \Delta) : \overline{Y} \text{ is not interval}\}. \tag{2}$$

Then $|S| \leq \ell(X) - 2$.

Let π be a vertex coloring of X. Given a pair $(\Delta, \Gamma) \in \pi \times \pi$, we define an equivalence relation $e_{\Delta, \Gamma}$ on Δ by setting

$$(\delta, \delta') \in e_{\Delta, \Gamma} \iff \delta \text{ and } \delta' \text{ are twins in } X_{\Delta, \Gamma}. \tag{3}$$

Note that the equivalence relation $e_{\Gamma, \Delta}$ is defined on Γ, and coincides with $e_{\Delta, \Gamma}$ only if $\Gamma = \Delta$. The sets of classes of $e_{\Delta, \Gamma}$ and $e_{\Gamma, \Delta}$ are denoted by $\Delta/e_{\Delta, \Gamma}$ and $\Gamma/e_{\Gamma, \Delta}$, respectively.

Lemma 5. *Let X be a chordal graph, π a stable coloring, and $\Delta, \Gamma \in \pi$. Assume that the graph X_Δ is complete. Then*

$$|\Delta/e_{\Delta, \Gamma}| \leq 2^\ell \quad \text{and} \quad |\Gamma/e_{\Gamma, \Delta}| \leq \ell, \tag{4}$$

where $\ell = \ell(X)$.

4   Critical Set of a Chordal Graph

Let X be a chordal graph and π a stable coloring. Denote by $\Omega^* = \Omega^*(X, \pi)$ the union of all $\Delta \in \pi$ such that

$$|\mathrm{Conn}(X_\Delta)| \leq \ell(X). \tag{5}$$

By Lemma 2(1), the graph X_Δ is a disjoint union of cliques; thus the above condition means that the number of them is at most $\ell(X)$. By Lemma 3, the set Ω^* is not empty if the graph X is connected.

Theorem 3. *Let X be a chordal graph and $\Omega^* = \Omega^*(X, \pi)$. Then one of the following statements holds:*

(i) for every $Y \in \mathrm{Conn}(X - \Omega^*)$, the graph $\overline{Y}$ is interval,
(ii) there is a invariant stable coloring $\pi' > \pi$.

Moreover, in case (ii), the coloring π' can be found in polynomial time in $|\Omega|$.

We say that Ω^* is a *critical set* of X (with respect to π) if statement (i) of Theorem 3 holds. In the rest of the section we define a hypergraph $\mathcal{H}^*$ associated with the critical set Ω^* and show that the groups $\mathrm{Aut}(\mathcal{H}^*)^{\Omega^*}$ and $\mathrm{Aut}(X)^{\Omega^*}$ are closely related.

The vertices of $\mathcal{H}^*$ are set to be the elements of the disjoint union

$$V = \bigcup_{\Delta \in \pi_{\Omega^*}} \bigcup_{\Gamma \in \pi} \Delta / e_{\Delta, \Gamma},$$

where $e_{\Delta, \Gamma}$ is the equivalence relation on Δ, defined by formula (3). Thus any vertex of $\mathcal{H}^*$ is a class of some $e_{\Delta, \Gamma}$. Taking the disjoint union means, in particular, that if Λ is a class of $e_{\Delta, \Gamma}$ and $e_{\Delta, \Gamma'}$, then V contains two vertices corresponding to Λ. The partition

$$\overline{\pi} = \{ \Delta / e_{\Delta, \Gamma} : \ \Delta \in \pi_{\Omega^*}, \ \Gamma \in \pi \}$$

of the set V is treated as a coloring of V.

Let us define the hyperedges of $\mathcal{H}^*$. First, let $\alpha \in \Omega^*$. Denote by Δ the class of π, containing α. Then $\Delta \in \pi_{\Omega^*}$. Moreover, for every $\Gamma \in \pi$, there is a unique class $\Lambda_\alpha(\Delta, \Gamma)$ of the equivalence relation $e_{\Delta, \Gamma}$, containing α. Put

$$\overline{\alpha} = \{ \Lambda_\alpha(\Delta, \Gamma) : \ \Gamma \in \pi \},$$

in particular, $\overline{\alpha} \subseteq V$. It is easily seen that $\overline{\alpha} = \overline{\beta}$ if and only if the vertices α and β are twins in X, lying in the same class of π. Next, let $\beta \in \Omega^*$ be adjacent to α in X, and Γ the class of π, containing β. Then every vertex in $\Lambda_\alpha(\Delta, \Gamma)$ is adjacent to every vertex of $\Lambda_\beta(\Gamma, \Delta)$. Put

$$\overline{\{\alpha, \beta\}} = \{ \Lambda_\alpha(\Delta, \Gamma), \Lambda_\beta(\Gamma, \Delta) \},$$

again $\overline{\{\alpha, \beta\}} \subseteq V$. With this notation, the hyperedge set of $\mathcal{H}^*$ is defined as the union:

$$E^* = \{ \overline{\alpha} : \ \alpha \in \Omega^* \} \cup \{ \overline{\{\alpha, \beta\}} : \ \alpha, \beta \in \Omega^*, \ \beta \in \alpha X \}.$$

As we are interested only in the automorphisms of E^* that stabilize the two parts $\{ \overline{\alpha} : \ \alpha \in \Omega^* \}$ and $\{ \overline{\{\alpha, \beta\}} : \ \alpha, \beta \in \Omega^*, \ \beta \in \alpha X \}$, we can color the hyperedges in E^* using two distinct colors to ensure this. Clearly, the hypergraph $\mathcal{H}^* = (V, E^*)$ and the coloring $\overline{\pi}$ can be constructed in polynomial time in $|\Omega|$.

Theorem 4. *Let X be a chordal graph, π an invariant stable vertex coloring of X, $\Omega^* = \Omega^*(X, \pi)$ the critical set, and $\mathcal{H}^* = (V, E^*)$ is the above hypergraph with vertex coloring $\overline{\pi}$. Then*

(i) $\max\{ |\Delta| : \ \Delta \in \overline{\pi} \} \le \ell 2^\ell$, *where $\ell = \ell(X)$,*
(ii) *if X is twinless, then the mapping $f : \Omega^* \to E^*$, $\alpha \mapsto \overline{\alpha}$, is an injection,*
(iii) *if X is twinless and $G = G(\mathcal{H}^*)$ is the group induced by the natural action of $\mathrm{Aut}(\mathcal{H}^*)$ on $\mathrm{Im}(f) = \{ \overline{\alpha} \mid \alpha \in \Omega^* \} \subseteq E^*$, then*

$$\mathrm{Aut}(X)^{\Omega^*} \le G^{f^{-1}} \le \mathrm{Aut}(X_{\Omega^*}), \tag{6}$$

where $G^{f^{-1}} = fGf^{-1}$.[2]

---

[2] Note that the composition fGf^{-1} is defined from left to right.

5   The Hypergraph Associated with the Complement of the Critical Set

The goal of this section is to provide some tools related to the critical set that will help design the algorithm for computing the automorphism group of a chordal graph in $\mathfrak{K}_\ell$.

Suppose X is a chordal graph on Ω and π an invariant stable coloring of X. Further, let Ω^* denote the critical set of X with respect to π. Let $G^\diamond = G^\diamond(X)$ denote the kernel of the restriction homomorphism $\mathrm{Aut}(X) \to \mathrm{Aut}(X)^{\Omega^*}$. We claim that a generating set for $G^\diamond$ can be efficiently computed.

Theorem 5. *A generating set for the kernel $G^\diamond \le \mathrm{Sym}(\Omega)$ of the restriction homomorphism from $\mathrm{Aut}(X)$ to $\mathrm{Aut}(X)^{\Omega^*}$ can be found in polynomial time in $|\Omega|$.*

In what follows, X is a chordal graph, π a stable coloring of X, Ω^* the critical set of X with respect to π, and $\Omega^\diamond = \Omega \setminus \Omega^*$. Recall that by the definition of critical set, every graph $\overline{Y}$, $Y \in \mathrm{Conn}(X_{\Omega^\diamond})$, is interval and

$$\partial Y = \Omega(\overline{Y}) \cap \Omega^*.$$

Lemma 6. *For every $Y \in \mathrm{Conn}(X_{\Omega^\diamond})$, there is a colored hypergraph $H = H_Y$ whose vertex set is ∂Y colored by $\pi_{\partial Y}$, and such that*

$$\mathrm{Iso}(H_Y, H_{Y'}) = \mathrm{Iso}(\overline{Y}, \overline{Y'})^{\partial Y}, \quad Y' \in \mathrm{Conn}(X_{\Omega^\diamond}). \tag{7}$$

Moreover, in time polynomial in $|\overline{Y}|$ one can

(a) *construct the hypergraph H_Y,*
(b) *given $\overline{g} \in \mathrm{Iso}(H_Y, H_{Y'})$, find $g \in \mathrm{Iso}(\overline{Y}, \overline{Y'})$ such that $g^{\partial Y} = \overline{g}$.*

Let us define a colored order-2 hypergraph $\mathcal{H}^\diamond$ with vertex set Ω^* and hyper-edge set $\mathcal{E}_1 \cup \mathcal{E}_2$, where

$$\mathcal{E}_1 = \bigcup_{Y \in \mathrm{Conn}(X_{\Omega^\diamond})} E(H_Y) \quad \text{and} \quad \mathcal{E}_2 = \{E(H_Y) : Y \in \mathrm{Conn}(X_{\Omega^\diamond})\}.$$

The vertex coloring of $\mathcal{H}^\diamond$ is set to be π_{Ω^*}. Note that the union in the definition of $\mathcal{E}_1$ is not disjoint; the color $\pi^\diamond(e)$ of a hyperedge $e \in \mathcal{E}_1$ is defined to be the multiset of the colors of e in $\mathcal{H}_Y$, where Y runs over all graphs $Y \in \mathrm{Conn}(X_{\Omega^\diamond})$ such that $e \in E(H_Y)$.

To define a coloring of $\mathcal{E}_2$, denote by $\sim$ the equivalence relation on $\mathrm{Conn}(X_{\Omega^\diamond})$ by setting

$$Y \sim Y' \quad \Leftrightarrow \quad H_Y = H_{Y'}.$$

Condition (7) implies that $Y \sim Y'$ if and only if there exists an isomorphism $g \in \mathrm{Iso}(\overline{Y}, \overline{Y'})$ such that the bijection $g^{\partial Y}$ is identical. The color $\pi^\diamond(e)$ of the hyperedge $e \in \mathcal{E}_2$ is defined to be so that if $e = \{E(H_Y)\}$ and $e' = \{E(H_{Y'})\}$, then

$$\pi^\diamond(e) = \pi^\diamond(e') \quad \Leftrightarrow \quad \mathrm{Iso}(\overline{Y}, \overline{Y'}) \ne \varnothing \quad \text{and} \quad n_Y = n_{Y'}, \tag{8}$$

where n_Y and $n_{Y'}$ are the cardinalities of the classes of the equivalence relation $\sim$, containing Y and Y', respectively.

Remark 1. Let $e \in \mathcal{E}_2$ and $Y \in \mathrm{Conn}(X - \Omega^*)$ be such that $e = E(H_Y)$. In general, the coloring π_e of the hyperedges of $\mathcal{E}_1$, contained in e, is different from the coloring π_Y of the corresponding hyperedges of H_Y. However, $\pi_e \geq \pi_Y$ and π_Y is uniquely determined by π_e.

Lemma 7. *Let X' be a colored graph obtained from X by deleting all edges of the induced subgraph X_{Ω^*}. Then*

$$\mathrm{Aut}(\mathcal{H}^\circ) = \mathrm{Aut}(X')^{\Omega^*}.$$

Moreover, given $\overline{g} \in \mathrm{Aut}(\mathcal{H}^\circ)$ one can construct $g \in \mathrm{Aut}(X')$ such that $g^{\Omega^*} = \overline{g}$ in polynomial time in $|\Omega|$.

The following theorem is the main result of the section, which together with Theorem 5 essentially provides a polynomial-time reduction of finding the group $\mathrm{Aut}(X)$ to finding the groups $\mathrm{Aut}(\mathcal{H}^*)$ and $\mathrm{Aut}(\mathcal{H}^\circ)$.

Theorem 6. *In the conditions and notation of Theorem 4, set $G^* = G(\mathcal{H}^*)^{f^{-1}}$. Then*

$$\mathrm{Aut}(X)^{\Omega^*} = \mathrm{Aut}(\mathcal{H}^\circ) \cap G^*.$$

Moreover, every permutation $\overline{g} \in \mathrm{Aut}(\mathcal{H}^\circ) \cap G^*$ can be lifted in polynomial time to an automorphism $g \in \mathrm{Aut}(X)$ such that $g^{\Omega^*} = g$.

6   Order-k Hypergraph Isomorphism: Bounded Color Classes

As stated in the theorem below, we show that the problem of testing isomorphism of colored order-k hypergraphs, in which the sizes of vertex color classes are bounded by a fixed parameter, is fixed parameter tractable; no assumption is made on the hyperedge color class sizes. Our algorithm is a generalization of the one for usual hypergraphs [3]. The detailed proof with the algorithm can be found in the arXiv version [2].

Theorem 7. *Let $k \geq 1$. Given two colored order-k hypergraphs H and H', the isomorphism coset $\mathrm{Iso}(H, H')$ can be computed in time $(b!\, s)^{O(k)}$, where b is the maximal size of a vertex color class of H and s is the size of H. In particular, the group $\mathrm{Aut}(H)$ can be found within the same time.*

Remark 2. More recently, we learned about Schweitzer and Wiebking's work [24]. They study computing canonical forms (under permutation group action) of an expressive class of combinatorial objects called heriditarily finite, and obtain algorithms to compute canonical labeling cosets for such objects. In particular, Theorem 17 of [24] can be applied to compute automorphism groups of order-k hypergraphs, and significantly improves on the time bound of our algorithm in Theorem 7 (in fact, their algorithm removes k from the exponent). This, however, does not improve the bound in our main theorem, because we use Theorem 7 for $k = 3$ only.

7   Main Algorithm and the Proof of Theorem 2

Based on the results obtained in the previous sections, we present an algorithm that constructs the automorphism group of a chordal twinless graph.

Main Algorithm

Input: a chordal twinless graph X and vertex coloring π of X.

Output: the group $\mathrm{Aut}(X, \pi)$.

Step 1. Construct $\pi = \mathrm{WL}(X, \pi)$ and $\Omega^* = \Omega^*(X, \pi)$.

Step 2. While the set Ω^* is not critical with respect to π, find $\pi := \mathrm{WL}(X, \pi')$ and set $\Omega^* := \Omega^*(X, \pi)$, where π' is the coloring from Theorem 3(ii).

Step 3. If $\Omega^* = \varnothing$, then X is interval and we output the group $\mathrm{Aut}(X, \pi)$ found by the algorithm from [19, Theorem 5].

Step 4. Construct the mapping f and colored hypergraph $\mathcal{H}^*$ on $(\Omega^*)^f$, defined in Sect. 4, and the colored hypergraph H° on Ω^*, defined in Sect. 5.

Step 5. Using the algorithm from Theorem 7, find a generating set $\overline{S}$ of the automorphism group of the colored order-3 hypergraph $\mathcal{H}^* \uparrow (\mathcal{H}^\circ)^f$.

Step 6. For each $\overline{g} \in \overline{S}$ find a lifting $g \in \mathrm{Aut}(X, \pi)$ of $f\overline{g}f^{-1} \in \mathrm{Sym}(\Omega^*)$ by the algorithm from Theorem 6; let S be the set of all these automorphisms g's.

Step 7. Output the group $\mathrm{Aut}(X, \pi) = \langle G^\circ, S \rangle$, where G° is the group defined in Theorem 5.

Theorem 8. *The Main Algorithm correctly finds the group* $\mathrm{Aut}(X, \pi)$ *in time* $t(\ell) \cdot n^{O(1)}$, *where* $n = |\Omega(X)|$, t *is a function independent of* n, *and* $\ell = \ell(X)$.

Proof. Note that the number of iterations of the loop at Step 2 is at most n, because $|\pi| \leq n$ and $|\pi'| > |\pi|$. Next, the running time at each other step, except for Step 5, is bounded by a polynomial in n, see the time bounds in the used statements. On the other hand, at Step 5, the cardinality of each vertex color class of the order-3 hypergraph $\mathcal{H}^* \uparrow (\mathcal{H}^\circ)^f$ is at most $\ell 2^\ell$ (Theorem 4(i)). By Theorem 7 for $b = \ell 2^\ell$ and $k = 3$, the running time of the Main Algorithm is at most $t(\ell) \cdot n^{O(1)}$ with $t(\ell) = ((\ell 2^\ell)!)^{O(1)}$.

To prove the correctness of the algorithm, we exploit the natural restriction homomorphism
$$\varphi : \mathrm{Aut}(X) \to \mathrm{Sym}(\Omega^*), \quad g \mapsto g^{\Omega^*}.$$
Given a generating set S' of the group $\mathrm{Im}(\varphi)$, we have $\mathrm{Aut}(X) = \langle \ker(\varphi), S \rangle$, where $S \subseteq \mathrm{Aut}(X)$ is a set of cardinality $|\overline{S}|$ such that $S' = \{\varphi(g) : g \in S\}$.

According to Step 7, $\ker(\varphi) = G^\circ$. Thus, it suffices to verify that as the set S' one can take the set $\{f\overline{g}f^{-1} : \overline{g} \in \overline{S}\}$, where f is the bijection found at Step 4 and $\overline{S}$ is the generating set of the group $\mathrm{Aut}(\mathcal{H}^* \uparrow (\mathcal{H}^\circ)^f)$, found at Step 5. By Theorem 6, we need to check that

$$\mathrm{Aut}(\mathcal{H}^* \uparrow (\mathcal{H}^\circ)^f)^{f^{-1}} = G^* \cap \mathrm{Aut}(\mathcal{H}^\circ). \tag{9}$$

Notice that

$$h \in \operatorname{Aut}(\mathcal{H}^* \uparrow (\mathcal{H}^\circ)^f) \Leftrightarrow h \in \operatorname{Aut}(\mathcal{H}^*) \quad \text{and} \quad (E(\mathcal{H}^\circ)^f)^h = E(\mathcal{H}^\circ)^f$$
$$\Leftrightarrow fhf^{-1} \in G^* \quad \text{and} \quad fhf^{-1} \in \operatorname{Aut}(\mathcal{H}^\circ)$$
$$\Leftrightarrow fhf^{-1} \in G^* \cap \operatorname{Aut}(\mathcal{H}^\circ),$$

which proves equality (9).

Proof (Theorem 2). Denote by e_X the equivalence relation on $\Omega = \Omega(X)$ such that $(\alpha, \beta) \in e_X$ if and only if the vertices α and β are twins in X. Since e_X is $\operatorname{Aut}(X)$-invariant, there is a natural homomorphism

$$\varphi : \operatorname{Aut}(X) \to \operatorname{Sym}(\Omega/e_X).$$

To find the group $\operatorname{Aut}(X)$, it suffices to construct generating sets of the groups $\ker(\varphi)$ and $\operatorname{Im}(\varphi)$, and then to lift every generator of the latter to an automorphism of X.

First, we note that every class of the equivalence relation e_X consists of twins of X. Consequently,

$$\ker(\varphi) = \prod_{\Delta \in \Omega/e_X} \operatorname{Sym}(\Delta),$$

and this group can efficiently be found.

Now let X' be the graph with vertex set Ω/e, in which the classes Δ and Γ are adjacent if and only if some (and hence each) vertex in Δ is adjacent to some (and hence each) vertex of Γ. Note that X' is isomorphic to an induced subgraph of X, and hence belongs to the class $\mathfrak{K}_\ell$. Let π' be the vertex coloring of X' such that $\pi'(\Delta) = \pi'(\Gamma)$ if and only if X_Δ and X_Γ are isomorphic, which is easy to check because each of X_Δ and X_Γ is either empty or complete. Then

$$\operatorname{Im}(\varphi) = \operatorname{Aut}(X', \pi'),$$

and this group can efficiently be found in time $t(\ell) \cdot n^{O(1)}$ by Theorem 8.

To complete the proof, we need to show that given $g' \in \operatorname{Aut}(X', \pi')$, one can efficiently find $g \in \operatorname{Aut}(X)$ such that $\varphi(g) = g'$. To this end, choose an arbitrary bijection $g_\Delta : \Delta \to \Delta^{\bar{g}}$; recall that $\pi'(\Delta) = \pi'(\Delta^{\bar{g}})$ and so $|\Delta| = |\Delta^{\bar{g}}|$. Then the mapping g taking a vertex $\alpha \in \Omega$ to the vertex α^{g_Δ}, where Δ is the class of e_X, containing α is a permutation of Ω. Moreover, from the definition of e_X, it follows that $g \in \operatorname{Aut}(X)$. It remains to note that g can efficiently be constructed.

8   Concluding Remarks

In this paper we have presented an isomorphism testing algorithm for n-vertex chordal graphs of leafage ℓ which has running time $t(\ell) \cdot n^{O(1)}$, where $t(\ell)$ is a double exponential function not depending on n. A natural question is to improve the running time dependence on the leafage.

References

1. Agaoglu, D., Hlinený, P.: Isomorphism problem for S_d-graphs. In: Esparza, J., Král', D. (eds.) 45th International Symposium on Mathematical Foundations of Computer Science, MFCS 2020, Prague, Czech Republic, 24–28 August 2020. LIPIcs, vol. 170, pp. 4:1–4:14. Schloss Dagstuhl - Leibniz-Zentrum für Informatik (2020). https://doi.org/10.4230/LIPIcs.MFCS.2020.4
2. Arvind, V., Nedela, R., Ponomarenko, I., Zeman, P.: Testing isomorphism of chordal graphs of bounded leafage is fixed-parameter tractable (2021). https://arxiv.org/abs/2107.10689
3. Arvind, V., Das, B., Köbler, J., Toda, S.: Colored hypergraph isomorphism is fixed parameter tractable. Algorithmica **71**(1), 120–138 (2015). https://doi.org/10.1007/s00453-013-9787-y
4. Babai, L.: Graph isomorphism in quasipolynomial time [extended abstract]. In: Wichs, D., Mansour, Y. (eds.) Proceedings of the 48th Annual ACM SIGACT Symposium on Theory of Computing, STOC 2016, Cambridge, MA, USA, 18–21 June 2016, pp. 684–697. ACM (2016). https://doi.org/10.1145/2897518.2897542
5. Biro, M., Hujter, M., Tuza, Z.: Precoloring extension I. Interval graphs. Discret. Math. **100**(1–3), 267–279 (1992)
6. Brand, N.: Isomorphisms of cyclic combinatorial objects. Discret. Math. **78**(1-2), 73–81 (1989). https://doi.org/10.1016/0012-365X(89)90162-3
7. Çagirici, D.A., Hlinený, P.: Isomorphism testing for T-graphs in FPT. In: Mutzel, P., Rahman, M.S., Slamin (eds.) WALCOM 2022. LNCS, vol. 13174, pp. 239–250. Springer, Cham (2022). https://doi.org/10.1007/978-3-030-96731-4_20
8. Chaplick, S., Töpfer, M., Voborník, J., Zeman, P.: On H-topological intersection graphs. In: Bodlaender, H.L., Woeginger, G.J. (eds.) WG 2017. LNCS, vol. 10520, pp. 167–179. Springer, Cham (2017). https://doi.org/10.1007/978-3-319-68705-6_13
9. Chaplick, S., Zeman, P.: Combinatorial problems on H-graphs. Electron. Notes Discret. Math. **61**, 223–229 (2017). https://doi.org/10.1016/j.endm.2017.06.042
10. Chaplick, S., Zeman, P.: Isomorphism-completeness for H-graphs (2021). https://kam.mff.cuni.cz/pizet/gic.pdf
11. Chen, G., Ponomarenko, I.: Coherent Configurations. Central China Normal University Press, Wuhan (2019). http://www.pdmi.ras.ru/~inp/ccNOTES.pdf
12. Fomin, F.V., Golovach, P.A., Raymond, J.-F.: On the tractability of optimization problems on H-graphs. Algorithmica **82**(9), 2432–2473 (2020). https://doi.org/10.1007/s00453-020-00692-9
13. Gavril, F.: The intersection graphs of subtrees in trees are exactly the chordal graphs. J. Comb. Theory Series B **16**(1), 47–56 (1974). https://doi.org/10.1016/0095-8956(74)90094-X. https://www.sciencedirect.com/science/article/pii/009589567490094X
14. Grohe, M., Neuen, D., Schweitzer, P.: A faster isomorphism test for graphs of small degree. In: 59th IEEE Annual Symposium on Foundations of Computer Science, FOCS 2018, Paris, France, 7–9 October 2018, pp. 89–100. IEEE Computer Society (2018). https://doi.org/10.1109/FOCS.2018.00018
15. Grohe, M., Wiebking, D., Neuen, D.: Isomorphism testing for graphs excluding small minors. In: 61st IEEE Annual Symposium on Foundations of Computer Science, FOCS 2020, Durham, NC, USA, 16–19 November 2020, pp. 625–636. IEEE (2020). https://doi.org/10.1109/FOCS46700.2020.00064

16. Habib, M., Stacho, J.: Polynomial-time algorithm for the leafage of chordal graphs. In: Fiat, A., Sanders, P. (eds.) ESA 2009. LNCS, vol. 5757, pp. 290–300. Springer, Heidelberg (2009). https://doi.org/10.1007/978-3-642-04128-0_27

17. Lin, I., McKee, T.A., West, D.B.: The leafage of a chordal graph. Discuss. Math. Graph Theory **18**(1), 23–48 (1998). https://doi.org/10.7151/dmgt.1061

18. Lokshtanov, D., Pilipczuk, M., Pilipczuk, M., Saurabh, S.: Fixed-parameter tractable canonization and isomorphism test for graphs of bounded treewidth. SIAM J. Comput. **46**(1), 161–189 (2017). https://doi.org/10.1137/140999980

19. Lueker, G.S., Booth, K.S.: A linear time algorithm for deciding interval graph isomorphism. J. ACM **26**(2), 183–195 (1979). https://doi.org/10.1145/322123.322125

20. Luks, E.M.: Isomorphism of graphs of bounded valence can be tested in polynomial time. J. Comput. Syst. Sci. **25**(1), 42–65 (1982). https://doi.org/10.1016/0022-0000(82)90009-5

21. Neuen, D.: Hypergraph isomorphism for groups with restricted composition factors. In: Czumaj, A., Dawar, A., Merelli, E. (eds.) 47th International Colloquium on Automata, Languages, and Programming, ICALP 2020, Saarbrücken, Germany, 8–11 July 2020 (Virtual Conference). LIPIcs, vol. 168, pp. 88:1–88:19. Schloss Dagstuhl - Leibniz-Zentrum für Informatik (2020). https://doi.org/10.4230/LIPIcs.ICALP.2020.88

22. Ponomarenko, I.N.: The isomorphism problem for classes of graphs closed under contraction. J. Sov. Math. **55**(2), 1621–1643 (1991)

23. Ponomarenko, I.: Polynomial isomorphism algorithm for graphs which do not pinch to $K_{3,g}$. J. Sov. Math. **34**(4), 1819–1831 (1986)

24. Schweitzer, P., Wiebking, D.: A unifying method for the design of algorithms canonizing combinatorial objects. In: Charikar, M., Cohen, E. (eds.) Proceedings of the 51st Annual ACM SIGACT Symposium on Theory of Computing, STOC 2019, Phoenix, AZ, USA, 23–26 June 2019, pp. 1247–1258. ACM (2019)

25. Stacho, J.: On 2-subcolourings of chordal graphs. In: Laber, E.S., Bornstein, C., Nogueira, L.T., Faria, L. (eds.) LATIN 2008. LNCS, vol. 4957, pp. 544–554. Springer, Heidelberg (2008). https://doi.org/10.1007/978-3-540-78773-0_47

26. Weisfeiler, B., Leman, A.: The reduction of a graph to canonical form and the algebra which appears therein. NTI Series **2**, 12–16 (1968). https://www.iti.zcu.cz/wl2018/pdf/wl_paper_translation.pdf. The URL links to an English translation

Twin-Width and Transductions of Proper
k-Mixed-Thin Graphs

Jakub Balabán⬥, Petr Hliněný$^{(\boxtimes)}$⬥, and Jan Jedelský⬥

Faculty of Informatics, Masaryk University, Botanická 68a, Brno, Czech Republic
`jakbal@mail.muni.cz`, `hlineny@fi.muni.cz`

Abstract. The new graph parameter twin-width, recently introduced
by Bonnet et al., allows for an FPT algorithm for testing all FO properties of graphs. This makes classes of efficiently bounded twin-width
attractive from the algorithmic point of view. In particular, such classes
(of small twin-width) include proper interval graphs, and (as digraphs)
posets of width k. Inspired by an existing generalization of interval graphs
into so-called k-thin graphs, we define a new class of *proper k-mixed-thin
graphs* which largely generalizes proper interval graphs. We prove that
proper k-mixed-thin graphs have twin-width linear in k, and that a certain subclass of k-mixed-thin graphs is transduction-equivalent to posets
of width k' such that there is a quadratic relation between k and k'.

Keywords: twin-width · proper interval graph · proper mixed-thin
graph · transduction equivalence

1 Introduction

The notion of twin-width (of graphs, digraphs, or matrices) was introduced quite
recently, in 2020, by Bonnet, Kim, Thomassé and Watrigant [7], and yet has
already found many very interesting applications. These applications span from
efficient parameterized algorithms and algorithmic metatheorems, through finite
model theory, to classical combinatorial questions. See also the (still growing)
series of follow-up papers [3–6,8].

We leave formal definitions for the next section. Informally, in simple graphs,
twin-width measures how diverse the neighbourhoods of the graph vertices are.
E.g., *cographs* (the graphs which can be built from singleton vertices by repeated
operations of a disjoint union and taking the complement) have the lowest possible value of twin-width, 0, which means that the graph can be brought down
to a single vertex by successively identifying twin[1] vertices. Hence the name,
twin-width, for the parameter, and the term *contraction sequence* referring to
the described identification process of vertices.

---

[1] Two vertices x and y are called *twins* in a graph G if they have the same neighbours
in $V(G) \setminus \{x, y\}$.

Supported by the Czech Science Foundation, project no. 20-04567S.

M. A. Bekos and M. Kaufmann (Eds.): WG 2022, LNCS 13453, pp. 43–55, 2022.
https://doi.org/10.1007/978-3-031-15914-5_4

Twin-width is particularly useful in the algorithmic metatheorem area. Namely, Bonnet et al. [7] proved that classes of binary relational structures (such as graphs and digraphs) of bounded twin-width have efficient first-order (FO) model checking algorithms, given a witness of the boundedness (a "good" contraction sequence). In one of the previous studies on algorithmic metatheorems for dense structures, Gajarský et al. [10] proved that posets of bounded width (the *width of a poset* is the maximum size of an antichain) admit efficient FO model checking algorithms. In this regard, [7] generalizes [10] since posets of bounded width have bounded twin-width. The original proof of the latter in [7] was indirect (via so-called mixed minors, but this word 'mixed' has nothing to do with our 'mixed-thin') and giving a loose bound, and Balabán and Hliněný [2] have recently proved a straightforward linear upper bound (with an efficient construction of a contraction sequence) on the twin-width of posets in terms of width.

Following this research direction further, we propose a new graph class of *proper k-mixed-thin graphs* (Definition 1) which is related to previous generalizations of interval graphs to thin [12] and proper thin [9] graphs. We show some basic properties and relations of our new class, and prove that proper k-mixed-thin graphs have the twin-width linearly bounded in k. Moreover, a contraction sequence can be constructed efficiently if the "geometric" representation of the graph is given. This result brings new possibilities of proving boundedness of twin-width for various graph classes in a direct and efficient way. The aspect of an efficient construction of the relevant contraction sequence is quite important from the algorithmic point of view; the exact twin-width is NP-hard to determine [3], and no efficient approximations of it are known in general.

Another interesting aspect of our research stems from the following deep result of [7]: the property of a class to have bounded twin-width is preserved under *FO transductions* which are, roughly explaining, expressions (or logical *interpretations*) of another graph in a given graph using formulae of FO logic with help of arbitrary additional parameters in the form of vertex labels. E.g., to prove that the class of interval graphs has unbounded twin-width, it suffices to show that they interpret in FO all graphs. In this regard we prove that a subclass of our new class, of the *inversion-free* proper k-mixed-thin graphs, is transduction-equivalent to the class of posets of width k' (with a quadratic dependence between k and k'). So, our results can be seen as a generalization of [2] and, importantly for possible applications, they target undirected graphs instead of special digraphs in the poset case.

1.1 Outline of the Paper

⋄ In Sect. 2 we give an overview of the necessary concepts from graph theory and FO logic; namely about intersection graphs, the twin-width and its basic properties, and FO transductions.

⋄ In Sect. 3 we define the new classes of k-mixed-thin and proper k-mixed-thin graphs, and their inversion-free subclasses (Definition 1). We also state the following results;

- comparing proper k-mixed-thin to k-thin graphs (Propositions 2 and 3),
- proving that multidimensional full grids (i.e., strong products of paths), and the proper intersection graphs of subpaths in a subdivision of a given graph, are proper k-mixed-thin for suitable k (Theorems 4 and 5).

◇ Sect. 4 brings the first core result composed of
  - an efficient constructive proof that the class of proper k-mixed thin graphs has twin-width at most $9k$ (Theorem 6), and an example in which this bound cannot be improved below a linear function (Proposition 9),
  - followed by a consequence that FO properties on these graphs can be tested in FPT, given the representation (Corollary 8).

◇ Sect. 5 then states the second core result – the transduction equivalence.
  - The class of inversion-free proper k-mixed-thin graphs is a transduction of the class of posets of width at most $5 \cdot \binom{k}{2} + 5k$ (Theorem 10), and
  - the class of posets of width at most k is a transduction of the class of inversion-free proper $(2k+1)$-mixed-thin unordered graphs (Theorem 11).

◇ We conclude our findings, state some open questions and outline future research directions in the final Sect. 6.

We leave proofs of the *-marked statements for the full preprint [1].

2 Preliminaries and Formal Definitions

A (simple) *graph* is a pair $G = (V, E)$ where $V = V(G)$ is the *finite* vertex set and $E = E(G)$ is the edge set – a set of unordered pairs of vertices $\{u, v\}$, shortly uv. For a set $Z \subseteq V(G)$, we denote by $G[Z]$ the subgraph of G induced on the vertices of Z. A *subdivision* of an edge uv of a graph G is the operation of replacing uv with a new vertex x and two new edges ux and xv.

A *poset* is a pair $P = (X, \leq)$ where the binary relation $\leq$ is an ordering on X. We represent posets also as special digraphs (directed graphs with ordered edges). The *width of a poset* P is the maximum size of an antichain in P, i.e., the maximum size of an independent set in the digraph P. We say that $(x, y) \in X^2$ is a *cover pair* if $x \lneq y$ and there is no $z \in X$ such that $x \lneq z \lneq y$.

2.1 Intersection Graphs

The *intersection graph* G of a finite collection of sets $\{S_1, \ldots, S_n\}$ is a graph in which each set S_i is associated with a vertex $v_i \in V(G)$ (then S_i is the *representative* of v_i), and each pair v_i, v_j of vertices is joined by an edge if and only if the corresponding sets have a non-empty intersection, i.e. $v_i v_j \in E(G) \iff S_i \cap S_j \neq \emptyset$. We say that an intersection graph G is *proper* if G is the intersection graph of $\{S_1, \ldots, S_n\}$ such that $S_i \not\subseteq S_j$ for all $i \neq j \in \{1, \ldots, n\}$.

A nice example of intersection graphs are *interval graphs*, which are the intersection graphs of intervals on the real line. More generally, for a fixed graph H, if H' is a subdivision of H, then an *H-graph* is the intersection graph of the vertex sets of connected subgraphs of H'. Such an intersection representation is also called an H-representation. For instance, interval graphs coincide with K_2-graphs. We can speak also about proper interval or proper H-graphs.

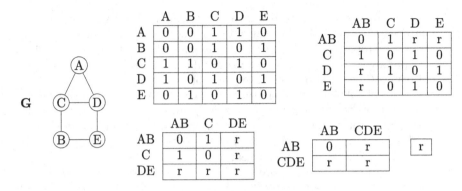

Fig. 1. An example of a graph G (left), and a symmetric contraction sequence of its adjacency matrix (right), which certifies that the symmetric twin-width of the adjacency matrix of G is at most 3, and so is the twin-width of G.

2.2 Twin-Width

We present the definition of twin-width focusing on matrices, as taken from [7, Section 5]. Later in the paper, we will restrict ourselves only to the *symmetric* twin-width because the more general version is not relevant for graphs.

Let A be a symmetric square matrix with entries from a finite set (here $\{0, 1, r\}$ for graphs) and let X be the set indexing both rows and columns of A. The entry r is called a *red entry*, and the *red number* of a matrix A is the maximum number of red entries over all columns and rows in A.

Contraction of the rows (resp. columns) k and ℓ results in the matrix obtained by deleting the row (resp. column) ℓ, and replacing entries of the row (resp. column) k by r whenever they differ from the corresponding entries in the row (resp. column) ℓ. Informally, if A is the adjacency matrix of a graph, the red entries ("errors") in a contraction of rows k and ℓ record where the graph neighbourhoods of the vertices k and ℓ differ.

A sequence of matrices $A = A_n, \dots, A_1$ is a *contraction sequence* of the matrix A, whenever A_1 is (1×1) matrix and for all $1 \le i < n$, the matrix A_i is a contraction of matrix A_{i+1}. A contraction sequence is *symmetric* if every contraction of a pair of rows (resp. columns) is immediately followed by a contraction of the corresponding pair of columns (resp. rows).

The *twin-width* of a matrix A is the minimum integer d, such that there is a contraction sequence $A = A_n, \dots, A_1$, such that for all $1 \le i \le n$, the red number of the matrix A_i is at most d. The *symmetric twin-width* of a matrix A is defined analogously, requiring that the contraction sequence is symmetric, and we only count the red number after both symmetric row and column contractions are performed. See Fig. 1. The *twin-width of a graph* G is then the symmetric twin-width of its adjacency matrix $A(G)$.[2]

2.3   FO Logic and Transductions

A *relational signature* Σ is a finite set of relation symbols R_i, each with associated arity r_i. A *relational structure* A with signature Σ (or shortly a Σ-structure) is defined by a *domain* A and relations $R_i(A) \subseteq A^{r_i}$ for each relation symbol $R_i \in \Sigma$ (the relations *interpret* the relational symbols). For example, graphs can be viewed as relational structures with the set of vertices as the domain and a single relation symbol E with arity 2 in the relational signature.

Let Σ and Γ be relational signatures. An *interpretation* I of Γ-structures in Σ-structures is a function from Σ-structures to Γ-structures defined by a formula $\varphi_0(x)$ and a formula $\varphi_R(x_1, \ldots, x_k)$ for each relation symbol $R \in \Gamma$ with arity k (these formulae may use the relational symbols of Σ).

Given a Σ-structure A, $I(A)$ is a Γ-structure whose domain B contains all elements $a \in A$ such that $\varphi_0(a)$ holds in A, and in which every relation symbol $R \in \Gamma$ of arity k is interpreted as the set of tuples $(a_1, \ldots, a_k) \in B^k$ satisfying $\varphi_R(a_1, \ldots, a_k)$ in A.

A transduction T from Σ-structures to Γ-structures is defined by an interpretation I_T of Γ-structures in Σ^+-structures where Σ^+ is Σ extended by a finite number of unary relation symbols (called *marks*). Given a Σ-structure A, the transduction $T(A)$ is a set of all Γ-structures B such that $B = I_T(A')$ where A' is A with arbitrary elements of A marked by the unary marks. If C is a class of Σ-structures, then we define $T(C) = \bigcup_{A \in C} T(A)$. A class $\mathcal{D}$ of Γ-structures is a *transduction* of C if there exists a transduction T such that $\mathcal{D} \subseteq T(C)$.

For simplicity, our transductions are non-copying.

3   Generalizing Proper k-Thin Graphs

So-called k-thin graphs (as defined below) have been proposed and studied as a generalization of interval graphs by Mannino et al. [12]. Likewise, proper interval graphs have been naturally generalized into proper k-thin graphs [9]. As forwarded in the introduction, we further generalize these classes into the classes of (proper) k-mixed-thin graphs as follows.

Definition 1 (Mixed-thin and Proper mixed-thin). Let $G = (V, E)$ be a graph and $k > 0$ an integer. Let $\bar{E} = \binom{V}{2} \setminus E$ be the complement of its edge set. For two linear orders $\leq$ and $\leq'$ on the same set, we say that $\leq$ and $\leq'$ are *aligned* if they are the same or one is the inverse of the other.

---

[2] Note that one can also define the "natural" twin-width of graphs which, informally, ignores the red entries on the main diagonal (as there are no loops in a simple graph). The natural twin-width is never larger, but possibly by one lower, than the symmetric matrix twin-width. For instance, for the sequence in Fig. 1, the natural twin-width would be only 2.

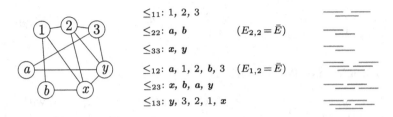

$\leq_{11}$: 1, 2, 3

$\leq_{22}$: a, b ($E_{2,2} = \bar{E}$)

$\leq_{33}$: x, y

$\leq_{12}$: a, 1, 2, b, 3   ($E_{1,2} = \bar{E}$)

$\leq_{23}$: x, b, a, y

$\leq_{13}$: y, 3, 2, 1, x

Fig. 2. An illustration of Definition 1. **Left**: a proper 3-mixed-thin graph G, with the vertex set partitioned into $V_1 = \{1, 2, 3\}$, $V_2 = \{a, b\}$ and $V_3 = \{x, y\}$. **Middle**: the six linear orders $\leq_{ij}$, and the sets $E_{i,j}$ defaulting to $E_{i,j} = E(G)$, except for $E_{1,2}$ and $E_{2,2}$. **Right**: a "geometric" proper interval representation of the orders $\leq_{ij}$ (notice – separately for each pair i, j), such that the edges between V_i and V_j belonging to $E_{i,j}$ are represented by intersections between intervals of colour i and colour j.

The graph G is *proper k-mixed-thin* if there exists a partition $\mathcal{V} = (V_1, \ldots, V_k)$ of V, and for each $1 \leq i \leq j \leq k$ a linear order $\leq_{ij}$ on $V_i \cup V_j$ and a choice of $E_{i,j} \in \{E, \bar{E}\}$ (Fig. 2), such that, again for every $1 \leq i \leq j \leq k$,

(a) the restriction of $\leq_{ij}$ to V_i (resp. to V_j) is aligned with $\leq_{ii}$ (resp. $\leq_{jj}$), and
(b) for every triple u, v, w such that $(\{u, v\} \subseteq V_i$ and $w \in V_j)$ or $(\{u, v\} \subseteq V_j$ and $w \in V_i)$, we have that if $u \leq_{ij} v \leq_{ij} w$ and $uw \in E_{i,j}$, then $vw \in E_{i,j}$.
(c) for every triple u, v, w such that $(\{v, w\} \subseteq V_i$ and $u \in V_j)$ or $(\{v, w\} \subseteq V_j$ and $u \in V_i)$, we have that if $u \leq_{ij} v \leq_{ij} w$ and $uw \in E_{i,j}$, then $uv \in E_{i,j}$.

General (not proper) k-mixed-thin graphs do not have to satisfy (c). A (proper) k-mixed-thin graph G is *inversion-free* if, above, (a) is replaced with

(a') the restriction of $\leq_{ij}$ to V_i (resp. to V_j) is equal to $\leq_{ii}$ (resp. $\leq_{jj}$).

We remark that aforementioned *(proper) k-thin graphs* are those (proper) k-mixed-thin graphs for which the orders $\leq_{ij}$ (for $1 \leq i \leq j \leq k$) in the definition can be chosen as the restrictions of the same linear order on V, and all $E_{i,j} = E$ ('inversion-free' is insignificant in such case).

The class of k-mixed-thin graphs is thus a superclass of the class of k-thin graphs, and the same holds in the 'proper' case. On the other hand, the class of interval graphs is 1-thin, but it is not proper k-mixed-thin for any finite k; the latter follows, e.g., easily from further Theorem 6.

Bonomo and de Estrada [9, Theorem 2] showed that given a (proper) k-thin graph G and a suitable ordering $\leq$ of $V(G)$, a partition of $V(G)$ into k parts compatible with $\leq$ can be found in polynomial time. On the other hand [9, Theorem 5], given a partition $\mathcal{V}$ of $V(G)$ into k parts, the problem of deciding whether there is an ordering of $V(G)$ compatible with $\mathcal{V}$ in the proper sense is NP-complete. These results do not answer whether the recognition of k-thin graphs is efficient or not, and neither can we at this stage say whether the recognition of proper k-mixed-thin graphs is efficient.

3.1   Comparing (Proper) k-Mixed-Thin to Other Classes

We illustrate use of our Definition 1 by comparing it to ordinary thinness on some natural graph classes. Recall that the (square) $(r \times r)$-grid is the Cartesian product of two paths of length r. Denote by $\overline{tK_2}$ the complement of the matching with t edges. We show that thinness and proper mixed-thinness are incomparable.

Proposition 2 (Mannino et al. [12], Bonomo and de Estrada [9])

a) For every $t \geq 1$, the graph $\overline{tK_2}$ is t-thin but not $(t-1)$-thin.
b) The $(r \times r)$-grid has thinness linear in r.
c) The thinness of the complete m-ary tree $(m > 1)$ is linear in its height.

Proposition 3.* a) For every $t \geq 1$, $\overline{tK_2}$ is inversion-free proper 1-mixed-thin.
b) For all m, n the $(m \times n)$-grid is inversion-free proper 3-mixed-thin.
c) Every tree T is inversion-free proper 3-mixed-thin.

Proposition 3(b) can be extended much further. A d-dimensional grid is the Cartesian product of $d \geq 1$ paths, and the d-dimensional full grid is the strong product of $d \geq 1$ paths (also known as the "grid with all diagonals"). We have:

Theorem 4.* Let $d \geq 1$ be an arbitrary integer. Both d-dimensional grids and d-dimensional full grids are inversion-free proper 3^{d-1}-mixed-thin.

To further illustrate the strength of the new concept, we show that k-mixed-thin graphs generalize the following class [11], which itself can be viewed as a natural generalization of proper interval graphs and k-fold proper interval graphs (a subclass of interval graphs whose representation can be decomposed into k proper interval subrepresentations):

Theorem 5.* Let $G = (V, E)$ be a proper intersection graph of paths in some subdivision of a fixed connected graph H with m edges, and let k be the number of (all) distinct paths in H. Then G is a proper $(m^2 k)$-mixed-thin graph.

4   Proper k-Mixed-Thin Graphs Have Bounded Twin-Width

In the founding series of papers, Bonnet et al. [4–7] proved that many common graph classes (in addition to aforementioned posets of bounded width) are of bounded twin-width. Their proof methods have usually been indirect (using other technical tools such as 'mixed minors'), but for a few classes including proper interval graphs and multidimensional grids and full grids (cf. Theorem 4) they provided a direct construction of a contraction sequence.

We have shown [2] that a direct and efficient construction of a contraction sequence is possible also for posets of width k. Stepping further in this direction, our proper k-mixed-thin graphs, which largely generalize proper interval graphs, still have bounded twin-width, as we are now going to show with a direct and efficient construction of a contraction sequence for them.

Before stating the result, we mention that 1-thin graphs coincide with interval graphs which have unbounded twin-width by [4], and hence the assumption of 'proper' in the coming statement is necessary.

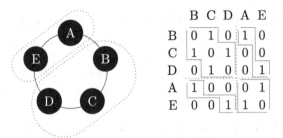

$$
\begin{array}{c|ccccc}
 & \text{B} & \text{C} & \text{D} & \text{A} & \text{E} \\
\hline
\text{B} & 0 & 1 & 0 & 1 & 0 \\
\text{C} & 1 & 0 & 1 & 0 & 0 \\
\text{D} & 0 & 1 & 0 & 0 & 1 \\
\text{A} & 1 & 0 & 0 & 0 & 1 \\
\text{E} & 0 & 0 & 1 & 1 & 0 \\
\end{array}
$$

Fig. 3. On the left, there is a partition of C_5 into parts $(\{B, C, D\}, \{A, E\})$, which with the ordering e.g. $A \leq B \leq C \leq D \leq E$ certifies that C_5 is proper 2-thin, therefore proper 2-mixed-thin as well. On the right, there is an adjacency matrix of C_5, together with eight blue diagonal boundaries obtained by the process described in Lemma 7.

Theorem 6. *Let G be a proper k-mixed-thin graph. Then the twin-width of G, i.e., the symmetric twin-width of $A(G)$, is at most $9k$. The corresponding contraction sequence for G can be computed in polynomial time from the vertex partition $(V_1, \ldots, V_k)$ and the orders $\leq_{ij}$ for G from Definition 1.*

In the course of proving Theorem 6, an adjacency matrix $A(G)$ of G is always obtained by ordering the k parts arbitrarily, and then inside each part using the order $\leq_{ii}$. Furthermore, we denote $A_{i,j}(G)$ the submatrix with rows from V_i and columns from V_j.

We would like to talk about parts ("areas") of a $(p \times q)$ matrix M. To do so, we embed such a matrix into the plane as a $((p + 1) \times (q + 1))$-grid, where entries of the matrix are represented by labels of the bounded square faces of the grid. We call a *boundary* any path in the grid, which is also a separator of the grid. In this view, we say that a matrix entry a is *next to* a boundary if at least one of the vertices of the face of a lies on the boundary.

Note that the grid has four corner vertices of degree 2, and a *diagonal boundary* is a shortest (i.e., geodesic) path going either between the top-left and the bottom-right corners, or between the top-right and the bottom-left corners. We say that two diagonal boundaries b_1 and b_2 are *crossing* if b_1 contains two grid vertices v and v' not contained in b_2, such that v and v' belong to different parts of the matrix separated by b_2. We call a matrix M *diagonally trisected* if M contains two non-crossing diagonal boundaries with the same ends which separate the matrix into three *parts*. The part bounded by both diagonal boundaries is called the *middle part*. See Fig. 3.

Lemma 7.* *Let G be a proper k-mixed-thin graph. For all $1 \leq i, j \leq k$, the submatrix $A_{i,j}(G)$ is diagonally trisected, such that each part has either all entries 0 or all entries 1, with the exception of entries on the main diagonal of $A(G)$. Furthermore, the diagonal boundaries of the submatrix $A(G)_{ii}$ are symmetric (w. r. to the main diagonal).*

Proof (of Theorem 6). For each $1 \leq i \neq j \leq k$, by Lemma 7, the submatrix $A_{i,j}(G)$ of $A(G)$ is diagonally trisected such that each part has all entries equal (i.e., all 0 or all 1). The case of $i = j$ is similar, except that the entries on the main diagonal might differ from the remaining entries in the same area.

Furthermore, since the matrix $\boldsymbol{A}$ is symmetric, we can assume that the diagonal boundaries are symmetric as well.

We generalize this setup to matrices with *red* entries r; these come from contractions of non-equal entries in $\boldsymbol{A}(G)$, cf. Subsect. 2.2. Considering a matrix $\boldsymbol{M} = (m_{uv})_{u,v}$ obtained by symmetric contractions from $\boldsymbol{A}(G)$, we assume that

- $\boldsymbol{M}$ is *consistent* with the partition $(V_1, \ldots, V_k)$, meaning that only rows and columns from the same part have been contracted in $\boldsymbol{A}(G)$,
- $\boldsymbol{M}$ is *red-aligned*, meaning that each submatrix $\boldsymbol{M}_{i,j}$ obtained from $\boldsymbol{A}_{i,j}(G)$ by row contractions in V_i and column contractions in V_j, is diagonally trisected such that (again with the possible *exception* of entries on the main diagonal of $\boldsymbol{M}$): each of the three parts has all entries either from $\{0, r\}$ or from $\{1, r\}$, and moreover, the entries r are only in the middle part and next to one of the diagonal boundaries, and
- the diagonal boundaries of $\boldsymbol{M}$ are also symmetric, that is, there is a boundary between m_{uv} and m_{uw} iff there is a boundary between m_{vu} and m_{wu}.

We are going to show that there is a symmetric matrix-contraction sequence starting from $\boldsymbol{M}^0 := \boldsymbol{A}(G)$ down to an $(8k \times 8k)$ matrix $\boldsymbol{M}^t$, such that all square matrices $\boldsymbol{M}^m$, $0 \leq m \leq t$, in this sequence are consistent with $(V_1, \ldots, V_k)$, red-aligned, and have red value at most $9k$. Furthermore, the matrices in our sequence are symmetric, and so are the diagonal boundaries. Hence we only need to observe the red values of the rows. Then, once we get to $\boldsymbol{M}^t$, we may finish the contraction sequence arbitrarily while not exceeding the red value of $8k$.

Assume we have got to a matrix $\boldsymbol{M}^m$, $m \geq 0$, of the claimed properties in our sequence, and $\boldsymbol{M}^m$ has more than $8k$ rows. The induction step to the next matrix $\boldsymbol{M}^{m+1}$ consists of two parts:

(i) We find a pair of consecutive rows from (some) one part of $(V_1, \ldots, V_k)$, such that their contraction does not yield more than $7k$ red entries.
(ii) After we do this row contraction followed by the symmetric column contraction to $\boldsymbol{M}^{m+1}$ (which may add one red entry up to each other row of $\boldsymbol{M}^{m+1}$), we show that the red value of any other row does not exceed $7k + 2k = 9k$.

Part (i) importantly uses the property of $\boldsymbol{M}^m$ being red-aligned, and is given separately in the next claim:

Claim. If a matrix $\boldsymbol{M}^m$ satisfies the above claimed properties and is of size more than $8k$, then there exists a pair of consecutive rows from one part in $\boldsymbol{M}^m$, such that their contraction gives a row with at most $7k$ red entries (a technical detail; this number includes the entry coming from the main diagonal of $\boldsymbol{M}^m$). After this contraction in $\boldsymbol{M}^m$, the matrix will be again red-aligned.

In part (ii) of the induction step, we fix any row $i \in \{1, \ldots, k\}$ of $\boldsymbol{M}^{m+1}$. Row i initially (in $\boldsymbol{M}^0$) has no red entry, and it possibly got up to $7k$ red entries in the previous last contraction involving it. After that, row i has possibly gained additional red entries only through column contractions, and such a contraction leading to a new red entry in row i (except on the main diagonal which has been accounted for in Claim 4) may happen only if the two non-red contracted entries lied on two sides of the same diagonal boundary. Since we have $2k$ such boundaries throughout our sequence, we get that the number of red entries in M^{m+1} is indeed at most $7k + 2k = 9k$.

We have finished the induction step, and so the whole proof by the above outline. Note that all steps are efficient, including Claim 4 since at every step there is at most a linear number of contractions which we are choosing from. $\square$

Corollary 8 (based on [7]**).** *Assume a proper k-mixed-thin graph G, given alongside with the vertex partition and the orders from Definition 1. Then FO model checking on G is solvable in FPT time with respect to k.* $\square$

Furthermore, the bound in Theorem 6 cannot be substantially improved (below linear dependence) due to the following:

Proposition 9. *For every integer $k \geq 1$, there exists an inversion-free proper $(2k + 1)$-mixed-thin graph G such that the twin-width of G is at least k.*

5   Transductions Between Inversion-Free Proper k-Mixed-Thin Graphs and Posets

In relation to the deep fact [7] that the class property of having bounded twin-width is preserved under FO transductions (cf. Sect. 2), it is interesting to look at how our class of proper k-mixed-thin graphs relates to other studied classes of bounded twin-width. In this regard we show that our class is nearly (note the inversion-free assumption!) transduction equivalent to the class of posets of bounded width. We stress that the considered transductions here are always non-copying (i.e., not "expanding" the ground set of studied structures).

Theorem 10. *The class of inversion-free proper k-mixed-thin graphs is a transduction of the class of posets of width at most $5 \cdot \binom{k}{2} + 5k$. For a given graph, together with the vertex partition and the orders as from Definition 1, the corresponding poset and its transduction parameters can be computed in polytime.*

Proof. Let $G = (V, E)$ be an inversion-free proper k-mixed-thin graph. Let $\mathcal{V} = (V_1, \ldots, V_k)$ be the partition of V and $\leq_{ij}$ for $1 \leq i \leq j \leq k$ be the orders given by Definition 1. On a suitable ground set $X \supseteq V$ defined below, we are going to construct a poset $P = (X, \preceq)$ equipped with vertex labels (*marks*), such that the edges of G will be interpreted by a binary FO formula within P. To simplify notation, we will also consider posets as special digraphs, and naturally use digraph terms for them.

For start, let $P_0 = (V, \preceq_0)$ be the poset formed by (independent) chains $V_1, \ldots, V_k$, where each chain V_i is ordered by $\leq_{ii}$. Let us denote by $V_{i,j} := V_i \cup V_j$.

In order to define set X, we first introduce the notion of *connectors*. Consider $1 \leq i \leq j \leq k$, $X \supsetneq V$, a vertex $x \in X \setminus V$ and a pair $l_x \in V_i$ and $u_x \in V_j$.

If $i = j$, we additionally demand $l_x \leq_{ii} u_x$. If $\sqsubseteq_x$ is a binary relation (on X) defined by $l_x \sqsubseteq_x x \sqsubseteq_x u_x$, then we call $(x, \sqsubseteq_x)$ a *connector* with the *centre* x and the *joins* $l_x x$ *and* $x u_x$. (Note that it will be important to have u_x from V_j and not from V_i, wrt. $i \leq j$.) We also order the connector centres $x \neq y$ with joins to V_i and V_j by $x \sqsubseteq_{ij} y$, if and only if $l_x \leq_{ii} l_y$ and $u_x \leq_{jj} u_y$. There may be more that one connector connecting the same pair of vertices.

Our construction relies on the following observation which, informally, tells us that connectors can (all together) encode some information about pairs of vertices of V in an unambiguous way.

Claim. Recall $P_0 = (V, \preceq_0)$. Let $X \supsetneq V$ be such that each $x \in X \setminus V$ is the centre of a connector, as defined above. Let $\preceq_1$ be a binary relation on $X \supsetneq V$ defined as the reflexive and transitive closure of $(\preceq_0 \cup \sqsubseteq)$ where $\sqsubseteq := \left(\bigcup_{x \in X \setminus V} \sqsubseteq_x \right) \cup \left(\bigcup_{1 \leq i \leq j \leq k} \sqsubseteq_{ij} \right)$. Then $P_1 = (X, \preceq_1)$ is a poset, and each join of every connector $(x, \sqsubseteq_x)$ from $x \in X \setminus V$ is a cover pair in P_1.

We continue with the construction of the poset P encoding G; this is done by adding suitable connectors to P_0, and marks $\boldsymbol{S}$, $\boldsymbol{V_i}$, $\boldsymbol{B_{ij}}$, or $\boldsymbol{C_{ij}}$. To explain, $\boldsymbol{S}$ stands for successor (cf. $\leq_{ij}$), $\boldsymbol{V_i}$ stands for the part V_i, $\boldsymbol{B_{ij}}$ means a border-pair (to be defined later in $G[V_{i,j}]$), and $\boldsymbol{C_{ij}}$ stands for complement (cf. $E_{i,j} = \bar{E}$).

1. We apply the mark $\boldsymbol{V_i}$ to every vertex of each part $V_i \in \mathcal{V}$.
2. For each $1 \leq i < j \leq k$, and every pair $(v, w) \in V_i \times V_j$ such that w is the immediate successor of v in $\leq_{ij}$, we add a connector with a new vertex x marked $\boldsymbol{S}$ and joins to $l_x = v$ and $u_x = w$. Note that one could think about symmetrically adding connectors for w being the immediate predecessor, but these can be uniquely recovered from the former connectors.
3. For $1 \leq i \leq j \leq k$ and $v, w \in V_{i,j}$, let $V_{i,j}[v, w] := \{x \in V_{i,j} : v \leq_{ij} x \leq_{ij} w\}$ be a consecutive subchain, and call the set $V_{i,j}[v, w]$ *homogeneous* if, moreover, every pair of vertices between $V_i \cap V_{i,j}[v, w]$ and $V_j \cap V_{i,j}[v, w]$ is an edge in $E_{i,j}$. (In particular, for $i = j$, homogeneous $V_{i,i}[v, w]$ means a clique in G if $E = E_{i,i}$ or an independent set of G otherwise.) If $V_{i,j}[v, w]$ is an inclusion-maximal homogeneous set in $V_{i,j}$, then we call (v, w) a *border pair* in $V_{i,j}$, and we add a connector with a new vertex x marked $\boldsymbol{B_{ij}}$ and joins to v and w. Specifically, it is $l_x = v$ and $u_x = w$, unless $v \in V_j$ and $w \in V_i$ in which case $l_x = w$ and $u_x = v$.
4. For $1 \leq i \leq j \leq k$, if $E_{i,j} = \bar{E}$, then we mark just any vertex by $\boldsymbol{C_{ij}}$.

Now we define the poset $P = (X, \preceq)$, where the set $X \supseteq V$ results from adding all marked connector centres defined above to $P_0 = (V, \preceq_0)$, and $\preceq$ is the transitive closure of $(\preceq_0 \cup \sqsubseteq)$ as defined in Claim 5 for the added connectors.

First, we claim that P with the applied marks uniquely determines our starting graph G. Notice that, for each connector centre $x \in X \setminus V$, the (unique) cover pairs of x to and from respective V_i and V_j, by Claim 5, determine the joins of x.

The vertex set of G is determined by the marks $\boldsymbol{V_i}$, $i = 1, \ldots, k$. For $1 \leq i \leq j \leq k$, the linear order $\leq_{ij}$ is directly determined by $\preceq$ if $i = j$, and otherwise the following holds. For $v \in V_i$ and $w \in V_j$, we have $v \leq_{ij} w$ if and only if there exists a connector x marked $\boldsymbol{S}$ with joins to $l_x \in V_i$ and $u_x \in V_j$ such that $v \preceq l_x$ and $u_x \preceq w$. For $v \in V_j$ and $w \in V_i$, we have $v \leq_{ij} w$ if and only if $w \not\leq_{ij} v$.

To determine the edge set of G, we observe that Definition 1 shows that every edge f (resp. non-edge) of $G[V_{i,j}]$ is contained in some homogeneous consecutive subchain of $\leq_{ij}$. Hence f is contained in some maximal such subchain, and so determined by some border pair in $V_{i,j}$ which we recover from its connector marked $\boldsymbol{B}_{ij}$ using the already determined order $\leq_{ij}$. We then determine whether f means an edge or a non-edge in G using the mark $\boldsymbol{C}_{ij}$.

Finally, we verify that the above-stated definition of the graph G within P can be expressed in FO logic. We leave the technical details for the next claim:

Claim. The transduction described in the proof of Theorem 10 can be defined by FO formulae on the marked poset P.

Second, we compute the width of P. In fact, we show that P can be covered by a small number of chains. There are the k chains of $V_1, \ldots, V_k$. Then, for each pair $1 \leq i \leq j \leq k$, we have one chain of the connector centres marked $\boldsymbol{S}$ from V_i to V_j (only $i < j$), and four chains of the connector centres marked $\boldsymbol{B}_{ij}$, sorted by how their border pairs fall into the sets V_i or V_j (they are indeed chains because border pairs demarcate maximal homogeneous sets). To summarize, there are $k + 5 \cdot \binom{k}{2} + 4k$ chains covering whole P.

Efficiency of the construction of marked poset P from given (already partitioned and with the orders) graph G is self-evident. The whole proof of Theorem 10 is now finished.   □

In the converse direction to Theorem 10 we can straightforwardly prove:

Theorem 11. *The class of posets of width at most k is a transduction of the class of inversion-free proper $(2k + 1)$-mixed-thin graphs. For a given poset, a corresponding inv.-free proper $(2k+1)$-mixed-thin graph is computed in polytime.*

6   Conclusions

Regarding the results in Sect. 5, we remark that it is considered very likely that the classes of graphs of bounded twin-width are not transductions of the classes of posets of bounded width (although we are not aware of a published proof of this). We think that the proper k-mixed-thin graph classes are, in the "transduction hierarchy", positioned strictly between the classes of posets of bounded width and the classes of bounded twin-width, meaning that they are not transductions of posets of bounded width and they do not transduce all graphs of bounded twin-width. We plan to further investigate this question.

Furthermore, Bonnet et al. [8] proved that the classes of structures of bounded twin-width are transduction-equivalent to the classes of permutations with a forbidden pattern. It would be very nice to find an analogous asymptotic characterization with permutations replaced by the graphs of some natural graph property. As a step forward, we would like to further generalize proper k-mixed-thin graphs while keeping the property of bounded twin-width.

References

1. Balabán, J., Hlinený, P., Jedelský, J.: Twin-width and transductions of proper k-mixed-thin graphs. CoRR abs/2202.12536 (2022)
2. Balabán, J., Hliněný, P.: Twin-width is linear in the poset width. In: IPEC. LIPIcs, vol. 214, pp. 6:1–6:13. Schloss Dagstuhl - Leibniz-Zentrum für Informatik (2021)
3. Bergé, P., Bonnet, É., Déprés, H.: Deciding twin-width at most 4 is NP-complete. In: ICALP. LIPIcs, vol. 229, pp. 18:1–18:20. Schloss Dagstuhl - Leibniz-Zentrum für Informatik (2022)
4. Bonnet, É., Geniet, C., Kim, E.J., Thomassé, S., Watrigant, R.: Twin-width II: small classes. In: SODA, pp. 1977–1996. SIAM (2021)
5. Bonnet, É., Geniet, C., Kim, E.J., Thomassé, S., Watrigant, R.: Twin-width III: max independent set, min dominating set, and coloring. In: ICALP. LIPIcs, vol. 198, pp. 35:1–35:20. Schloss Dagstuhl - Leibniz-Zentrum für Informatik (2021)
6. Bonnet, É., Giocanti, U., de Mendez, P.O., Simon, P., Thomassé, S., Torunczyk, S.: Twin-width IV: ordered graphs and matrices. In: STOC, pp. 924–937. ACM (2022)
7. Bonnet, É., Kim, E.J., Thomassé, S., Watrigant, R.: Twin-width I: tractable FO model checking. In: FOCS, pp. 601–612. IEEE (2020)
8. Bonnet, É., Nesetril, J., de Mendez, P.O., Siebertz, S., Thomassé, S.: Twin-width and permutations. CoRR abs/2102.06880 (2021)
9. Bonomo, F., de Estrada, D.: On the thinness and proper thinness of a graph. Discret. Appl. Math. **261**, 78–92 (2019)
10. Gajarský, J., et al.: FO model checking on posets of bounded width. In: FOCS, pp. 963–974. IEEE Computer Society (2015)
11. Jedelský, J.: Classes of bounded and unbounded twin-width [online] (2021). https://is.muni.cz/th/utyga/. Bachelor thesis, Masaryk University, Faculty of Informatics, Brno
12. Mannino, C., Oriolo, G., Ricci-Tersenghi, F., Chandran, L.S.: The stable set problem and the thinness of a graph. Oper. Res. Lett. **35**(1), 1–9 (2007)

Token Sliding on Graphs of Girth Five

Valentin Bartier[1], Nicolas Bousquet[1] (ID), Jihad Hanna[2],
Amer E. Mouawad[2,3]([☒]) (ID), and Sebastian Siebertz[3]

[1] CNRS, LIRIS, Université de Lyon, Université Claude Bernard Lyon 1, Lyon, France
`valentin.bartier@grenoble-inp.fr`, `nicolas.bousquet@univ-lyon1.fr`
[2] American University of Beirut, Beirut, Lebanon
`jgh20@mail.aub.edu.lb`, `aa368@aub.edu.lb`
[3] University of Bremen, Bremen, Germany
`siebertz@uni-bremen.de`

Abstract. In the TOKEN SLIDING problem we are given a graph G and
two independent sets I_s and I_t in G of size $k \geq 1$. The goal is to decide
whether there exists a sequence $\langle I_1, I_2, \ldots, I_\ell \rangle$ of independent sets such
that for all $i \in \{1, \ldots, \ell\}$ the set I_i is an independent set of size k, $I_1 = I_s$,
$I_\ell = I_t$ and $I_i \triangle I_{i+1} = \{u, v\} \in E(G)$. Intuitively, we view each independent set as a collection of tokens placed on the vertices of the graph.
Then, the problem asks whether there exists a sequence of independent
sets that transforms I_s into I_t where at each step we are allowed to slide
one token from a vertex to a neighboring vertex. In this paper, we focus
on the parameterized complexity of TOKEN SLIDING parameterized by
k. As shown by Bartier et al. [2], the problem is W[1]-hard on graphs of
girth four or less, and the authors posed the question of whether there
exists a constant $p \geq 5$ such that the problem becomes fixed-parameter
tractable on graphs of girth at least p. We answer their question positively and prove that the problem is indeed fixed-parameter tractable on
graphs of girth five or more, which establishes a full classification of the
tractability of TOKEN SLIDING based on the girth of the input graph.

Keywords: token sliding · independent set · girth · combinatorial
reconfiguration · parameterized complexity

1 Introduction

Many algorithmic questions present themselves in the following form: Given the
description of a system state and the description of a state we would prefer the
system to be in, is it possible to transform the system from its current state
into the more desired one without "breaking" certain properties of the system

This work is supported by PHC Cedre project 2022 "PLR".

V. Bartier—Supported by ANR project GrR (ANR-18-CE40-0032).

N. Bousquet—Supported by ANR project GrR (ANR-18-CE40-0032).

A. E. Mouawad—Research supported by the Alexander von Humboldt Foundation
and partially supported by URB project "A theory of change through the lens of
reconfiguration".

© Springer Nature Switzerland AG 2022
M. A. Bekos and M. Kaufmann (Eds.): WG 2022, LNCS 13453, pp. 56–69, 2022.
https://doi.org/10.1007/978-3-031-15914-5_5

in the process? Such questions, with some generalizations and specializations, have received a substantial amount of attention under the so-called *combinatorial reconfiguration framework* [9,27,29].

Historically, the study of reconfiguration questions predates the field of computer science, as many classic one-player games can be formulated as reachability questions [19,21], e.g., the 15-puzzle and Rubik's cube. More recently, reconfiguration problems have emerged from computational problems in different areas such as graph theory [10,16,17], constraint satisfaction [14,25], computational geometry [24], and even quantum complexity theory [13]. We refer the reader to the surveys by van den Heuvel [27] and Nishimura [26] for extensive background on combinatorial reconfiguration.

Independent Set Reconfiguration. In this work, we focus on the reconfiguration of independent sets. Given a simple undirected graph G, a set of vertices $S \subseteq V(G)$ is an *independent set* if the vertices of this set are pairwise non-adjacent. Finding an independent set of size k, i.e., the INDEPENDENT SET problem, is known to be NP-hard, but also W[1]-hard[1] parameterized by solution size k and not approximable within $O(n^{1-\epsilon})$, for any $\epsilon > 0$, unless $\mathsf{P} = \mathsf{NP}$ [30]. Moreover, INDEPENDENT SET remains W[1]-hard on graphs excluding C_4 (the cycle on four vertices) as an induced subgraph [7].

We view an independent set as a collection of tokens placed on the vertices of a graph such that no two tokens are placed on adjacent vertices. This gives rise to two natural adjacency relations between independent sets (or token configurations), also called *reconfiguration steps*. These reconfiguration steps, in turn, give rise to two combinatorial reconfiguration problems.

In the TOKEN SLIDING problem, introduced by Hearn and Demaine [15], two independent sets are adjacent if one can be obtained from the other by removing a token from a vertex u and immediately placing it on another vertex v with the requirement that $\{u, v\}$ must be an edge of the graph. The token is then said to *slide* from vertex u to vertex v along the edge $\{u, v\}$. Generally speaking, in the TOKEN SLIDING problem, we are given a graph G and two independent sets I_s and I_t of G. The goal is to decide whether there exists a sequence of slides (a *reconfiguration sequence*) that transforms I_s to I_t. The problem has been extensively studied under the combinatorial reconfiguration framework [6,8,11,12,18,20,23]. It is known that the problem is PSPACE-complete, even on restricted graph classes such as graphs of bounded bandwidth (and hence pathwidth) [28], planar graphs [15], split graphs [4], and bipartite graphs [22]. However, TOKEN SLIDING can be decided in polynomial time on trees [11], interval graphs [6], bipartite permutation and bipartite distance-hereditary graphs [12], and line graphs [16].

In the TOKEN JUMPING problem, introduced by Kamiński et al. [20], we drop the restriction that the token should move along an edge of G and instead we allow it to move to any vertex of G provided it does not break the independence of the set of tokens. That is, a single reconfiguration step consists of first removing a token on some vertex u and then immediately adding it back on

---

[1] Informally, this means that it is unlikely to be fixed-parameter tractable.

any other vertex v, as long as no two tokens become adjacent. The token is said to *jump* from vertex u to vertex v. TOKEN JUMPING is also PSPACE-complete on graphs of bounded bandwidth [28] and planar graphs [15]. Lokshtanov and Mouawad [22] showed that, unlike TOKEN SLIDING, which is PSPACE-complete on bipartite graphs, the TOKEN JUMPING problem becomes NP-complete on bipartite graphs. On the positive side, it is "easy" to show that TOKEN JUMPING can be decided in polynomial-time on trees (and even on split/chordal graphs) since we can simply jump tokens to leaves (resp. vertices that only appear in the bag of a leaf in the clique tree) to transform one independent set into another.

In this paper we focus on the parameterized complexity of the TOKEN SLID-ING problem on graphs where cycles with prescribed lengths are forbidden. Given an NP-hard problem, parameterized complexity permits to refine the notion of hardness; does the hardness come from the whole instance or from a small param-eter? A problem Π is FPT (fixed-parameter tractable) parameterized by k if one can solve it in time $f(k) \cdot poly(n)$, for some computable function f. In other words, the combinatorial explosion can be restricted to the parameter k. In the rest of the paper, our parameter k will be the size of the independent set (i.e. the number of tokens). TOKEN SLIDING is known to be W[1]-hard param-eterized by k on general [23] and bipartite [2] graphs. It remains W[1]-hard on $\{C_4, \ldots, C_p\}$-free graphs for any $p \in \mathbb{N}$ [2] and becomes FPT parameterized by k on bipartite C_4-free graphs. The TOKEN JUMPING problem is W[1]-Hard on general graphs [18] and is FPT when parameterized by k on graphs of girth five or more [2]. For graphs of girth four, it was shown that TOKEN JUMPING being FPT would imply that *Gap-ETH*, an unproven computational hardness hypothe-sis, is false [1]. Both TOKEN JUMPING and TOKEN SLIDING were recently shown to be XL-complete [5].

Our Result. The parameterized complexity of the TOKEN JUMPING problem parameterized by k is settled with regard to the girth of the graph, i.e., the problem is unlikely to be FPT for graphs of girth four or less and FPT for graphs of girth five or more. For TOKEN SLIDING, it was only known that the problem is W[1]-hard for graphs of girth four or less and the authors in [2] posed the question of whether there exists a constant p such that the problem becomes fixed-parameter tractable on graphs of girth at least p. We answer their question positively and prove that the problem is indeed fixed-parameter tractable on graphs of girth five or more, which establishes a full classification of the tractability of TOKEN SLIDING parameterized by the number of tokens based on the girth of the input graph.

Our Methods. Our result extends and builds on the recent *galactic reconfiguration* framework introduced by Bartier et al. [3] to show that TOKEN SLIDING is FPT on graphs of bounded degree, chordal graphs of bounded clique number, and planar graphs. Let us briefly describe the intuition behind the framework and how we adapt it for our use case. One of the main reasons why the TOKEN SLIDING problem is believed to be "harder" than the TOKEN JUMPING problem is due to what the authors in [3] call the *bottleneck effect*. Indeed, if we consider TOKEN SLIDING on trees, there might be a lot of empty leaves/subtrees in the tree but there might be a bottleneck in the graph that prevents any other tokens from reaching these vertices. For instance, if we consider a star with one long subdivided branch, then one cannot move any tokens from the leaves of the star to the long branch while there are at least two tokens on leaves. That being said, if the long branch of the star is "long enough" with respect to k then it *should* be possible to reduce parts of it; as some part would be irrelevant. In fact, this observation can be generalized to many other cases. For instance, when we have a large grid minor, then whenever a token slides into the structure it should then be able to slide freely within the structure (while avoiding conflicts with any other tokens in that structure). However, proving that a structure can be reduced in the context of reconfiguration is usually a daunting task due to the many moving parts. To overcome this problem, the authors in [3] introduce a new type of vertices called *black holes*, which can simulate the behavior of a large grid minor by being able to *absorb* as many tokens as they see fit; and then *project* them back as needed.

Since we need to maintain the girth property, we do not use the notion of black holes and instead show that when restricted to graphs of girth five or more we can efficiently find structures that behave like large grid minors (from the discussion above) and replace them with subgraphs of size bounded by a function of k that can absorb/project tokens in a similar fashion (and do not decrease the girth of the graph). We note that our strategy for reducing such structures is not limited to graphs of high girth and could in principle apply to any graph. At a high level, our FPT algorithm can then be summarized as follows. We let (G, k, I_s, I_t) denote an instance of the problem, where G has girth five or more. In a first stage, we show that we can always find a reconfiguration sequence from I_s to I'_s and from I_t to I'_t such that each vertex $v \in I'_s \cup I'_t$ has degree bounded by some function of k. This immediately implies that we can bound the size of $L_1 \cup L_2$, where $L_1 = I'_s \cup I'_t$ and $L_2 = N_G(I'_s \cup I'_t)$. In a second stage, we show that every connected component C of $L_3 = V(G) \backslash (L_1 \cup L_2)$ can be classified as either a *degree-safe component*, a *diameter-safe component*, a *bad component*, or a *bounded component*. The remainder of the proof consists in showing that degree-safe and diameter-safe components behave like large grid minors and can be replaced by bounded-size gadgets. We then show that bounded components and bad components will eventually have bounded size and we then conclude the algorithm by showing how to bound the total number of components in L_3.

Finally, we note that many interesting questions remain open. In particular, it remains open whether TOKEN SLIDING admits a (polynomial) kernel on graphs of girth five or more and whether the problem remains tractable if we forbid cycles of length $p \mod q$, for integers p and q, or if we exclude odd cycles.

2   Preliminaries

We denote the set of natural numbers by $\mathbb{N}$. For $n \in \mathbb{N}$ we let $[n] = \{1, 2, \ldots, n\}$.

Graphs. We assume that each graph G is finite, simple, and undirected. We let $V(G)$ and $E(G)$ denote the vertex set and edge set of G, respectively. The *open neighborhood* of a vertex v is denoted by $N_G(v) = \{u \mid \{u, v\} \in E(G)\}$ and the *closed neighborhood* by $N_G[v] = N_G(v) \cup \{v\}$. For a set of vertices $Q \subseteq V(G)$, we define $N_G(Q) = \{v \notin Q \mid \{u, v\} \in E(G), u \in Q\}$ and $N_G[Q] = N_G(Q) \cup Q$. The subgraph of G induced by Q is denoted by $G[Q]$, where $G[Q]$ has vertex set Q and edge set $\{\{u, v\} \in E(G) \mid u, v \in Q\}$. We let $G - Q = G[V(G)\backslash Q]$.

A *walk* of length ℓ from v_0 to v_ℓ in G is a vertex sequence $v_0, \ldots, v_\ell$, such that for all $i \in \{0, \ldots, \ell - 1\}$, $\{v_i, v_{i+1}\} \in E(G)$. It is a *path* if all vertices are distinct. It is a *cycle* if $\ell \geq 3$, $v_0 = v_\ell$, and $v_0, \ldots, v_{\ell-1}$ is a path. A path from vertex u to vertex v is also called a *uv-path*. For a pair of vertices u and v in $V(G)$, by $\mathsf{dist}_G(u, v)$ we denote the *distance* or length of a shortest uv-path in G (measured in number of edges and set to ∞ if u and v belong to different connected components). The *eccentricity* of a vertex $v \in V(G)$, $\mathsf{ecc}(v)$, is equal to $\max_{u \in V(G)}(\mathsf{dist}_G(u, v))$. The *diameter* of G, $\mathsf{diam}(G)$, is equal to $\max_{v \in V(G)}(\mathsf{ecc}(v))$. The *girth* of G, $\mathsf{girth}(G)$, is the length of a shortest cycle contained in G. If the graph does not contain any cycles (that is, it is a forest), its girth is defined to be infinity.

Reconfiguration. In the TOKEN SLIDING problem we are given a graph $G = (V, E)$ and two independent sets I_s and I_t of G, each of size $k \geq 1$. The goal is to determine whether there exists a sequence $\langle I_0, I_1, \ldots, I_\ell \rangle$ of independent sets of size k such that $I_s = I_0$, $I_\ell = I_t$, and $I_i \Delta I_{i+1} = \{u, v\} \in E(G)$ for all $i \in \{0, \ldots, \ell - 1\}$. In other words, if we view each independent set as a collection of tokens placed on a subset of the vertices of G, then the problem asks for a sequence of independent sets which transforms I_s to I_t by individual token slides along edges of G which maintain the independence of the sets. Note that TOKEN SLIDING can be expressed in terms of a *reconfiguration graph* $\mathcal{R}(G, k)$. $\mathcal{R}(G, k)$ contains a node for each independent set of G of size exactly k. We add an edge between two nodes whenever the independent set corresponding to one node can be obtained from the other by a single reconfiguration step. That is, a single token slide corresponds to an edge in $\mathcal{R}(G, k)$. The TOKEN SLIDING problem asks whether $I_s, I_t \in V(\mathcal{R}(G, k))$ belong to the same connected component of $\mathcal{R}(G, k)$.

3   The Algorithm

Let (G, k, I_s, I_t) be an instance of TOKEN SLIDING, where G has girth five or more. The aim of this section is to bound the size of the graph by a function of k. We start with a very simple reduction rule that allows us to get rid of most twin vertices in the graph. Two vertices $u, v \in V(G)$ are said to be *twins* if u and v have the same set of neighbours, that is, if $N(u) = N(v)$.

Lemma 1. *Assume* $u, v \in V(G)\backslash(I_s \cup I_t)$ *and* $N(u) = N(v)$. *Then* (G, k, I_s, I_t) *is a yes-instance if and only if* $(G - \{v\}, k, I_s, I_t)$ *is a yes-instance.*

Note that in a graph of girth at least five twins can have degree at most one. Given Lemma 1, we assume in what follows that twins have been reduced. In other words, we let (G, k, I_s, I_t) be an instance of TOKEN SLIDING where G has girth five or more and twins not in $I_s \cup I_t$ have been removed. We now partition our graph into three sets $L_1 = I_s \cup I_t$, $L_2 = N_G(L_1)$, and $L_3 = V(G)\backslash(L_1 \cup L_2)$.

Lemma 2. *If $u \in L_2 \cup L_3$, then u has at most $|L_1| \leq 2k$ neighbors in $L_1 \cup L_2$, i.e., $|N_{L_1 \cup L_2}(u)| \leq 2k$.*

Proof. Assume u_1 is a vertex in L_2 and $u_2 \in N_{L_2}(u_1)$ is a neighbor of u_1 in L_2. If u_1 and u_2 have a common neighbor $u_3 \in L_1$, then this would imply the existence of a C_3 (triangle) in G, a contradiction. Now assume $u_1 \in L_3$ and assume $u_2, u_3 \in N_{L_2}(v_1)$ are two neighbors of u_1 in L_2. If u_2 and u_3 have a common neighbor $u_4 \in L_1$ this would imply the existence of a C_4 in G, a contradiction. Hence, for any vertex $u \in L_2 \cup L_3$ we have $N_{L_1}(v) \cap N_{L_1}(w) = \emptyset$ for all $v, w \in N_{L_2}[u]$. Since each vertex in L_2 has at least one neighbor in L_1 by definition, each vertex $u \in L_2 \cup L_3$ can have at most one neighbor in L_2 for each non-neighbor in L_1, for a total of $|L_1| \leq 2k$ neighbors in $L_1 \cup L_2$. $\qquad \square$

3.1 Safe, Bounded, and Bad Components

Given G and the partition $L_1 = I_s \cup I_t$, $L_2 = N_G(L_1)$, and $L_3 = V(G) \backslash (L_1 \cup L_2)$ we now classify components of $G[L_3]$ into four different types.

Definition 1. *Let C be a maximal connected component in $G[L_3]$.*

- *We call C a diameter-safe component whenever $diam(G[V(C)]) > k^3$.*
- *We call C a degree-safe component whenever $G[V(C)]$ has a vertex u with at least $k^2 + 1$ neighbors X in C and at least k^2 vertices of X have degree two in $G[V(C)]$.*
- *We call C a bounded component whenever $diam(G[V(C)]) \leq k^3$ and no vertex of C has degree more than k^2 in $G[V(C)]$.*
- *We call C a bad component otherwise.*

Note that every component of $G[L_3] = G - (L_1 \cup L_2)$ is *safe* (degree- or diameter-safe), bad, or bounded.

Lemma 3. *A bounded component C in $G[L_3]$ has at most k^{2k^3} vertices.*

Proof. Let T be a spanning tree of C and let $u \in V(C)$ denote the root of T. Each vertex in T has at most k^2 children given the degree bound of C and the height of the tree is at most k^3 given the diameter bound of C. Hence the total number of vertices in C is at most k^{2k^3}. $\qquad \square$

We now describe a crucial property of degree-safe and diameter-safe components, which we call the *absorption-projection property*. We note that this notion is similar to the notion of black holes introduced in [3]. The key (informal) insight is that for a safe component C we can show the following:

1. If there exists a reconfiguration sequence $\mathcal{S} = \langle I_0, I_1, \ldots, I_{\ell-1}, I_\ell \rangle$ from I_s to I_t, then we may assume that $I_i \cap N_G(V(C)) \leq 1$, for $0 \leq i \leq \ell$.
2. A safe component can *absorb* all k tokens, i.e., a safe component contains an independent set of size at least k and whenever a token reaches $N_G(V(C))$ then we can (but do not have to) absorb it into C (regardless of how many tokens are already in C). Moreover, a safe component can then *project* the tokens back into its neighborhood as needed.

Let us start by proving the absorption-projection property for degree-safe components. An s-*star* is a vertex with s pairwise non-adjacent neighbors, which are called the leaves of the s-star. A *subdivided s-star* is an s-star where each edge is subdivided (replaced by a new vertex of degree two adjacent to the endpoints of the edge) any number of times. We say that each leaf of a subdivided star belongs to a *branch* of the star.

Lemma 4. *Let C be a degree-safe component in $G[L_3]$. Then C contains an induced subdivided k-star where all k branches have length more than one.*

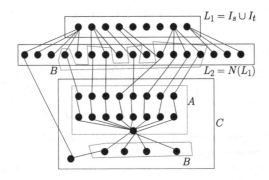

Fig. 1. An illustration of a degree-safe component C.

Lemma 5. *Let C be a degree-safe component in $G[L_3]$ and let A be an induced subdivided k-star contained in C where all branches have length exactly two. Let $B = N_G(A)$. If (G, k, I_s, I_t) is a yes-instance, then there exists a reconfiguration sequence from I_s to I_t in G where we have at most one token on a vertex of B at all times.*

Proof. First, note that the existence of A follows from Lemma 4 and that it is indeed the case that $I_s \cap B = I_t \cap B = \emptyset$. Let r denote the root of the induced subdivided k-star and let N_1 and N_2 denote the first and second levels of subdivided the star, respectively. Let us explain how we can adapt a transformation $\mathcal{S}$ from I_s to I_t into a transformation containing at most one token on a vertex of B at all times and such that, at any step, the number of tokens in $A \cup B$ in both transformations is the same and the positions of the tokens in $V(G)\backslash(A \cup B)$ are the same. Assume that, in the transformation $\mathcal{S}$, a token is about to reach a vertex $b \in B$, that is, we consider the step right before a token is about to slide into B. We first move all tokens residing in A, if any, to the second level of their branches, i.e., to N_2. This is possible as A is an induced subdivided star and there are no other tokens on B. Note that we can assume that there is no token on r (and hence every token is on a branch and "the branch" of a token is well defined) since we can otherwise slide this token to one of the empty branches while B is still empty of tokens. Then we proceed as follows:

- If b is a neighbor of the root r of the subdivided star, then b is not a neighbor of any vertex at the second level of A, since otherwise this would create a cycle of length four. Hence, we can slide the token into b and then r and then some empty branch of A (which is possible since we have k branches in A).
- Otherwise, if b has no neighbors in the first level N_1 of A, we choose a branch that has a neighbor a of b in N_2 (which exists since b is not adjacent to r nor N_1). Then, if the branch of a already contains a token, we can safely slide the token into another branch by going to the first level, then the root r, then to another empty branch of A. Now we slide all tokens in A to the first level of their branch and finally we slide the initial token to b and then to a.
- Finally, if b has neighbors in the first level of A, note that it cannot have more than one neighbor in N_1 since that would imply the existence of a cycle of length four. Let a denote the unique neighbor of b in N_1. If the branch of a has a token on it, then we safely slide it into another empty branch. Now we slide all tokens in A to the first level of their branch and finally we slide the initial token to b and then to a.

Note that all of above slides are reversible and we can therefore use a similar strategy to project tokens from A to B. If, in S, a token is about to leave the vertex $b \in B$, then we can similarly move a token from A to b and then perform the same move. Finally, if a reconfiguration step in S consists of moving tokens in $A \cup B$ to $A \cup B$, we ignore that step. And, if it consists of moving a token from $V(G) \backslash (A \cup B)$ to $V(G) \backslash (A \cup B)$ we perform the same step. It follows from the previous procedure that whenever (G, k, I_s, I_t) is a yes-instance we can find a reconfiguration sequence from I_s to I_t in G where we have at most one token in B at all times, as claimed (see Fig. 1). $\square$

Corollary 1. *Let C be a degree-safe component. If (G, k, I_s, I_t) is a yes-instance, then there exists a reconfiguration sequence from I_s to I_t in G where we have at most one token in $N(C) \subseteq L_2$ at all times.*

Proof. Assume a token slides to a vertex $c \in N(C)$ (for the first time). If $c \in B$, then the result follows from Lemma 5. Otherwise, we can follow a path P contained in C that leads to the root of the induced k-subdivided star (such a path exists since $c \in N(C)$ and C is connected) and right before we reach B we then again can apply Lemma 5. Note that, regardless of whether c is in B or not, once the token reaches $N(C)$ we can assume that it is immediately absorbed by the degree-safe component (and later projected as needed). This implies that we can always find a path P to slide along (i.e., having no tokens in the way). $\square$

We now turn our attention to diameter-safe components and show that they exhibit a similar absorption-projection behavior as degree-safe components. Given a component C we say that a path A in C is a *diameter path* if A is a longest shortest path in C.

Lemma 6. *Let C be a diameter-safe component, let A be a diameter path of C, and let $B = N_G(V(A))$. If (G, k, I_s, I_t) is a yes-instance, then there exists a reconfiguration sequence from I_s to I_t in G where we have at most one token on vertices of B at all times.*

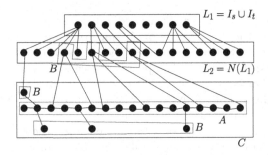

Fig. 2. An illustration of a diameter-safe component C.

Proof. As in the proof of Lemma 5, the goal will consist in proving that we can adapt a transformation $\mathcal{S}$ from I_s to I_t into a transformation containing at most one token on a vertex of B at all times and such that, at any step, the number of tokens in $A \cup B$ in both transformations is the same and the positions of the tokens in $V(G) \backslash (A \cup B)$ are the same. As in the proof of Lemma 5, all the tokens in $A \cup B$ will be absorbed into A (and later projected back as needed) and it suffices to explain how we can move the tokens on A when a new token wants to enter in B or leave into B. We know that two non-consecutive vertices in A cannot be adjacent by minimality of the path. Now assume a token t is about to reach a vertex $b \in B$. Note that neighbors of b in A are pairwise at distance at least three in A, since otherwise that would create a cycle of length less than five. We call the intervals between consecutive neighbor of b *gap intervals* (with respect to b). If b has more than k neighbors in A, then we can put the already in A tokens (at most $k - 1$ of them) in the at most $k - 1$ first gap intervals. Indeed, since there is no token on B and A is an induced path, we can freely move tokens where we want. Then we can slide the token t to b, since none of its neighbors in A have a token on them, and then slide it to the next neighbor of b in A since it has more than k neighbors. Otherwise, b has at most k neighbors in A. Hence there are at most $k + 1$ gap intervals in A (with respect to b). The average number of vertices in the gap intervals (assuming $k \geq 4$) is $\alpha = \frac{\text{diam}(C) - |N_A(b)|}{|N_A(b)| + 1} \geq \frac{k^3 - k}{k + 1} \geq 2k$. Hence at least one gap interval has length at least α and therefore we can slide all tokens currently in A (at most $k - 1$ of them) into this gap interval in such a way no token is on the border of the gap interval (since the gap interval contains an independent set of size at least $k - 1$ which does not contain an endpoint of the gap interval). Now we can simply slide the token t onto b and then onto any of the neighbors of b in A. Combined with the fact that the above strategy can also be applied to project a token from A to B, it then follows that whenever (G, k, I_s, I_t) is a yes-instance we can find a reconfiguration sequence from I_s to I_t in G where we have at most one token in B at all times, as claimed (see Fig. 2). □

Corollary 2. *Let C be a diameter-safe component. If (G, k, I_s, I_t) is a yes-instance then there exists a reconfiguration sequence from I_s to I_t where we have at most one token in $N(C) \subseteq L_2$ at all times.*

Putting Corollary 1 and Corollary 2 together, we know that if (G, k, I_s, I_t) is a yes-instance, then there exists a reconfiguration sequence from I_s to I_t where we have at most one token in $N(C) \subseteq L_2$ at all times, where C is either a degree-safe or a diameter-safe component. We now show how to reduce a safe component C by replacing it by another smaller subgraph that we denote by H.

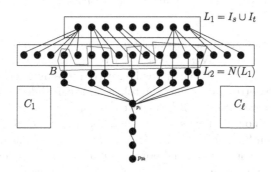

Fig. 3. An illustration of the replacement gadget for a safe component C.

Lemma 7. *Let C be a safe component in $G[L_3]$ and let G' be the graph obtained from G as follows:*

- *Delete all vertices of C (and their incident edges).*
- *For each vertex $v \in N(C) \subseteq L_2$ add two new vertices v' and v'' and add the edges $\{v, v'\}$ and $\{v', v''\}$.*
- *Add a path of length $3k$ consisting of new vertices p_1 to p_{3k}.*
- *Add an edge $\{p_1, v''\}$ for every vertex v''.*

Note that this new component has size $3k + |2N(C)|$ (see Fig. 3). We claim that (G, k, I_s, I_t) is a yes-instance if and only if (G', k, I_s, I_t) is a yes-instance.

Proof. First, we note that replacing C with this new component, H, cannot create cycles of length less than five. This follows from the fact that all the vertices at distance one or two from p_1 have distinct neighbors. Assume (G, k, I_s, I_t) is a yes-instance. Then, by Corollary 1 and Corollary 2, we know that there exists a reconfiguration sequence from I_s to I_t in G where we have at most one token in $N(C) \subseteq L_2$ at all times, where C is either a degree-safe or a diameter-safe component. Hence, we can mimic the reconfiguration sequence from I_s to I_t in G' by simply projecting tokens onto the path of length $3k$ in each of the safe components that we replaced. Now assume that (G', k, I_s, I_t) is a yes-instance. By the same arguments, and combined with the fact that a safe component C can absorb/project the same number of tokens as its replacement component H, we can again mimic the reconfiguration sequence of G' in G. $\quad\square$

3.2 Bounding the Size of L_2

Having classified the components in L_3 and the edges between L_2 and L_3, our next goal is to bound the size of L_2, which until now could be arbitrarily large. We know that vertices in L_2 are the neighbors of vertices in L_1, hence the size of L_2 will grow whenever there are vertices in L_1 with arbitrarily large degrees. Bounding L_2 will therefore be done by first proving the following lemma.

Lemma 8. *Assume a vertex u in $L_1 = I_s \cup I_t$ has degree greater than $2k^2$. Moreover, assume, without loss of generality, that $u \in I_s$. Then, there exists I'_s such that $I_s \triangle I'_s = \{u, u'\}$, u' has degree at most $2k^2$, and the token on u can slide to u'.*

Proof. First note that from such a vertex $u \in I_s$ we can always slide to a vertex in L_2. Indeed, for every v, $|N(u) \cap N(v)| \leq 1$ by the assumption on the girth of the graph. Thus, since the degree of u is larger than the number of tokens, there exists at least one vertex in L_2 that the token on u can slide to. If we slide to a vertex $v \in L_2$ of degree at most $2k^2$, then we are done (we set $u' = v$). Otherwise, by Lemma 2, we know that most of the neighbors of v are in L_3; since v has degree greater than $2k^2$ and at most $2k$ of its neighbors are in $L_1 \cup L_2$. Hence, we are guaranteed at least one neighbor w of v in some component of L_3. If we reach a bounded component C, i.e., if w belongs to a bounded component, then all vertices of C (including w) have at most k^2 neighbors in C and have at most $2k$ neighbors in L_2 (by Lemma 2) and thus we can set $u' = w$. If we reach a bad component C, then we know that C has a vertex b with at least $k^2 + 1$ neighbors in C and at most $k^2 - 1$ of those neighbors have other neighbors in C. Let z denote a vertex in the neighborhood of b that does not have other neighbors in C. By Lemma 2, z will have degree at most $2k + 1$ and we can therefore let $u' = z$. Finally, if we reach a safe component, then after our replacement such components contain a lot of vertices of degree exactly two and we can therefore slide to any such vertex, which completes the proof. □

After exhaustively applying Lemma 8, each time relabeling vertices in L_1, L_2 and L_3 and replacing safe components as described in Lemma 7, we get an equivalent instance where the maximum degree in L_1 is at most $2k^2$ and hence we get a bound on $|L_2|$. We conclude this section with the following lemma.

Lemma 9. *Let (G, k, I_s, I_t) be an instance of* TOKEN SLIDING, *where G has girth at least five. Then we can compute an equivalent instance (G', k, I'_s, I'_t), where G' has girth at least five, $|L_1 \cup L_2| \leq 2k + 4k^3 = O(k^3)$, and each safe component of G is replaced in G' by at most $3k + 8k^3 = O(k^3)$ vertices.*

3.3 Bounding the Size of L_3

We have proved that the number of vertices in L_1 and L_2 is bounded by a function of k, namely $|L_1 \cup L_2| = O(k^3)$. We have also showed that every safe or bounded component in L_3 has a bounded number of vertices, namely safe components have $O(k^3)$ vertices and bounded components have k^{2k^3} vertices. We still need to show that L_3 is bounded. We start by showing that bad components become bounded after bounding L_2:

Lemma 10. *Let (G, k, I_s, I_t) be an instance where G has girth at least five, $|L_1 \cup L_2| \leq 2k + 4k^3 = O(k^3)$, and each safe component has at most $3k + 8k^3 = O(k^3)$ vertices. Then, every bad component has at most $k^{O(k^3)}$ vertices.*

Proof. Let C be a bad component, hence $\text{diam}(C) \leq k^3$ since C is not diameter-safe. Let $v \in V(C)$ be a vertex in C whose degree is $d > k^2$. Since C is not a degree-safe component v can have at most $k^2 - 1$ neighbors in C that have other neighbors in C. Hence, at least $d - (k^2 - 1) = d - k^2 + 1$ neighbors of v will have only v as a neighbor in C and all their other neighbors must be in L_2. Since, by Lemma 1, we can assume that L_3 contains no twin vertices, $d - k^2$ of the neighbors of v in C must have at least one neighbor in L_2. But we know that L_2 has size $O(k^3)$ and if two neighbors of v had a common neighbor in L_2, this would imply the existence of a cycle of length four. Therefore, d must be at most $O(k^3)$. Having bounded the degree and diameter of bad components, we can now apply the same argument as in the proof of Lemma 3.   □

Since bounded and bad components now have the same asymptotic number of vertices, in what follows we refer to both of them as bounded components. What remains to show is that the number of safe and bounded components is also bounded by a function of k and hence L_3 and the whole graph will have size bounded by a function of k.

Definition 2. *Let C_1 and C_2 be two components in $G[L_3]$ and B_1 and B_2 be their respective neighborhoods in L_2. We say C_1 and C_2 are equivalent whenever $B_1 = B_2 = B$ and $G[V(C_1) \cup B]$ is isomorphic to $G[V(C_2) \cup B]$ by an isomorphism that fixes B point-wise. We let $\beta(G)$ and $\sigma(G)$ denote the number of equivalence classes of bounded components and safe components, respectively.*

Lemma 11. *Let S_1 and S_2 be equivalent safe components and let $B_1, \ldots, B_{k+1}$ be equivalent bounded components. Then, (G, k, I_s, I_t), $(G - V(S_2), k, I_s, I_t)$ and $(G - V(B_{k+1}), k, I_s, I_t)$ are equivalent instances.*

After exhaustively removing equivalent components as described in Lemma 11 we obtain the following corollary, which leads to the final lemma.

Corollary 3. *There are at most $k\beta(G)$ bounded components and $\sigma(G)$ safe components.*

Lemma 12. *We have $\beta(G) = 2^{k^{O(k^3)}}$, $\sigma(G) = 2^{O(k^6)}$, $|L_3| \leq k^{O(k^3)} 2^{k^{O(k^3)}} + k^3 2^{O(k^6)} = 2^{k^{O(k^3)}}$, and $|V(G)| = |L_1| + |L_2| + |L_3| = 2^{k^{O(k^3)}}$.*

Theorem 1. TOKEN SLIDING *is fixed-parameter tractable when parameterized by k on graphs of girth five or more.*

References

1. Agrawal, A., Allumalla, R.K., Dhanekula, V.T.: Refuting FPT algorithms for some parameterized problems under Gap-ETH. In: Golovach P.A., Zehavi M. (eds.) 16th International Symposium on Parameterized and Exact Computation. IPEC, vol. 214 of LIPIcs, pp. 2:1–2:12. Schloss Dagstuhl - Leibniz-Zentrum für Informatik (2021)
2. Bartier, V., Bousquet, N., Dallard, C., Lomer, K., Mouawad, A.E.: On girth and the parameterized complexity of token sliding and token jumping. Algorithmica **83**(9), 2914–2951 (2021)
3. Bartier, V., Bousquet, N., Mouawad, A.E.: Galactic token sliding. CoRR, abs/2204.05549 (2022)
4. Belmonte, R., Kim, E.J., Lampis, M., Mitsou, V., Otachi, Y., Sikora, F.: Token sliding on split graphs. Theory Comput. Syst. **65**(4), 662–686 (2021)
5. Bodlaender, H.L., Groenland, C., Swennenhuis, C.M. F.: Parameterized complexities of dominating and independent set reconfiguration. In: Golovach P.A., Zehavi M. (eds.) 16th International Symposium on Parameterized and Exact Computation. IPEC, 8–10 September Lisbon. LIPIcs, vol. 214, pp. 9:1–9:16. Schloss Dagstuhl - Leibniz-Zentrum für Informatik , Portugal (2021)
6. Bonamy, M., Bousquet, N.: Token sliding on chordal graphs. In: Bodlaender, H.L., Woeginger, G.J. (eds.) WG 2017. LNCS, vol. 10520, pp. 127–139. Springer, Cham (2017). https://doi.org/10.1007/978-3-319-68705-6_10
7. Bonnet, É., Bousquet, N., Charbit, P., Thomassé, S., Watrigant, R.: Parameterized complexity of independent set in h-free graphs. Algorithmica **82**(8), 2360–2394 (2020). https://doi.org/10.1007/s00453-020-00730-6
8. Bonsma, P., Kamiński, M., Wrochna, M.: Reconfiguring independent sets in claw-free graphs. In: Ravi, R., Gørtz, I.L. (eds.) SWAT 2014. LNCS, vol. 8503, pp. 86–97. Springer, Cham (2014). https://doi.org/10.1007/978-3-319-08404-6_8
9. Brewster, R.C., McGuinness, S., Moore, B., Noel, J.A.: A dichotomy theorem for circular colouring reconfiguration. Theor. Comput. Sci. **639**, 1–13 (2016)
10. Cereceda, L., van den Heuvel, J., Johnson, M.: Connectedness of the graph of vertex-colourings. Discret. Math. **308**(5–6), 913–919 (2008)
11. Demaine, E.D., Demaine, M.L., Fox-Epstein, E., Hoang, D.A., Ito, T., Ono, H., Otachi, Y., Uehara, R., Yamada, T.: Polynomial-time algorithm for sliding tokens on trees. In: Ahn, H.-K., Shin, C.-S. (eds.) ISAAC 2014. LNCS, vol. 8889, pp. 389–400. Springer, Cham (2014). https://doi.org/10.1007/978-3-319-13075-0_31
12. Fox-Epstein, E., Hoang, D.A., Otachi, Y., Uehara, R.: Sliding token on bipartite permutation graphs. In: Elbassioni, K., Makino, K. (eds.) ISAAC 2015. LNCS, vol. 9472, pp. 237–247. Springer, Heidelberg (2015). https://doi.org/10.1007/978-3-662-48971-0_21
13. Gharibian, S., Sikora, J.: Ground state connectivity of local hamiltonians. ACM Trans. Comput. Theory **10**(2), 8:1–8:28 (2018)
14. Gopalan, P., Kolaitis, P.G., Maneva, E.N., Papadimitriou, C.H.: The connectivity of boolean satisfiability: computational and structural dichotomies. SIAM J. Comput. **38**(6), 2330–2355 (2009)
15. Hearn, R.A., Demaine, E.D.: PSPACE-completeness of sliding-block puzzles and other problems through the nondeterministic constraint logic model of computation. Theor. Comput. Sci. **343**(1–2), 72–96 (2005)
16. Ito, T., et al.: On the complexity of reconfiguration problems. Theor. Comput. Sci. **412**(12–14), 1054–1065 (2011)

17. Ito, T., Kaminski, M., Demaine, E.D.: Reconfiguration of list edge-colorings in a graph. Discret. Appl. Math. **160**(15), 2199–2207 (2012)
18. Ito, T., Kamiński, M., Ono, H., Suzuki, A., Uehara, R., Yamanaka, K.: On the parameterized complexity for token jumping on graphs. In: Gopal, T.V., Agrawal, M., Li, A., Cooper, S.B. (eds.) TAMC 2014. LNCS, vol. 8402, pp. 341–351. Springer, Cham (2014). https://doi.org/10.1007/978-3-319-06089-7_24
19. Johnson, W.W., Story, W.E.: Notes on the "15" puzzle. Am. J. Math. **2**(4), 397–404 (1879)
20. Kaminski, M., Medvedev, P., Milanic, M.: Complexity of independent set reconfigurability problems. Theor. Comput. Sci. **439**, 9–15 (2012)
21. Kendall, G., Parkes, A.J., Spoerer, K.: A survey of NP-complete puzzles. J. Int. Comput. Games Assoc. **31**(1), 13–34 (2008)
22. Lokshtanov, D., Mouawad, A.E.: The complexity of independent set reconfiguration on bipartite graphs. ACM Trans. Algorithms **15**(1), 1–19 (2019)
23. Lokshtanov, D., Mouawad, A.E., Panolan, F., Ramanujan, M.S., Saurabh, S.: Reconfiguration on sparse graphs. J. Comput. Syst. Sci. **95**, 122–131 (2018)
24. Lubiw, A., Pathak, V.: Flip distance between two triangulations of a point set is NP-complete. Comput. Geom. **49**, 17–23 (2015)
25. Mouawad, A.E., Nishimura, N., Pathak, V., Raman, V.: Shortest reconfiguration paths in the solution space of boolean formulas. SIAM J. Discret. Math. **31**(3), 2185–2200 (2017)
26. Nishimura, N.: Introduction to reconfiguration. Algorithms **11**(4), 52 (2018)
27. Heuvel, J.V.D.: The complexity of change. In: Blackburn S.R., Gerke S., Wildon, M. (eds.) Surveys in Combinatorics 2013. London Mathematical Society Lecture Note Series, vol. 409, pp. 127–160. University Press, Cambridge (2013)
28. Wrochna, M.: Reconfiguration in bounded bandwidth and tree-depth. J. Comput. Syst. Sci. **93**, 1–10 (2018)
29. Wrochna, M.: Homomorphism reconfiguration via homotopy. SIAM J. Discret. Math. **34**(1), 328–350 (2020)
30. Zuckerman, D.: Linear degree extractors and the inapproximability of max clique and chromatic number. Theo. Comput. **3**(1), 103–128 (2007)

Recognition of Linear and Star Variants of Leaf Powers is in P

Bergougnoux Benjamin, Svein Høgemo, Jan Arne Telle[✉],
and Martin Vatshelle

Department of Informatics, University of Bergen, 5020 Bergen, Norway
{benjamin.bergougnoux,svein.hogemo,jan.arne.telle,
martin.vatshelle}@uib.no

Abstract. A k-leaf power of a tree T is a graph G whose vertices are the leaves of T and whose edges connect pairs of leaves whose distance in T is at most k. A graph is a leaf power if it is a k-leaf power for some k. Over 20 years ago, Nishimura et al. [J. Algorithms, 2002] asked if recognition of leaf powers was in P. Recently, Lafond [SODA 2022] showed an XP algorithm when parameterized by k, while leaving the main question open. In this paper, we explore this question from the perspective of two alternative models of leaf powers, showing that both a linear and a star variant of leaf powers can be recognized in polynomial-time.

Keywords: Leaf power · Co-threshold tolerance graphs · Interval graphs

1 Introduction

Leaf powers were introduced by Nishimura et al. in [22], and have enjoyed a steady stream of research. Leaf powers are related to the problem of reconstructing *phylogenetic trees*. For an integer k, a graph G is a k-leaf power if there exists a tree T – called a *leaf root* – with a one-to-one correspondence between $V(G)$ and the leaves of T, such that two vertices u and v are neighbors in G iff the distance between the two corresponding leaves in T is at most k. G is a leaf powers if it is a k-leaf power for some k. The most important open problem in the field is whether leaf powers can be recognized in polynomial time.

Most of the results on leaf powers have followed two main lines, focusing either on the distance values k or on the relation of leaf powers to other graph classes, see e.g. the survey by Calamoneri et al. [7]. For the first approach, steady research for increasing values of k has shown that k-leaf powers for any $k \leqslant 6$ is recognizable in polytime [4,5,8,11,12,22]. Moreover, the recognition of k-leaf powers is known to be FPT parameterized by k and the degeneracy of the graph [13]. Recently Lafond [19] gave a polynomial time algorithm to

---

Omitted proofs and a conclusion can be found in the full version of this paper available on https://arxiv.org/abs/2105.12407.

© Springer Nature Switzerland AG 2022
M. A. Bekos and M. Kaufmann (Eds.): WG 2022, LNCS 13453, pp. 70–83, 2022.
https://doi.org/10.1007/978-3-031-15914-5_6

recognize k-leaf powers for any constant value of k. For the second approach, we can mention that interval graphs [5] and rooted directed path graphs [3] are leaf powers, and also that leaf powers have mim-width one [17] and are strongly chordal. Moreover, an infinite family of strongly chordal graphs that are not leaf powers has been identified [18]; see also Nevris and Rosenke [21].

To decide if leaf powers are recognizable in polynomial time, it may be better not to focus on the distance values k. Firstly, the specialized algorithms for k-leaf powers for $k \leqslant 6$ do not seem to generalize. Secondly, the recent XP algorithm of Lafond [19] uses techniques that will not allow removing k from the exponent. In this paper we therefore take a different approach, and consider alternative models for leaf powers that do not rely on a distance bound. In order to make progress towards settling the main question, we consider fundamental restrictions on the shape of the trees, in two distinct directions: subdivided caterpillars (linear) and subdivided stars, in both cases showing polynomial-time recognizability. We use two models: weighted leaf roots for the linear case and NeS models for the stars.

The first model uses rational edge weights between 0 and 1 in the tree T which allows to fix a bound of 1 for the tree distance. It is not hard to see that this coincides with the standard definition of leaf powers using an unweighted tree T and a bound k on distance. Given a solution of the latter type we simply set all edge weights to $1/k$, while in the other direction we let k be the least common denominator of all edge weights and then subdivide each edge a number of times equal to its weight times k.

The second model arises by combining the result of Brandstädt et al. that leaf powers are exactly the fixed tolerance NeST graphs [3, Theorem 4] with the result of Bibelnieks et al. [1, Theorem 3.3] that these latter graphs are exactly those that admit what they call a "neighborhood subtree intersection representation", that we choose to call a NeS model. NeS models are a generalization of interval models: by considering intervals of the line as having a center that stretches uniformly in both directions, we can generalize the line to a tree embedded in the plane, and the intervals to embedded subtrees with a center, stretching uniformly in all directions from the center, along tree edges. Thus a NeS model of a graph G consists of an embedded tree and one such subtree for each vertex, such that two vertices are adjacent in G iff their subtrees have non-empty intersection. Precise definitions are given later. The leaf powers are exactly the graphs having a NeS model. Some results are much easier to prove using NeS models, to illustrate this, we show that leaf powers are closed under several operations such as the addition of a universal vertex (see Lemma 6).

We show that fundamental constraints on these models allow polynomial-time recognition. Using the first model, we restrict to edge-weighted caterpillars, i.e. trees with a path containing all nodes of degree 2 or more. We call linear leaf power a graph with such model. Brandstädt et al. [2] considered leaf roots restricted to caterpillars (see also [6]) in the unweighted setting, showing that unit interval graphs are exactly the k-leaf powers for some k with a leaf root being an unweighted caterpillar. In the unweighted setting, linear leaf powers are graphs with a leaf root that is a *subdivision* of a caterpillar. We show that linear leaf powers are exactly the *co-threshold tolerance graphs* [20], and combined with the algorithm of Golovach et al. [14] this implies that we can recognize linear leaf powers in $O(n^2)$ time. Our proof goes via the equivalent concept of blue-red interval graphs introduced by [15].

The recognition of linear leaf powers in polynomial time could have practical applications for deciding whether the most-likely evolutionary tree associated with a set of organisms has a linear topology. Answering this question might find particular relevance inside the field of tumor phylogenetics where, under certain model assumptions, linear topologies are considered more likely [9,24].

For NeS models, we restrict to graphs having a NeS model where the embedding tree is a star, and show that they can be recognized in polynomial time. Note that allowing the embedding tree to be a subdivided star will result in the same class of graphs. Our algorithm uses the fact that the input graph must be a chordal graph, and for each maximal clique X we check if G admits a star NeS model where the set of vertices having a subtree containing the central vertex of the star is X. To check this we use a combinatorial characterization, that we call a "good partition", of a star NeS model.

2   Preliminaries

For positive integer k, denote by $[k]$ the set $\{1, 2, \ldots, k\}$. A partition of a set S is a collection of non-empty disjoint subsets $B_1, \ldots, B_t$ of S – called *blocks* – such that $S = B_1 \cup \cdots \cup B_t$. Given two partitions $\mathcal{A}, \mathcal{B}$ of S, we say $\mathcal{A} \sqsubseteq \mathcal{B}$ if every block of $\mathcal{A}$ is included in a block of $\mathcal{B}$, i.e. $\sqsubseteq$ is the *refinement* relation.

Graph. Our graph terminology is standard and we refer to [10]. The vertex set of a graph G is denoted by $V(G)$ and its edge set by $E(G)$. An edge between two vertices x and y is denoted by xy or yx. The set of vertices that is adjacent to x is denoted by $N(x)$. A vertex x is simplicial if $N(x)$ is a clique. Two vertices x, y are true twins if $xy \in E(G)$ and $N(x)\backslash\{y\} = N(y)\backslash\{x\}$. Given $X \subseteq V(G)$, we denote by $G[X]$ the graph induced by X. Given a vertex $v \in V(G)$, we denote the subgraph $G[V(G)\backslash\{v\}]$ by $G - v$. We denote by $\mathsf{CC}(G)$ the partition of G into its connected components. Given a tree T and an edge-weight function $\mathsf{w} : E(T) \to \mathbb{Q}$, the distance between two vertices x and y is denoted by $d_T(x, y)$ is $\sum_{e \in E(P)} \mathsf{w}(e)$ with P is the unique path between x and y. A caterpillar is a tree in which there exists a path that contains every vertex of degree two or more

Leaf Power. In the Introduction we have already given the standard definition of leaf powers and leaf roots, and also we argued the equivalence with the following. Given a graph G, a leaf root of G is a pair (T, w) of a tree T and a rational-valued weight function $\mathsf{w} : E(T) \to [0, 1]$ such that the vertices of G are the leaves of T and for every $u, v \in V(G)$, u and v are adjacent iff $d_T(u, v) \leqslant 1$. Moreover, if T is a caterpillar we call (T, w) a *linear leaf root*. A graph is a *leaf power* if it admits a leaf root and it is a linear leaf power if it admits a linear leaf root. Since we manipulate both the graphs and the trees representing them, the vertices of trees will be called *nodes* to avoid confusion.

Interval Graphs. A graph G is an *interval graph* if there exists a set of intervals in $\mathbb{Q}$, $\mathcal{I} = (I_v)_{v \in V(G)}$, such that for every pair of vertices $u, v \in V(G)$, the intervals I_v and I_u intersect iff $uv \in E(G)$. We call $(I_v)_{v \in V(G)}$ an *interval model* of G. For an interval $I = [\ell, r]$, we define the midpoint of I as $(\ell + r)/2$ and its length as $r - \ell$.

Clique Tree. A Chordal graph is a graph in which every induced cycle have exactly three vertices. For a chordal graph G, a *clique tree* CT of G is a tree whose vertices are the maximal cliques of $V(G)$ and for every vertex $v \in V(G)$, the set of maximal cliques of G containing v induces a subtree of CT. Figure 2 gives an example of clique tree. Every chordal graph admits $O(n)$ maximal cliques and given a graph G, in time $O(n+m)$ we can construct a clique tree of G or confirm that G is not chordal [16,23]. When a clique tree is a path, we call it a *clique path*. We denote by $(K_1, \ldots, K_k)$ the clique path whose vertices are $K_1, \ldots, K_k$ and where K_i is adjacent to K_{i+1} for every $i \in [k-1]$.

3 Linear Leaf Powers

In this section we show that linear leaf powers are exactly the *co-threshold tolerance graphs* (co-TT graphs). Combined with the algorithm in [14], this implies that we can recognize linear leaf powers in $O(n^2)$ time.

Co-TT graphs were defined by Monma, Trotter and Reed in [20]; we will not define them here as we do not use this characterization. Rather, we work with the equivalent class of *blue-red interval graphs* [15, Proposition 3.3].

Definition 1 (Blue-red interval graph). *A graph G is a* blue-red interval graph *if there exists a bipartition (B, R) of $V(G)$ and an interval model $\mathcal{I} = (I_v)_{v \in V(G)}$ (with $(B, R, \mathcal{I})$ called a* blue-red interval model*) such that $E(G) = \{b_1 b_2 : b_1, b_2 \in B \text{ and } I_{b_1} \cap I_{b_2} \neq \varnothing\} \cup \{rb : r \in R, b \in B \text{ and } I_r \subseteq I_b\}$.*

The red vertices induce an independent set, $(I_b)_{b \in B}$ is an interval model of $G[B]$, and we have a blue-red edge for each red interval contained in a blue interval. The following fact can be easily deduced from Fig. 1.

Fact 2 *Consider two intervals I_1, I_2 with lengths ℓ_1, ℓ_2 and midpoints m_1, m_2 respectively. We have $I_1 \cap I_2 \neq \varnothing$ iff $|m_1 - m_2| \leqslant \frac{\ell_1 + \ell_2}{2}$. Moreover, we have $I_2 \subseteq I_1$ iff $|m_1 - m_2| \leqslant \frac{\ell_1 - \ell_2}{2}$.*

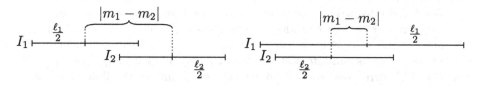

Fig. 1. Example of two intervals overlapping and one interval containing another one.

To prove that linear leaf powers are exactly blue-red interval graphs, we use a similar construction as the one used in [2, Theorem 6] to prove that every interval graph is a leaf power, but in our setting, we have to deal with red vertices and this complicates things quite a bit.

Theorem 3. *G is a blue-red interval graph iff G is a linear leaf power.*

Proof (Sketch of proof). ($\Rightarrow$) Let $(B, R, (I_v)_{v \in V(G)})$ be a blue-red interval model of a graph G. For each $v \in V(G)$, we denote by m_v and ℓ_v the midpoint and the length of the interval I_v. We assume w.l.o.g. that G is connected and the lengths of the intervals I_v's are not 0 and at most 1. We prove that we can construct a linear leaf root (T, w) of G. For doing so, we construct (T, w) as follows. The inner path of T represents the midpoints of the intervals: each node of this path is associated with a midpoint and the weights on the edges of this path are the distances between consecutive midpoints (since G is connected by assumption and the length are at most 1, these distances are at most 1). Each vertex v of G is adjacent through an edge e to the node representing the midpoint m_v, the weight of e is $\frac{1-\ell_v}{2}$ if $v \in B$ and $\frac{1+\ell_v}{2}$ if $v \in R$. With these weights, the distance between two red vertices in T is strictly greater than 1. Moreover, for every $v \in B$ and $u \in V(G)$, we have ($\spadesuit$) $d_T(u, v) = 1 + |m_u - m_v| - \left(\frac{\ell_u + \ell_v}{2}\right)$ if $u \in B$ and $1 + |m_v - m_u| - \left(\frac{\ell_v - \ell_u}{2}\right)$ if $u \in R$. From Fact 2, one easily proves that (T, w) is a linear leaf root of G.

($\Leftarrow$) Let (T, w) be a linear leaf root of a graph G with $(u_1, \ldots, u_t)$ the path induced by the internal vertices of T. The previous construction can easily be reversed to construct a blue-red interval model $(B, R, (I_v)_{v \in V(G)})$ of G. For each vertex $v \in V(G)$ whose neighbor in T is u_i, we associate v with an interval I_v and a color as follows. We color v blue if $\mathsf{w}(u_i v) \leqslant 1/2$, otherwise we color it red. The midpoint m_v of I_v is the distance between u_1 and u_i in T. We define the length ℓ_v of I_v to be $1 - 2\mathsf{w}(u_i v)$ if v is blue and $2\mathsf{w}(u_i v) - 1$ if v is red. By construction, $\mathsf{w}(u_i v) = \frac{1-\ell_v}{2}$ if $v \in R$ and $\frac{1+\ell_v}{2}$ if $v \in B$. Observe that two red vertices cannot be adjacent since their distance in T is strictly greater than 1. We can prove that ($\spadesuit$) holds also for this direction and with Fact 2 one concludes that $(B, R, (I_v)_{v \in V(G)})$ is a blue-red interval model of G. $\square$

4   Star NeS Model

In this section, we first present an alternative definition of leaf powers through the notion of NeS models. We then show that we can recognize in polynomial time graphs with a *star NeS model*: a NeS model whose embedding tree is a star (considering subdivided stars instead of stars does not make a difference).

For each tree T, we consider a corresponding tree $\mathcal{T}$ embedded in the Euclidean plane so that each edge of T corresponds to a line segment of $\mathcal{T}$, these lines segments can intersect one another only at their endpoints, and the vertices of T correspond (one-to-one) to the endpoints of the lines. Each line segment of $\mathcal{T}$ has a positive Euclidean length. These embedded trees allow us to consider $\mathcal{T}$ as the infinite set of points on the line segments of $\mathcal{T}$. The notion of tree embedding used here is consistent with that found in [25]. The line segments of $\mathcal{T}$ and their endpoints are called respectively the edges and the nodes of $\mathcal{T}$. The distance between two points x, y of $\mathcal{T}$ denoted by $d_{\mathcal{T}}(x, y)$ is the length of the unique path in $\mathcal{T}$ between x and y (the distance between two vertices of T and their corresponding endpoints in $\mathcal{T}$ are the same).

Definition 4 (Neighborhood subtree, NeS-model). *Let $\mathcal{T}$ be an embedding tree. For some point $c \in \mathcal{T}$ and non-negative rational w, we define the neighborhood subtree with center c and radius w as the set of points $\{p \in \mathcal{T} : d_{\mathcal{T}}(p, c) \leqslant w\}$. A NeS model $(\mathcal{T}, (T_v)_{v \in V(G)})$ of a graph G is a pair of an embedding tree $\mathcal{T}$ and a collection of neigbhorhood subtrees of $\mathcal{T}$ associated with each vertex of G such that for every $u, v \in V(G)$, we have $uv \in E(G)$ iff $T_u \cap T_v \neq \varnothing$.*

Theorem 5. *A graph is a leaf power iff it admits a NeS model.*

Proof. Brandstädt et al. showed that leaf powers correspond to the graph class fixed tolerance NeST graph [3, Theorem 4]. Bidelnieks and Dearing showed that G is a fixed tolerance NeST graph iff G has a NeS model [1, Theorem 3.3]. □

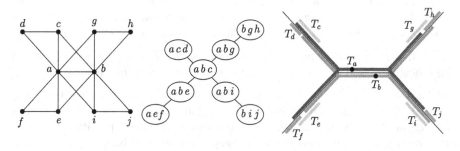

Fig. 2. A graph with a clique tree and NeS model. The dots are centers of T_a and T_b.

See Fig. 2 for a NeS model. From now, we consider that the intervals of an interval models are segments of a line in the plane rather than intervals in $\mathbb{Q}$. Observe that every interval graph has a NeS model where the embedding tree is a single edge. Moreover, if a graph G admits a NeS model $(\mathcal{T}, (T_v)_{v \in V(G)})$, then for every embedding path $\mathcal{L}$ of $\mathcal{T}$, $(\mathcal{L}, (T_v \cap \mathcal{L})_{v \in X})$ is an interval model of $G[X]$ with X the set of vertices v such that T_v intersects $\mathcal{L}$. As illustrated by the proofs of the following two lemmata, some results are easier to prove with NeS models than with other characterizations.

Lemma 6. *For a graph G and $u \in V(G)$ such that either (1) u is universal, or (2) u has degree 1, or (3) $N(u)$ is a minimal separator in $G - u$, or (4) $N(u)$ is a maximal clique in $G - u$. Then G is a leaf power iff $G - u$ is a leaf power.*

Lemma 7. *Let G be a graph and $u \in V(G)$ a cut vertex. Then G is a leaf power iff for every component C of $G - u$, $G[V(C) \cup \{u\}]$ is a leaf power.*

We now give the algorithm for recognizing graphs having a star NeS model. Our result is based on the purely combinatorial definition of *good partition*, and we show that a graph admits a star NeS model iff it admits a good partition. Given a good partition, we compute a star NeS model in polynomial time. Finally, we prove that our Algorithm 1 in polynomial time constructs a good partition of the input graph or confirms that it does not admit one.

Consider a star NeS model $(\mathcal{T}, (T_v)_{v \in V(G)})$ of a graph G. Observe that $\mathcal{T}$ is the union of line segments $L_1, \ldots, L_\beta$ with a common endpoint c that is the center of $\mathcal{T}$. Let X be the set of vertices whose neighborhood subtrees contain c. For each $i \in [\beta]$, we let B_i be the set of all vertices in $V(G) \backslash X$ whose neighborhood subtrees are subsets of L_i. The family $\mathcal{B} = \{B_i : i \in [\beta]\}$ must then constitute a partition of $V(G) \backslash X$. We will show in Theorem 12 that the pair $(X, \mathcal{B})$ has the properties of a good partition.

Fact 8 *Let G, $(\mathcal{T}, (T_v)_{v \in V(G)})$ and $\mathcal{B}$ be as defined above. We then have:*

- *There is no edge between B_i and B_j for $i \neq j$ and thus $\mathsf{CC}(G - X) \sqsubseteq \mathcal{B}$.*
- *For every $i \in [\beta]$ the NeS model $(L_i, (T_v \cap L_i)_{v \in B_i \cup X})$ is an interval model of $G[X \cup B_i]$.*
- *For each $x \in X$ the neighborhood subtree T_x is the union of the β intervals $L_1 \cap T_x, \ldots, L_\beta \cap T_x$ and there exist positive rationals ℓ_x and h_x with $\ell_x \leqslant h_x$ such that one interval among these intervals has length h_x and the other $\beta - 1$ intervals have length ℓ_x. If $\ell_x = h_x$, then the center of T_x is c.*

Claim. If G has a star NeS model, it has a one, $(\mathcal{T}, (T_v)_{v \in V(G)})$, where vertices whose neighborhood subtrees contain the center of $\mathcal{T}$ is a maximal clique.

Claim 4 follows since we can always stretch some intervals to make X a maximal clique. So far we have described a good partition as it arises from a star NeS model. Now we introduce the properties of a good partition that will allow to abstract away from geometrical aspects while still being equivalent, i.e. so that a graph has a good partition $(X, \mathcal{B})$ iff it has a star NeS model. The first property is $\mathsf{CC}(G - X) \sqsubseteq \mathcal{B}$ and the second is that for every $B \in \mathcal{B}$ the graph $G[X \cup B]$ is an interval graph having a model where the intervals of X contain the last point used in the interval representation.

Definition 9 (X-interval graph). *Let X be a maximal clique of G. We say that G is an X-interval graph if G admits a clique path ending with X.*

The third property is the existence of an elimination order for the vertices of X based on the lengths ℓ_x in the last item of Fact 8, namely the permutation $(x_1, \ldots, x_t)$ of X such that $\ell_{x_1} \leqslant \ell_{x_2} \leqslant \ldots \leqslant \ell_{x_t}$. This permutation has the property that for any $i \in [t]$, among the vertices $x_i, x_{i+1}, \ldots, x_t$ the vertex x_i must have the minimal neighborhood in at least $\beta - 1$ of the blocks of $\mathcal{B}$; we say that x_i is *removable* from $\{x_i, \ldots, x_t\}$ for $\mathcal{B}$.

Definition 10 (Removable vertex). *Let $X \subseteq V(G)$, $Y \subseteq X$ and let $\mathcal{B}$ be a partition of $V(G) \backslash X$. Given a block B of $\mathcal{B}$ and $x \in Y$, we say $N(x)$ is minimal in B for Y if $N(x) \cap B \subseteq N(y)$ for every $y \in Y$. We say that a vertex $x \in Y$ is removable from Y for $\mathcal{B}$ if $N(x)$ is minimal in at least $|\mathcal{B}| - 1$ blocks of $\mathcal{B}$ for Y.*

Definition 11 (Good partition). *A good partition of a graph G is a pair $(X, \mathcal{B})$ where X is a maximal clique of G and $\mathcal{B}$ a partition of $V(G) \backslash X$ satisfying:*

1. *$\mathsf{CC}(G - X) \sqsubseteq \mathcal{B}$, i.e. every $C \in \mathsf{CC}(G - X)$ is contained in a block of $\mathcal{B}$.*
2. *For each block $B \in \mathcal{B}$, $G[X \cup B]$ is an X-interval graph.*
3. *There exists an elimination order $(x_1, \ldots, x_t)$ on X such that for every $i \in [t]$, x_i is removable from $\{x_i, \ldots, x_t\}$ for $\mathcal{B}$.*

X is the central clique of $(X, \mathcal{B})$ and $(x_1, \ldots, x_t)$ a good permutation of $(X, \mathcal{B})$.

Theorem 12. *A graph G admits a good partition iff it admits a star NeS model. Moreover, given the former we can compute the latter in polynomial time.*

Proof (Sketch of proof). ($\Leftarrow$) Let $(\mathcal{T}, (T_v)_{v \in V(G)})$ be a star NeS model of a graph G and $(X, \mathcal{B})$ be the pair that we defined above Fact 8. Properties 1 and 2 follows immediately from the first two items of Fact 8. Take the permutation $(x_1, \ldots, x_t)$ defined above Definition 10. As argued there, for every $i \in [t]$, $N(x_i)$ is minimal in $\{x_i, \ldots, x_t\}$ for at least $|\mathcal{B}| - 1$ blocks of $\mathcal{B}$, and thus $(x_1, \ldots, x_t)$ is a good permutation and Property 3 is satisfied.

($\Rightarrow$) Let $(X, \mathcal{B})$ be a good partition of a graph G with $\mathcal{B} = \{B_1, \ldots, B_\beta\}$ and $(x_1, \ldots, x_t)$ be a good permutation of $(X, \mathcal{B})$. Take $\mathcal{T}$, the embedding of a star with center c that is the union of β line segments $L_1, \ldots, L_\beta$ whose intersection is $\{c\}$. We start by constructing the neighborhood subtrees of the vertices in X. For doing so, we associate each $x_i \in X$ and each line segment L_j with a rational $\ell(x_i, L_j)$ and define T_{x_i} as the union over $j \in [\beta]$ of the points on L_j at distance at most $\ell(x_i, L_j)$ from c. We set $\ell(x_i, L_j) = i$ if $N(x_i)$ is minimal in $\{x_i, \ldots, x_t\}$ for B_j, and otherwise, we set $\ell(x_i, L_j)$ to a value computed from the lengths associated with the vertices $x_{i+1}, \ldots, x_t$ on L_j such that ($\clubsuit$) for every $x, y \in X$, if $N(x) \cap B_j \subset N(y)$ then $\ell(x, L_j) < \ell(y, L_j)$.

Without going into details of how we compute these lengths, roughly what we do is the following: for each i and each j, if we have some vertices $\hat{x}$ in X "non-minimal" on B_j such that $N(x_i) \cap B_j \subseteq N(\hat{x}) \cap B_j \subset N(x_{i+1}) \cap B_j$, then we place each $\hat{x}$ spaced between i and $i+1$. Since $(x_1, \ldots, x_t)$ is a good permutation, we deduce that for each i the lengths $\ell(x_i, L_j)$ respect the last item of Fact 8 and each T_{x_i} is a neighborhood subtree. For the vertices not in X, with every $j \in [\beta]$, we associate each vertex $v \in B_j$ to an interval of L_j so that $(L_i, (T_v \cap L_i)_{v \in B_i \cup X})$ is an interval model of $G[X \cup B_i]$, which can always be done since we know that ($\clubsuit$) holds. This allows us to obtain a NeS model in polynomial time. $\square$

It is easy to see that every graph that admits a star NeS model has a clique-tree that is a subdivided star. The converse is not true. In fact, for every graph G with a clique tree that is a subdivided star with center X, the pair $(X, \mathsf{CC}(G-X))$ satisfies Properties 1 and 2 of Definition 11 but Property 3 might not be satisfied. See for example the graph in Fig. 2 and note that the pair $(\{a, b, c\}, \mathsf{CC}(G - \{a, b, c\}))$ does not satisfy Property 3, as after removing the vertex c neither a nor b is removable from $\{a, b\}$.

We now describe Algorithm 1 that decides whether a graph G admits a good partition. Clearly G must be chordal, so we start by checking this. A chordal graph has $O(n)$ maximal cliques, and for each maximal clique X we try to construct a good partition $(X, \mathcal{A})$ of G. We start with $\mathcal{A} \leftarrow \mathsf{CC}(G - X)$ and note that $(X, \mathcal{A})$ trivially satisfies Property 1 of Definition 11. Moreover, if G admits a good partition with central clique X, then $(X, \mathcal{A})$ must satisfy Property 2 of Definition 11, and we check this in Line 3. Then, Algorithm 1 iteratively in a while loop tries to construct a good permutation $(w_1, \ldots, w_{|X|})$ of X, while possibly merging some blocks of $\mathcal{A}$ along the way so that it satisfies Property 3, or discover that there is no good partition with central clique X.

For doing so, at every iteration of the while loop, Algorithm 1 searches for a vertex w in W (the set of unprocessed vertices) such that $\mathsf{notmin}(w, W, \mathcal{A})$ – the union of the blocks of $\mathcal{A}$ where $N(w)$ is not minimal for W (see Definition 13) – induces an X-interval graph with X. If such a vertex w exists, then Algorithm 1 sets w_r to w, increments r and merges the blocks of $\mathcal{A}$ contained in $\mathsf{notmin}(w_r, W, \mathcal{A})$ (Line 8) to make w_r removable in $\mathcal{A}$ for W. Otherwise, when no such vertex w exists, Algorithm 1 stops the while loop (Line 7) and tries another candidate for X. For the graph in Fig. 2, with $X = \{a, b, c\}$ the first iteration of the while loop will succeed and set $w_1 = c$, but in the second iteration neither a nor b satisfy the condition of Line 6.

At the start of an iteration of the while loop, the algorithm has already choosen the vertices $w_1, w_2, \ldots, w_{r-1}$ and $W = X \backslash \{w_1, \ldots, w_{r-1}\}$. For every $i \in [r-1]$, w_i is removable from $X \backslash \{w_1, \ldots, w_{i-1}\}$ for $\mathcal{A}$. According to Definition 10 the next vertex w_r to be removed should have $N(w_r)$ non-minimal for W in at most one block of the good partition we want to construct. However, the neighborhood $N(w_r)$ may be non-minimal for W in several blocks of the current partition $\mathcal{A}$, since these blocks may be (unions of) separate components of $G \backslash X$ that should live on the same line segment of a star NeS model and thus actually be a single block which together with X induces an X-interval graph. An example of this merging is given in Fig. 3. The following definition captures, for each $w \in W$, the union of the blocks of $\mathcal{A}$ where $N(w)$ is not minimal for W.

Definition 13 (notmin). *For $W \subseteq X$, $x \in W$ and partition $\mathcal{A}$ of $V(G) \backslash X$, we denote by $\mathsf{notmin}(x, W, \mathcal{A})$ the union of the blocks $A \in \mathcal{A}$ where $N(x)$ is not minimal in A for W.*

As we already argued, when Algorithm 1 starts a while loop, $(X, \mathcal{A})$ satisfies Properties 1 and 2, and it is not hard to argue that in each iteration, for every $i \in [r-1]$, w_i is removable from $X \backslash \{w_1, \ldots, w_{i-1}\}$ for $\mathcal{A}$. Hence, when $W = \varnothing$, then $(w_1, \ldots, w_{|X|})$ is a good permutation of $(X, \mathcal{A})$ and Property 3 is satisfied.

Lemma 14. *If Algorithm 1 returns $(X, \mathcal{B})$, then $(X, \mathcal{B})$ is a good partition.*

To prove the opposite direction, namely that if G has a good partition $(X, \mathcal{B})$ associated with a good permutation $(x_1, \ldots, x_t)$, then Algorithm 1 finds a good partition, we need two lemmata. The easy case is when Algorithm 1 chooses consecutively $w_1 = x_1, \ldots, w_t = x_t$, and we can use Lemma 15 to prove that it will not return no. However, Algorithm 1 does not have this permutation as input and at some iteration with $w_1 = x_1, \ldots, w_{r-1} = x_{r-1}$, the algorithm might

Algorithm 1:

Input: A graph G.

Output: A good partition of G or "no".

1   Check if G is chordal and if so, compute its set of maximal cliques $\mathcal{X}$, otherwise **return** no;

2   **for** *every $X \in \mathcal{X}$* **do**

3   **if** *there exists $C \in \mathsf{CC}(G - X)$ such that $G[X \cup C]$ is not an X-interval graph* **then continue**;

4      $\mathcal{A} \leftarrow \mathsf{CC}(G - X)$, $W \leftarrow X$ and $r \leftarrow 1$;

5   **while** $W \neq \varnothing$ **do**

6   **if** *there exists $w \in W$ such that $G[X \cup \mathsf{notmin}(w_r, W, \mathcal{A})]$ is an X-interval graph* **then** $w_r \leftarrow w$;

7   **else break**;

8   Replace the blocks of $\mathcal{A}$ contained in $\mathsf{notmin}(w_r, W, \mathcal{A})$ by $\mathsf{notmin}(w_r, W, \mathcal{A})$;

9         $W \leftarrow W \backslash \{w_r\}$ and $r \leftarrow r + 1$;

10   **end**

11   **if** $W = \varnothing$ **then return** $(X, \mathcal{A})$

12   **end**

13   **return** no;

stop to follow the permutation $(x_1, \ldots, x_t)$ and choose a vertex $w_r = x_i$ with $r < i$ because x_r may not be the only vertex satisfying the condition of Line 6. In Lemma 16 we show that choosing $w_r = x_i$ is then not a mistake as it implies the existence of another good partition and another good permutation that starts with $(x_1, \ldots, x_{r-1}, w_r = x_i)$. See Fig. 3 for an example of a very simple graph with several good permutations leading to quite distinct star NeS models.

We need some definitions. Given permutation $P = (x_1, \ldots, x_\ell)$ of a subset of X and $i \in [\ell]$, define $\mathcal{A}_0^P = \mathsf{CC}(G - X)$ and $\mathcal{A}_i^P$ the partition of $V(G) \backslash X$ obtained from $\mathcal{A}_{i-1}^P$ by merging the blocks contained in $\mathsf{notmin}(x_i, X \backslash \{x_1, \ldots, x_{i-1}\}, \mathcal{A}_{i-1}^P)$. Observe that when Algorithm 1 treats X and we have $w_1 = x_1, \ldots, w_\ell = x_\ell$, then the values of $\mathcal{A}$ are successively $\mathcal{A}_0^P, \ldots, \mathcal{A}_\ell^P$. The following lemma proves that if there exists a good permutation $P = (x_1, \ldots, x_t)$ and at some iteration we have $w_1 = x_1, \ldots, w_{r-1} = x_{r-1}$, then the vertex x_r satisfies the condition of Line 6 and Algorithm 1 does not return no during this iteration. Thus, as long as Algorithm 1 follows a good permutation, it will not return no.

Lemma 15. *Let G be a graph with good partition $(X, \mathcal{B})$ and $P = (x_1, \ldots, x_t)$ be a good permutation of $(X, \mathcal{B})$. For every $i \in [t]$, we have $\mathcal{A}_i^P \sqsubseteq \mathcal{B}$ and the graph $G[X \cup \mathsf{notmin}(x_i, \{x_i, \ldots, x_t\}, \mathcal{A}_{i-1}^P)]$ is an X-interval graph.*

Lemma 16. *Let $P = (x_1, \ldots, x_t)$ be a good permutation of X and $i \in [t]$. For every $w \in \{x_i, \ldots, x_t\}$ such that $G[X \cup \mathsf{notmin}(w, \{x_i, \ldots, x_t\}, \mathcal{A}_{i-1}^P)]$ is an X-interval graph, there exists a good permutation of X starting with $(x_1, \ldots, x_{i-1}, w)$.*

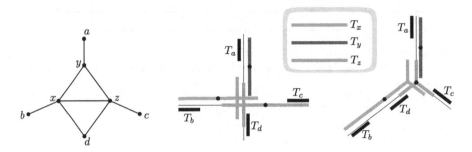

Fig. 3. A graph G and two star NeS models. The dots are the centers of the subtrees. $X = \{x, y, z\}$ is a maximal clique with four components in $G - X$. The left NeS model corresponds to permutations (y, x, z) or (y, z, x) and the right to permutation (x, y, z). If Algorithm 1 chooses $w_1 = x$, then the components $\{b\}$ and $\{d\}$ will be merged in Line 8. There is a third star NeS model, corresponding to permutation (z, y, x), similar to the one on the right. The last two permutations of X are not good permutations.

Proof. (Sketch of proof) For every $A \subseteq V(G)\backslash X$, we define $X\text{-max}_\cap(A) = N(A) \cap X$ and $X\text{-min}_\cap(A) = \cap_{a \in A} N(a) \cap X$. Given $A_1, A_2 \subseteq V(G)\backslash X$, we say that $A_1 \leqslant_X A_2$ if $X\text{-max}_\cap(A_1) \subseteq X\text{-min}_\cap(A_2)$. In Fig. 3, we have $\{b\} \leqslant_{\{x,y,z\}} \{d\}$. We denote by X_i the set $\{x_i, \ldots, x_t\}$. Let $w \in X_i$ such that the graph $G[X \cup \mathsf{notmin}(w, X_i, \mathcal{A}_{i-1}^P)]$ is an X-interval graph. Let $(X, \mathcal{B})$ be a good partition of G such that $(x_1, \ldots, x_t)$ is a good permutation of $(X, \mathcal{B})$. If $w = x_p$ is removable from $\mathcal{B}$ for X_i, then we are done since $(x_1, \ldots, x_{i-1}, w, x_i, \ldots, x_{p-1}, x_{p+1}, \ldots, x_t)$ is a good permutation of $(X, \mathcal{B})$. In particular, w is removable from $\mathcal{B}$ if we have $|\mathsf{notmin}(w, X_i, \mathcal{A}_{i-1}^P)| \leqslant 1$.

We supose from now that $|\mathsf{notmin}(w, X_i, \mathcal{A}_{i-1}^P)| \geqslant 2$. We construct a good partition $(X, \mathcal{B}_{\mathsf{new}})$ that admits a good permutation starting with $(x_1, \ldots, x_{i-1}, w)$. Let $A_1, \ldots, A_k$ be the blocks of $\mathcal{A}_{i-1}^P$ such that $\mathsf{notmin}(w, X_i, \mathcal{A}_{i-1}^P) = A_1 \cup A_2 \cup \cdots \cup A_k$. We prove a separate claim implying that as $G[X \cup \mathsf{notmin}(w, X_i, \mathcal{A}_{i-1}^P)]$ is an X-interval graph, the blocks $A_1, \ldots, A_k$ are pairwise comparable for $\leqslant_X$. Suppose w.l.o.g. that $A_1 \leqslant_X \ldots \leqslant_X A_k$.

In Fig. 3 with $\mathsf{notmin}(x, \{x, y, z\}, \{\{a\}, \{b\}, \{c\}, \{d\}\}) = \{b\} \cup \{d\}$ we have $A_1 = \{b\}$ and $A_2 = \{d\}$. By Lemma 15, we have $\mathcal{A}_{i-1}^P \sqsubseteq \mathcal{B}$. Thus, there exists a block B_1 of $\mathcal{B}$ containing A_1 and B_1 is a union of blocks of $\mathcal{A}_{i-1}^P$. See Fig. 4. Let B_1^{max} be the union of all the blocks A of $\mathcal{A}_{i-1}^P$ included in B_1 such that A is not contained in $\mathsf{notmin}(w, X_i, \mathcal{A}_{i-1}^P)$ and $A_1 \leqslant_X A$. Note that for every block $A \in B_1^{\mathsf{max}}$, we have $A_k \leqslant_X A$ because otherwise we have $A_1 \leqslant_X A \leqslant_X A_k$ and that implies $A \subseteq \mathsf{notmin}(w, X_i, \mathcal{A}_{i-1}^P)$.

We prove a separate claim allowing us to assume that $B_1^{\mathsf{max}} = \varnothing$. We then construct the partition $\mathcal{B}_{\mathsf{new}}$ as follows. Recall that $A_1 \subseteq B_1$. We create a new block $\widehat{B}_1 = B_1 \cup A_2 \cup \cdots \cup A_k$, and for every block $B \in \mathcal{B}$ such that $B \neq B_1$, we create a new block $\widehat{B} = B\backslash(A_2 \cup \cdots \cup A_k)$. We define $\mathcal{B}_{\mathsf{new}} = \{\widehat{B}_1\} \cup \{\widehat{B} : B \in \mathcal{B}\backslash\{B_1\}$ and $\widehat{B} \neq \varnothing\}$. The construction of $\mathcal{B}_{\mathsf{new}}$ is illustrated in Fig. 4. Finally, we

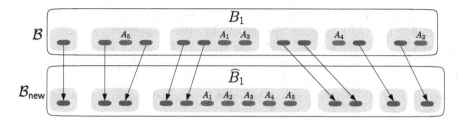

Fig. 4. Construction of $\mathcal{B}_{\text{new}}$ from $\mathcal{B}$ with $k = 5$. The blocks of $\mathcal{B}$ and $\mathcal{B}_{\text{new}}$ are in blue, the blocks $A_1, \ldots, A_5$ of $\mathcal{A}_{i-1}^P$ contained in $\mathsf{notmin}(w, X_i, \mathcal{A}_{i-1}^P)$ are in red, the other blocks of $\mathcal{A}_{i-1}^P$ are in purple. In each block of $\mathcal{B}$ or $\mathcal{B}_{\text{new}}$, the blocks of $\mathcal{A}_{i-1}^P$ are ordered w.r.t. $\leqslant_X$ from left to right. (Color figure online)

prove 4 separate claims that allow us to show that $(X, \mathcal{B}_{\text{new}})$ is a good partition for a good permutation starting with $(x_1, \ldots, x_{i-1}, w)$. □

Lemma 17. *If G has a good partition then Algorithm 1 returns one.*

Proof. Suppose G admits a good partition with central clique X. We prove the following invariant holds at the end of the i-th iteration of the while loop for X.

Invariant. G admits a good permutation starting with $w_1, \ldots, w_i$.

By assumption, X admits a good permutation and thus the invariant holds before the algorithm starts the first iteration of the while loop. By induction, assume the invariant holds when Algorithm 1 starts the i-th iteration of the while loop. Let $L = (w_1, \ldots, w_{i-1})$ be the consecutive vertices chosen at Line 6 before the start of iteration i (observe that L is empty when $i = 1$). The invariant implies that there exists a good permutation $P = (w_1, \ldots, w_{i-1}, x_i, x_{i+1}, \ldots, x_t)$ of G starting with L. By Lemma 15, the graph $G[X \cup \mathsf{notmin}(x_i, \{x_i, \ldots, x_t\}, \mathcal{A}_{i-1}^P)]$ is an X-interval graph. Observe that $\mathcal{A}_{i-1}^P$ and $\{x_i, \ldots, x_t\}$ are the values of the variables $\mathcal{A}$ and W when Algorithm 1 starts the i-th iteration. Thus, at the start of the i-th iteration, there exists a vertex w_i such that $G[X \cup \mathsf{notmin}(w_i, W, \mathcal{A})]$ is an X-interval graph. Consequently, the algorithm does not return no at the i-th iteration and chooses a vertex $w_i \in \{x_i, \ldots, x_t\}$ such that the graph $G[X \cup \mathsf{notmin}(w_i, \{x_i, \ldots, x_t\}, \mathcal{A}_i^P)]$ is an X-interval graph. By Lemma 16, G admits a good permutation starting with $(w_1, \ldots, w_{i-1}, w_i)$. Thus, the invariant holds at the end of the i-th iteration. If at the end of the i-th iteration W is empty, then the while loop stops and the algorithm returns a pair $(X, \mathcal{A})$. Otherwise, the algorithm starts an $i + 1$-st iteration and the invariant holds at the start of this new iteration. By induction, the invariant holds at every step and Algorithm 1 returns a pair $(X, \mathcal{A})$ which is a good partition by Lemma 14. □

Theorem 18. *Algorithm 1 decides in polynomial time if G admits a star NeS model.*

Proof. By Theorem 12 G admits a star NeS model iff G admits a good partition, and by Lemma 14 and 17 Algorithm 1 finds a good partition iff the input graph has a good partition. Let us argue for the runtime. Checking that G is chordal and finding the $O(n)$ maximal cliques can be done in polynomial time [16,23]. Given $X, Y \subseteq V(G)$ we check whether $G[X \cup Y]$ is an X-interval graph, as follows. Take G' the graph obtained from $G[X \cup Y]$ by adding u and v such that $N(u) = \{v\}$ and $N(v) = \{u\} \cup X$. It is easy to see that $G[X \cup Y]$ is an X-interval graph iff G' is an interval graph, which can be checked in polynomial time.   □

References

1. Bibelnieks, E., Dearing, P.M.: Neighborhood subtree tolerance graphs. Discret. Appl. Math. **43**(1), 13–26 (1993)
2. Brandstädt, A., Hundt, C.: Ptolemaic graphs and interval graphs are leaf powers. In: Laber, E.S., Bornstein, C., Nogueira, L.T., Faria, L. (eds.) LATIN 2008. LNCS, vol. 4957, pp. 479–491. Springer, Heidelberg (2008). https://doi.org/10.1007/978-3-540-78773-0_42
3. Brandstädt, A., Hundt, C., Mancini, F., Wagner, P.: Rooted directed path graphs are leaf powers. Discret. Math. **310**(4), 897–910 (2010)
4. Brandstädt, A., Le, V.B.: Structure and linear time recognition of 3-leaf powers. Inf. Process. Lett. **98**(4), 133–138 (2006)
5. Brandstädt, A., Le, V.B., Sritharan, R.: Structure and linear-time recognition of 4-leaf powers. ACM Trans. Algorithms **5**(1), 11:1-11:22 (2008). https://doi.org/10.1145/1435375.1435386
6. Calamoneri, T., Frangioni, A., Sinaimeri, B.: Pairwise compatibility graphs of caterpillars. Comput. J. **57**(11), 1616–1623 (2014)
7. Calamoneri, T., Sinaimeri, B.: Pairwise compatibility graphs: a survey. SIAM Rev. **58**(3), 445–460 (2016)
8. Chang, M.-S., Ko, M.-T.: The 3-Steiner root problem. In: Brandstädt, A., Kratsch, D., Müller, H. (eds.) WG 2007. LNCS, vol. 4769, pp. 109–120. Springer, Heidelberg (2007). https://doi.org/10.1007/978-3-540-74839-7_11
9. Davis, A., Gao, R., Navin, N.: Tumor evolution: Linear, branching, neutral or punctuated? Biochim. Biophys. Acta (BBA) - Rev. Cancer **1867**(2), 151–161 (2017). https://doi.org/10.1016/j.bbcan.2017.01.003, evolutionary principles - heterogeneity in cancer?
10. Diestel, R.: Graph Theory, 4th edn, Graduate Texts in Mathematics, vol. 173. Springer, London (2012)
11. Dom, M., Guo, J., Hüffner, F., Niedermeier, R.: Extending the tractability border for closest leaf powers. In: Kratsch, D. (ed.) WG 2005. LNCS, vol. 3787, pp. 397–408. Springer, Heidelberg (2005). https://doi.org/10.1007/11604686_35
12. Ducoffe, G.: The 4-Steiner root problem. In: Sau, I., Thilikos, D.M. (eds.) WG 2019. LNCS, vol. 11789, pp. 14–26. Springer, Cham (2019). https://doi.org/10.1007/978-3-030-30786-8_2
13. Eppstein, D., Havvaei, E.: Parameterized leaf power recognition via embedding into graph products. Algorithmica **82**(8), 2337–2359 (2020)
14. Golovach, P.A., et al.: On recognition of threshold tolerance graphs and their complements. Discret. Appl. Math. **216**, 171–180 (2017)
15. Golumbic, M.C., Weingarten, N.L., Limouzy, V.: Co-TT graphs and a characterization of split co-TT graphs. Discret. Appl. Math. **165**, 168–174 (2014)
16. Habib, M., McConnell, R.M., Paul, C., Viennot, L.: Lex-BFS and partition refinement, with applications to transitive orientation, interval graph recognition and consecutive ones testing. Theor. Comput. Sci. **234**(1–2), 59–84 (2000)

17. Jaffke, L., Kwon, O., Strømme, T.J.F., Telle, J.A.: Mim-width III. graph powers and generalized distance domination problems. Theor. Comput. Sci. **796**, 216–236 (2019). https://doi.org/10.1016/j.tcs.2019.09.012

18. Lafond, M.: On strongly chordal graphs that are not leaf powers. In: Graph-Theoretic Concepts in Computer Science - 43rd International Workshop, WG 2017, Eindhoven, The Netherlands, 21–23 June 2017, Revised Selected Papers, pp. 386–398 (2017). https://doi.org/10.1007/978-3-319-68705-6_29

19. Lafond, M.: Recognizing k-leaf powers in polynomial time, for constant k. In: Proceedings of the 2022 Annual ACM-SIAM Symposium on Discrete Algorithms (SODA), pp. 1384–1410. SIAM (2022). https://doi.org/10.1137/1.9781611977073.58

20. Monma, C.L., Reed, B.A., Trotter, W.T.: Threshold tolerance graphs. J. Graph Theor. **12**(3), 343–362 (1988)

21. Nevries, R., Rosenke, C.: Towards a characterization of leaf powers by clique arrangements. Graphs Combin. **32**(5), 2053–2077 (2016). https://doi.org/10.1007/s00373-016-1707-x

22. Nishimura, N., Ragde, P., Thilikos, D.M.: On graph powers for leaf-labeled trees. J. Algorithms **42**(1), 69–108 (2002)

23. Rose, D.J., Tarjan, R.E., Lueker, G.S.: Algorithmic aspects of vertex elimination on graphs. SIAM J. Comput. **5**(2), 266–283 (1976)

24. Azer, E.S., Ebrahimabadi, M.H., Malikić, S., Khardon, R., Sahinalp, S.C.: Tumor phylogeny topology inference via deep learning. iScience **23**(11), 101655 (2020). https://doi.org/10.1016/j.isci.2020.101655

25. Tamir, A.: A class of balanced matrices arising from location problems. Siam J. Algebraic Discrete Methods **4**, 363–370 (1983)

Problems Hard for Treewidth but Easy for Stable Gonality

Hans L. Bodlaender[1]([⊠])[iD], Gunther Cornelissen[2][iD],
and Marieke van der Wegen[2][iD]

[1] Department of Information and Computing Sciences, Utrecht University,
Princetonplein 5, 3584CC Utrecht, The Netherlands
h.l.bodlaender@uu.nl
[2] Department of Mathematics, Utrecht University, P.O. Box 80010, 3508 TA Utrecht,
The Netherlands
{g.cornelissen,.vanderwegen}@uu.nl

Abstract. We show that some natural problems that are XNLP-hard (hence W[t]-hard for all t) when parameterized by pathwidth or treewidth, become FPT when parameterized by stable gonality, a novel graph parameter based on optimal maps from graphs to trees. The problems we consider are classical flow and orientation problems, such as UNDIRECTED FLOW WITH LOWER BOUNDS, MINIMUM MAXIMUM OUTDEGREE, and capacitated optimization problems such as CAPACITATED (RED-BLUE) DOMINATING SET. Our hardness claims beat existing results. The FPT algorithms use a new parameter "treebreadth", associated to a weighted tree partition, as well as DP and ILP.

Keywords: Parameterized complexity · Graph algorithms · Network flow · Graph orientation · Capacitated dominating set · Tree partitions · Stable gonality

1 Introduction

The Parameterization Paradigm. Problems on finite (multi-)graphs that are NP-hard may become polynomial by restricting a specific graph parameter k. If there exists an algorithm that solves the problem in time bounded by a computable function of the parameter k times a power of the input size, we say that the problem becomes *fixed parameter tractable* (FPT) for the parameter k [15, 1.1]. Despite the fact that computing the parameter itself can often be shown to be NP-hard or NP-complete, the FPT-paradigm, originating in the work of Downey and Fellows [18], has shown to be very fruitful in both theory and practice.

One successful approach is to consider graph parameters that measure how far a given graph is from being acyclic; e.g. how the graph may be decomposed into "small" pieces, such that the interrelation of the pieces is described by a tree-like structure. A prime example of such a parameter is the *treewidth* tw(G) of a graph G ([15, Ch. 7]).

Other parameters have been considered (see, e.g., [21,27], [15, 7.9]), but for some famous graph orientation and graph flow problems, as well as capacitated

M. A. Bekos and M. Kaufmann (Eds.): WG 2022, LNCS 13453, pp. 84–97, 2022.
https://doi.org/10.1007/978-3-031-15914-5_7

version of classical problems, many of these parameters did not succumb to the FPT paradigm. As shown by Ganian et al. [22], the parameter "tree-cut width" of Wolan [32] is successful in dealing with several such problems. We propose a new parameter, based on *mapping* the graph to a tree, rather than decomposing the graph, that gives FPT-algorithms for a larger collection of graphs.

A Novel Parameter: Stable Gonality. The new multigraph parameter, based on "tree-likeness", is the so-called *stable gonality* sgon(G) of a multigraph G, introduced in [13, §3], and originating in algebraic geometry, where a similar construction has been used since the 19th century. One replaces tree *decompositions* of a graph G by graph *morphisms* from G to trees, and the "width" of the decomposition by the "degree" of the morphism, where lower degree maps correspond to less complex graphs. For example, connected graphs of stable gonality 1 are trees [7, Example 2.13], those of stable gonality 2 are so-called hyperelliptic graphs, i.e., graphs that admit, after refinement, a graph automorphism of order two such that the quotient graph is a tree (decidable in quasilinear time [7, Thm. 6.1]). The formal definition, given in Sect. 2.2, requires taking care of two technicalities, related to harmonicity of the map and refinement of the graph.

It has been shown that $\mathrm{tw}(G) \leq \mathrm{sgon}(G)$ [16, §6], that sgon(G) is computable, and NP-complete [24,26]. One attractive point of stable gonality as parameter for weighted problems stems from the fact that it is sensitive to multigraph properties, whereas the treewidth is not. Given an undirected weighted graph $G = (V, E, w)$ where $w: E \to \mathbb{Z}_{>0}$ denotes the edge weights, we have an associated (unweighted) multigraph $\tilde{G}$, with the same vertex set, but where each simple edge $e = uv$ in G is replaced by $w(e)$ parallel edges between the vertices u and v. The stable gonality of the weighted graph G is then by definition $\mathrm{sgon}(G) := \mathrm{sgon}(\tilde{G})$.

Three Sample Problems. We now introduce three problems that are exemplary for our work. We later discuss a few additional variants of these problems. Throughout, we assume that all integers are given in unary.

A typical orientation problem is the following.

MINIMUM MAXIMUM OUTDEGREE (cf. Szeider [31])
Given: Undirected weighted graph $G = (V, E, w)$ with a weight function
$w: E \to \mathbb{Z}_{>0}$; integer r
Question: Is there an orientation of G such that for each $v \in V$, the total weight of all edges directed out of v is at most r?

A *flow network* (see, e.g., [1]) is a *directed* graph $D = (N, A)$, given with two nodes s (source) and t (target) in N, and a capacity $c(e) \in \mathbb{Z}_{>0}$ for each arc $e \in A$. Given a function $f: A \to \mathbb{Z}_{\geq 0}$ and a node v, we call $\sum_{wv \in A} f(wv)$ the *flow to v* and $\sum_{vw \in A} f(vw)$ the *flow out of v*. We say f is an *s-t-flow* if for each arc $a \in A$, the flow over the arc is nonnegative and at most its capacity (i.e., $0 \leq f(a) \leq c(a)$), and for each node $v \in N \backslash \{s, t\}$, the flow conservation law holds: the flow to v equals the flow out of v. The *value* val(f) of a flow is the flow out of s minus the flow to s.

UNDIRECTED FLOW WITH LOWER BOUNDS ([23, Problem ND37])[1]

Given: Undirected graph $G = (V, E)$, for each edge $e \in E$ a capacity $c(e) \in \mathbb{Z}_{>0}$ and a lower bound $\ell(e) \in \mathbb{Z}_{\geq 0}$, vertices s (source) and t (target), a value $R \in \mathbb{Z}_{>0}$

Question: Is there an orientation of G such that the resulting directed graph D allows an s-t-flow f that meets capacities and lower bounds (i.e., $\ell(a) \leq f(a) \leq c(a)$ for all arcs a in D), with value R?

Capacitated versions of classical graph problems impose a limitation on the available "resources", placing them closer to real-world situations. The following is a well-studied such graph problem, that can be viewed as an abstract form of facility location questions.

CAPACITATED DOMINATING SET

Given: Undirected graph $G = (V, E)$, for each vertex $v \in V$ a positive integer capacity $c(v) \in \mathbb{Z}_{>0}$, integer k

Question: Are there a set $D \subset V$ of size $|D| \leq k$ and a function $f : V \backslash D \to D$ such that $vf(v) \in E$ for all $v \in V \backslash D$ and $|f^{-1}(v)| \leq c(v)$ for all $v \in D$?

Main Results: Hard Problems for Treewidth but Easy for Stable Gonality. To specify the (parameterized) hardness of problems, we use the parameterized complexity class XNLP from Elberfeld et al. [20]: problems that can be solved non-deterministically in time $O(f(k)n^c)$ $(c \geq 0)$ and space $O(f(k)\log(n))$ where n is the input size, k the parameter, and f is a computable function. We note that, in terms of the more familiar W-hierarchy of Downey and Fellows [15, 13.3], XNLP-hardness implies W[t]-hardness for all t [11, Lemma 2.2].

Theorem 1. MINIMUM MAXIMUM OUTDEGREE (MMO), UNDIRECTED FLOW WITH LOWER BOUNDS (UFLB) *and* CAPACITATED DOMINATING SET (CDS) *are* XNLP-*complete for pathwidth, and* XNLP-*hard for treewidth* (*given a path or tree decomposition realising the path- or treewidth*), *but are* FPT *for stable gonality* (*given a refinement and graph morphism from the associated multigraph to a tree realising the stable gonality*).

Our proof that UFLB is XNLP-hard for pathwidth is by reduction from ACCEPTING NON-DETERMINISTIC CHECKING COUNTER MACHINE from [11]. XNLP-completeness of CDS for pathwidth was shown in [10, Thm. 8]. Hardness for the other problems follows by easy transformations from UFLB. Membership in XNLP follows each time by observing that a known dynamic programming algorithm can be transformed to a non-deterministic algorithm with bounded space. Details are given in the full paper [8]. The condition that a path decomposition realising the pathwidth is given as part of the input may be removed when an FPT algorithm is known that finds such decompositions and uses logarithmic space (see [11, 19, 20]).

---

[1] In [23] it is required that val$(f) \geq R$ rather than val$(f) = R$, but the problems are of the same complexity, cf. [28].

Itai [28, Thm. 4.1] showed that UFLB is strongly NP-complete. DOMINATING SET is $W[2]$-complete for the size of the dominating set [15, Thm. 13.28], and FPT for treewidth [15, Thm. 7.7]. CDS was shown to be W[1]-hard for treewidth (more precisely, for treedepth plus the size of the dominating set) by Dom et al. [17]. Szeider [31] showed that CDS is W[1]-hard for treewidth, which was improved to W[1]-hardness for vertex cover by Gima et al. [25].

For proving FPT under stable gonality, we revive an older idea of Seese on tree-partite graphs and their widths [30]; in contrast to the tree decompositions used in defining treewidth, we partition the original graph vertices into *disjoint* sets ('bags') labelled by vertices of a tree, such that adjacent vertices are in the same bag or in bags labelled by adjacent vertices in the tree. Seese introduced *tree partition width* to be the maximal size of a bag in such a partition. We consider weighted graphs and define a new parameter, *breadth*, given as the maximum of the bag size *and* the sum of the weights of edges between adjacent bags. The *treebreadth* of a graph G is the minimum breadth of a tree partition of G. This allows us to divide the proof in two parts: (a) show that, given a graph morphism from the associated multigraph to a tree, one can compute in polynomial time a tree partition of the weighted graph of breadth upper bounded by the stable gonality of the associated multigraph; (b) provide an FPT-algorithm, given a tree partition of bounded breadth. By reductions, the two algorithms we specify are the following, for the indicated parameters.

Theorem 2. MMO *is* FPT *for treebreadth* (*given a tree partition realising the treebreadth*), *and* CDS *is* FPT *for tree partition width* (*given a tree partition with bounded width*).

The technique to obtain Theorem 2 is similar to one used by Ganian et al. [22] who obtained FPT algorithms for a number of problems with *tree-cut width* as parameter, including CDS. Our results show that the technique from [22] can be extended to a wider class of graphs: with an upper bound on the weight of all edges, tree partition width and treebreadth are bounded by a polynomial in the tree-cut width, while stable tree partition width and tree partition width are polynomially related; see the discussion in [9, §5]. The second half of Theorem 1 is obtained from Theorem 2 by transforming the data required for sgon (a refinement with to a harmonic morphism to a tree) into a tree partition. We also prove that MMO and UFLB are W[1]-hard for vertex cover number by reduction from BIN PACKING [29]. In Sect. 3, we list some related problems for which algorithmic and hardness results hold as well. Due to space considerations, several details and all hardness proofs are omitted and can be found in the full version [8].

2 Preliminaries

2.1 Conventions and Notations

We will consider *multigraphs* $G = (V, E)$ that consist of a finite set V of vertices, as well as a finite multiset E of unoriented (unweighted) edges, i.e., a set of pairs of (possibly equal) vertices, with finite multiplicity on each such pair. We denote such an edge between vertices $u, v \in V$ as uv. For $v \in V$, E_v denotes the edges incident with v, and for two disjoint subsets $X, Y \subset V$, $E(X, Y)$ is the collection of edges from any vertex in X to any vertex in Y. We also consider *weighted simple graphs*, where edges have positive integer weights. We will make repeated use of the correspondence between integer weighted simple graphs and multigraphs given by replacing every edge with weight k by k parallel edges. All graphs we consider are connected. (If a graph is not connected, we can solve the problem at hand separately on each connected component.) For convenience, we use the terminology "vertex" and "edge" for undirected graphs, and "arc" and "node" for either directed graphs, or for trees that occur in graph morphisms or tree partitions. We write $\mathbb{Z}$ for all integers, with unique subscripts indicating ranges (so $\mathbb{Z}_{>0}$ is the positive integers and $\mathbb{Z}_{\geq 0}$ the non-negative integers). We use interval notation for sets of integers, e.g., $[2, 5] = \{2, 3, 4, 5\}$.

2.2   Stable Gonality and Treebreadth

Stable Gonality. A *graph homomorphism* between two multigraphs G and H, denoted $\phi\colon G \to H$ consists of two (not necessarily surjective) maps $\phi\colon V(G) \to V(H)$ and $\phi\colon E(G) \to E(H)$ such that $\phi(uv) = \phi(u)\phi(v) \in E(H)$ for all $uv \in E(G)$. One would like to define the "degree" of such a graph homomorphism as the number of pre-images of any vertex or edge, but in general, this obviously depends on the chosen vertex or edge. However, by introducing certain weights on the edges via an additional index function, we get a large collection of "indexed" maps for which the degree can be defined as the sum of the indices of the pre-image of a given edge, as long as the indices satisfy a certain condition of "harmonicity" above every vertex in the target. We make this precise.

Definition 1. *A finite morphism ϕ between two loopless multigraphs G and H consists of a graph homomorphism $\phi\colon G \to H$ (denoted by the same letter), and an index function $r\colon E(G) \to \mathbb{Z}_{>0}$ (hidden from notation). The* index *of $v \in V(G)$ in the direction of $e \in E(H)$, where e is incident to $\phi(v)$, is defined by*

$$m_e(v) := \sum_{\substack{e' \in E_v, \\ \phi(e')=e}} r(e').$$

We call ϕ harmonic *if this index is independent of $e \in E(H)$ for any given vertex $v \in V(G)$. We call this simply the* index *of v, and denote it by $m(v)$. The degree of a finite harmonic morphism ϕ is*

$$\deg(\phi) = \sum_{\substack{e' \in E(G), \\ \phi(e')=e}} r(e') = \sum_{\substack{v' \in V(G), \\ \phi(v')=v}} m(v').$$

where $e \in E(H)$ is any edge and $v \in V(H)$ is any vertex. Since ϕ is harmonic, this number does not depend on the choice of e or v, and both expressions are indeed equal.

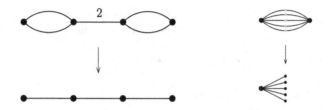

Fig. 1. Two examples of a finite harmonic morphisms of degree 2. The edges without label have index 1. The small grey vertices represent refinements of the graph. (Color figure online)

The second ingredient in defining stable gonality is the notion *refinement*.

Definition 2. *Let G be a multigraph. A* refinement *of G is a graph obtained using the following two operations iteratively finitely often: (a) add a leaf (i.e., a vertex of degree one), (b) subdivide an edge.*

Definition 3. *Let G be a multigraph. The* stable gonality *of G is the minimum degree of a finite harmonic morphism from a refinement of G to a tree.*

Two examples are found in Fig. 1. The left-hand side illustrates the need for an index function (the middle edge needs label 2), and the right hand side shows the effect of subdivision. Stable gonality is well-defined, as each graph $G = (V, E)$ has a refinement that maps to $K_{1,|E|}$: refine each edge once, map each original vertex to the center, and each refinement vertex to a unique leaf.

Tree Partitions and Their Breadth. The existence of a harmonic morphism to a tree imposes a special structure on the graph that we can exploit in designing algorithms. To capture this structure, we define the "breadth" of *tree partitions* of weighted graphs. The notion resembles that of "tree-partite graphs" from [30].

Definition 4. *A* tree partition $\mathcal{T}$ *of a weighted graph $G = (V, E, w)$ is a pair*

$$\mathcal{T} = (\{X_i \mid i \in I\}, \ T = (I, F))$$

where each X_i is a (possibly empty) subset of the vertex set V and $T = (I, F)$ is a tree, such that $\{X_i \mid i \in I\}$ forms a partition of V (i.e., for each $v \in V$, there is exactly one $i \in I$ with $v \in X_i$); and adjacent vertices are in the same set X_i or in sets corresponding to adjacent nodes (i.e., for each $uv \in E$, there exists an $i \in I$ such that $\{u, v\} \subseteq X_i$ or there exists $ij \in F$ with $\{u, v\} \subseteq X_i \cup X_j$). The breadth *of a tree partition $\mathcal{T}$ of G is defined as*

$$b(\mathcal{T}) := \max\left\{ \max_{i \in I} |X_i|, \max_{jk \in F} w(X_j, X_k) \right\},$$

with $w(X_j, X_k) = \sum\limits_{e \in E(X_j, X_k)} w(e)$ the total weight of the edges connecting vertices in X_j to vertices in X_k. The treebreadth $\mathrm{tb}(G)$ *of a weighted graph G is the minimum breadth of a tree partition of G.*

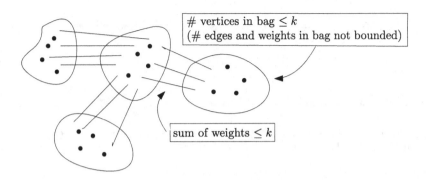

Fig. 2. Schematic representation of a tree partition of a graph of breadth $\leq k$

We refer to Fig. 2 for a schematic view of a tree partition with weights and bounded breadth. If we have a tree partition of a weighted graph G using a tree T, for convenience we will call the vertices of T *nodes* and the edges of T *arcs*. We call the sets X_i *bags*. Observe that in a tree partition of breadth k, if the total weight of edges between two vertices u and v is more than k, then u and v will be in the same bag.

Remark 1. In Seese's work [30], the structure/weights of edges between bags does not contribute to the total width; Seese's tree-partition-width tpw(G) of a simple graph G, defined as the minimum over all tree partitions of G of the maximum bag size in the tree partition, is thus a lower bound for the treebreadth tb(G) (in particular, for sgon, see below). For any G, tpw(G) is lower bounded in terms of tw(G), but also upper bounded in terms of tw(G) and the maximal degree in G, cf. [33].

From Morphisms to Tree Partitions. The existence of a finite harmonic morphism ϕ of some degree k from a multigraph to a tree implies the existence of a tree partition of breadth k for the associated weighted simple graph. The basic idea is to use the pre-images of vertices in T as partitioning sets.

Theorem 3. *Suppose G is a weighted simple graph, and $\phi\colon H \to T$ is a finite harmonic morphism of degree $\deg(\phi) = k$, where H is a loopless refinement of the multigraph corresponding to G and T is a tree. Then one can construct in time $O(k \cdot |V(T)|)$ a tree partition $\mathcal{T} = (X, T')$ for a subdivision of G such that $\mathrm{b}(\mathcal{T}) \leq k$, and $|V(T')| \leq 2|V(G)|$.*

For the proof, construct a tree partition $\mathcal{T} = (X, T)$ as follows. For every node $t \in V(T)$, define $X_t = \phi^{-1}(t) \cap V(G)$. For every edge $uv \in E(G)$, do the following. Let $i \in V(T)$ be such that $u \in X_i$ and let $j \in V(T)$ be such that $v \in X_j$. Let $i, t_1, t_2, \ldots, t_l, j$ be the path between i and j in T. Subdivide the edge uv into a path $u, s_1, s_2, \ldots, s_l, v$ and add the vertex s_r to the set X_{t_r} for each r. To get a bound on the size of T, remove all vertices t from T for which

$X_t = \emptyset$, and, for every degree 2 vertex t of T' for which X_t does not contain a vertex of $V(G)$, contract t with one of its neighbours, and contract all vertices in X_t with a neighbour as well.

Example 1. For the multigraphs in Fig. 1, the constructed tree partitions have breadth two, equal to the stable gonality: for (a), each vertex forms an individual bag, and bags are connected by edges of weight 2; for (b), there is one bag containing both non-subdivision vertices, and no edges.

Thus, to prove that a graph problem is FPT for sgon (given a morphism of a refinement of the corresponding multigraph to a tree of the correct degree), it suffices to prove that it is FPT for the breadth of a given tree-partition of a subdivision of the weighted graph.

3 Related Problems and Reductions

We consider variations of the problems MMO and UFLB.

- CIRCULATING ORIENTATION (CO): given an undirected weighted graph (V, E, w), is there an orientation such that for all vertices, the total weight of outgoing edges equals that of incoming edges?
- OUTDEGREE RESTRICTED ORIENTATION (ORO): given an interval for each vertex, is there an orientation such that for every vertex, the total weight of outgoing edges belongs to the given interval?
- TARGET OUTDEGREE ORIENTATION (TOO): given an integer m_v for each vertex v, is there an orientation such that for every vertex v, the total weight of outgoing edges equals m_v?
- CHOSEN MAXIMUM OUTDEGREE (CMO): given an integer m_v for each vertex v, is there an orientation such that for every vertex v, the total weight of outgoing edges is at most m_v?
- ALL OR NOTHING FLOW (AoNF): Given a directed graph with a positive capacity for each arc, two nodes s, t and a value R, is there a flow with value R whose value on each arc is either zero or the given capacity? (Cf. [2].)

We claim that these problems can be transformed into one another according to the diagrams in Fig. 3 preserving parameterized complexity for the indicated parameters. All complexity statements are then reduced to the following claims: (a) ORO is FPT for treebreadth; (b) AoNF is XNLP-complete for pathwidth; (c) TOO is W[1]-hard for vertex cover number.

The proof of (a) is outlined in Sect. 4.1; a full proof of (a) and the hardness proofs in (b) and (c) are given in the full version [8].

4 Algorithms for ORO and CDS for Graphs with Bounded Treebreadth

4.1 Outdegree Restricted Orientations

We give the main ideas for an algorithm for ORO, when we are given a tree partition of a subdivision of G of bounded breadth. Subdivisions can be handled

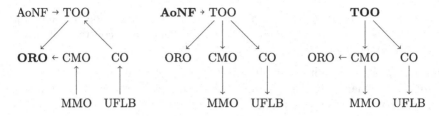

Fig. 3. Transformation between different problems with respect to parameter (a) sgon and treebreadth, for which ORO is FPT, (b) pathwidth, for which AoNF is XNLP-complete, and (c) vertex cover number, for which TOO is W[1]-hard. AonF with vertex cover number has a separate $W[1]$-hardness proof (see [8], based on [2].)

by replacing each subdivision of an edge e by a vertex x_e with $D_{x_e} = [w(e), w(e)]$. This gives an equivalent instance with a corresponding tree partition of the same breadth.

We can now assume that we have a tree partition T of G, i.e., adjacent vertices are in the same or neighbouring bags. Let k be the breadth of T.

We add a new root node r to T, and set $X_r = \emptyset$. For each node i, we let V_i be the union of all bags j with $j = i$ or j is a descendant of i. For an arc $a = ii'$ in T with i the parent of i', we let E_a be the set of all edges with either both endpoints in $V_{i'}$, or with one endpoint in X_i and one endpoint in $X_{i'}$.

A *partial solution* for the arc $a = ii'$ (with again i the parent of i') is an orientation of all edges in E_a such that for all vertices $v \in V_{i'}$, its total outdegree in this orientation is in D_v. Vertices in X_i can have oriented incident edges (namely to neighbours in $X_{i'}$) and incident edges that are not yet oriented (namely to neighbours in other bags than $X_{i'}$). The *fingerprint* of a partial solution is the function $\delta : X_i \rightarrow [0, k]$, that maps each vertex in X_i to its total outdegree in this partial solution, i.e., for $v \in X_i$, the sum of the weights of the edges vw with $w \in X_{i'}$ and vw is oriented from v to w. These sums are bounded by the breadth, and thus, the total number of possible fingerprints for an arc is bounded by a function of k.

The algorithm to solve ORO uses dynamic programming: for each arc in T, we compute the table A_a of all fingerprints of partial solutions for that arc. This is done bottom-up in the tree.

It is straightforward to compute the table A_a for an arc a to a leaf of T, by enumerating all orientations of edges in E_a.

Now, suppose we have a node i' with children $j_1, \ldots, j_q$ and we have already computed the tables $A_{i'j_1}, \ldots, A_{i'j_q}$. To compute the table $A_{ii'}$, for the arc ii' from i' to its parent i, we express in an Integer Linear Program (with number of variables bounded by a function of k), the property that we can extend an orientation of the edges between X_i and $X_{i'}$ and between pairs of vertices in $X_{i'}$ to a partial solution. Some details are given below.

Define an equivalence relation $\sim$ on the children of i', with two children equivalent if they have precisely the same set of fingerprints. As the number of different fingerprints is a function of k, the number of equivalence classes of $\sim$ is a (double exponential) function of k.

We write Γ for the set of equivalence classes of $\sim$, and Δ for the set of all possible fingerprints of partial solutions for arcs from i' to a child.

We then enumerate all orientations ρ of the edges in $X_i \times X_{i'}$ and in $X_{i'} \times X_{i'}$. Each such orientation would fix a fingerprint for ii'—what needs to be done is checking whether there is actually a partial solution for ii' that extends ρ.

To do this, we introduce yet another concept: the *blueprint* of a partial solution for ii'. The blueprint is a function that maps a pair (γ, f) of an equivalence class $\gamma \in \Gamma$ and a fingerprint $f \in \Delta$ to the number of children j_α of i' with the following two properties: (1) the restriction of the partial solution to $E_{i' j_\alpha}$ has fingerprint f, and (2) j_α is in equivalence class γ.

Note that ρ and the blueprint contain all that is needed to compute the outdegrees of vertices in $X_{i'}$: from it, we can see, for each $v \in X_{i'}$ and for each weight in $[0, k]$, how many edges with that weight are directed from v to a vertex in a child bag of i'.

This allows us to formulate an ILP that expresses the property that there exists a blueprint of a partial solution that extends ρ. We have a non-negative integer variable $x_{\gamma, f}$ for each pair $\gamma \in \Gamma$ and $f \in \Delta$ that should give the value of this pair in the blueprint.

The ILP has no objective function, and the following constraints:

- For each γ, $\sum_f x_{\gamma, f}$ equals the number of children in equivalence class γ.
- If f is not a fingerprint for children in equivalence class γ, then $x_{\gamma, f} = 0$.
- For all $v \in X_{i'}$, we have a condition that checks that the outdegree of v in the orientation belongs to D_v. Let $D_v = [d_{\min, v}, d_{\max, v}]$. Let α be the total weight of all edges in ρ that have v as endpoint and are directed out of v. Now, add the inequalities:

$$d_{\min, v} \leq \alpha + \sum_{\gamma, f} f(v) \cdot x_{\gamma, f} \leq d_{\max, v}$$

We sum over all $\gamma \in \Gamma$, and $f \in \gamma$.

The first two conditions guarantee that we can choose for each child a fingerprint, such that for each equivalence class γ and each fingerprint f we have $x_{\gamma, f}$ children in the class γ with fingerprint f; the first condition ensures that we have the right amount of fingerprints per class, and the second that we do not assign fingerprints to children that have no corresponding partial solution in that subtree.

The third condition ensures that each vertex in $X_{i'}$ has an outdegree in its interval: we have $x_{\gamma, f}$ children in the equivalence class γ from which we take fingerprint f, and here v gets outdegree $f(v)$ for the edges between v and a vertex in such a child bag.

Note that the number of variables of the ILP is bounded by a function of k. Thus, the ILPs can be solved by an FPT algorithm, see [15, Theorem 6.4].

Once we have the table A_{a_r} for the arc a_r to the root, we can decide: the instance of ORO has a solution if and only if this table A_{a_r} is non-empty, as any partial solution for this arc is actually an orientation that fulfils the requirements for G. Thus, by processing the bags of $\mathcal{T}$ in bottom-up order, we finally obtain the table for the root and can decide the problem.

4.2   Capacitated Dominating Set

We now also sketch some main ideas for the FPT algorithm for CAPACITATED DOMINATING SET. The algorithm again uses dynamic programming, with an ILP that determines the number of children with a partial solution having a fingerprint of a certain type (compare with [22]).[2]

We define *a partial solution* for an arc ii' with i the parent of i' as a set S of vertices in $V_{i'}$ together with a mapping that maps all vertices in $V_{i'} \backslash X_{i'}$ and a subset $D \subseteq X_i \cup X_{i'}$ to a neighbour in S, such that no vertex in S has more than its capacity number of neighbours mapped to it. At this point, vertices in $X_{i'}$ do not need to be dominated yet, and they can be used to dominate vertices in the parent bag X_i. All vertices in bags that are a descendant of i' must be dominated. The *fingerprint* of a partial solution is the set D: the dominated vertices in $X_i \cup X_{i'}$.

In a dynamic programming algorithm, we compute for each arc ii' and for each fingerprint $D \subseteq X_i \cup X_{i'}$ the minimum size of a set S that gives a partial solution with this fingerprint. Let $B_{ii'}(D) \in \mathbb{Z}_{>0} \cup \{\infty\}$ be the minimum such size for a fingerprint D. Using the classical theory of matchings in graphs and inspiration from [12], we find the following.

Lemma 1. *If the instance of* CDS *has a solution, then $B_{ii'}(\emptyset) \in \mathbb{Z}_{>0}$. If there is a partial solution with fingerprint $D \subseteq X_i \cup X_{i'}$, then $B_{ii'}(\emptyset) \leq B_{ii'}(D) \leq B_{ii'}(\emptyset) + 2k$.*

In the step where we attempt to compute the table $B_{ii'}$ given such tables for the children of i', we add up all values $B_{i'j_\alpha}(\emptyset)$ and create tables $B'_{i'j_\alpha}$ by setting

$$B'_{i'j_\alpha}(D) = B_{i'j_\alpha}(D) - B_{i'j_\alpha}(\emptyset)$$

Now, $B'_{i'j_\alpha}$ is a function that maps subsets of $X_i \cup X_{i'}$ to values in $[0, 2k]$, and thus, the number of possible such functions is bounded by a function of k. This is, however, not sufficient to build an equivalence relation on the children of i', as the non-dominated vertices in such children still must be dominated by vertices in X_i. Instead, we look to extensions of partial solutions, where we also dominate vertices in $X_{i'}$ by vertices in X_i, and prescribe how much capacity each vertex in X_i uses to dominate vertices in $X_{i'}$. This gives a number of equivalence classes that is bounded by a function of k. Once we built an

---

[2] In the full version [8], we in fact give a detailed, slightly different, algorithm for CAPACITATED RED-BLUE DOMINATING SET and then deduce the result for CAPACITATED DOMINATING SET.

equivalence relation on the children, the algorithm proceeds in a similar fashion as for ORO: an ILP is constructed that expresses for each class in the equivalence relation and fingerprint of a partial solution how many children in that class have that fingerprint. The ILP has an objective function which gives the size of the partial solution built (which is a sum of B'-values).

5   Conclusion

We showed that various classical instances of flow, orientation and capacitated graph problems are XNLP-hard when parameterized by treewidth (and even pathwidth), but FPT for a novel graph parameter, stable gonality. Following Goethe's motto "Das Schwierige leicht behandelt zu sehen, gibt uns das Anschauen des Unmöglichen", we venture into stating some open problems.

Is stable gonality fixed parameter tractable? Can multigraphs of fixed stable gonality be recognized efficiently (this holds for treewidth; for sgon = 2 this can be done in quasilinear time [7])? Given the stable gonality of a graph, can a refinement and morphism of that degree to a tree be constructed in reasonable time (the analogous problem for treewidth can be done in linear time)? Can we find a tree partition of a subdivision with bounded treebreadth? The same question can be asked in the approximate sense.

Find a multigraph version of Courcelle's theorem (that provides a logical characterisation of problems that are FPT for treewidth, see [14]), using stable gonality instead of treewidth: give a logical description of the class of multigraph problems that are FPT for stable gonality.

Stable gonality and (stable) treebreadth seem useful parameters for more edge-weighted or multigraph problems that are hard for treewidth. Find other problems that become FPT for such a parameter. Here, our proof technique of combining tree partitions with ILP with a bounded number of variables becomes relevant.

Conversely, find problems that are hard for treewidth and remain hard for stable gonality or (stable) treebreadth. We believe candidates to consider are in the realm of problems concerning "many" neighbours of given vertices (where our use of ILP seems to break down), such as DEFENSIVE ALIANCE and SECURE SET, proven to be W[1]-hard for treewidth (but FPT for solution size) [5,6]. For such problems, it is also interesting to upgrade known W[1]-hardness to XNLP.

Other flavours of graph gonality (untied to stable gonality) exist, based on the theory of divisors on graphs (cf. [3,4].) Investigate whether such 'divisorial' gonality is a useful parameter for hard graph problems.

Acknowledgements. We thank Carla Groenland and Hugo Jacob for various discussions, and in particular for suggestions related to the capacitated dominating set problems, and the relations between tree cut-width, tree partition width, and stable tree-partition width.

References

1. Ahuja, R.K., Magnanti, T.L., Orlin, J.B.: Network Flows – Theory, Algorithms and Applications. Prentice Hall (1993)
2. Alexandersson, P.: NP-complete variants of some classical graph problems. CoRR, abs/2001.04120 (2020). arXiv:2001.04120
3. Baker, M.: Specialization of linear systems from curves to graphs. Algebra Number Theory **2**(6), 613–653 (2008). https://doi.org/10.2140/ant.2008.2.613. With an appendix by Brian Conrad
4. Baker, M., Norine, S.: Harmonic morphisms and hyperelliptic graphs. Int. Math. Res. Not. IMRN **15**, 2914–2955 (2009). https://doi.org/10.1093/imrn/rnp037
5. Bliem, B., Woltran, S.: Complexity of secure sets. Algorithmica **80**(10), 2909–2940 (2017). https://doi.org/10.1007/s00453-017-0358-5
6. Bliem, B., Woltran, S.: Defensive alliances in graphs of bounded treewidth. Discrete Appl. Math. **251**, 334–339 (2018). https://doi.org/10.1016/j.dam.2018.04.001
7. Bodewes, J.M., Bodlaender, H.L., Cornelissen, G., van der Wegen, M.: Recognizing hyperelliptic graphs in polynomial time. Theoret. Comput. Sci. **815**, 121–146 (2020). https://doi.org/10.1016/j.tcs.2020.02.013
8. Bodlaender, H.L., Cornelissen, G., van der Wegen, M.: Problems hard for treewidth but easy for stable gonality. CoRR, abs/2202.06838 (2022). arXiv:2202.06838
9. Bodlaender, H.L., Groenland, C., Jacob, H.: On the parameterized complexity of computing tree-partitions. CoRR, abs/2206.11832 (2022). arXiv:2206.11832
10. Bodlaender, H.L., Groenland, C., Jacob, H.: XNLP-completeness for parameterized problems on graphs with a linear structure. CoRR, abs/2201.13119 (2022). arXiv:2201.13119
11. Bodlaender, H.L., Groenland, C., Nederlof, J., Swennenhuis, C.M.F.: Parameterized problems complete for nondeterministic FPT time and logarithmic space. In: Proceedings 62nd IEEE Annual Symposium on Foundations of Computer Science, FOCS 2021, pp. 193–204 (2021). https://doi.org/10.1109/FOCS52979.2021.00027
12. Bodlaender, H.L., van Antwerpen-de Fluiter, B.: Reduction algorithms for graphs of small treewidth. Inf. Comput. **167**(2), 86–119 (2001). https://doi.org/10.1006/inco.2000.2958
13. Cornelissen, G., Kato, F., Kool, J.: A combinatorial Li-Yau inequality and rational points on curves. Math. Ann. (10), 211–258 (2014). https://doi.org/10.1007/s00208-014-1067-x
14. Courcelle, B.: The monadic second-order logic of graphs. I. Recognizable sets of finite graphs. Inform. and Comput. **85**(1), 12–75 (1990). https://doi.org/10.1016/0890-5401(90)90043-H
15. Cygan, M., et al.: Parameterized Algorithms. Springer, Cham (2015). https://doi.org/10.1007/978-3-319-21275-3
16. van Dobben de Bruyn, J., Gijswijt, D.: Treewidth is a lower bound on graph gonality. Algebr. Comb. **3**(4), 941–953 (2020). https://doi.org/10.5802/alco.124
17. Dom, M., Lokshtanov, D., Saurabh, S., Villanger, Y.: Capacitated domination and covering: a parameterized perspective. In: Grohe, M., Niedermeier, R. (eds.) IWPEC 2008. LNCS, vol. 5018, pp. 78–90. Springer, Heidelberg (2008). https://doi.org/10.1007/978-3-540-79723-4_9
18. Downey, R.G., Fellows, M.R.: Parameterized Complexity. Monographs in Computer Science, Springer, New York (1999). https://doi.org/10.1007/978-1-4612-0515-9

19. Elberfeld, M., Jakoby, A., Tantau, T.: Logspace versions of the theorems of Bodlaender and Courcelle. In: Proceedings 51th Annual IEEE Symposium on Foundations of Computer Science, FOCS 2010, pp. 143–152. IEEE Computer Society (2010). https://doi.org/10.1109/FOCS.2010.21

20. Elberfeld, M., Stockhusen, C., Tantau, T.: On the Space and Circuit Complexity of Parameterized Problems: classes and Completeness. Algorithmica **71**(3), 661–701 (2014). https://doi.org/10.1007/s00453-014-9944-y

21. Fiala, J., Golovach, P.A., Kratochvíl, J.: Parameterized complexity of coloring problems: treewidth versus vertex cover. Theoret. Comput. Sci. **412**(23), 2513–2523 (2011). https://doi.org/10.1016/j.tcs.2010.10.043

22. Ganian, R., Kim, E.J., Szeider, S.: Algorithmic applications of tree-cut width. In: Italiano, G.F., Pighizzini, G., Sannella, D.T. (eds.) MFCS 2015. LNCS, vol. 9235, pp. 348–360. Springer, Heidelberg (2015). https://doi.org/10.1007/978-3-662-48054-0_29

23. Garey, M.R., Johnson, D.S.: Computers and Intractability: A Guide to the Theory of NP-Completeness. W. H. Freeman (1979)

24. Gijswijt, D., Smit, H., van der Wegen, M.: Computing graph gonality is hard. Discret. Appl. Math. **287**, 134–149 (2020). https://doi.org/10.1016/j.dam.2020.08.013

25. Gima, T., Hanaka, T., Kiyomi, M., Kobayashi, Y., Otachi, Y.: Exploring the gap between treedepth and vertex cover through vertex integrity. Theoret. Comput. Sci. **918**, 60–76 (2022). https://doi.org/10.1016/j.tcs.2022.03.021

26. Koerkamp, R.G., van der Wegen, M.: Stable gonality is computable. Discrete Math. Theor. Comput. Sci. **21**(1), 14 (2019). https://doi.org/10.23638/DMTCS-21-1-10. Paper No. 10

27. Hliněný, P., Oum, S., Seese, D., Gottlob, G.: Width parameters beyond tree-width and their applications. Comput. J. **51**(3), 326–362 (2007). https://doi.org/10.1093/comjnl/bxm052

28. Itai, A.: Two-commodity flow. J. ACM **25**(4), 596–611 (1978). https://doi.org/10.1145/322092.322100

29. Jansen, K., Kratsch, S., Marx, D., Schlotter, I.: Bin packing with fixed number of bins revisited. J. Comput. Syst. Sci. **79**(1), 39–49 (2013). https://doi.org/10.1016/j.jcss.2012.04.004

30. Seese, D.: Tree-partite graphs and the complexity of algorithms. In: Budach, L. (ed.) FCT 1985. LNCS, vol. 199, pp. 412–421. Springer, Heidelberg (1985). https://doi.org/10.1007/BFb0028825

31. Szeider, S.: Not so easy problems for tree decomposable graphs. In: Advances in Discrete Mathematics and Applications: Mysore, 2008. Ramanujan Mathematical Society Lecture Note Series, vol. 13, pp. 179–190. Ramanujan Mathematical Society, Mysore (2010). arXiv:1107.1177

32. Wollan, P.: The structure of graphs not admitting a fixed immersion. J. Comb. Theory Ser. B **110**, 47–66 (2015). https://doi.org/10.1016/j.jctb.2014.07.003

33. Wood, D.R.: On tree-partition-width. Eur. J. Combin. **30**(5), 1245–1253 (2009). https://doi.org/10.1016/j.ejc.2008.11.010

Edge-Cut Width: An Algorithmically Driven Analogue of Treewidth Based on Edge Cuts

Cornelius Brand[ID], Esra Ceylan[ID], Robert Ganian[(✉)][ID], Christian Hatschka[ID], and Viktoriia Korchemna[ID]

Algorithms and Complexity Group, TU Wien, Vienna, Austria
{cbrand,rganian,vkorchemna}@ac.tuwien.ac.at,
{e1526801,e1525634}@student.tuwien.ac.at

Abstract. Decompositional parameters such as treewidth are commonly used to obtain fixed-parameter algorithms for NP-hard graph problems. For problems that are W[1]-hard parameterized by treewidth, a natural alternative would be to use a suitable analogue of treewidth that is based on edge cuts instead of vertex separators. While tree-cut width has been coined as such an analogue of treewidth for edge cuts, its algorithmic applications have often led to disappointing results: out of twelve problems where one would hope for fixed-parameter tractability parameterized by an edge-cut based analogue to treewidth, eight were shown to be W[1]-hard parameterized by tree-cut width.

As our main contribution, we develop an edge-cut based analogue to treewidth called edge-cut width. Edge-cut width is, intuitively, based on measuring the density of cycles passing through a spanning tree of the graph. Its benefits include not only a comparatively simple definition, but mainly that it has interesting algorithmic properties: it can be computed by a fixed-parameter algorithm, and it yields fixed-parameter algorithms for all the aforementioned problems where tree-cut width failed to do so.

Keywords: tree-cut width · parameterized complexity · graph parameters

1 Introduction

While the majority of computational problems on graphs are intractable, in most cases it is possible to exploit the structure of the input graphs to circumvent this intractability. This basic fact has led to the extensive study of a broad hierarchy of decompositional graph parameters (see, e.g., Fig. 1 in [3]), where for individual problems of interest the aim is to pinpoint which parameters can be used to develop fixed-parameter algorithms for the problem. Treewidth [31] is by far the most prominent parameter in the hierarchy, and it is known that many problems

Cornelius Brand, Robert Ganian and Viktoriia Korchemna gratefully acknowledge support from the Austria Science Foundation (FWF, Project Y1329).

M. A. Bekos and M. Kaufmann (Eds.): WG 2022, LNCS 13453, pp. 98–113, 2022.
https://doi.org/10.1007/978-3-031-15914-5_8

of interest are fixed-parameter tractable when parameterized by treewidth; some of these problem can even be solved efficiently on more general parameters such as rank-width [13,30] or other decompositional parameters above treewidth in the hierarchy [4]. However, in this article we will primarily be interested in problems that lie on the other side of this spectrum: those which remain intractable when parameterized by treewidth.

Aside from non-decompositional parameters[1] such as the vertex cover number [10,12] or feedback edge number [1,18,21], the most commonly applied parameters for problems which are not fixed-parameter tractable with respect to treewidth are tied to the existence of small vertex separators. One example of such a parameter is treedepth [29], which has by now found numerous applications in diverse areas of computer science, e.g., [17,23,28]. An alternative approach is to use a decompositional parameter that is inherently tied to edge-cuts—in particular, tree-cut width [27,34].

Tree-cut width was discovered by Wollan, who described it as a variation of tree decompositions based on edge cuts instead of vertex separators [34]. But while it is true that "tree-cut decompositions share many of the natural properties of tree decompositions" [27], from the perspective of algorithmic design tree-cut width seems to behave differently than an edge-cut based alternative to treewidth. To illustrate this, we note that tree-cut width is a parameter that lies between treewidth and treewidth plus maximum degree (which may be seen as a "heavy-handed" parameterization that enforces small edge cuts) in the parameter hierarchy [14,24]. There are numerous problems which are W[1]-hard (and sometimes even NP-hard) w.r.t. treewidth but fixed-parameter tractable w.r.t. the latter parameterization, and the aim would be to have an edge-cut based parameter that can lift this fixed-parameter tractability towards graphs of unbounded degree.

Unfortunately, out of twelve problems with these properties where a tree-cut width parameterization has been pursued so far, only four are fixed-parameter tractable [14,15] while eight turn out to be W[1]-hard [5,14,16,18,22]. The most appalling example of the latter case is the well-established EDGE DISJOINT PATHS (EDP) problem: VERTEX DISJOINT PATHS is a classical example of a problem that is FPT parameterized by treewidth [33], and one should by all means expect a similar outcome for EDP parameterized by the analogue of treewidth based on edge cuts [18,19]. But if EDP is W[1]-hard parameterized by tree-cut width, what is the algorithmic analogue of treewidth for edge cuts? Here, we attempt to answer to this question through the notion of edge-cut width.

Contribution. Edge-cut width is an edge-cut based decompositional parameter which has a surprisingly streamlined definition: instead of specialized decompositions such as those employed by treewidth, clique-width or tree-cut width, the "decompositions" for edge-cut width are merely spanning trees (or, in case of

---

[1] We view a parameter as decompositional if it is tied to a well-defined graph decomposition; all decompositional parameters are closed under the disjoint union operation of graphs.

disconnected graphs, maximum spanning forests). To define edge-cut width of a spanning tree T, we observe that for each edge in $G - E(T)$ there is a unique path in T connecting its endpoints, and the edge-cut width of T is merely the maximum number of such paths that pass through any particular vertex in T; as usual, the edge-cut width of G is then the minimum width of a spanning tree (i.e., decomposition).

After introducing edge-cut width, establishing some basic properties of the parameter and providing an in-depth comparison to tree-cut width, we show that the parameter has surprisingly useful algorithmic properties. As our first task, we focus on the problem of computing edge-cut width along with a suitable decomposition. This is crucial, since we will generally need to compute an edge-cut width decomposition before we can use the parameter to solve problems of interest. As our first algorithmic result, we leverage the connection of edge-cut width to spanning trees of the graph to obtain an explicit fixed-parameter algorithm for computing edge-cut width decompositions. This compares favorably to tree-cut width, for which only an explicit 2-approximation fixed-parameter algorithm [24] and a non-constructive fixed-parameter algorithm [20] are known.

Finally, we turn to the algorithmic applications of edge-cut width. Recall that among the twelve problems where a parameterization by tree-cut width had been pursued, eight were shown to be W[1]-hard parameterized by tree-cut width: LIST COLORING [14], PRECOLORING EXTENSION [14], BOOLEAN CONSTRAINT SATISFACTION [14], EDGE DISJOINT PATHS [18], BAYESIAN NETWORK STRUCTURE LEARNING [16], POLYTREE LEARNING [16], MINIMUM CHANGEOVER COST ARBORESCENCE [22], and MAXIMUM STABLE ROOMMATES WITH TIES AND INCOMPLETE LISTS [5][2]. Here, we follow up on previous work by showing that *all* of these problems are fixed-parameter tractable when parameterized by edge-cut width. We obtain our algorithms using a new dynamic programming framework for edge-cut width, which can also be adapted for other problems of interest.

Related Work. The origins of edge-cut width lie in the very recent work of Ganian and Korchemna on learning polytrees and Bayesian networks [16], who discovered an equivalent parameter when attempting to lift the fixed-parameter tractability of these problems to a less restrictive parameter than the feedback edge number[3]. That same work also showed that computing edge-cut width can be expressed in Monadic Second Order Logic which implies fixed-parameter tractability, but obtaining an explicit fixed-parameter algorithm for computing optimal decompositions was left as an open question.

As far as the authors are aware, there are only four problems for which it is known that fixed-parameter tractability can be lifted from the parameterization by "maximum degree plus treewidth" to tree-cut width. These are CAPACITATED VERTEX COVER [14], CAPACITATED DOMINATING SET [14], IMBALANCE [14] and BOUNDED DEGREE VERTEX DELETION [15]. Additionally, Gozu-

---

[2] We remark that besides establishing W[1]-hardness for this problem, the authors also showed that the problem becomes FPT w.r.t. tree-cut width under additional restrictions.

[3] The authors originally used the name "local feedback edge number".

pek et al. [22] showed that the MINIMUM CHANGEOVER COST ARBORESCENCE problem is fixed-parameter tractable when parameterized by a special, restricted version of tree-cut width where one essentially requires the so-called *torsos* to be stars.

2 Preliminaries

We use standard terminology for graph theory, see for instance [7], and assume basic familiarity with the parameterized complexity paradigm including, in particular, the notions of *fixed-parameter tractability* and W[1]-*hardness* [6,8]. Let $\mathbb{N}$ denote the set of natural numbers including zero. We use $[i]$ to denote the set $\{0, 1, \ldots, i\}$.

Given two graph parameters $\alpha, \beta : G \to \mathbb{N}$, we say that α *dominates* β if there exists a function p such that for each graph G, $\alpha(G) \leq p(\beta(G))$. For a vertex set Y, we use $N(Y)$ to denote the set of all vertices that are outside of Y and have a neighbor in Y.

Treewidth. *Treewidth* [31] is a fundamental graph parameter that has found a multitude of algorithmic applications throughout computer science.

Definition 1. *A* tree decomposition *of a graph G is a pair $(T, \{\beta_t\}_{t \in V(T)})$, where T is a tree, and each node $t \in V(T)$ is associated with a bag $\beta_t \subseteq V(G)$, satisfying the following conditions:*

1. *Every vertex of G appears in some bag of T.*
2. *Every edge of G is contained as a subset in some bag of T.*
3. *For every vertex $v \in V(G)$, the set of nodes $t \in V(T)$ such that $v \in \beta_t$ holds is connected in T.*

The width *of a tree decomposition is defined as $\max_t |\beta_t| - 1$, and the* treewidth $\mathrm{tw}(G)$ *of G is defined as the minimum width of any of its tree decompositions.*

Tree-Cut Width. The notion of tree-cut decompositions was introduced by Wollan [34], see also [27]. A family of subsets $X_1, \ldots, X_k$ of X is a *near-partition* of X if they are pairwise disjoint and $\bigcup_{i=1}^{k} X_i = X$, allowing the possibility of $X_i = \emptyset$.

Definition 2. *A* tree-cut decomposition *of G is a pair $(T, \mathcal{X})$ which consists of a rooted tree T and a near-partition $\mathcal{X} = \{X_t \subseteq V(G) : t \in V(T)\}$ of $V(G)$. A set in the family $\mathcal{X}$ is called a* bag *of the tree-cut decomposition.*

For any node t of T other than the root r, let $e(t) = ut$ be the unique edge incident to t on the path to r. Let T_u and T_t be the two connected components in $T - e(t)$ which contain u and t, respectively. Note that $(\bigcup_{q \in T_u} X_q, \bigcup_{q \in T_t} X_q)$ is a near-partition of $V(G)$, and we use E_t to denote the set of edges with one endpoint in each part. We define the *adhesion* of t ($\mathrm{adh}(t)$) as $|E_t|$; we explicitly set $\mathrm{adh}(r) = 0$ and $E_r = \emptyset$.

The *torso* of a tree-cut decomposition $(T, \mathcal{X})$ at a node t, written as H_t, is the graph obtained from G as follows. If T consists of a single node t, then the torso of $(T, \mathcal{X})$ at t is G. Otherwise, let $T_1, \ldots, T_\ell$ be the connected components of $T - t$. For each $i = 1, \ldots, \ell$, the vertex set $Z_i \subseteq V(G)$ is defined as the set $\bigcup_{b \in V(T_i)} X_b$. The torso H_t at t is obtained from G by *consolidating* each vertex set Z_i into a single vertex z_i (this is also called *shrinking* in the literature). Here, the operation of consolidating a vertex set Z into z is to substitute Z by z in G, and for each edge e between Z and $v \in V(G) \setminus Z$, adding an edge zv in the new graph. We note that this may create parallel edges.

The operation of *suppressing* (also called *dissolving* in the literature) a vertex v of degree at most 2 consists of deleting v, and when the degree is two, adding an edge between the neighbors of v. Given a connected graph G and $X \subseteq V(G)$, let the *3-center* of (G, X) be the unique graph obtained from G by exhaustively suppressing vertices in $V(G) \setminus X$ of degree at most two. Finally, for a node t of T, we denote by $\tilde{H}_t$ the 3-center of (H_t, X_t), where H_t is the torso of $(T, \mathcal{X})$ at t. Let the *torso-size* $\mathrm{tor}(t)$ denote $|\tilde{H}_t|$.

Definition 3. *The width of a tree-cut decomposition $(T, \mathcal{X})$ of G is $\max_{t \in V(T)} \{\mathrm{adh}(t), \mathrm{tor}(t)\}$. The tree-cut width of G, or $\mathrm{tcw}(G)$ in short, is the minimum width of $(T, \mathcal{X})$ over all tree-cut decompositions $(T, \mathcal{X})$ of G.*

We refer to previous work [14,24,27,34] for a more detailed comparison of tree-cut width to other parameters. Here, we mention only that tree-cut width is dominated by treewidth and dominates treewidth plus maximum degree, which we denote $\mathrm{degtw}(G)$.

Lemma 1 ([14,27,34]). *For every graph G, $\mathrm{tw}(G) \leq 2\,\mathrm{tcw}(G)^2 + 3\,\mathrm{tcw}(G)$ and $\mathrm{tcw}(G) \leq 4\,\mathrm{degtw}(G)^2$.*

3   Edge-Cut Width

Let us begin by considering a maximal spanning forest T of a graph G, and recall that $E(G) - T$ forms a minimum feedback edge set in G; the size of this set is commonly called the *feedback edge number* [1,18,21], and it does not depend on the choice of T. We will define our parameter as the maximum number of edges from the feedback edge set that form cycles containing some particular vertex $v \in V(G)$.

Formally, for a graph G and a maximal spanning forest T of G, let the *local feedback edge set* at $v \in V$ be $E_{\mathrm{loc}}^{G,T}(v) = \{uw \in E(G) \setminus E(T) \mid$ the unique path between u and w in T contains $v\}$; we remark that this unique path forms a so-called *fundamental cycle* with the edge uw. The *edge-cut width* of (G, T) (denoted $\mathrm{ecw}(G, T)$) is then equal to $1 + \max_{v \in V} |E_{\mathrm{loc}}^{G,T}(v)|$, and the *edge-cut width* of G is the smallest edge-cut width among all possible maximal spanning forests of G.

Notice that the definition increments the edge-cut width of T by 1. This "cosmetic" change may seem arbitrary, but it matches the situation for treewidth (where the width is the bag size minus one) and allows trees to have a width of 1. Moreover, defining edge-cut width in this way provides a more concise description of the running times for our algorithms, where the records will usually depend on a set that is one larger than $|E_{\text{loc}}^{G,T}(v)|$. We note that the predecessor to edge-cut width, called the *local feedback edge number* [16], was defined without this cosmetic change and hence is equal to edge-cut width minus one.

While it is obvious that $\text{ecw}(G)$ is upper-bounded by (and hence dominates) the feedback edge number of G ($\text{fen}(G)$), we observe that graphs of constant $\text{ecw}(G)$ can have unbounded feedback edge number—see Fig. 1. We also note that Ganian and Korchemna established that edge-cut width is dominated by tree-cut width.

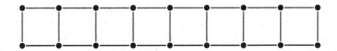

Fig. 1. Example of a graph G with a spanning tree T (marked in red) such that $\text{ecw}(G) = \text{ecw}(G,T) = 3$. The feedback edge number of G, i.e., its edge deletion distance to acyclicity, is exactly the number of black edges and can be made arbitrarily large in this fashion while preserving $\text{ecw}(G) = 3$. (Color figure online)

Proposition 1 ([16]). *For every graph G, $\text{tcw}(G) \leq \text{ecw}(G) \leq \text{fen}(G) + 1$.*

As for the converse, we already have conditional evidence that edge-cut width cannot dominate tree-cut width: BAYESIAN NETWORK STRUCTURE LEARNING is W[1]-hard w.r.t. the latter, but fixed-parameter tractable w.r.t. the former [16]. We conclude our comparisons with a construction that not only establishes this relationship unconditionally, but—more surprisingly—implies that edge-cut width is incomparable to degtw.

Lemma 2. *For each $m \in \mathbb{N}$, there exists a graph G_m of maximum degree at most 3, tree-cut width at most 2, and edge-cut width at least $m + 1$.*

Proof (Sketch). We start from two regular binary trees Y and Y' of depth m, i.e., rooted binary trees where every node except leaves has precisely two children and the path from any leaf to the root contains m edges. We glue Y and Y' together by identifying each leaf of Y with a unique leaf of Y' (see the left part of Fig. 2 for an illustration). It remains to show that the resulting graph, which we denote G_m, has the desired properties. □

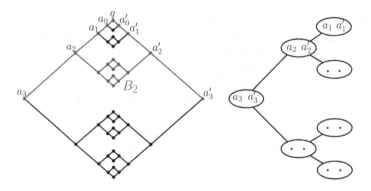

Fig. 2. Left: Graph G_4, where the roots of Y and Y' are a_3 and a_3'. A path between the two roots in the spanning tree is marked in green, and this path separates a copy of G_2 (marked in purple and denoted B_2) from the rest of the graph. Crucially, the purple subgraph must contribute a weight of at least 1 to q, and the same situation applies for the other graphs G_i, $i \leq 2$. **Right**: Fragment of the tree-cut decomposition (Y^*, χ) of G_4.

Since it is known that treewidth dominates tree-cut width (see Lemma 1), Lemma 2 implies that edge-cut width does not dominate degtw. Conversely, it is easy to build graphs with unbounded degtw and bounded edge-cut width (e.g., consider the class of stars). Hence, we obtain that edge-cut width is incomparable to degtw. An illustration of the parameter hierarchy including edge-cut width is provided in Fig. 3.

Fig. 3. Position of edge-cut width in the hierarchy of graph parameters. Here an arrow from parameter β to parameter α represents the fact that α dominates β, i.e., there exists a function p such that for each graph G, $\alpha(G) \leq p(\beta(G))$. We use fen to denote the feedback edge number.

Next, we note that even though Lemma 1 and Proposition 1 together imply that $\mathrm{tw}(G) \leq 2\,\mathrm{ecw}(G)^2 + 3\,\mathrm{ecw}(G)$, one can in fact show that the gap is linear. This will also allow us to provide a better running time bound in Sect. 4.

Lemma 3. *For every graph G, $\mathrm{tw}(G) \leq \mathrm{ecw}(G)$.*

Last but not least, we show that—also somewhat surprisingly—edge-cut width is not closed under edge or vertex deletion. (see Fig. 4).

Corollary 1. *There exist graphs G and H such that* $\mathrm{ecw}(G - e) > \mathrm{ecw}(G)$ *and* $\mathrm{ecw}(H - v) > \mathrm{ecw}(H)$ *for some* $e \in E(G)$ *and* $v \in V(H)$.

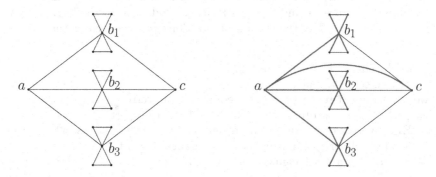

Fig. 4. Left: Graph $G - ac$ of $\mathrm{ecw}(G - ac) \geq 5$. **Right**: Green tree witnessing that $\mathrm{ecw}(G) \leq 4$.

4   Computing Edge-Cut Width

Before we proceed to the algorithmic applications of edge-cut width, we first consider the question of computing the parameter along with an optimal "decomposition" (i.e., spanning tree). Here, we provide an explicit fixed-parameter algorithm for this task.

By Lemma 3, the treewidth of G can be linearly bounded by $\mathrm{ecw}(G)$. The algorithm uses this to perform dynamic programming on a tree decomposition $(T, \{\beta_t\}_{t \in V(T)})$ of G. For a node $t \in V(T)$, we let Y_t be the union of all bags β_s such that s is either t itself or a descendant of t in T, and let G_t be the subgraph $G[Y_t]$ of G induced by Y_t.

Lemma 4. *Given an n-vertex graph G of treewidth k and a bound w, it is possible to decide whether G has edge-cut width at most w in time $k^{\mathcal{O}(wk^2)} \cdot n$. If the answer is positive, we can also output a spanning tree of G of edge-cut width at most w.*

Using the relation between treewidth and edge-cut width above, we immediately obtain:

Theorem 1. *Given a graph G, the edge-cut width $\mathrm{ecw}(G)$ can be computed time $\mathrm{ecw}(G)^{\mathrm{ecw}(G)^3} \cdot n$.*

Proof (Proof Sketch of Lemma 4). Without loss of generality, we assume that G is connected. Using state-of-the-art approximation algorithms [2,25], we first compute a "nice" tree decomposition $(T, \{\beta_t\}_{t \in V(T)})$ with root $r \in V(T)$ of width $k = \mathcal{O}(\mathrm{tw}(G))$ in time $2^{\mathcal{O}(k)} \cdot n$.

On a high level, the algorithm relies on the fact that if G has edge-cut width at most w, then at each bag β_t the number of unique paths contributing to the edge-cut width of vertices in β_t is upper-bounded by $|\beta_t|w \leq kw$. Otherwise, at least one of the vertices in β_t would lie on more than w cycles. We can use this to branch on how these at most kw paths are routed through the bag.

At each vertex $t \in T$ of the tree decomposition, we store *records* that consist of:

- an acyclic subset F of edges of $G[\beta_t]$,
- a partition $\mathcal{C}$ of β_t, and
- two multisets `future, past` of sequences of vertex-pairs in the form $((u_1, v_1), \ldots, (u_\iota, v_\iota))$ from β_t, with the following property:
  - Every vertex of β_t appears on at most w distinct u-v paths, where (u, v) is a pair of vertices in a sequence in `future` or `past`.
  - v_i and u_{i+1} are not connected by an edge in β_t.

The semantics of these records are as follows: For every spanning tree of width at most w, the record describes the intersection of the solution with $G[\beta_t]$, and the intersection of every fundamental cycle of this solution with $G[\beta_t]$. We encode the path that a cycle takes through $G[\beta_t]$ via a sequence of vertex pairs that indicate where the path leaves and enters $G[\beta_t]$ from the outside (it may be that these are the same vertex). More precisely, `past` contains those cycles that correspond to an edge that has already appeared in G_t, whereas `future` corresponds to those cycles that correspond to an edge not in G_t. In particular, this allows to reconstruct on how many cycles a vertex of β_t lies. The partition $\mathcal{C}$ says which vertices of β_t are connected via the solution in G_t.

To be more precise, let $t \in T$ and let S be an acyclic subset of edges of G that has width at most w on G_t (that is, each vertex of S lies on at most w fundamental cycles of S in G_t). We call such S *partial solutions* at t. Then, we let the *t-projection* of S be defined as $(F, \mathcal{C}, \texttt{future}, \texttt{past})$, where

- $F = S \cap E(G[\beta_t])$.
- $\mathcal{C}$ is a partition of F according to the connected components of S in G_t.
- Let C_e be a fundamental cycle of S in G corresponding to the edge $e \in G - S$. Then, there is a sequence $P_e = ((u_1, v_1), \ldots, (u_t, v_t))$ in either `future` or `past` of vertex pairs such that the intersection of C_e with S traverses F along the unique u_i-v_i paths in the order they appear in P_e (note that $u_i = v_i$ is possible, in which case the path contains just the vertex u_i).
- For each fundamental cycle C_e of S in G, if $e \in E(G_t)$, then $P_e \in \texttt{past}$, otherwise, $P_e \in \texttt{future}$.

Note that P_e can (and often will) be the empty sequence $P_e = \emptyset$. Moreover, we assume that the correspondence between $\texttt{future} \cup \texttt{past}$ and the edges in $G - S$ is bijective, in the sense that if two edges e, e' produce the same sequence $P_e = P_{e'}$, then P_e and $P_{e'}$ occur as two separate copies in $\texttt{future} \cup \texttt{past}$.

The encoding length of a single record is $\mathcal{O}(wk^2 \log k)$, dominated by the at most kw sequences P_e of k pairs of vertices each, with indices having $\mathcal{O}(\log k)$ bits. Overall, the number of records is hence bounded by $2^{\mathcal{O}(wk^2 \log k)}$.

For each $t \in T$, we store a set of records $\mathcal{R}(t)$ that has the property that $\mathcal{R}(t)$ contains the set of all t-projections of spanning trees of width at most w (that is, projections of solutions of the original instance). In addition, we require for every record in $\mathcal{R}(t)$ that there is a partial solution S of G_t of width at most w that agrees with F, C and past of the record. In this case, we call $\mathcal{R}(t)$ *valid*. Supposing correctness of this procedure, G is a YES-instance if and only if $(F_r, C_r, \text{past}_r, \text{future}_r) \in \mathcal{R}(r)$, with $F_r = C_r = \text{future}_r = \emptyset$, $\text{past}_r = \{\emptyset^{m-n}\}$, and a NO-instance otherwise.

To conclude the proof, it now suffices to compute $\mathcal{R}(t)$ in a leaf-to-root fashion. □

5 Algorithmic Applications of Edge-Cut Width

Here we obtain algorithms for the following five NP-hard problems (where a sixth problem mentioned in the introduction, PRECOLORING EXTENSION, is a special case of LIST COLORING, and the fixed-parameter tractability of BAYESIAN NETWORK STRUCTURE LEARNING and POLYTREE LEARNING follows from previous work [16]). In all of these, we will parameterize either by the edge-cut width of the input graph or of a suitable graph representation of the input. Recall that all problems are known to be W[1]-hard when parameterized by tree-cut width [5,14,18,22], and here we will show they are all fixed-parameter tractable w.r.t. edge-cut width.

As a unified starting point for all algorithms, we will apply Theorem 1 to compute a minimum-width spanning tree T of the input graph (or the graph representation of the input) G; the running time of Theorem 1 is also an upper-bound for the running time of all algorithms except for MAXSRTI, which has a quadratic dependence on the input size. Let r be an arbitrarily chosen root in T. For each node $v \in V(T)$, we will use T_v to denote the subtree of T rooted at v. Without loss of generality, in all our problems we will assume that G is connected.

The central notion used in our dynamic programming framework is that of a *boundary*, which fills a similar role as the bags in tree decompositions. Intuitively, the boundary contains all the edges which leave T_v (including the vertices incident to these edges).

Definition 4. *For each $v \in V(T)$, the boundary $\partial(v)$ of T_v is the edge-induced subgraph of G induced by those edges which have precisely one endpoint in T_v.*

Observe that for each $v \in V(T)$, $|E(\partial(v))| \leq \text{ecw}(G)$ and $|V(\partial(v))| \leq 2\,\text{ecw}(G)$. It will also sometimes be useful to speak of the graph induced by the vertices that are "below" v in T, and so we set $\mathcal{Y}_v = \{w \mid w \text{ is a descendant of } v \text{ in } T\}$ and $\mathcal{G}_v = G[\mathcal{Y}_v]$; we note that $v \in \mathcal{Y}_v$. Observe that $\partial(v)$ acts as a separator between vertices outside of $\mathcal{Y}_v \cup V(\partial(v))$ and vertices in $\mathcal{Y}_v \setminus V(\partial(v))$.

Edge Disjoint Paths. We start with the classical EDGE DISJOINT PATHS problem, which has been extensively studied in the literature. While its natural counterpart, the VERTEX DISJOINT PATHS problem, is fixed-parameter tractable when parameterized by treewidth, EDGE DISJOINT PATHS is W[1]-hard not only when parameterized by tree-cut width [18] but also by the vertex cover number [11].

EDGE DISJOINT PATHS (EDP)

Input: A graph G and a set P of *terminal pairs*, i.e., a set of subsets of $V(G)$ of size two.

Question: Is there a set of pairwise edge disjoint paths connecting every set of terminal pairs in P?

A vertex which occurs in a terminal pair is called a *terminal* and a set of pairwise edge disjoint paths connecting every set of terminal pairs in P is called a *solution*.

Theorem 2. EDP *is fixed-parameter tractable when parameterized by the edge-cut width of the input graph.*

Proof (Sketch). We start by defining the syntax of the records we will use in our dynamic program. For $v \in V(G)$, let a record be a tuple of the form (S, D, R), where:

- $S = \{(t_0, e_0), \ldots, (t_i, e_i)\}$ where for each $j \in [i]$, $t_j \in \mathcal{Y}_v$ is a terminal whose counterpart is not in $\mathcal{G}_v$, $e_j \in E(\partial(v))$, and where each terminal without a partner in $\mathcal{Y}_v$ appears in exactly one pair,
- D, R are sets of unordered pairs of elements from $E(\partial(v))$, and
- each edge of $E(\partial(v))$ may only appear in at most one tuple over all of these sets.

We refer to the edges in S, D, R as *single, donated* and *received* edges, respectively, in accordance with how they will be used in the algorithm. Let $\mathcal{R}(v)$ be a set of records for v. From the syntax, it follows that $|\mathcal{R}(v)| \leq 2^{\mathcal{O}(k \log k)}$ for each $v \in V(G)$.

Let $P_v \subseteq (\mathcal{Y}_v \cup V(\partial(v)))^2$ be a set that can be obtained from P by the following three operations:

- for some $\{a, b\} \in P$ where $a \in \mathcal{Y}_v$, $b \notin \mathcal{Y}_v$, replacing b by some $c \in V(\partial(v))$, and
- for some $a', b' \in V(\partial(v)) \setminus \mathcal{Y}_v$, adding $\{a', b'\}$ to P_v, and
- for each $\{a, b\} \in P$ where $a, b \notin \mathcal{Y}_v$, remove $\{a, b\}$.

To define a partial solution we need the following graph H_v:

- First, we add $\mathcal{G}_v \cup \partial(v)$ to H_v, where $\mathcal{G}_v \cup \partial(v)$ is the (non-disjoint) union of these two graphs.
- Next, we create for each edge $e \in E(\partial(v))$ a pendant vertex v_e adjacent to the endpoint of e that is outside of $\mathcal{Y}_v$. Let V_∂ denote the set of these new vertices.

– Finally, we add edges to $E(H_v)$ such that V_∂ is a clique.

Let a partial solution at v be a solution to the instance (H_v, P_v) for some P_v defined as above. Obviously, since at the root r we have that $\partial(r)$ is empty, $P_r = P$ and $H_r = G$. Notice that a partial solution at the root is a solution.

Consider then the set $\mathcal{W}$ containing all partial solutions at v. The v-*projection* of a partial solution $W \in \mathcal{W}$ at v is a record (S_W, D_W, R_W) where:

– $(t, e) \in S_W$ if and only if t is a terminal in $\mathcal{Y}_v$ whose counterpart t' is not in $\mathcal{Y}_v$ and e is the first edge in $E(\partial(v))$ encountered by the t-t' path in W,
– $\{e_i, e_j\} \in D_W$ if and only if there is a path $Q \in W$ with $Q = e_i, e_{i+1}, \ldots, e_{j-1}, e_j$ such that the edges in $Q \backslash \{e_i, e_j\}$ are contained in $E(\mathcal{G}_v)^4$, and
– $\{e_i, e_j\} \in R_W$ if and only if there is some s-t path $Q \in W$ such that s, t in $\mathcal{Y}_v$, e_i is the first edge in $E(\partial(v))$ that occurs in Q, and e_j is the last edge in $E(\partial(v))$ that occurs in Q.

We say that $\mathcal{R}(v)$ is *valid* if and only if it contains all v-projections of partial solutions in $\mathcal{W}$, and in addition, for every record in $\mathcal{R}(v)$, there is a partial solution such that its v-projection yields this record.

Observe that if $\mathcal{R}(r) = \emptyset$, then (G, P) is a NO-instance, while if $\mathcal{R}(r) = \{(\emptyset, \emptyset, \emptyset)\}$, then $R(r)$ is a YES-instance. To complete the proof, it now suffices to dynamically compute a set of valid records in a leaf-to-root fashion along T. We note that if at any stage we obtain that a vertex v has no records (i.e., $\mathcal{R}(v) = \emptyset$), we immediately reject.   □

List Coloring. The second problem we consider is LIST COLORING [9, 14]. It is known that this problem is W[1]-hard parameterized by tree-cut width. A *coloring* col is a mapping from the vertex set of a graph to a set of colors; a coloring is *proper* if for every pair of adjacent vertices a, b, it holds that $\text{col}(a) \neq \text{col}(b)$.

LIST COLORING

Input: A graph $G = (V, E)$ and for each vertex $v \in V$ a list $L(v)$ of permitted colors.

Question: Does G admit a proper coloring col where for each vertex v it holds $\text{col}(v) \in L(v)$?

Theorem 3. LIST COLORING *is fixed-parameter tractable when parameterized by the edge-cut width of the input graph.*

Proof (Sketch). We start by defining the syntax of the records we will use in our dynamic program. For $v \in V(G)$, let a record for a vertex v consist of tuples of the form (u, c), where (1) $u \in V(\partial(v)) \cap \mathcal{Y}_v$, (2) $c \in L(u) \cup \{\delta\}$, and (3) each vertex of $V(\partial(v)) \cap \mathcal{Y}_v$ appears exactly once in a record.

---

[4] Note that by the syntax, it follows that e_i and e_j are both contained in $\partial(v)$.

To introduce the semantics of the records, consider the set $\mathcal{W}$ containing all partial solutions (i.e., all proper colorings) at v to the instance $(\mathcal{G}_v, (L(u))_{u \in \mathcal{Y}_v})$. The v-*projection* of a partial solution $\mathtt{col} \in \mathcal{W}$ is a set $R_{\mathtt{col}} = \{(u, c) \mid u \in V(\partial(v)) \cap \mathcal{Y}_v, c \in L(u)\}$ where $(u, c) \in R_{\mathtt{col}}$ if and only if $\mathtt{col}(u) = c$.

Let $\mathcal{R}(v)$ be a set of records for v. For two records $R_1, R_2 \in \mathcal{R}(v)$ we say $R_1 \preceq R_2$ if and only if for each $u \in V(\partial(v)) \cap \mathcal{Y}_v$ the following holds:

- Either $(u, c) \in R_1 \cap R_2$ with $c \in L(u)$,
- Or $(u, c) \in R_1$ with $c \in L(u)$ and $(u, \delta) \in R_2$.

We say that $\mathcal{R}(v)$ is *valid* if for each v-projection $R_{\mathtt{col}}$ of a partial solution $\mathtt{col} \in \mathcal{W}$ there is a record $R \in \mathcal{R}(v)$ which satisfies $R_{\mathtt{col}} \preceq R$, and in addition, for every record $R \in \mathcal{R}(v)$, there is a partial solution $\mathtt{col} \in \mathcal{W}$ such that its v-projection fulfills $R_{\mathtt{col}} \preceq R$. Observe that if $\mathcal{R}(r) = \emptyset$, then $(G, (L(v))_{v \in V(G)})$ is a NO-instance, while if $\mathcal{R}(r) = \{\emptyset\}$, then $R(r)$ is a YES-instance.

If a record in $\mathcal{R}(v)$ contains a tuple (u, δ), then this means that there is always a possible coloring for the vertex u, e.g., if $|L(u)| > d_G(u)$; the symbol δ is introduced specifically to bound $|L(v)|$. Therefore, it follows that $|\mathcal{R}(v)| \leq 2^{\mathcal{O}(k \log k)}$ for each $v \in V(G)$. To complete the proof, it now suffices to dynamically compute a set of valid records in a leaf-to-root fashion along T.   □

Boolean CSP. Next, we consider the classical constraint satisfaction problem [32].

BOOLEAN CSP

Input: A set of variables X and a set of constraints $\mathcal{C}$.
Question: Is there an assignment $\sigma : X \to \{0, 1\}$ such that all constraints in $\mathcal{C}$ are satisfied?

We represent this problem via the *incidence graph*, whose vertex set is $X \cup \mathcal{C}$ and which contains an edge between a variable and a constraint if and only if the variable appears in the scope of the constraint.

Theorem 4. BOOLEAN CSP *is fixed-parameter tractable when parameterized by the edge-cut width of the incidence graph.*

Proof (Sketch). For this problem, we do not need to consider all the vertices in the boundary. Instead, for a vertex $v \in V$, let $B(v) = V(\partial(v)) \cap \mathcal{Y}_v \cap X$. Hence, we will consider only the vertices in the boundary inside of the current subtree, which correspond to variables in the input instance. Note that $|B(v)| \leq |V(\partial(v))| \leq 2k$.

We continue with defining the syntax of the records we will use in our dynamic program. For $v \in V(G)$, let a record for a vertex v be a set of functions of the form $\varphi : B(v) \to \{0, 1\}$. Let $\mathcal{R}(v)$ be a set of records for v. From the syntax, it follows that $|\mathcal{R}(v)| \leq 2^{\mathcal{O}(k)}$ for each $v \in V(G)$. To introduce the semantics of the records, consider the set $\mathcal{W}$ containing all partial solutions (i.e., all assignments of the variables such that every constraint is fulfilled) at v for the instance $(\mathcal{Y}_v \cap X, \mathcal{Y}_v \cap \mathcal{C})$.

The function φ is a *v-projection* of a solution $\sigma \in \mathcal{W}$ if and only if $\sigma|_{B(v)} = \varphi$. This means, that the functions in a record represent the assignments of variables, which are compatible with $\mathcal{Y}_v$.

We say that $\mathcal{R}(v)$ is *valid* if it contains all v-projections of partial solutions in $\mathcal{W}$, and in addition, for every record in $\mathcal{R}(v)$, there is a partial solution such that its v-projection yields this record. Observe that if $\mathcal{R}(r) = \emptyset$, then $(X, \mathcal{C})$ is a NO-instance, while if $\mathcal{R}(r) = \{\emptyset\}$, then $R(r)$ is a YES-instance. To complete the proof, it now suffices to dynamically compute a set of valid records in a leaf-to-root fashion along T.   □

Further Problems. As our final two results, we use the algorithmic framework developed above to also establish the fixed-parameter tractability for the remaining two problems which were shown to be W[1]-hard w.r.t. tree-cut width. These are MAXIMUM STABLE ROOMMATES WITH TIES AND INCOMPLETE LISTS (MaxSRTI) [5] and MINIMUM CHANGEOVER COST ARBORESCENCE (MinCCA) [22].

Theorem 5. MaxSRTI *is fixed-parameter tractable when parameterized by the edge-cut width of the acceptability graph.*

Theorem 6. MinCCA *is fixed-parameter tractable when parameterized by the edge-cut width of the input graph.*

6   Conclusion

The parameter developed in this paper, edge-cut width, is aimed at mitigating the algorithmic shortcomings of tree-cut width and filling the role of an "easy-to-use" edge-based alternative to treewidth. We show that edge-cut width essentially has all the desired properties one would wish for as far as algorithmic applications are concerned: it is easy to compute, uses a natural structure as its decomposition, and yields fixed-parameter tractability for all problems that one would hope an edge-based alternative to treewidth could solve.

Last but not least, we note that a preprint exploring a different parameter that is aimed at providing an edge-based alternative to treewidth appeared shortly after the results presented in our paper were obtained [26]. While it is already clear that the two parameters are not equivalent, it would be interesting to explore the relationship between them in future work.

References

1. Bentert, M., Haag, R., Hofer, C., Koana, T., Nichterlein, A.: Parameterized complexity of min-power asymmetric connectivity. Theory Comput. Syst. **64**(7), 1158–1182 (2020)

2. Bodlaender, H.L., Drange, P.G., Dregi, M.S., Fomin, F.V., Lokshtanov, D., Pilipczuk, M.: A c^k n 5-approximation algorithm for treewidth. SIAM J. Comput. **45**(2), 317–378 (2016)
3. Bodlaender, H.L., Jansen, B.M.P., Kratsch, S.: Preprocessing for treewidth: a combinatorial analysis through kernelization. SIAM J. Discret. Math. **27**(4), 2108–2142 (2013)
4. Bonnet, É., Kim, E.J., Thomassé, S., Watrigant, R.: Twin-width I: tractable FO model checking. In: 61st IEEE Annual Symposium on Foundations of Computer Science, FOCS 2020, Durham, NC, USA, 16–19 November 2020, pp. 601–612. IEEE (2020)
5. Bredereck, R., Heeger, K., Knop, D., Niedermeier, R.: Parameterized complexity of stable roommates with ties and incomplete lists through the lens of graph parameters. In: Lu, P., Zhang, G. (eds.) 30th International Symposium on Algorithms and Computation, ISAAC 2019, 8–11 December 2019, Shanghai University of Finance and Economics, Shanghai, China. LIPIcs, vol. 149, pp. 44:1–44:14. Schloss Dagstuhl - Leibniz-Zentrum für Informatik (2019)
6. Cygan, M., et al.: Lower bounds for kernelization. In: Parameterized Algorithms, pp. 523–555. Springer, Cham (2015). https://doi.org/10.1007/978-3-319-21275-3_15
7. Diestel, R.: Graph Theory. Graduate Texts in Mathematics, vol. 173, 4th edn. Springer, Heidelberg (2012)
8. Downey, R.G., Fellows, M.R.: Fundamentals of Parameterized Complexity. Texts in Computer Science, Springer, Heidelberg (2013). https://doi.org/10.1007/978-1-4471-5559-1
9. Fellows, M.R., et al.: On the complexity of some colorful problems parameterized by treewidth. Inf. Comput. **209**(2), 143–153 (2011)
10. Fellows, M.R., Lokshtanov, D., Misra, N., Rosamond, F.A., Saurabh, S.: Graph layout problems parameterized by vertex cover. In: Hong, S.-H., Nagamochi, H., Fukunaga, T. (eds.) ISAAC 2008. LNCS, vol. 5369, pp. 294–305. Springer, Heidelberg (2008). https://doi.org/10.1007/978-3-540-92182-0_28
11. Fleszar, K., Mnich, M., Spoerhase, J.: New algorithms for maximum disjoint paths based on tree-likeness. Math. Program. **171**, 433–461 (2017). https://doi.org/10.1007/s10107-017-1199-3
12. Ganian, R.: Improving vertex cover as a graph parameter. Discret. Math. Theor. Comput. Sci. **17**(2), 77–100 (2015). http://dmtcs.episciences.org/2136
13. Ganian, R., Hlineny, P.: On parse trees and Myhill-Nerode-type tools for handling graphs of bounded rank-width. Discret. Appl. Math. **158**(7), 851–867 (2010)
14. Ganian, R., Kim, E.J., Szeider, S.: Algorithmic applications of tree-cut width. In: Italiano, G.F., Pighizzini, G., Sannella, D.T. (eds.) MFCS 2015. LNCS, vol. 9235, pp. 348–360. Springer, Heidelberg (2015). https://doi.org/10.1007/978-3-662-48054-0_29
15. Ganian, R., Klute, F., Ordyniak, S.: On structural parameterizations of the bounded-degree vertex deletion problem. Algorithmica **83**(1), 297–336 (2021)
16. Ganian, R., Korchemna, V.: The complexity of Bayesian network learning: Revisiting the superstructure. In: Proceedings of NeurIPS 2021, The Thirty-Fifth Conference on Neural Information Processing Systems (2021, to appear)
17. Ganian, R., Ordyniak, S.: The complexity landscape of decompositional parameters for ILP. Artif. Intell. **257**, 61–71 (2018)
18. Ganian, R., Ordyniak, S.: The power of cut-based parameters for computing edge-disjoint paths. Algorithmica **83**(2), 726–752 (2021)

19. Ganian, R., Ordyniak, S., Ramanujan, M.S.: On structural parameterizations of the edge disjoint paths problem. Algorithmica **83**(6), 1605–1637 (2021)
20. Giannopoulou, A.C., Kwon, O., Raymond, J., Thilikos, D.M.: Lean tree-cut decompositions: obstructions and algorithms. In: Niedermeier, R., Paul, C. (eds.) 36th International Symposium on Theoretical Aspects of Computer Science, STACS 2019, Berlin, Germany, 13–16 March 2019. LIPIcs, vol. 126, pp. 32:1–32:14. Schloss Dagstuhl - Leibniz-Zentrum für Informatik (2019)
21. Golovach, P.A., Komusiewicz, C., Kratsch, D., Le, V.B.: Refined notions of parameterized enumeration kernels with applications to matching cut enumeration. J. Comput. Syst. Sci. **123**, 76–102 (2022)
22. Gözüpek, D., Özkan, S., Paul, C., Sau, I., Shalom, M.: Parameterized complexity of the MINCCA problem on graphs of bounded decomposability. Theor. Comput. Sci. **690**, 91–103 (2017)
23. Gutin, G.Z., Jones, M., Wahlström, M.: The mixed Chinese postman problem parameterized by pathwidth and treedepth. SIAM J. Discret. Math. **30**(4), 2177–2205 (2016)
24. Kim, E.J., Oum, S., Paul, C., Sau, I., Thilikos, D.M.: An FPT 2-approximation for tree-cut decomposition. Algorithmica **80**(1), 116–135 (2018)
25. Korhonen, T.: Single-exponential time 2-approximation algorithm for treewidth. CoRR abs/2104.07463 (2021). https://arxiv.org/abs/2104.07463
26. Magne, L., Paul, C., Sharma, A., Thilikos, D.M.: Edge-treewidth: algorithmic and combinatorial properties. CoRR abs/2112.07524 (2021)
27. Marx, D., Wollan, P.: Immersions in highly edge connected graphs. SIAM J. Discret. Math. **28**(1), 503–520 (2014)
28. Nederlof, J., Pilipczuk, M., Swennenhuis, C.M.F., Węgrzycki, K.: Hamiltonian cycle parameterized by treedepth in single exponential time and polynomial space. In: Adler, I., Müller, H. (eds.) WG 2020. LNCS, vol. 12301, pp. 27–39. Springer, Cham (2020). https://doi.org/10.1007/978-3-030-60440-0_3
29. Nešetřil, J., Ossona de Mendez, P.: Sparsity. Algorithms and Combinatorics, vol. 28. Springer, Heidelberg (2012). https://doi.org/10.1007/978-3-642-27875-4
30. Oum, S.: Approximating rank-width and clique-width quickly. ACM Trans. Algorithms **5**(1), 1–20 (2008)
31. Robertson, N., Seymour, P.D.: Graph minors. II. Algorithmic aspects of tree-width. J. Algorithms **7**(3), 309–322 (1986)
32. Samer, M., Szeider, S.: Constraint satisfaction with bounded treewidth revisited. J. Comput. Syst. Sci. **76**(2), 103–114 (2010)
33. Scheffler, P.: Practical linear time algorithm for disjoint paths in graphs with bounded tree-width. Technical report TR 396/1994. FU Berlin, Fachbereich 3 Mathematik (1994)
34. Wollan, P.: The structure of graphs not admitting a fixed immersion. J. Comb. Theory Ser. B **110**, 47–66 (2015)

An Algorithmic Framework for Locally Constrained Homomorphisms

Laurent Bulteau[1] , Konrad K. Dabrowski[2(✉)] , Noleen Köhler[3] ,
Sebastian Ordyniak[4] , and Daniël Paulusma[5]

[1] LIGM, CNRS, Université Gustave Eiffel, Champs-sur-Marne, France
laurent.bulteau@univ-eiffel.fr
[2] School of Computing, Newcastle University, Newcastle upon Tyne, UK
konrad.dabrowski@newcastle.ac.uk
[3] LAMSADE, CNRS, Université Paris-Dauphine, PSL University, Paris, France
noleen.kohler@dauphine.psl.eu
[4] School of Computing, University of Leeds, Leeds, UK
sordyniak@gmail.com
[5] Department of Computer Science, University of Durham, Durham, UK
daniel.paulusma@durham.ac.uk

Abstract. A homomorphism ϕ from a guest graph G to a host graph H is locally bijective, injective or surjective if for every $u \in V(G)$, the restriction of ϕ to the neighbourhood of u is bijective, injective or surjective, respectively. The corresponding decision problems, LBHom, LIHom and LSHom, are well studied both on general graphs and on special graph classes. We prove a number of new FPT, W[1]-hard and para-NP-complete results by considering a hierarchy of parameters of the guest graph G. For our FPT results, we do this through the development of a new algorithmic framework that involves a general ILP model. To illustrate the applicability of the new framework, we also use it to prove FPT results for the ROLE ASSIGNMENT problem, which originates from social network theory and is closely related to locally surjective homomorphisms.

Keywords: (locally constrained) graph homomorphism ·
parameterized complexity · fracture number

1 Introduction

A *homomorphism* from a graph G to a graph H is a mapping $\phi : V(G) \to V(H)$ such that $\phi(u)\phi(v) \in E(H)$ for every $uv \in E(G)$. Graph homomorphisms generalise graph colourings (using a complete graph for H) and have been intensively studied over a long period of time. We refer to the textbook of Hell and Nešetřil [34] for a further introduction.

The second and fourth authors acknowledge support from the Engineering and Physical Sciences Research Council (EPSRC, project EP/V00252X/1).

M. A. Bekos and M. Kaufmann (Eds.): WG 2022, LNCS 13453, pp. 114–128, 2022.
https://doi.org/10.1007/978-3-031-15914-5_9

We write $G \to H$ if there exists a homomorphism from G to H; here, G is called the *guest graph* and H is the *host graph*. We denote the corresponding decision problem by HOM, and if H is fixed, that is, not part of the input, we write H-HOM. The renowned Hell-Nešetřil dichotomy [33] states that H-HOM is polynomial-time solvable if H is bipartite, and NP-complete otherwise. We denote the vertices of H by $1, \ldots, |V(H)|$ and call them *colours*.

Instead of fixing the host graph H, one can also restrict the structure of the guest graph G by bounding some graph parameter. Here, it is known that if FPT $\neq$ W[1], then HOM can be solved in polynomial time if and only if the so-called core of the guest graph has bounded treewidth [31].

Locally Constrained Homomorphisms. We are interested in three well-studied variants of graph homomorphisms that occur after placing constraints on the neighbourhoods of the vertices of the guest graph G. Consider a homomorphism ϕ from a graph G to a graph H. We say that ϕ is locally injective, locally bijective or locally surjective for $u \in V(G)$ if restricting ϕ to a function $\phi_u : N_G(u) \to N_H(\phi(u))$ is injective, bijective or surjective, respectively. Here, $N_G(u) = \{v \mid uv \in E(G)\}$ denotes the (open) neighbourhood of a vertex u in a graph G. We say that ϕ is *locally injective, locally bijective* or *locally surjective* if it is locally injective, locally bijective or locally surjective for every $u \in V(G)$. We denote existence of these *locally constrained* homomorphisms by $G \xrightarrow{B} H$, $G \xrightarrow{I} H$ and $G \xrightarrow{S} H$, respectively.

Locally injective homomorphisms are also known as *partial graph coverings* and are used in telecommunications [23], in distance constrained labelling [22] and as indicators of the existence of homomorphisms of derivative graphs [46]. Locally bijective homomorphisms originate from topological graph theory [4,45] and are more commonly known as *graph coverings*. They are used in distributed computing [2,3,7] and in constructing highly transitive regular graphs [5]. Locally surjective homomorphisms are sometimes called *colour dominations* [41]. They have applications in distributed computing [11,12] and in social science [20,50,53,54]. In the latter context they are known as *role assignments*.

Let LBHOM, LIHOM and LSHOM be the three problems of deciding, for two graphs G and H, whether $G \xrightarrow{B} H$, $G \xrightarrow{I} H$ or $G \xrightarrow{S} H$ holds, respectively. As before, we write H-LBHOM, H-LIHOM and H-LSHOM in the case when the host graph H is fixed. Out of the three problems, only the complexity of H-LSHOM has been completely classified, both for general graphs and bipartite graphs [26]. We refer to a series of papers [1,6,23,25,38,39,44] for polynomial-time solvable and NP-complete cases of H-LBHOM and H-LIHOM; see also the survey by Fiala and Kratochvíl [24]. Some more recent results include sub-exponential algorithms for H-LBHOM, H-LIHOM and H-LSHOM on string graphs [48] and complexity results for H-LBHOM for host graphs H that are multigraphs [40] or that have semi-edges [9].

In our paper we assume that both G and H are part of the input. We note a fundamental difference between locally injective homomorphisms on the one hand and locally bijective and surjective homomorphisms on the other. Namely, for connected graphs G and H, we must have $|V(G)| \geq |V(H)|$ if $G \xrightarrow{B} H$ or

$G \xrightarrow{S} H$ (this is a consequence of Observation 1), whereas H might be arbitrarily larger than G if $G \xrightarrow{I} H$ holds. For example, if we let G be a complete graph and H be a graph without self-loops, then $G \xrightarrow{I} H$ holds if and only if H contains a clique on at least $|V(G)|$ vertices.

The above difference is also reflected in the complexity results for the three problems under input restrictions. In fact, LIHOM is closely related to the SUBGRAPH ISOMORPHISM problem and is usually the hardest problem. For example, LBHOM is GRAPH ISOMORPHISM-complete on chordal guest graphs, but polynomial-time solvable on interval guest graphs and LSHOM is NP-complete on chordal guest graphs, but polynomial-time solvable on proper interval guest graphs [32]. In contrast, LIHOM is NP-complete even on complete guest graphs G, which follows from a reduction from the CLIQUE problem via the aforementioned equivalence: $G \xrightarrow{I} H$ holds if and only if H contains a clique on at least $|V(G)|$ vertices.

The aforementioned polynomial-time result on HOM for guest graphs G with a core of bounded treewidth [15,30] does not carry over to any of the three locally constrained homomorphism problems. Indeed, LBHOM, LSHOM and LIHOM are NP-complete for guest graphs G of path-width at most 5, 4 and 2, respectively [14] (all three problems are polynomial-time solvable if G is a tree [14,27]). It is also known that LBHOM [37], LSHOM [41] and LIHOM [23] are NP-complete even if G is cubic and H is the complete graph K_4 on four vertices, but polynomial-time solvable if G has bounded treewidth and one of the two graphs G or H has bounded maximum degree [14].

An Application. Locally surjective homomorphisms from a graph G to a graph H are known as H-role assignments in social network theory. Role assignments were introduced by White and Reitz [54]. A connected graph G has an h-role assignment if and only if $G \xrightarrow{S} H$ for some connected graph H with $|V(H)| = h$, as long as we allow H to have self-loops (while we assume that G is a graph with no self-loops). The ROLE ASSIGNMENT problem is to decide, for a graph G and an integer h, whether G has an h-role assignment. If h is fixed, we denote the problem h-ROLE ASSIGNMENT. h-ROLE ASSIGNMENT is NP-complete for planar graphs ($h \geq 2$) [51], cubic graphs ($h \geq 2$) [52], bipartite graphs ($h \geq 3$) [49], chordal graphs ($h \geq 3$) [35] and split graphs ($h \geq 4$) [16].

Our Focus. We continue the line of study in [14] and focus on the following research question: *For which parameters of the guest graph do LBHOM, LSHOM and LIHOM become fixed-parameter tractable?*

We will also apply our new techniques towards answering this question for the ROLE ASSIGNMENT problem. In order to address our research question, we need some additional terminology. A graph parameter p *dominates* a parameter q if there is a function f such that $p(G) \leq f(q(G))$ for every graph G. If p dominates q but q does not dominate p, then p is *more powerful* than q. We denote this by $p \triangleright q$. If neither p dominates q nor q dominates p, then p and q are *incomparable (orthogonal)*. Given the para-NP-hardness results on LBHOM, LSHOM and LIHOM for graph classes of bounded path-width [14], we will consider a range of graph parameters that are less powerful than path-width. In this way we aim to increase our understanding of the (parameterized) complexity of LBHOM, LSHOM and LIHOM.

For an integer $c \geq 1$, a *c-deletion set* of a graph G is a subset $S \subseteq V(G)$ such that every connected component of $G \setminus S$ has at most c vertices. The *c-deletion set number* $\text{ds}_c(G)$ of a graph G is the minimum size of a c-deletion set in G. If $c = 1$ we obtain the *vertex cover number* $\text{vc}(G)$ of G. The c-deletion set number is closely related to the *fracture number* $\text{fr}(G)$, introduced in [19], which is the minimum k such that G has a k-deletion set on at most k vertices. Both these parameters are also closely related to *vertex integrity* [18]. Note that $\text{fr}(G) \leq \max\{c, \text{ds}_c(G)\}$ holds for every integer c. The *feedback vertex set number* $\text{fv}(G)$ of a graph G is the size of a smallest set S such that $G \setminus S$ is a forest. We write $\text{tw}(G)$, $\text{pw}(G)$ and $\text{td}(G)$ for the treewidth, path-width and tree-depth of a graph G, respectively; see [47] for more information. It is known that $\text{tw}(G) \rhd \text{pw}(G) \rhd \text{td}(G) \rhd \text{fr}(G) \rhd \text{ds}_c(G)$(fixed c) $\rhd \text{vc}(G) \rhd |V(G)|$, where the second relationship is proven in [8] and the others follow immediately from their definitions (see also Sect. 2). It is readily seen that $\text{tw}(G) \rhd \text{fv}(G) \rhd \text{ds}_2(G)$ and that $\text{fv}(G)$ is incomparable with the parameters $\text{pw}(G)$, $\text{td}(G)$, $\text{fr}(G)$ and $\text{ds}_c(G)$ for every fixed $c \geq 3$ (consider e.g. a tree of large path-width and the disjoint union of many triangles).

Our Results. We prove a number of new parameterized complexity results for LBHOM, LSHOM and LIHOM by considering some property of the guest graph G as the parameter. In particular, we consider the graph parameters above. Our two main results, which are proven in Sect. 4, show that LBHOM and LSHOM are fixed-parameter tractable parameterized by the fracture number of G. These two results cannot be strengthened to the tree-depth of the guest graph, for which we prove para-NP-completeness. Note that the latter results imply the known para-NP-completeness results for path-width of the guest graph [14]. We also prove that LBHOM and LSHOM are para-NP-complete when parameterized by the feedback vertex set number of the guest graph. This result and the para-NP-hardness for tree-depth motivated us to consider the fracture number as a natural remaining graph parameter for obtaining an fpt algorithm.

Concerning LIHOM, we prove that it is in XP and W[1]-hard when parameterized by the vertex cover number, or equivalently, the c-deletion set number for $c = 1$. We then show that the XP-result for LIHOM cannot be generalised to hold for $c \geq 2$. In fact, in Sect. 4, we will determine the complexity of LIHOM on graphs with c-deletion set number at most k for every fixed pair of integers c and k. Our results for LBHOM, LSHOM and LIHOM are summarised, together with the known results, in Table 1.

Algorithmic Framework. The fpt algorithms for LBHOM and LSHOM are proven via a new algorithmic framework (described in detail in Sect. 3) that involves a reduction to an integer linear program (ILP) that has a wider applicability. To illustrate this, in Sect. 4 we also use our general framework to prove that ROLE ASSIGNMENT is in FPT when parameterized by $c + \text{ds}_c$, or equivalently by fracture number.

Table 1. Table of results. The results marked with a ($\star$) are the new results in this paper. The remaining results are either known results, some of which are now also implied by our new results, or follow immediately from other results in the table; in particular, for a graph G, $\mathrm{ds}_c(G) \geq \mathrm{fr}(G)$ if $c \leq \mathrm{fr}(G) - 1$, and $\mathrm{ds}_c(G) \leq \mathrm{fr}(G)$ if $c \geq \mathrm{fr}(G)$. Also note that LIHom is W[1]-hard when parameterized by $|V(G)|$, as Clique is W[1]-hard when parameterized by the clique number [17], so as before, we can let G be the complete graph in this case.

guest graph parameter	LIHom	LBHom	LSHom
$|V(G)|$	XP, W[1]-hard [17]	FPT	FPT
vertex cover number	XP ($\star$), W[1]-hard	FPT	FPT
c-deletion set number (fixed c)	para-NP-c ($c \geq 2$) ($\star$)	FPT	FPT
fracture number	para-NP-c	FPT (Theorem 4) ($\star$)	FPT (Theorem 4) ($\star$)
tree-depth	para-NP-c	para-NP-c ($\star$)	para-NP-c ($\star$)
path-width	para-NP-c [14]	para-NP-c [14]	para-NP-c [14]
treewidth	para-NP-c	para-NP-c	para-NP-c
maximum degree	para-NP-c [23]	para-NP-c [37]	para-NP-c [41]
treewidth plus maximum degree	XP, W[1]-hard	XP [14]	XP [14]
feedback vertex set number	para-NP-c	para-NP-c ($\star$)	para-NP-c ($\star$)

Techniques. The main ideas behind our algorithmic ILP framework are as follows. Let G and H be the guest and host graphs, respectively. First, we observe that if G has a c-deletion of size at most k and there is a locally surjective homomorphism from G to H, then H must also have a c-deletion set of size at most k. However it does not suffice to compute c-deletion sets D_G and D_H for G and H, guess a partial homomorphism h from D_G to D_H, and use the structural properties of c-deletion sets to decide whether h can be extended to a desired homomorphism from G to H. This is because a homomorphism from G to H does not necessarily map D_G to D_H. Moreover, even if it did, vertices in $G \setminus D_G$ can still be mapped to vertices in D_H. Consequently, components of $G \setminus D_G$ can still be mapped to more than one component of $H \setminus D_H$. This makes it difficult to decompose the homomorphism from G to H into small independent parts. To overcome this challenge, we prove that there are small sets D_G and D_H of vertices in G and H, respectively, such that every locally surjective homomorphism from G to H satisfies:

1. the pre-image of D_H is a subset of D_G,
2. D_H is a c'-deletion set for H for some c' bounded in terms of only $c + k$, and
3. all but at most k components of $G \setminus D_G$ have at most c vertices and, while the remaining components can be arbitrarily large, their treewidth is bounded in terms of $c + k$.

As D_G and D_H are small, we can enumerate all possible homomorphisms from some subset of D_G to D_H. Condition 2 allows us to show that any locally surjective homomorphism from G to H can be decomposed into locally surjective homomorphisms from a small set of components of $G \setminus D_G$ (plus D_G) to one component of $H \setminus D_H$ (plus D_H). This enables us to formulate the question of whether a homomorphism from a subset of D_G to D_H can be extended to a desired homomorphism from G to H in terms of an ILP. Finally, Condition 3 allows us to efficiently compute the possible parts of the decomposition, that is, which (small) sets of components of $G \setminus D_G$ can be mapped to which components of $H \setminus D_H$.

2   Preliminaries

Let G be a graph. We denote the vertex set and edge set of G by $V(G)$ and $E(G)$, respectively. Let $X \subseteq V(G)$ be a set of vertices of G. The *subgraph of G induced by X*, denoted $G[X]$, is the graph with vertex set X and edge set $\{uv \in E(G) \mid u, v \in X\}$. When the underlying graph is clear from the context, we will sometimes refer to an induced subgraph simply by its set of vertices. We use $G \setminus X$ to denote the subgraph of G induced by $V(G) \setminus X$. Similarly, for $Y \subseteq E(G)$ we let $G \setminus Y$ be the subgraph of G obtained by deleting all edges in Y from G. For a graph G and a vertex $u \in V(G)$, we let $N_G(u) = \{v \mid uv \in E(G)\}$ and $N_G[v] = N_G(v) \cup \{v\}$ denote the open and closed neighbourhood of v in G, respectively. Recall that we assume that the guest graph G does not contain self-loops, while the host graph H is permitted to have self-loops. In this case, by definition, $u \in N_H(u)$ if $uu \in E(H)$. We need the following well-known fact:

Proposition 1 ([42]). *Let G be a graph and let k and c be natural numbers. Then, deciding whether G has a c-deletion set of size at most k is fixed-parameter tractable parameterized by $k + c$.*

A (k, c)-*extended deletion set* for G is a set $D \subseteq V(G)$ such that: (1) every component of $G \setminus D$ either has at most c vertices or has a c-deletion set of size at most k and (2) at most k components of $G \setminus D$ have more than c vertices.

Locally Constrained Homomorphisms. Here we show some basic properties of locally constrained homomorphisms.

Observation 1. *Let G and H be non-empty connected graphs and let ϕ be a locally surjective homomorphism from G to H. Then ϕ is surjective.*

Observation 2. *Let G and H be graphs, let $D \subseteq V(G)$, and let ϕ be a homomorphism from G to H. Then, for every component C_G of $G \setminus D$ such that $\phi(C_G) \cap \phi(D) = \emptyset$, there is a component C_H of $H \setminus \phi(D)$ such that $\phi(C_G) \subseteq C_H$. Moreover, if ϕ is locally injective/surjective/bijective, then $\phi|_{D \cup C_G}$ is a homomorphism from $G' = G[D \cup C_G]$ to $H' = H[\phi(D) \cup C_H]$ that is locally injective/surjective/bijective for every $v \in V(C_G)$.*

Lemma 1. *Let G and H be non-empty connected graphs, let $D \subseteq V(G)$ be a c-deletion set for G, and let ϕ be a locally surjective homomorphism from G to H. Then $\phi(D)$ is a c-deletion set for H.*

Integer Linear Programming. Given a set $\mathcal{X}$ of variables and a set $\mathcal{C}$ of linear constraints (i.e. inequalities) over the variables in $\mathcal{X}$ with integer coefficients, the task in the feasibility variant of *integer linear programming* (ILP) is to decide whether there is an assignment $\alpha : \mathcal{X} \to \mathbb{Z}$ of the variables satisfying all constraints in $\mathcal{C}$. We will use the following well-known result by Lenstra [43].

Proposition 2 ([21,29,36,43]).   ILP *is fpt parameterized by the number of variables.*

3   Our Algorithmic Framework

Here we present our main algorithmic framework that will allow us to show that LSHOM, LBHOM and ROLE ASSIGNMENT are fpt parameterized by $k + c$ when the guest graph has c-deletion set number at most k. To illustrate the main ideas behind our framework, let us first explain these ideas for the examples of LSHOM and LBHOM. In this case we are given G and H and we know that G has a c-deletion set of size at most k. Because of Lemma 1, it then follows that if (G, H) is a yes-instance of LSHOM or LBHOM, then H also has a c-deletion set of size at most k. Informally, our next step is to compute a small set Φ of partial locally surjective homomorphisms such that (1) every locally surjective homomorphism from G to H augments some $\phi_P \in \Phi$ and (2) for every $\phi_P \in \Phi$, the domain of ϕ_P is a (k, c)-extended deletion set of G and the co-domain of ϕ_P is a c'-deletion set of H, where c' is bounded by a function of $k + c$. Here and in what follows, we say that a function $\phi : V(G) \to V(H)$ *augments* (or is an *augmentation* of) a partial function $\phi_P : V_G \to V_H$, where $V_G \subseteq V(G)$ and $V_H \subseteq V(H)$ if $v \in V_G \Leftrightarrow \phi(v) \in V_H$ and $\phi|_{V_G} = \phi_P$. This allows us to reduce our problems to (boundedly many) subproblems of the following form: Given a (k, c)-extended deletion set D_G for G, a c'-deletion set D_H for H, and a locally surjective (respectively bijective) homomorphism ϕ_P from D_G to D_H, find a locally surjective homomorphism ϕ from G to H that augments ϕ_P. We will then show how to formulate this subproblem as an integer linear program and how this program can be solved efficiently. Importantly, our ILP formulation will allow us to solve a much more general problem, where the host graph H is not explicitly given, but defined in terms of a set of linear constraints, which will allow us to solve the ROLE ASSIGNMENT problem.

Partial Homomorphisms for the Deletion Set. For a graph G and $m \in \mathbb{N}$ we let $D_G^m := \{v \in V(G) \mid \deg_G(v) \geq m\}$. We will show in Lemma 4 that there is a small set Φ of partial homomorphisms such that every locally surjective (respectively bijective) homomorphism from G to H augments some $\phi_P \in \Phi$ and, for every $\phi_P \in \Phi$, the domain of ϕ_P is a (k, c)-extended deletion set for G of size at most k and its co-domain is a c'-deletion set of size at most k for H. The main idea behind finding this set Φ is to consider the set of high degree vertices in G and H, i.e. the sets D_G^{k+c} and D_H^{k+c}. As it turns out (see Lemma 2), for every subset $D \subseteq D_G^{k+c}$, D is a $(k - |D|, c)$-extended deletion set for G of size at most k and D_H^{k+c} is a c'-deletion set for H of size at most k, where $c' = kc(k + c)$. Moreover, as we will show in Lemma 3, every locally surjective (respectively bijective) homomorphism from G to H has to augment a locally surjective (respectively bijective) homomorphism from some induced subgraph of $G[D_G^{k+c}]$ to $D_H = D_H^{k+c}$. Intuitively, this holds because for every locally surjective homomorphism, only vertices of high degree in G can be mapped to a vertex of high degree in H and every vertex in H must have a pre-image in G.

Lemma 2. *Let G be a graph. If G has a c-deletion set of size at most k, then the set D_G^{k+c} is a $kc(k+c)$-deletion set of size at most k. Furthermore, every subset $D \subseteq D_G^{k+c}$ is a $(k - |D|, c)$-extended deletion set of G.*

Lemma 3. *Let G and H be non-empty connected graphs such that G has a c-deletion set of size at most k. If there is a locally surjective homomorphism ϕ from G to H, then there is a set $D \subseteq D_G^{k+c}$ and a locally surjective homomorphism ϕ_P from $G[D]$ to $H[D_H^{k+c}]$ such that ϕ augments ϕ_P. If ϕ is locally bijective, then $D = D_G^{k+c}$ and ϕ_P is a locally bijective homomorphism.*

Proof. By Lemma 2, D_G^{k+c} is a $kc(k+c)$-deletion set of size at most k. Furthermore, observe that for a locally surjective homomorphism ϕ from G to H, the inequality $\deg_G(v) \geq \deg_H(\phi(v))$ holds for every $v \in V(G)$ ($\deg_G(v) = \deg_H(\phi(v))$ holds in the locally bijective case). Since ϕ is surjective by Observation 1, this implies that $\phi(D_G^{k+c}) \supseteq D_H^{k+c}$ (and if ϕ is locally bijective, then $\phi(D_G^{k+c}) = D_H^{k+c}$). By Lemma 1, $\phi(D_G^{k+c})$ is a $kc(k+c)$-deletion set for H. Let $D = \phi^{-1}(D_H^{k+c})$, so $D \subseteq D_G^{k+c}$ (note that $D = D_G^{k+c}$ if ϕ is locally bijective). Now $\phi|_D$ is a surjective map from D to D_H^{k+c}. Furthermore, $\phi(D_G^{k+c} \setminus D) \cap \phi(D) = \phi(D_G^{k+c} \setminus D) \cap D_H^{k+c} = \emptyset$. Moreover, for every $v \in V(G) \setminus D_G^{k+c}$, $\phi(v) \notin D_H^{k+c} = \phi|_D(D)$, since $\deg_G(v) \geq \deg_H(\phi(v))$. Furthermore, $\phi|_D$ is a homomorphism from $G[D]$ to $H[D_H^{k+c}]$ because ϕ is a homomorphism. We argue that $\phi|_D$ is locally surjective (respectively bijective) by contradiction. Suppose $\phi|_D$ is not locally surjective. Then there is a vertex $u \in D$ and a neighbour $v \in D_H^{k+c}$ of $\phi|_D(u)$ such that $v \notin \phi|_D(N_G(u) \cap D)$. Since ϕ is locally surjective, there must be $w \in N_G(u) \setminus D$ such that $\phi(w) = v$. This contradicts the fact that $\phi(V(G) \setminus D) \cap D_H^{k+c} = \emptyset$. Hence $\phi|_D$ is a locally surjective homomorphism. In the bijective case we just need to additionally observe that $\phi|_D$ restricted to the neighbourhood of any vertex $v \in D$ must be injective. This completes the proof. $\square$

Lemma 4. *Let G and H be non-empty connected graphs and let k, c be non-negative integers. For any $D \subseteq D_G^{k+c}$, we can compute the set Φ_D of all locally surjective (respectively bijective) homomorphisms ϕ_P from $G[D]$ to $H[D_H^{k+c}]$ in $\mathcal{O}(|D|^{|D|+2})$ time. Furthermore, $|\Phi_D| \leq |D|^{|D|}$.*

ILP Formulation. We will show how to formulate the subproblem obtained in the previous subsection in terms of an ILP instance. More specifically, we will show that the following problem can be formulated in terms of an ILP: given a partial locally surjective (respectively bijective) homomorphism ϕ_P from some induced subgraph D_G of G to some induced subgraph D_H of H, can this be augmented to a locally surjective (respectively bijective) homomorphism from G

to H? Moreover, we will actually show that for this to work, the host graph H does not need to be given explicitly, but can instead be defined by a certain system of linear constraints.

The main ideas behind our translation to ILP are as follows. Suppose that there is a locally surjective (respectively bijective) homomorphism ϕ from G to H that augments ϕ_P. Because ϕ augments ϕ_P, Observation 2 implies that ϕ maps every component C_G of $G \setminus V(D_G)$ entirely to some component C_H of $H \setminus V(D_H)$, moreover, $\phi|_{V(D_G) \cup V(C_G)}$ is already locally surjective (respectively bijective) for every vertex $v \in V(C_G)$. Our aim now is to describe ϕ in terms of its parts consisting of locally surjective (respectively bijective) homomorphisms from *extensions* of D_G in G, i.e. sets of components of $G \setminus D_G$ plus D_G, to *simple extensions* of D_H in H, i.e. single components of $H \setminus D_H$ plus D_H. Note that the main difficulty comes from the fact that we need to ensure that ϕ is locally surjective (respectively bijective) for every $d \in D_G$ and not only for the vertices within the components of $G \setminus D_G$. This is why we need to describe the parts of ϕ using sets of components of $G \setminus D_G$ and not just single components. However, as we will show, it will suffice to consider only minimal extensions of D_G in G, where an extension is minimal if no subset of it allows for a locally surjective (respectively bijective) homomorphism from it to some simple extension of D_H in H. The fact that we only need to consider minimal extensions is important for showing that we can compute the set of all possible parts of ϕ efficiently. Having shown this, we can create an ILP that has one variable $x_{\text{Ext}_G \text{Ext}_H}$ for every minimal extension Ext_G and every simple extension Ext_H such that there is a locally surjective (respectively bijective) homomorphism from Ext_G to Ext_H that augments ϕ_P. The value of the variable $x_{\text{Ext}_G \text{Ext}_H}$ now corresponds to the number of parts used by ϕ that map minimal extensions isomorphic to Ext_G to simple extensions isomorphic to Ext_H that augment ϕ_P. We can then use linear constraints on these variables to ensure that:

(SB2') H contains exactly the right number of extensions isomorphic to Ext_H required by the assignment for $x_{\text{Ext}_G \text{Ext}_H}$,
(B1') G contains exactly the right number of minimal extensions isomorphic to Ext_G required by the assignment for $x_{\text{Ext}_G \text{Ext}_H}$ (if ϕ is locally bijective),
(S1') G contains at least the number of minimal extensions isomorphic to Ext_G required by the assignment for $x_{\text{Ext}_G \text{Ext}_H}$ (if ϕ is locally surjective),
(S3') for every simple extension Ext_G of G that is not yet used in any part of ϕ, there is a homomorphism from Ext_G to some simple extension of D_H in H that augments ϕ_P and is locally surjective for every vertex in $\text{Ext}_G \setminus D_G$ (if ϕ is locally surjective).

Together, these constraints ensure that there is a locally surjective (respectively bijective) homomorphism ϕ from G to H that augments ϕ_P. To do so, we need the following additional notation.

Given a graph D, an *extension* for D is a graph E containing D as an induced subgraph. It is *simple* if $E \setminus D$ is connected, and *complex* in general. Given two extensions $\text{Ext}_1, \text{Ext}_2$ of D, we write $\text{Ext}_1 \sim_D \text{Ext}_2$ if there is an isomorphism τ from Ext_1 to Ext_2 with $\tau(d) = d$ for every $d \in D$. Then $\sim_D$ is an equivalence relation. Let the *types* of D, denoted $\mathcal{T}_D$, be the set of equivalence classes of $\sim_D$ of simple extensions of D. We write $\mathcal{T}_D^c$ to denote the set of types of D of size at most $|D| + c$, so $|\mathcal{T}_D^c| \leq (|D| + c)2^{\binom{|D|+c}{2}}$.

Given a complex extension E of D, let C be a connected component of $E \setminus D$. Then C has type $T \in \mathcal{T}_D$ if $E[D \cup C] \sim_D T$ (depending on the context, we also say that the extension $E[D \cup C]$ has type T). The *type-count* of E is the function $\mathrm{tc}_E : \mathcal{T}_D \to \mathbb{N}$ such that $\mathrm{tc}_E(T)$ for $T \in \mathcal{T}_D$ is the number of connected components of $E \setminus D$ with type T (in particular if E is simple, the type-count is 1 for E and 0 for other types). Note that two extensions are equivalent if and only if they have the same type-counts; this then also implies that there is an isomorphism τ between the two extensions satisfying $\tau(d) = d$ for every $d \in D$. We write $E \preceq E'$ if $\mathrm{tc}_E(T) \leq \mathrm{tc}_{E'}(T)$ for all types $T \in \mathcal{T}_D$. If E is an extension of D, we write $\mathcal{T}_D(E) = \{T \in \mathcal{T}_D \mid \mathrm{tc}_E(T) \geq 1\}$ for the *set of types of E* and $\mathcal{E}_D(E)$ for the set of simple extensions of E. Moreover, for $T \in \mathcal{T}_D$, we write $\mathcal{E}_D(E, T)$ for the set of simple extensions in E having type T.

A *target description* is a tuple (D_H, c, CH) where D_H is a graph, c is an integer and CH is a set of linear constraints over variables x_T, $T \in \mathcal{T}_{D_H}^c$. A type-count for D_H is an integer assignment of the variables x_T. A graph H satisfies the target description (D_H, c, CH) if it is an extension of D_H, $\mathrm{tc}_H(T) = 0$ for $T \notin \mathcal{T}_{D_H}^c$, and setting $x_T = \mathrm{tc}_H(T)$ for all $T \in \mathcal{T}_{D_H}^c$ satisfies all constraints in CH.

In what follows, we assume that the following are given: the graphs D_G, D_H, an extension G of D_G, a target description $\mathcal{D} = (D_H, c, \mathrm{CH})$, and a locally surjective (respectively bijective) homomorphism $\phi_P : D_G \to D_H$. Let Ext_G be an extension of D_G with $\mathrm{Ext}_G \preceq G$ and let $T_H \in \mathcal{T}_{D_H}^c$; note that we only consider $T_H \in \mathcal{T}_{D_H}^c$, because we assume that T_H is a type of a simple extension of a graph H that satisfies the target description $\mathcal{D}$. We say that Ext_G can be *weakly ϕ_P-S-mapped* to a type T_H if there exists an augmentation $\phi : \mathrm{Ext}_G \to T_H$ of ϕ_P such that ϕ is locally surjective for every $v \in \mathrm{Ext}_G \setminus D_G$. We say that Ext_G can be *ϕ_P-S-mapped* (respectively *ϕ_P-B-mapped*) to a type T_H if there exists an augmentation $\phi : \mathrm{Ext}_G \to T_H$ of ϕ_P such that ϕ is locally surjective (respectively locally bijective). Furthermore, Ext_G can be *minimally ϕ_P-S-mapped* (respectively *minimally ϕ_P-B-mapped*) to T_H if Ext_G can be ϕ_P-S-mapped (respectively ϕ_P-B-mapped) to T_H and no other extension Ext_G' with $\mathrm{Ext}_G' \preceq \mathrm{Ext}_G$ can be ϕ_P-S-mapped (respectively ϕ_P-B-mapped) to T_H. Let $\mathrm{wSM}(G, D_G, \mathcal{D}, \phi_P)$ be the set of all pairs (T_G, T_H) such that $T_G \in \mathcal{T}_{D_G}(G)$ can be weakly ϕ_P-S-mapped to T_H. Let $\mathrm{SM}(G, D_G, \mathcal{D}, \phi_P)$ be the set of all pairs (Ext_G, T_H) with $\mathrm{Ext}_G \preceq G$, $T_H \in \mathcal{T}_{D_H}^c$ such that Ext_G can be minimally ϕ_P-S-mapped to T_H and let $\mathrm{BM}(G, D_G, \mathcal{D}, \phi_P)$ be the set of all pairs (Ext_G, T_H) with $\mathrm{Ext}_G \preceq G$, $T_H \in \mathcal{T}_{D_H}^c$ such that Ext_G can be minimally ϕ_P-B-mapped to T_H.

We now build a set of linear constraints. To this end, besides variables x_T for $T \in \mathcal{T}_H$, we introduce variables $x_{\mathrm{Ext}_G T_H}$ for each $(\mathrm{Ext}_G, T_H) \in \mathrm{SM}$ (respectively BM), where here and in what follows $\mathrm{wSM} = \mathrm{wSM}(G, D_G, \mathcal{D}, \phi_P)$, $\mathrm{SM} = \mathrm{SM}(G, D_G, \mathcal{D}, \phi_P)$ and $\mathrm{BM} = \mathrm{BM}(G, D_G, \mathcal{D}, \phi_P)$.

(S1) $\sum_{(\mathrm{Ext}_G, T_H) \in \mathrm{SM}} \mathrm{tc}_{\mathrm{Ext}_G}(T_G) * x_{\mathrm{Ext}_G T_H} \leq \mathrm{tc}_G(T_G)$ for every $T_G \in \mathcal{T}_{D_G}(G)$,

(B1) $\sum_{(\mathrm{Ext}_G, T_H) \in \mathrm{BM}} \mathrm{tc}_{\mathrm{Ext}_G}(T_G) * x_{\mathrm{Ext}_G T_H} = \mathrm{tc}_G(T_G)$ for every $T_G \in \mathcal{T}_{D_G}(G)$,

(S2) $\sum_{\mathrm{Ext}_G:(\mathrm{Ext}_G, T_H) \in \mathrm{SM}} x_{\mathrm{Ext}_G, T_H} = x_{T_H}$ for every $T_H \in \mathcal{T}_{D_H}$,

(B2) $\sum_{\mathrm{Ext}_G:(\mathrm{Ext}_G, T_H) \in \mathrm{BM}} x_{\mathrm{Ext}_G, T_H} = x_{T_H}$ for every $T_H \in \mathcal{T}_{D_H}$,

(S3) $\sum_{(T_G, T_H) \in \mathrm{wSM}} x_{T_H} \geq 1$ for every $T_G \in \mathcal{T}_{D_G}(G)$.

Lemma 5. *Let D_G and D_H be graphs, let G be an extension of D_G and let $\mathcal{D} = (D_H, c, \mathrm{CH})$ be a target description. Moreover, let $\phi_P : V(D_G) \rightarrow V(D_H)$ be a locally surjective (respectively bijective) homomorphism from D_G to D_H. There exists a graph H satisfying $\mathcal{D}$ and a locally surjective (respectively bijective) homomorphism ϕ augmenting ϕ_P if and only if the equation system (CH, S1, S2, S3) (respectively (CH, B1, B2)) admits a solution.*

Constructing and Solving the ILP. We show the following theorem.

Theorem 3. *Let G be a graph, let D_G be a (k, c)-extended deletion set (respectively a c-deletion set) of size at most k for G, let $\mathcal{D} = (D_H, c', \mathrm{CH})$ be a target description and let $\phi_P : D_G \rightarrow D_H$ be a locally surjective (respectively bijective) homomorphism from D_G to D_H. Then, deciding whether there is a locally surjective (respectively bijective) homomorphism that augments ϕ_P from G to any graph satisfying CH is fpt parameterized by $k + c + c'$.*

To prove Theorem 3, we need to show that we can construct and solve the ILP instance given in the previous section. The main ingredient for the proof of Theorem 3 is Lemma 7, which shows that we can efficiently compute the sets wSM, SM, and BM. A crucial insight for its proof is that if $(\mathrm{Ext}_G, \mathrm{Ext}_H) \in$ SM (or $(\mathrm{Ext}_G, \mathrm{Ext}_H) \in$ BM), then Ext_G consists of only boundedly many (in terms of some function of the parameters) components, which will allow us to enumerate all possibilities for Ext_G in fpt-time. We start by showing that the set $\mathcal{T}_{D_G}(G)$ can be computed efficiently and has small size.

Lemma 6. *Let G be a graph and let D_G be a (k, c)-extended deletion set of size at most k for G. Then, $\mathcal{T}_{D_G}(G)$ has size at most $k + (|D_G| + c)2^{\binom{|D_G|+c}{2}}$ and computing $\mathcal{T}_{D_G}(G)$ and tc_G is fpt parameterized by $|D_G| + k + c$.*

Lemma 7. *Let G be a graph, let D_G be a (k, c)-extended deletion set (respectively a c-deletion set) of size at most k for G, let $\mathcal{D} = (D_H, c', \mathrm{CH})$ be a target description and let ϕ_P be a locally surjective (respectively bijective) homomorphism from D_G to D_H. Then, the sets wSM $= \mathrm{wSM}(G, D_G, \mathcal{D}, \phi_P)$ and SM $= \mathrm{SM}(G, D_G, \mathcal{D}, \phi_P)$ (respectively the set BM $= \mathrm{BM}(G, D_G, \mathcal{D}, \phi_P)$) can be computed in fpt-time parameterized by $k + c + c'$ and $|\mathrm{SM}|$ (respectively $|\mathrm{BM}|$) is bounded by a function depending only on $k + c + c'$. Moreover, the number of variables in the equation system (CH, S1, S2, S3) (respectively (CH, B1, B2)) is bounded by a function depending only on $k + c + c'$.*

4   Applications of Our Algorithmic Framework

Here we show the main results of our paper, which are simple applications of our framework from the previous section. Our first result implies that LSHOM and LBHOM are fpt parameterized by the fracture number of the guest graph.

Theorem 4. LSHOM *and* LBHOM *are fpt parameterized by* $k+c$, *where* k *and* c *are such that the guest graph* G *has a* c-*deletion set of size at most* k.

Proof. Let G and H be non-empty connected graphs such that G has a c-deletion set of size at most k. Let $D_H = H[D_H^{k+c}]$. We first verify whether H has a c-deletion set of size at most k using Proposition 1. Because of Lemma 1, we can return that there is no locally surjective (and therefore also no bijective) homomorphism from G to H if this is not the case. Therefore, we can assume in what follows that H also has a c-deletion set of size at most k, which together with Lemma 2 implies that $V(D_H)$ is a $kc(k+c)$-deletion set of size at most k for H. Therefore, using Lemma 6, we can compute tc_H in fpt-time parameterized by $k + c$. This now allows us to obtain a target description $\mathcal{D} = (D_H, c', \mathrm{CH})$ with $c' = kc(k+c)$ for H, i.e. $\mathcal{D}$ is satisfied only by the graph H, by adding the constraint $x_T = \mathrm{tc}_H(T_H)$ to CH for every simple extension type $T_H \in \mathcal{T}_{D_H}^{c'}$; note that $\mathcal{T}_{D_H}^{c'}$ can be computed in fpt-time parameterized by $k + c$ by Lemma 6.

Because of Lemma 3, we obtain that there is a locally surjective (respectively bijective) homomorphism ϕ from G to H if and only if there is a set $D \subseteq D_G^{k+c}$ and a locally surjective (respectively bijective) homomorphism ϕ_P from $D_G = G[D]$ to D_H such that ϕ augments ϕ_P. Therefore, we can solve LSHOM by checking, for every $D \subseteq D_G^{k+c}$ and every locally surjective homomorphism ϕ_P from $D_G = G[D]$ to D_H, whether there is a locally surjective homomorphism from G to H that augments ϕ_P. Note that there are at most 2^k subsets D and because of Lemma 4, we can compute the set Φ_D for every such subset in $\mathcal{O}(k^{k+2})$ time. Furthermore, due to Lemma 2, D is a $(k - |D|, c)$-extended deletion set of size at most k for G. Therefore, for every $D \subseteq D_G^{k+c}$ and $\phi_p \in \Phi_D$, we can use Theorem 3 to decide in fpt-time parameterized by $k+c$ (because $c' = kc(k+c)$), if there is a locally surjective (resp. bijective) homomorphism from G to a graph satisfying $\mathcal{D}$ that augments ϕ_P. As H is the only graph satisfying $\mathcal{D}$, we proved the theorem.   □

The proof of our next theorem is similar to that of Theorem 4. The difference is that H is not given. Instead, we use Theorem 3 for a selected set of target descriptions. Each target description enforces that graphs satisfying it have to be connected and have precisely h vertices, where h is part of the input for ROLE ASSIGNMENT. We ensure that every graph H satisfying the requirements of ROLE ASSIGNMENT satisfies at least one of the selected target descriptions. The size of the set of considered target descriptions depends only on c and k, as it suffices to consider any small graph D_H and types of small simple extensions of D_H.

Theorem 5. ROLE ASSIGNMENT *is fpt parameterized by* $k + c$, *where* k *and* c *are such that* G *has a* c-*deletion set of size at most* k.

We also obtain the following dichotomy, where the $c = 1$, $k \geq 1$ case (vertex cover number case) follows from our ILP framework: we first find, in XP time, a partial mapping from a vertex cover of the host graph G to the guest graph H and then use our ILP framework to map the remaining vertices in FPT-time.

Theorem 6. *Let* $c, k \geq 1$*. Then* LIHOM *is polynomial-time solvable on guest graphs with a c-deletion set of size at most* k *if either* $c = 1$ *and* $k \geq 1$ *or* $c = 2$ *and* $k = 1$*; otherwise, it is* NP-*complete.*

5 Conclusions

We aim to extend our ILP-based framework. If successful, this will then also enable us to address the parameterized complexity of other graph homomorphism variants such as quasi-covers [28] and pseudo-covers [10,12,13]. We also recall the open problem from [14]: are LBHOM and LSHOM in FPT when parameterized by the treewidth of the guest graph plus the maximum degree of the guest graph?

References

1. Abello, J., Fellows, M.R., Stillwell, J.: On the complexity and combinatorics of covering finite complexes. Australas. J. Comb. **4**, 103–112 (1991)
2. Angluin, D.: Local and global properties in networks of processors (extended abstract). Proc. STOC **1980**, 82–93 (1980)
3. Angluin, D., Gardiner, A.: Finite common coverings of pairs of regular graphs. J. Comb. Theory Ser. B **30**, 184–187 (1981)
4. Biggs, N.J.: Algebraic Graph Theory. Cambridge University Press, Cambridge (1974)
5. Biggs, N.J.: Constructing 5-arc transitive cubic graphs. J. Lond. Math. Soc. **II**(26), 193–200 (1982)
6. Bílka, O., Lidický, B., Tesař, M.: Locally injective homomorphism to the simple weight graphs. In: Ogihara, M., Tarui, J. (eds.) TAMC 2011. LNCS, vol. 6648, pp. 471–482. Springer, Heidelberg (2011). https://doi.org/10.1007/978-3-642-20877-5_46
7. Bodlaender, H.L.: The classification of coverings of processor networks. J. Parallel Distrib. Comput. **6**, 166–182 (1989)
8. Bodlaender, H.L., Gilbert, J.R., Hafsteinsson, H., Kloks, T.: Approximating treewidth, pathwidth, frontsize, and shortest elimination tree. J. Algorithms **18**, 238–255 (1995)
9. Bok, J., Fiala, J., Hlinený, P., Jedličková, N., Kratochvíl, J.: Computational complexity of covering multigraphs with semi-edges: small cases. In: Proceedings of MFCS 2021. LIPIcs, vol. 202, pp. 21:1–21:15 (2021)
10. Chalopin, J.: Local computations on closed unlabelled edges: the election problem and the naming problem. In: Vojtáš, P., Bieliková, M., Charron-Bost, B., Sýkora, O. (eds.) SOFSEM 2005. LNCS, vol. 3381, pp. 82–91. Springer, Heidelberg (2005). https://doi.org/10.1007/978-3-540-30577-4_11
11. Chalopin, J., Métivier, Y., Zielonka, W.: Local computations in graphs: the case of cellular edge local computations. Fund. Inform. **74**, 85–114 (2006)
12. Chalopin, J., Paulusma, D.: Graph labelings derived from models in distributed computing: a complete complexity classification. Networks **58**, 207–231 (2011)
13. Chalopin, J., Paulusma, D.: Packing bipartite graphs with covers of complete bipartite graphs. Discret. Appl. Math. **168**, 40–50 (2014)

14. Chaplick, S., Fiala, J., van 't Hof, P., Paulusma, D., Tesař, M.: Locally constrained homomorphisms on graphs of bounded treewidth and bounded degree. Theor. Comput. Sci. **590**, 86–95 (2015)
15. Chekuri, C., Rajaraman, A.: Conjunctive query containment revisited. Theoret. Comput. Sci. **239**, 211–229 (2000)
16. Dourado, M.C.: Computing role assignments of split graphs. Theoret. Comput. Sci. **635**, 74–84 (2016)
17. Downey, R.G., Fellows, M.R.: Fixed-parameter tractability and completeness II: on completeness for W[1]. Theoret. Comput. Sci. **141**, 109–131 (1995)
18. Drange, P.G., Dregi, M.S., van 't Hof, P.: On the computational complexity of vertex integrity and component order connectivity. Algorithmica **76**, 1181–1202 (2016)
19. Dvořák, P., Eiben, E., Ganian, R., Knop, D., Ordyniak, S.: Solving integer linear programs with a small number of global variables and constraints. Proc. IJCAI **2017**, 607–613 (2017)
20. Everett, M.G., Borgatti, S.P.: Role colouring a graph. Math. Soc. Sci. **21**, 183–188 (1991)
21. Fellows, M.R., Lokshtanov, D., Misra, N., Rosamond, F.A., Saurabh, S.: Graph layout problems parameterized by vertex cover. In: Hong, S.-H., Nagamochi, H., Fukunaga, T. (eds.) ISAAC 2008. LNCS, vol. 5369, pp. 294–305. Springer, Heidelberg (2008). https://doi.org/10.1007/978-3-540-92182-0_28
22. Fiala, J., Kloks, T., Kratochvíl, J.: Fixed-parameter complexity of lambda-labelings. Discret. Appl. Math. **113**, 59–72 (2001)
23. Fiala, J., Kratochvíl, J.: Partial covers of graphs. Discuss. Math. Graph Theory **22**, 89–99 (2002)
24. Fiala, J., Kratochvíl, J.: Locally constrained graph homomorphisms - structure, complexity, and applications. Comput. Sci. Rev. **2**, 97–111 (2008)
25. Fiala, J., Kratochvíl, J., Pór, A.: On the computational complexity of partial covers of theta graphs. Discret. Appl. Math. **156**, 1143–1149 (2008)
26. Fiala, J., Paulusma, D.: A complete complexity classification of the role assignment problem. Theoret. Comput. Sci. **349**, 67–81 (2005)
27. Fiala, J., Paulusma, D.: Comparing universal covers in polynomial time. Theory Comput. Syst. **46**, 620–635 (2010)
28. Fiala, J., Tesař, M.: Dichotomy of the H-quasi-cover problem. In: Bulatov, A.A., Shur, A.M. (eds.) CSR 2013. LNCS, vol. 7913, pp. 310–321. Springer, Heidelberg (2013). https://doi.org/10.1007/978-3-642-38536-0_27
29. Frank, A., Tardos, É.: An application of simultaneous diophantine approximation in combinatorial optimization. Combinatorica **7**(1), 49–65 (1987)
30. Freuder, E.C.: Complexity of k-tree structured constraint satisfaction problems. Proc. AAAI **1990**, 4–9 (1990)
31. Grohe, M.: The complexity of homomorphism and constraint satisfaction problems seen from the other side. J. ACM **54**, 1:1-1:24 (2007)
32. Heggernes, P., van 't Hof, P., Paulusma, D.: Computing role assignments of proper interval graphs in polynomial time. J. Discret. Algorithms **14**, 173–188 (2012)
33. Hell, P., Nešetřil, J.: On the complexity of H-coloring. J. Comb. Theory Ser. B **48**, 92–110 (1990)
34. Hell, P., Nešetřil, J.: Graphs and Homomorphisms. Oxford University Press, Oxford (2004)
35. van 't Hof, P., Paulusma, D., van Rooij, J.M.M.: Computing role assignments of chordal graphs. Theoret. Comput. Sci. **411**, 3601–3613 (2010)

36. Kannan, R.: Minkowski's convex body theorem and integer programming. Math. Oper. Res. **12**(3), 415–440 (1987)
37. Kratochvíl, J.: Regular codes in regular graphs are difficult. Discret. Math. **133**, 191–205 (1994)
38. Kratochvíl, J., Proskurowski, A., Telle, J.A.: Covering regular graphs. J. Comb. Theory Ser. B **71**, 1–16 (1997)
39. Kratochvíl, J., Proskurowski, A., Telle, J.A.: On the complexity of graph covering problems. Nordic J. Comput. **5**, 173–195 (1998)
40. Kratochvíl, J., Telle, J.A., Tesař, M.: Computational complexity of covering three-vertex multigraphs. Theoret. Comput. Sci. **609**, 104–117 (2016)
41. Kristiansen, P., Telle, J.A.: Generalized H-coloring of graphs. In: Goos, G., Hartmanis, J., van Leeuwen, J., Lee, D.T., Teng, S.-H. (eds.) ISAAC 2000. LNCS, vol. 1969, pp. 456–466. Springer, Heidelberg (2000). https://doi.org/10.1007/3-540-40996-3_39
42. Kronegger, M., Ordyniak, S., Pfandler, A.: Backdoors to planning. Artif. Intell. **269**, 49–75 (2019)
43. Lenstra, H.W., Jr.: Integer programming with a fixed number of variables. Math. Oper. Res. **8**(4), 538–548 (1983)
44. Lidický, B., Tesař, M.: Complexity of locally injective homomorphism to the theta graphs. In: Iliopoulos, C.S., Smyth, W.F. (eds.) IWOCA 2010. LNCS, vol. 6460, pp. 326–336. Springer, Heidelberg (2011). https://doi.org/10.1007/978-3-642-19222-7_33
45. Massey, W.S.: Algebraic Topology: An Introduction. Harcourt, Brace and World (1967)
46. Nešetřil, J.: Homomorphisms of derivative graphs. Discret. Math. **1**, 257–268 (1971)
47. Nešetřil, J., Ossona de Mendez, P.: Sparsity: Graphs, Structures, and Algorithms, Algorithms and Combinatorics, vol. 28. Springer, Heidelberg (2012). https://doi.org/10.1007/978-3-642-27875-4
48. Okrasa, K., Rzążewski, P.: Subexponential algorithms for variants of the homomorphism problem in string graphs. J. Comput. Syst. Sci. **109**, 126–144 (2020)
49. Pandey, S., Sahlot, V.: Role coloring bipartite graphs. CoRR abs/2102.01124 (2021)
50. Pekeč, A., Roberts, F.S.: The role assignment model nearly fits most social networks. Math. Soc. Sci. **41**, 275–293 (2001)
51. Purcell, C., Rombach, M.P.: On the complexity of role colouring planar graphs, trees and cographs. J. Discret. Algorithms **35**, 1–8 (2015)
52. Purcell, C., Rombach, M.P.: Role colouring graphs in hereditary classes. Theoret. Comput. Sci. **876**, 12–24 (2021)
53. Roberts, F.S., Sheng, L.: How hard is it to determine if a graph has a 2-role assignment? Networks **37**, 67–73 (2001)
54. White, D.R., Reitz, K.P.: Graph and semigroup homomorphisms on networks of relations. Soc. Netw. **5**, 193–235 (1983)

s-Club Cluster Vertex Deletion on Interval and Well-Partitioned Chordal Graphs

Dibyayan Chakraborty[1] , L. Sunil Chandran[2] , Sajith Padinhatteeri[3] ,
and Raji R. Pillai[2(✉)]

[1] Laboratoire de l'Informatique du Parallélisme, ENS de Lyon, Lyon, France
[2] Department of Computer Science and Automation, Indian Institute of Science, Bengaluru, India
`rajipillai@iisc.ac.in`
[3] Department of Mathematics, BITS-Pilani, Hyderabad, India

Abstract. In this paper, we study the computational complexity of s-CLUB CLUSTER VERTEX DELETION. Given a graph, s-CLUB CLUSTER VERTEX DELETION (s-CVD) aims to delete the minimum number of vertices from the graph so that each connected component of the resulting graph has a diameter at most s. When $s = 1$, the corresponding problem is popularly known as CLUSTER VERTEX DELETION (CVD). We provide a faster algorithm for s-CVD on *interval graphs*. For each $s \geq 1$, we give an $O(n(n + m))$-time algorithm for s-CVD on interval graphs with n vertices and m edges. In the case of $s = 1$, our algorithm is a slight improvement over the $O(n^3)$-time algorithm of Cao *et al.* (Theor. Comput. Sci., 2018) and for $s \geq 2$, it significantly improves the state-of-the-art running time $\left(O\left(n^4\right)\right)$.

We also give a polynomial-time algorithm to solve CVD on *well-partitioned chordal graphs*, a graph class introduced by Ahn *et al.* (WG 2020) as a tool for narrowing down complexity gaps for problems that are hard on chordal graphs, and easy on split graphs. Our algorithm relies on a characterisation of the optimal solution and on solving polynomially many instances of the WEIGHTED BIPARTITE VERTEX COVER. This generalises a result of Cao *et al.* (Theor. Comput. Sci., 2018) on split graphs. We also show that for any even integer $s \geq 2$, s-CVD is NP-hard on well-partitioned chordal graphs.

1 Introduction

Detecting *"highly-connected"* parts or *"clusters"* of a complex system is a fundamental research topic in network science [28,38] with numerous applications in computational biology [7,12,30,34,35], machine learning [6], image processing [37], etc. In a graph-theoretic approach, a complex system or a network is often viewed as an undirected graph G that consists of a set of *vertices* $V(G)$ representing the atomic entities of the system and a set of *edges* $E(G)$ representing a binary relationship among the entities. A *cluster* is often viewed as a dense subgraph (often a *clique*) and *partitioning* a graph into such clusters is one of the main objectives of *graph-based data clustering* [7,13,33].

Ben-Dor *et al.* [7] and Shamir *et al.* [33] observed that the clusters of certain networks may be retrieved by making a small number of modifications in the network.

These modifications may be required to account for the errors introduced during the construction of the network. In graph-theoretic terms, the objective is to modify (*e.g.* edge deletion, edge addition, vertex deletion) a given input graph as little as possible so that each component of the resulting graph is a cluster. When deletion of vertices is the only valid operation on the input graph, the corresponding clustering problem falls in the category of *vertex deletion* problems, a core topic in algorithmic graph theory. Many classic optimization problems like MAXIMUM CLIQUE, MAXIMUM INDEPENDENT SET, VERTEX COVER are examples of vertex deletion problems. In this paper, we study popular vertex deletion problems called CLUSTER VERTEX DELETION and its generalisation s-CLUB CLUSTER VERTEX DELETION, both being important in the context of graph-based data clustering.

Given a graph G, the objective of CLUSTER VERTEX DELETION (CVD) is to delete a minimum number of vertices so that the remaining graph is a set of disjoint cliques. Below we give a formal definition of CVD.

CLUSTER VERTEX DELETION (CVD)

Input: An undirected graph G, and an integer k.

Output: YES, if there is a set S of vertices with $|S| \leq k$, such that each component of the graph induced by $V(G) \setminus S$ is a clique. NO, otherwise.

The term CLUSTER VERTEX DELETION was coined by Gramm *et al.* [19] in 2004. However NP-hardness of CVD, even on planar graphs and bipartite graphs, follows from the seminal works of Yannakakis [39] and Lewis & Yannakakis [24] from four decades ago. Since then many researchers have proposed *parameterized algorithms* and *approximation algorithms* for CVD on general graphs [4,9,15–18,20,31,36,40]. In this paper, we focus on polynomial-time solvability of CVD on special classes of graphs.

Cao *et al.* [10] gave polynomial-time algorithms for CVD on *interval* graphs (see Definition 2) and *split* graphs. Chakraborty *et al.* [11] gave a polynomial-time algorithm for CVD on *trapezoid* graphs. However, much remains unknown: Chakraborty *et al.* [11] pointed out that computational complexity of CVD on *planar bipartite* graphs and *cocomparability* graphs is unknown. Cao *et al.* [10] asked if CVD can be solved on chordal graphs (graphs with no induced cycle of length greater than 3) in polynomial-time. Ahn *et al.* [1] introduced *well-partitioned chordal* graphs (see Definition 1) as a tool for narrowing down complexity gaps for problems that are hard on chordal graphs, and easy on split graphs. Since several problems (for example: transversal of longest paths and cycles, tree 3-spanner problem, geodetic set problem) which are either hard or open on chordal graphs become polynomial-time solvable on well-partitioned chordal graphs [2], the computational complexity of CVD on well-partitioned chordal graphs is a well-motivated open question.

In this paper, we also study a generalisation of CVD known as s-CLUB CLUSTER VERTEX DELETION (s-CVD). In many applications the equivalence of cluster and clique is too restrictive [3,5,29]. For example, in protein networks where proteins are the vertices and the edges indicate the interaction between the proteins, a more appropriate notion of clusters may have a diameter of more than 1 [5]. Therefore researchers have defined the notion of *s-clubs* [5,26]. An s-club is a graph with *diameter* at most s. The objective of s-CLUB CLUSTER VERTEX DELETION (s-CVD) is to delete the minimum number of vertices from the input graph so that all connected components of the resultant graph is an s-club. Below we give a formal definition of s-CVD.

---

s-CLUB CLUSTER VERTEX DELETION (s-CVD)
Input: An undirected graph G, and integers k and s.
Output: YES, if there is a set S of vertices with $|S| \leq k$, such that each component of the graph induced by $V(G) \setminus S$ has diameter at most s. NO, otherwise.

Schäfer [32] introduced the notion of s-CVD and gave a polynomial-time algorithm for s-CVD on trees. Researchers have studied the particular case of 2-CVD as well [14, 25]. In general, s-CVD remains NP-hard on planar bipartite graphs for each $s \geq 2$, APX-hard on split graphs for $s = 2$ [11] (contrasting the polynomial-time solvability of CVD on split graphs). Combination of the ideas of Cao et al. [10] and Schäfer [32], provides an $O(n^8)$-time algorithm for s-CVD on a trapezoid graphs (intersection graphs of trapezoids between two horizontal lines) with n vertices [11]. This algorithm can be modified to give an $O(n^4)$-time algorithm for s-CVD on interval graphs with n vertices.

General Notations: For a graph G, let $V(G)$ and $E(G)$ denote the set of vertices and edges, respectively. For a vertex $v \in V(G)$, the set of vertices adjacent to v is denoted by $N(v)$ and $N[v] = N(v) \cup \{v\}$. For $S \subseteq V(G)$, let $G - S$ be an induced graph obtained by deleting the vertices in S from G. For two sets S_1, S_2, let $S_1 - S_2$ denotes the set obtained by deleting the elements of S_2 from S_1. The set $S_1 \Delta S_2$ denotes $(S_1 \cup S_2) - (S_1 \cap S_2)$.

2   Our Contributions

In this section, we state our results formally. We start with the definition of well-partitioned chordal graphs as given in [1].

Definition 1 ([1]). *A connected graph G is a well-partitioned chordal graph if there exists a partition $\mathscr{P}$ of $V(G)$ and a tree $\mathscr{T}$ having $\mathscr{P}$ as a vertex set such that the following hold.*

(a) *Each part $X \in \mathscr{P}$ is a clique in G.*
(b) *For each edge $XY \in E(\mathscr{T})$, there exist $X' \subseteq X$ and $Y' \subseteq Y$ such that edge set of the bipartite graph $G[X, Y]$ is $X' \times Y'$.*
(c) *For each pair of distinct $X, Y \in V(\mathscr{T})$ with $XY \notin E(\mathscr{T})$, there is no edge between a vertex in X and a vertex in Y.*

The tree $\mathscr{T}$ is called a partition tree *of G, and the elements of $\mathscr{P}$ are called its* bags *or* nodes *of $\mathscr{T}$.*

Our first result is on CVD for well-partitioned chordal graphs which generalises a result of Cao et al. [10] for split graphs. We prove the following theorem in Sect. 3.

Theorem 1. *Given a well-partitioned chordal graph G and its partition tree, there is an $O(m^2 n)$-time algorithm to solve CVD on G, where n and m are the number of vertices and edges.*

Since a partition tree of a well-partitioned chordal graph can be obtained in polynomial time [1], the above theorem adds CVD to the list of problems that are open on chordal graphs but admits polynomial-time algorithm on well-partitioned chordal graphs. Our algorithm relies on a characterisation of the solution set and we show that the optimal solution of a well-partitioned chordal graph with m edges can be obtained by finding weighted minimum vertex cover [23] of m many weighted bipartite graphs with weights at most n. Then standard *Max-flow* based algorithms [22,23,27] from the literature yields Theorem 1. On the negative side, we prove the following theorem in Sect. 4.

Theorem 2. *Unless the Unique Games Conjecture is false, for any even integer $s \geq 2$, there is no $(2 - \varepsilon)$-approximation algorithm for s-CVD on well-partitioned chordal graphs.*

Our third result is a faster algorithm for s-CVD on *interval graphs*.

Definition 2. *A graph G is an interval graph if there is a collection $\mathscr{I}$ of intervals on the real line such that each vertex of the graph can be mapped to an interval and two intervals intersect if and only if there is an edge between the corresponding vertices in G. The set $\mathscr{I}$ is an interval representation of G*

We prove the following theorem in Sect. 5.

Theorem 3. *For each $s \geq 1$, there is an $O(n(n + m))$-time algorithm to solve s-CVD on interval graphs with n vertices and m edges.*

We note that our techniques deviate significantly from the ones in the previous literature [10, 11, 32] and our result significantly improves the state-of-the-art running time $(O(n^4)$, See [11]) for s-CVD on interval graphs.

3 Polynomial Time Algorithm for CVD on Well-Partitioned Chordal Graphs

In this section, we shall give a polynomial-time algorithm to solve CVD on well-partitioned chordal graphs. We use the following notations extensively in the description of our algorithm and proofs.

Let G be a well-partitioned chordal graph with a partition tree $\mathscr{T}$ rooted at an arbitrary node. For a node X, let $\mathscr{T}_X$ be the subtree rooted at X and G_X be the subgraph of G induced by the vertices in the nodes of $\mathscr{T}_X$. For two adjacent nodes X, Y of $\mathscr{T}$, *the boundary of X with respect to Y* is the set $bd(X,Y) = \{x \in X : N(x) \cap Y \neq \emptyset\}$. For a node X, $P(X)$ denotes the parent of X in $\mathscr{T}$. We denote minimum CVD sets of G_X and $G_X - bd(X,P(X))$ as $OPT(G_X)$ and $OPT(G_X - bd(X,P(X)))$, respectively.

Our dynamic programming-based algorithm traverses $\mathscr{T}$ in a post-order fashion and for each node X of $\mathscr{T}$, computes $OPT(G_X)$ and $OPT(G_X - bd(X,P(X)))$. A set S of vertices is a CVD *set* of G if $G - S$ is disjoint union of cliques. At the heart of our algorithm lies a characterisation of CVD sets of G_X, showing that any CVD set of G_X can be exactly one of two types, namely, X-CVD *set* or (X,Y)-CVD *set* where Y is a child of X (See Definitions 3 and 4). Informally, for a node X, a CVD set is an X-CVD set if it contains X or removing it from G_X creates a cluster all of whose vertices are from X. On the contrary, a CVD set is an (X,Y)-CVD set if its removal creates a cluster intersecting both X and Y, where Y is a child of X. In Lemma 1, we formally show that any CVD set of G_X must be one of the above two types.

Now we introduce some definitions and prove the lemma that facilitates the construction of a polynomial-time algorithm for finding a minimum CVD set of well-partitioned chordal graphs. A *cluster* C of a graph G is a connected component that is isomorphic to a complete graph.

Definition 3. *Let G be a well-partitioned chordal graph, $\mathcal{T}$ be its partition tree, and X be the root node of $\mathcal{T}$. A CVD set S of G is an X-CVD set if either $X \subseteq S$ or $G - S$ contains a cluster $C \subseteq X$.*

Definition 4. *Let G be a well-partitioned chordal graph, $\mathcal{T}$ be its partition tree, X be the root node of $\mathcal{T}$. Let Y be a child of X. A CVD set S is a (X,Y)-CVD set if $G - S$ has a cluster C such that $C \cap X \neq \emptyset$ and $C \cap Y \neq \emptyset$.*

Lemma 1. *Let S be a CVD set of G. Then exactly one of the following holds.*

(a) The set S is a X-CVD set.
(b) There is exactly one child Y of X in $\mathcal{T}$ such that S is an (X,Y)-CVD set of G.

3.1 Finding Minimum X-CVD Sets

Theorem 4. *Let G be a well-partitioned chordal graph rooted at X and $\mathcal{T}$ be a partition tree of G. Assume for each node $Y \in V(\mathcal{T}) - \{X\}$ both $OPT(G_Y)$ and $OPT(G_Y - bd(Y,P(Y)))$ are given, where $P(Y)$ is the parent of Y in $\mathcal{T}$. Then a minimum X-CVD set of G can be computed in $O(|E(G)| \cdot |V(G)|)$ time.*

For the remainder of this section, we denote by G a fixed well-partitioned chordal graph rooted at X with a partition tree $\mathcal{T}$. Let $X_1, X_2, \ldots, X_t$ be the children of X. The main idea behind our algorithm for finding minimum X-CVD set of G is to construct an auxiliary vertex weighted bipartite graph $\mathcal{H}$ with at most $|V(G)|$ vertices such that the (minimum) vertex covers of $\mathcal{H}$ can be used to construct (minimum) X-CVD set. Below we describe the construction of $\mathcal{H}$.

Let $\mathcal{B} = \{bd(X_i,X): i \in [t]\}$. The vertex set of $\mathcal{H}$ is $X \cup \mathcal{B}$ and the edge set of $\mathcal{H}$ is defined as $E(\mathcal{H}) = \{uB: u \in X, B \in \mathcal{B}, \forall v \in B, uv \in E(G)\}$. The weight function on the vertices of $\mathcal{H}$ is defined as follows. For each vertex $u \in X$, define $w(u) = 1$ and for each set $B \in \mathcal{B}$ where $B = bd(X_j,X)$, define $w(B) = |B| + |OPT(G_{X_j} - B)| - |OPT(G_{X_j})|$. Note that, since $B \cup OPT(G_{X_j} - B)$ is a CVD set of G_{X_j}, we have $|OPT(G_{X_j})| \leq |B| + |OPT(G_{X_j} - B)|$ and therefore $w(B) \geq 0$.

Below we show how minimum weighted vertex covers of $\mathcal{H}$ can be used to compute minimum X-CVD set of G. For a vertex cover D of $\mathcal{H}$, define

$$S_1(D) = D \cap X, \quad S_2(D) = \bigcup_{\substack{B \in D \cap \mathcal{B} \\ B = bd(X_i,X)}} B \cup OPT(G_{X_i} - bd(X_i,X))$$

$$S_3(D) = \bigcup_{\substack{B \in \mathcal{B} - D \\ B = bd(X_i,X)}} OPT(G_{X_i}) \text{ and } Sol(D) = S_1(D) \cup S_2(D) \cup S_3(D)$$

Lemma 2. *Let D be a vertex cover of $\mathscr{H}$. Then $Sol(D)$ is an X-CVD set of G.*

A minimum weighted vertex cover D of $\mathscr{H}$ is also *minimal* if no proper subset of D is a vertex cover of $\mathscr{H}$. The restriction of minimality is to avoid the inclusion of redundant vertices with weight 0 in the minimum vertex cover. From now on D denotes a minimal minimum weighted vertex cover of $\mathscr{H}$ and Z denotes a fixed but arbitrary X-CVD set of G. Our goal is to show that $|Sol(D)| \leq |Z|$. We need some more notations and observations.

First we define four sets I_1, I_2, I_3, I_4 as follows. (Recall that $X_1, X_2, \ldots, X_t$ are children of the root X of the partition tree $\mathscr{T}$ of G.)

$$I_1 = \{i \in [t] : bd(X, X_i) \subseteq Sol(D) \text{ and } bd(X, X_i) \subseteq Z\} \tag{1}$$

$$I_2 = \{i \in [t] : bd(X, X_i) \subseteq Sol(D) \text{ and } bd(X, X_i) \not\subseteq Z\} \tag{2}$$

$$I_3 = \{i \in [t] - (I_1 \cup I_2) : bd(X_i, X) \subseteq Sol(D) \text{ and } bd(X_i, X) \subseteq Z\} \tag{3}$$

$$I_4 = \{i \in [t] - (I_1 \cup I_2) : bd(X_i, X) \subseteq Sol(D) \text{ and } bd(X_i, X) \not\subseteq Z\} \tag{4}$$

Note that $I_1 \cup I_2 \cup I_3 \cup I_4 = [t]$ and $(I_1 \cup I_2) \cap (I_3 \cup I_4) = \emptyset$. We have the following observations on the sets $I_i, 1 \leq i \leq 4$.

Observation A. *The sets I_1, I_2, I_3, I_4 form a partition of $[t]$.*

Based on the set I_1, we construct two sets $D_1 \subseteq Sol(D)$ and $Z_1 \subseteq Z$.
$$D_1 = \bigcup_{i \in I_1} bd(X, X_i) \cup (Sol(D) \cap G_{X_i}), \quad Z_1 = \bigcup_{i \in I_1} bd(X, X_i) \cup (Z \cap G_{X_i})$$

Based on the set I_2, we construct the following two sets $D_2 \subseteq Sol(D)$ and $Z_2 \subseteq Z$.
$$D_2 = \bigcup_{i \in I_2} bd(X, X_i) \cup (Sol(D) \cap G_{X_i}) - \bigcup_{i \in I_1} bd(X, X_i)$$
$$Z_2 = \bigcup_{i \in I_2} bd(X_i, X) \cup (Z \cap (G_{X_i} - bd(X_i, X)))$$

Based on the set I_3, we construct the following two sets $D_3 \subseteq Sol(D)$ and $Z_3 \subseteq Z$.
$$D_3 = \bigcup_{i \in I_3} bd(X_i, X) \cup OPT(G_{X_i} - bd(X_i, X))$$
$$Z_3 = \bigcup_{i \in I_3} bd(X_i, X) \cup (Z \cap (G_{X_i} - bd(X_i, X)))$$

Based on the set I_4, we construct the following two sets $D_4 \subseteq Sol(D)$ and $Z_4 \subseteq Z$.
$$D_4 = \bigcup_{i \in I_4} bd(X_i, X) \cup OPT(G_{X_i} - bd(X_i, X))$$
$$Z_4 = \bigcup_{i \in I_4} bd(X, X_i) \cup (Z \cap (G_{X_i})) - \bigcup_{i \in I_1} bd(X, X_i)$$

Observation B. *For each $i \in \{1, 2, 3, 4\}$, $|D_i| \leq |Z_i|$.*

Lemma 3. *$Sol(D) = \bigsqcup_{i=1}^{4} D_i$ and for each $i, j \subset [4]$, $Z_i \cap Z_j = \emptyset$.*

Proof of Theorem 4. Using Lemma 3, we have that $|Sol(D)| \leq |Z_1 \cup Z_2 \cup Z_3 \cup Z_4| \leq |Z|$. Hence, $Sol(D)$ is a minimum X-CVD set of G. Furthermore, $\mathscr{H}$ has at most $|V(G)|$ vertices and $|E(G)|$ edges. Therefore minimum weighted vertex cover of $\mathscr{H}$ can be found in $O(|V(G)| \cdot |E(G)|)$-time and $Sol(D)$ can be computed in total of $O(|V(G)| \cdot |E(G)|)$-time.

3.2   Finding Minimum (X,Y)-CVD Set of Well-partitioned Chordal Graphs

In this section, we prove the following theorem.

Theorem 5. *Let G be a well-partitioned chordal graph; $\mathscr{T}$ be a partition tree of G rooted at X; Y be a child of X. Moreover, for each $Z \in V(\mathscr{T}) - \{X\}$, assume both $OPT(G_Z)$ and $OPT(G_Z - bd(Z,P(Z)))$ are given ($P(Z)$ denotes the parent of Z in $\mathscr{T}$). Then a minimum (X,Y)-CVD set of G can be computed in $O\left(|E(G)|^2.|V(G)|\right)$ time.*

For the remainder of this section, the meaning of G, $\mathscr{T}$, X and Y will be as given in Theorem 5. For an (X,Y)-edge e, we say that a minimum (X,Y)-CVD set A *preserves* the edge e if $G - A$ contains the edge e. Let $e \in E(X,Y)$ be an (X,Y)-edges of G. Then to prove Theorem 5, we use Theorem 6. First we show how to construct a minimum (X,Y)-CVD set S_e that preserves the edge $e \in E(X,Y)$ and prove Theorem 6. Clearly, a minimum (X,Y)-CVD set S of G is the one that satisfies $|S| = \min\limits_{e \in E(X,Y)} |S_e|$. Therefore, Theorem 5 will follow directly from Theorem 6. The remainder of this section is devoted to prove Theorem 6.

Theorem 6. *Assuming the same conditions as in Theorem 5, for $e \in E(X,Y)$, a minimum (X,Y)-CVD set of G that preserves e can be computed in $O\left(|E(G)|.|V(G)|\right)$ time.*

First, we need the following observation about the partition trees of well-partitioned chordal graphs, which is easy to verify.

Observation C. *Let G be a well-partitioned chordal graph with a partition tree $\mathscr{T}$. Let X,Y be two adjacent nodes of $\mathscr{T}$ such that $X \cup Y$ induces a complete subgraph in G and $\mathscr{T}'$ be the tree obtained by contracting the edge XY in $\mathscr{T}$. Now associate the newly created node with the subset of vertices $(X \cup Y)$ and retain all the other nodes of $\mathscr{T}'$ and their associated subsets as in $\mathscr{T}$. Then $\mathscr{T}'$ is also a partition tree of G.*

Now we begin building the machinery to describe our algorithm for finding a minimum (X,Y)-CVD of G that preserves an (X,Y)-edge ab. Observe that any (X,Y)-CVD set that preserves the edge ab must contain the set $(N(a) \triangle N(b))$ as subset. (Otherwise, the connected component of $G - S$ containing ab would not be a cluster, a contradiction).

Let H denote the graph $G - (N(a) \triangle N(b))$. Now consider the partition $\mathscr{Q}$ defined as $\{Z - (N(a) \triangle N(b)) : Z \in V(\mathscr{T})\}$. Now construct a graph $\mathscr{F}$ whose vertex set is $\mathscr{Q}$ and two vertices Z_1, Z_2 are adjacent in $\mathscr{F}$ if there is an edge $uv \in E(H)$ such that $u \in Z_1$ and $v \in Z_2$. Observe that $\mathscr{F}$ is a forest.

Observation D. *There is a bijection f between the connected components of H and the connected components of $\mathscr{F}$, such that for a component C of H, $f(C)$ is the partition tree of C. Moreover, the vertices of the root node of $f(C)$ is subset of a node in $\mathscr{T}$.*

Consider the connected component H^* of H which contains a and b and let $\mathscr{F}' = f(H^*)$ where f is the function given by Observation D. Observe that the root R' of $\mathscr{F}'$ is actually $bd(X,Y)$. Moreover, R' has a child R'' which is actually $bd(Y,X)$. Observe that, $R' \cup R''$ induces a complete subgraph in H^*. Hence, due to Observation C, the tree $\mathscr{F}^*$ obtained by contracting the edge $R'R''$ is a partition tree of H^*. Moreover, $R^* = R' \cup R'' = bd(X,Y) \cup bd(Y,X)$ is the root node of $\mathscr{F}^*$. Recall that our objective is to find a minimum (X,Y)-CVD set that preserves the edge ab. We have the following lemma.

Lemma 4. *Let $H^*, H_1, H_2, \ldots, H_{k'}$ be the connected components of H. Let S^* be a minimum (R^*)-CVD set of H^*, $S_0 = (N(a) \triangle N(b))$, and for each $j \in [k']$, let S_j denote a minimum CVD set of H_j. Then $(S_0 \cup S_1 \cup S_2 \cup \ldots \cup S_{k'} \cup S^*)$ is a minimum (X,Y)-CVD set of G that preserves the edge ab.*

Lemma 4 provides a way to compute a minimum (X,Y)-CVD set of G that preserves the edge ab. Clearly, the set $S_0 = (N(a) \triangle N(b))$ can be computed in polynomial time. The following observation provides a way to compute a minimum CVD set of all connected components that are different from H^*.

Observation E. *Let A be a connected component of H which is different from H^*. Then a minimum CVD set of A can be computed in polynomial time.*

Let $H_1, H_2, \ldots, H_{k'}$ be the connected components of H, all different from H^*. Applying Observation E repeatedly on each component, it is possible to obtain, for each $j \in [k']$, a minimum CVD set S_j of H_j. The following observation provides a way to compute a minimum (R^*)-CVD set of H^*.

Observation F. *Let R be a child of R^* in $\mathscr{F}^*$. Then both $OPT(H_R^*)$ and $OPT(H_R^* - bd(R,R^*))$ are known.*

Due to Observation F and Theorem 4, it is possible to compute a minimum (R^*)-CVD set S^* of H^* in $O(|V(G)| \cdot |E(G)|)$ time. Now due to Lemma 4, we have that $(S_0 \cup S_1 \cup S_2 \cup \ldots \cup S_{k'} \cup S^*)$ is a minimum (X,Y)-CVD set of G that preserves the edge ab. This completes the proof of Theorem 6 and therefore of Theorem 5.

3.3   Main Algorithm

From now on G denote a fixed well-partitioned chordal graph with a partition tree $\mathscr{T}$ whose vertex set is $\mathscr{P}$, a partition of $V(G)$. We will process $\mathscr{T}$ in the post-order fashion and for each node X of $\mathscr{T}$, we give a dynamic programming algorithm to compute both $OPT(G_X)$ and $OPT(G_X - bd(X,P(X)))$ where $P(X)$ is the parent of X (when exists) in $\mathscr{T}$. Due to Observation C, we can assume that $bd(X,P(X)) \subsetneq X$. In the remaining section, X is a fixed node of $\mathscr{T}$, A has a fixed value (which is either $\emptyset$ or $bd(X,P(X))$), G_X^A denotes the graph $G_X - A$. Since well-partitioned chordal graphs are closed under vertex deletion, G_X^A is a well partitioned chordal graph which may be disconnected.

Since the vertices of $X - A$ induces a clique in G_X^A, there exists at most one component G^* in G_X^A that contains a vertex from $X - A$. Due to Observation D there exists a unique connected component $f(G^*) = \mathscr{T}^*$ of $\mathscr{T}'$ which is a partition tree of G^*. Let the remaining connected components of G_X^A be $G_1, G_2, \ldots, G_k$ and for each $i \in [k]$, let $f(G_i) = \mathscr{T}_i$ and X_i is the root of $\mathscr{T}_i$. Let X^* denote the root node of $\mathscr{T}^*$ and $X_1^*, X_2^*, \ldots, X_t^*$ be the children of X^* in $\mathscr{T}^*$. We have the following observation.

Observation G. *For each $j \in [t]$, there is a child Y_j of X in $\mathcal{T}$ such that $Y_j = X_j^*$ and $G_{Y_j} = G_{X_j^*}^*$.*

We have the following lemma.

Lemma 5. $OPT(G_X^A) = \left(\displaystyle\bigsqcup_{i=1}^{k} OPT(G_{X_i})\right) \sqcup OPT(G^*)$

Observe that $OPT(G_{X_i})$ is already known. Due to Lemma 4, any CVD set S of G^* is either a (X^*)-CVD set or there exists a unique child Y of X^*, such that S is a (X^*,Y)-CVD set of G^*. by Theorem 4, it is possible to compute a minimum (R^*)-CVD set S_0 of G^*. Due to Observation G, for any node Y of $\mathcal{T}^*$ which is different from X^*, both $OPT(G_Y)$ and $OPT(G_Y - bd(Y, P(Y)))$ are known, where $P(Y)$ is the parent of Y in $\mathcal{T}^*$. Hence, by Theorem 5 for each child X_i^*, $i \in [t]$, computing a minimum (X^*, X_i^*)-CVD set S_i is possible in $O(|V(G_{X_i^*}^*)| \cdot |E(G_{X_i^*}^*)|)$ time. Let $S^* \in \{S_0, S_1, S_2, \ldots, S_t\}$ be a set with the minimum cardinality. Due to Lemma 1, S^* is a minimum CVD set of G^* that can be obtained in $O(m^2n)$. Finally, due to Lemma 5, we have a minimum CVD set of G_X^A.

4   Hardness for Well-Partitioned Chordal Graphs

Observation H. *Let H be a well-partitioned chordal graph. Let H' be a graph obtained from H by adding a vertex of degree 1. Then H' is an well-partitioned chordal graph.*

Let $s \geq 2$ be an even integer and let $s = 2k$. We shall reduce MINIMUM VERTEX COVER (MVC) on general graphs to s-CVD on well partitioned graphs. Let $\langle G, k \rangle$ be an instance of MINIMUM VERTEX COVER such that maximum degree of G is at most $n - 3$. Let $\overline{G}$ denote the complement of G. Now construct an well-partitioned chordal graph G_{well} from G as follows. For each vertex of $v \in V(G)$, we introduce a new path P_v with $k - 1$ edges and let x_v, x_v' be the endpoints of P_v. For each edge $e \in E(\overline{G})$ we introduce a new vertex y_e in G_{well}. For each pair of edges $e_1, e_2 \in E(\overline{G})$ we introduce an edge between y_{e_1} and y_{e_2} in G_{well}. For each edge $e = uv \in E(\overline{G})$, we introduce the edges $x_u y_e$ and $x_v y_e$ in G_{well}. Observe that $C = \{y_e\}_{e \in E(\overline{G})}$ is a clique, $I = \{x_v\}_{v \in V(G)}$ is an independent set of G_{well}. Therefore $C \cup I$ induces a split graph, say G', in G_{well}. Since G_{well} can be obtained from G' by adding vertices of degree 1, due to Observation H, we have that G_{well} is an well-partitioned chordal graph. We show that G has a vertex cover of size k if and only if G_{well} has a s-CVD set of size k by the following Lemmas.

Lemma 6. *Let D be a subset of I and let $T = \{u \in V(G): x_u \in D\}$. The set D is a s-CVD set of G_{well} if and only if T is a vertex cover of G.*

Lemma 7. *There is a subset of I which is a minimum s-CVD set of G_{well}.*

Now Theorem 2 follows from a result of Khot and Regev [21], where they showed that unless the Unique Games Conjecture is false, there is no $(2 - \varepsilon)$-approximation algorithm for MINIMUM VERTEX COVER on general graphs, for any $\varepsilon > 0$.

5 $O(n(n+m))$-Time Algorithm for s-CVD on Interval Graphs

In this section we give an $O(n(n+m))$-time algorithm to solve s-CVD on interval graph G with n vertices and m edges. For a set $X \subseteq V(G)$, if each connected component of $G - X$ is an s-club, then we call X as an s-club vertex deleting set (s-CVD set). Below we state some definitions and the main idea behind our algorithm.

Let G denotes a connected interval graph with n vertices and m edges. The set $\mathscr{I}$ denotes a fixed interval representation of G where the endpoints of the representing intervals are distinct. Let $l(v)$ and $r(v)$ denote the left and right endpoints, respectively, of an interval corresponding to a vertex $v \in V(G)$. Then the interval assigned to the vertex v in $\mathscr{I}$ is denoted by $I(v) = [l(v), r(v)]$. Observe that, intervals on a real line satisfies the Helly property and hence for each maximal clique Q of G there is an interval $I = \bigcap_{v \in Q} I(v)$. We call I as the *Helly* region corresponding to the maximal clique Q. Let $Q_1, Q_2, \ldots, Q_k$ denote the set of maximal cliques of G ordered with respect to their *Helly* regions $I_a, 1 \le a \le k$ on the real line. That is, $I_1 < I_2 < \ldots < I_k$. Observe that, for any two integers a, b we have $I_a \cap I_b = \emptyset$ as both Q_a and Q_b are maximal cliques. Moreover, for any $a \le b \le c$ if a vertex $v \in Q_a \cap Q_c$, then $v \in Q_b$.

For a set $X \subseteq V(G)$, if each connected component of $G - X$ is an s-club, then we call X as an s-club vertex deleting set (s-CVD set). The key idea of the algorithm is to build a *minimum* s-CVD set for a given interval graph G by iteratively finding *minimum* s-CVD sets for a certain set of induced subgraphs $(Q_1 \cup Q_2 \cup \ldots \cup Q_a) - A, A \subseteq Q_a$ in each iteration. We will show that the total number of *subproblems* we solve is $O(n+m)$ and each subproblem can be solved in $O(n)$ time. First we introduce some definitions which we use extensively in the algorithm and proofs.

Definition 5. *(i) For integers a, b where $1 \le a < b \le k$, let $S_a^b = Q_a \cap Q_b$.*

(ii) For an integer a, let $\mathscr{S}(Q_a) = \left\{ S_a^b : a < b \le k \text{ and } S_a^b \ne S_a^{b'}, a < b' < b \right\} \cup \emptyset$ (Note that, the members of the set $\mathscr{S}(Q_a)$ are distinct.)

(iii) For $A \in \mathscr{S}(Q_a)$, let $Y_A^a = (Q_a - Q_{a-1}) - A$.

(iv) For a vertex $v \in V(G)$, the index $q_v^- = \min\{a : v \in Q_a\}$. That is, the minimum integer a such that v belongs to the maximal clique Q_a.

(v) For a vertex $v \in V(G)$, the index $q_v^+ = \max\{b : v \in Q_b\}$. That is, the maximum integer b such that v belongs to the maximal clique Q_b.

We use the following observation to prove our main lemma.

Observation I. *Let $X \subseteq V(G)$ and u, v be two vertices with $r(u) < l(v)$ such that u and v lie in different connected components in $G - X$. Then there exists an integer a with $q_u^+ \le a < q_v^-$, such that $S_a^{a+1} \subseteq X$.*

For two integers a, b with $1 \le a \le b \le k$, let $G[a,b]$ denotes the subgraph induced by the set $\{Q_a \cup Q_{a+1} \cup \ldots \cup Q_b\}$.

Definition 6. *For an induced subgraph H of G, a vertex $v \in V(H)$ and an integer a, let $L_H(a, v)$ denote the set of vertices in H that lie at distance a from v in H.*

Hereafter, we use the notation $L_H(s+1, v)$ where $H = G[1, a] - A$ for some integer a and $v \in Y_A^a$ (See Definition 5, (iii)) several times.

Definition 7. *For an integer $a, 1 \leq a \leq k-1$ and a set $A \in \mathscr{S}(Q_a)$ consider the induced subgraph $H = G[1,a] - A$ and the sub-interval representation $\mathscr{I}' \subseteq \mathscr{I}$ of H. We define the* frontal component *of the induced graph as the connected component of $G[1,a] - A$ containing the vertex with the rightmost endpoint in $\mathscr{I}'$.*

Note that for an integer a and $A \in \mathscr{S}(Q_a)$, the vertices of Y_A^a, if any, lies in the *frontal* component of $G[1,a] - A$. Below we categorize an s-CVD set X of $G[1,a] - A$ into four types.

Definition 8. *Consider an integer $a, 1 < a \leq k$ and a set $A \in \mathscr{S}(Q_a)$, an s-CVD set X of $G[1,a] - A$ is of*

1. *type-1: if $Y_A^a \subseteq X$.*
2. *type-2: if there is a vertex $v \in Y_A^a$ such that $L_H(s+1,v) \subseteq X$.*
3. *type-3: if there exists an integer $c, 1 \leq c < a$ such that $S_c^{c+1} - A \subseteq X$ and $G[c+1,a] - (S_c^{c+1} \cup A)$ is connected and has diameter at most s.*
4. *type-4: if there exists an integer $c, 1 \leq c < a$ such that $S_c^{c+1} - A \subseteq X$ and $G[c+1,a] - (S_c^{c+1} \cup A)$ is connected and has diameter exactly $s+1$.*

The following lemma is crucial for our algorithm.

Lemma 8 (Main Lemma). *Consider an integer $1 \leq a \leq k$ and a set $A \in \mathscr{S}(Q_a)$. Then at least one of the following holds:*

1. *Every connected component of $G[1,a] - A$ have diameter at most s.*
2. *Any s-CVD set of $G[1,a] - A$ is of some type-j where $j \in \{1,2,3,4\}$.*

5.1 The Algorithm

Our algorithm constructs a table Ψ iteratively whose cells are indexed by two parameters. For an integer $a, 1 \leq a \leq k$ and a set $A \in \mathscr{S}(Q_a)$, the cell $\Psi[a,A]$ contains a minimum s-CVD set of $G[1,a] - A$. Clearly, $\Psi[k,\emptyset]$ is a minimum s-CVD set of G.

Now we start the construction of Ψ. Since $G[1,1]$ is a clique, we set $\Psi[1,A] = \emptyset$ for all $A \in \mathscr{S}(Q_1)$:

Lemma 9. *For any $A \in \mathscr{S}(Q_1)$, $\Psi[1,A] = \emptyset$.*

From now on assume $a \geq 2$ and A be a set in $\mathscr{S}(Q_a)$. Let H be the graph $G[1,a] - A$ and F be the graph $G[1,a-1] - (A \cap Q_{a-1})$. Observe that for any two integers $a,b, 1 \leq a < b \leq k$ the set $S_{a-1}^b = S_a^b \cap Q_{a-1}$. Then, for any $A \in \mathscr{S}(Q_a)$ we have $(A \cap Q_{a-1}) \in \mathscr{S}(Q_{a-1})$ and $\Psi[a-1, A \cap Q_{a-1}]$ is defined. Note that $H - F = Y_A^a$. In the following lemma we show that $\Psi[a,A] = \Psi[a-1, A \cap Q_{a-1}]$ if the *frontal* component of H has diameter at most s.

Lemma 10. *Let $H = G[1,a] - A$, for $A \in \mathscr{S}(Q_a), 1 < a \leq k$. If the* frontal *component of H has diameter at most s, then $\Psi[a,A] = \Psi[a-1, A \cap Q_{a-1}]$.*

Now assume that the *frontal* component of $H = G[1,a] - A$ has diameter at least $s+1$. Recall that if $Y_A^a = \emptyset$, we have $\Psi[a,A] = \Psi[a-1,A \cap Q_{a-1}]$. Hence assume that $Y_A^a \neq \emptyset$. Due to Lemma 8, any s-CVD set of H has to be one of the four types defined above. First, for each $j \in \{1,2,3,4\}$, we find an s-CVD set of minimum cardinality, which is of type-j. We begin by showing how to construct a minimum cardinality s-CVD set X_1 which is of type-1 and defined as $X_1 = Y_A^a \cup \Psi[a-1,A \cap Q_{a-1}]$.

Lemma 11. *The set X_1 is a minimum cardinality s-CVD set of type-1 of $G[1,a] - A$.*

Let v be some vertex in Y_A^a and $b < a$ be the maximum integer such that $(Q_b \cap L_H(s+2,v)) \neq \emptyset$. We construct a minimum cardinality s-CVD set of type-2 for $G[1,a] - A$, which is defined as $X_2 = L_H(s+1,v) \cup \Psi[b, S_b^{b+1}]$.

Lemma 12. *The set X_2 is a minimum cardinality s-CVD set of type-2 of $G[1,a] - A$.*

Now we show how to construct a minimum cardinality s-CVD set X_3 of type-3 of $G[1,a] - A$. Let $B \subseteq \{1,2,\ldots,a-1\}$ be the set of integers such that for any $i \in B$ the graph $H_i = G[i+1,a] - (S_i^{i+1} \cup A)$ is connected and has diameter at most s. By definition, a type-3 s-CVD set X of H contains S_c^{c+1} for some $c \in B$. We call each such type-3 s-CVD set as type-3(c). Now we define minimum type-3(c) s-CVD set as follows. For each $c \in B$, $Z_c = (S_c^{c+1} - A) \cup \Psi[c, S_c^{c+1}]$ and $X_3 = \min\{Z_c : c \in B\}$.

Lemma 13. *The set X_3 is a minimum cardinality s-CVD set of type-3 of $G[1,a] - A$.*

Finally, we show the construction of a minimum cardinality s-CVD set X_4 of type-4 of $G[1,a] - A$. Let $C \subseteq \{1,2,\ldots,a-1\}$ be the set of integers such that for any $i \in C$ the graph $H_i = G[i+1,a] - (S_i^{i+1} \cup A)$ is connected and has diameter exactly $s+1$. By definition, a type-4 s-CVD set X of H contains S_i^{i+1} for some $i \in C$. We call each such type-4 s-CVD set as type-4(c). Now we define minimum type-4(c) s-CVD set as follows. Note that $Y_A^a \neq \emptyset$. Let v be some vertex in Y_A^a and $Y_i = L_{H_i}(s+1,v)$. For each $i \in C$, $Z_i = (S_i^{i+1} - A) \cup Y_i \cup \Psi[i, S_i^{i+1}]$ and we define $X_4 = \min\{Z_i : i \in C\}$.

Lemma 14. *The set X_4 is a minimum cardinality s-CVD set of type-4 of $G[1,a] - A$.*

Now we define a minimum s-CVD set of $G[1,a] - A$ as the one with minimum cardinality among the sets $X_i, 1 \leq i \leq 4$. That is, $\Psi[a,A] = \min\{X_1,X_2,X_3,X_4\}$. We formally summarize the above discussion in the following lemma.

Lemma 15. *For $1 < a \leq k$, if the diameter of the* frontal *component of $G[1,a] - A$ is at least $s+1$, then $\Psi[a,A] = \min\{X_1,X_2,X_3,X_4\}$.*

Proof. The proof follows from Lemma 8 and the above discussion on the minimality of the sets $X_i, 1 \leq i \leq 4$, in their respective types.

The proof of correctness of the algorithm follows from the Lemmas 9, 10 and 15.

5.2 Time Complexity

For a given interval graph G with n vertices and m edges, the algorithm first finds the ordered set of maximal cliques of G. Such an ordered list of the maximal cliques of G can be produced in linear time as a byproduct of the linear $(O(n + m))$ time recognition algorithm for interval graphs due to Booth and Leuker [8]. To compute the overall time complexity of our algorithm, we have the following claims.

Claim. Total number of subproblems computed by the algorithm is at most $O(|V| + |E|) = O(n + m)$.

Claim. For $H = G[1,a] - A, 1 \leq a \leq k, A \in \mathscr{S}(Q_a)$, the minimum cardinality s-CVD set of *4-types* can be computed in $O(n)$ time.

Therefore, it follows that s-CVD set of G can be computed in $O(n \cdot (n + m))$ and hence proved Theorem 3.

References

1. Ahn, J., Jaffke, L., Kwon, O., Lima, P.T.: Well-partitioned chordal graphs: obstruction set and disjoint paths. In: Adler, I., Müller, H. (eds.) WG 2020. LNCS, vol. 12301, pp. 148–160. Springer, Cham (2020). https://doi.org/10.1007/978-3-030-60440-0_12
2. Ahn, J., Jaffke, L., Kwon, O., Lima, P.T.: Three problems on well-partitioned chordal graphs. In: Calamoneri, T., Corò, F. (eds.) CIAC 2021. LNCS, vol. 12701, pp. 23–36. Springer, Cham (2021). https://doi.org/10.1007/978-3-030-75242-2_2
3. Alba, R.D.: A graph-theoretic definition of a sociometric clique. J. Math. Sociol. **3**(1), 113–126 (1973)
4. Aprile, M., Drescher, M., Fiorini, S., Huynh, T.: A tight approximation algorithm for the cluster vertex deletion problem. In: Singh, M., Williamson, D.P. (eds.) IPCO 2021. LNCS, vol. 12707, pp. 340–353. Springer, Cham (2021). https://doi.org/10.1007/978-3-030-73879-2_24
5. Balasundaram, B., Butenko, S., Trukhanov, S.: Novel approaches for analyzing biological networks. J. Comb. Optim. **10**(1), 23–39 (2005)
6. Bansal, N., Blum, A., Chawla, S.: Correlation clustering. Mach. Learn. **56**(1), 89–113 (2004)
7. Ben-Dor, A., Shamir, R., Yakhini, Z.: Clustering gene expression patterns. J. Comput. Biol. **6**(3–4), 281–297 (1999)
8. Booth, K.S., Lueker, G.S.: Testing for the consecutive ones property, interval graphs, and graph planarity using PQ-tree algorithms. J. Comput. Syst. Sci. **13**(3), 335–379 (1976)
9. Boral, A., Cygan, M., Kociumaka, T., Pilipczuk, M.: A fast branching algorithm for cluster vertex deletion. Theory Comput. Syst. **58**(2), 357–376 (2016)
10. Cao, Y., Ke, Y., Otachi, Y., You, J.: Vertex deletion problems on chordal graphs. Theoret. Comput. Sci. **745**, 75–86 (2018)
11. Chakraborty, D., Chandran, L.S., Padinhatteeri, S., Pillai, R.R.: Algorithms and complexity of s-club cluster vertex deletion. In: Flocchini, P., Moura, L. (eds.) IWOCA 2021. LNCS, vol. 12757, pp. 152–164. Springer, Cham (2021). https://doi.org/10.1007/978-3-030-79987-8_11
12. Dehne, F., Langston, M.A., Luo, X., Pitre, S., Shaw, P., Zhang, Y.: The cluster editing problem: implementations and experiments. In: Bodlaender, H.L., Langston, M.A. (eds.) IWPEC 2006. LNCS, vol. 4169, pp. 13–24. Springer, Heidelberg (2006). https://doi.org/10.1007/11847250_2

13. Fellows, M.R., Guo, J., Komusiewicz, C., Niedermeier, R., Uhlmann, J.: Graph-based data clustering with overlaps. Discrete Optim. **8**(1), 2–17 (2011)
14. Figiel, A., Himmel, A.-S., Nichterlein, A., Niedermeier, R.: On 2-clubs in graph-based data clustering: theory and algorithm engineering. In: Calamoneri, T., Corò, F. (eds.) CIAC 2021. LNCS, vol. 12701, pp. 216–230. Springer, Cham (2021). https://doi.org/10.1007/978-3-030-75242-2_15
15. Fiorini, S., Joret, G., Schaudt, O.: Improved approximation algorithms for hitting 3-vertex paths. In: Louveaux, Q., Skutella, M. (eds.) IPCO 2016. LNCS, vol. 9682, pp. 238–249. Springer, Cham (2016). https://doi.org/10.1007/978-3-319-33461-5_20
16. Fiorini, S., Joret, G., Schaudt, O.: Improved approximation algorithms for hitting 3-vertex paths. Math. Program. **182**(1), 355–367 (2019). https://doi.org/10.1007/s10107-019-01395-y
17. Fomin, F.V., Gaspers, S., Lokshtanov, D., Saurabh, S.: Exact algorithms via monotone local search. J. ACM (JACM) **66**(2), 1–23 (2019)
18. Fomin, F.V., Le, T.-N., Lokshtanov, D., Saurabh, S., Thomassé, S., Zehavi, M.: Subquadratic kernels for implicit 3-hitting set and 3-set packing problems. ACM Trans. Algorithms (TALG) **15**(1), 1–44 (2019)
19. Gramm, J., Guo, J., Hüffner, F., Niedermeier, R.: Automated generation of search tree algorithms for hard graph modification problems. Algorithmica **39**(4), 321–347 (2004)
20. Hüffner, F., Komusiewicz, C., Moser, H., Niedermeier, R.: Fixed-parameter algorithms for cluster vertex deletion. Theory Comput. Syst. **47**(1), 196–217 (2010)
21. Khot, S., Regev, O.: Vertex cover might be hard to approximate to within 2- ε. J. Comput. Syst. Sci. **74**(3), 335–349 (2008)
22. King, V., Rao, S., Tarjan, R.: A faster deterministic maximum flow algorithm. J. Algorithms **17**(3), 447–474 (1994)
23. Kleinberg, J., Tardos, E.: Algorithm Design. Pearson Education India (2006)
24. Lewis, J.M., Yannakakis, M.: The node-deletion problem for hereditary properties is NP-complete. J. Comput. Syst. Sci. **20**(2), 219–230 (1980)
25. Liu, H., Zhang, P., Zhu, D.: On editing graphs into 2-club clusters. In: Snoeyink, J., Lu, P., Su, K., Wang, L. (eds.) AAIM/FAW -2012. LNCS, vol. 7285, pp. 235–246. Springer, Heidelberg (2012). https://doi.org/10.1007/978-3-642-29700-7_22
26. Mokken, R.J.: Cliques, clubs and clans. Qual. Quant. **13**, 161–173 (1979)
27. Orlin, J.B.: Max flows in o (nm) time, or better. In: Proceedings of the Forty-Fifth Annual ACM Symposium on Theory of Computing, pp. 765–774 (2013)
28. Papadopoulos, S., Kompatsiaris, Y., Vakali, A., Spyridonos, P.: Community detection in social media. Data Min. Knowl. Disc. **24**(3), 515–554 (2012)
29. Pasupuleti, Srinivas: Detection of protein complexes in protein interaction networks using n-clubs. In: Marchiori, Elena, Moore, Jason H.. (eds.) EvoBIO 2008. LNCS, vol. 4973, pp. 153–164. Springer, Heidelberg (2008). https://doi.org/10.1007/978-3-540-78757-0_14
30. Rahmann, S., Wittkop, T., Baumbach, J., Martin, M., Truss, A., Böcker, S.: Exact and heuristic algorithms for weighted cluster editing. In: Computational Systems Bioinformatics, vol. 6, pp. 391–401. World Scientific (2007)
31. Sau, I., Souza, U.D.S.: Hitting forbidden induced subgraphs on bounded treewidth graphs. In: 45th International Symposium on Mathematical Foundations of Computer Science (MFCS 2020). Schloss Dagstuhl-Leibniz-Zentrum für Informatik (2020)
32. Schäfer, A.: Exact algorithms for s-club finding and related problems. Diploma thesis, Friedrich-Schiller-University Jena (2009)
33. Shamir, R., Sharan, R., Tsur, D.: Cluster graph modification problems. Discret. Appl. Math. **144**(1–2), 173–182 (2004)
34. Sharan, R., Shamir, R.: CLICK: a clustering algorithm with applications to gene expression analysis. In: Proceedings of the International Conference on Intelligent Systems in Molecular Biology, vol. 8, p. 16 (2000)

35. Spirin, V., Mirny, L.A.: Protein complexes and functional modules in molecular networks. Proc. Natl. Acad. Sci. **100**(21), 12123–12128 (2003)
36. Tsur, D.: Faster parameterized algorithm for cluster vertex deletion. Theory Comput. Syst. **65**(2), 323–343 (2021)
37. Zhenyu, W., Leahy, R.: An optimal graph theoretic approach to data clustering: theory and its application to image segmentation. IEEE Trans. Pattern Anal. Mach. Intell. **15**(11), 1101–1113 (1993)
38. Yang, Z., Algesheimer, R., Tessone, C.J.: A comparative analysis of community detection algorithms on artificial networks. Sci. Rep. **6**(1), 1–18 (2016)
39. Yannakakis, M.: Node-and edge-deletion NP-complete problems. In: Proceedings of the Tenth Annual ACM Symposium on Theory of Computing, pp. 253–264 (1978)
40. You, J., Wang, J., Cao, Y.: Approximate association via dissociation. Discret. Appl. Math. **219**, 202–209 (2017)

Polychromatic Colorings of Unions of Geometric Hypergraphs

Vera Chekan[1]([envelope]) [ORCID] and Torsten Ueckerdt[2] [ORCID]

[1] Humboldt-Universität zu Berlin, Berlin, Germany
Vera.Chekan@informatik.hu-berlin.de
[2] Institute of Theoretical Informatics, Karlsruhe Institute of Technology,
Karlsruhe, Germany
torsten.ueckerdt@kit.edu

Abstract. A polychromatic k-coloring of a hypergraph assigns to each vertex one of k colors in such a way that every hyperedge contains all the colors. A range capturing hypergraph is an m-uniform hypergraph whose vertices are points in the plane and whose hyperedges are those m-subsets of points that can be separated by some geometric object of a particular type, such as axis-aligned rectangles, from the remaining points. Polychromatic k-colorings of m-uniform range capturing hypergraphs are motivated by the study of weak ε-nets and cover decomposability problems.

We show that the hypergraphs in which each hyperedge is determined by a bottomless rectangle or by a horizontal strip in general do not allow for polychromatic colorings. This strengthens the corresponding result of Chen, Pach, Szegedy, and Tardos [Random Struct. Algorithms, 34:11–23, 2009] for axis-aligned rectangles, and gives the first explicit (not randomized) construction of non-2-colorable hypergraphs defined by axis-aligned rectangles of arbitrarily large uniformity.

In general we consider unions of range capturing hypergraphs, each defined by a type of unbounded axis-aligned rectangles. For each combination of types, we show that the unions of such hypergraphs either admit polychromatic k-colorings for $m = O(k)$, $m = O(k \log k)$, $m = O(k^{8.75})$, or do not admit in general polychromatic 2-colorings for any m.

Keywords: Hypergraph · Coloring · Polychromatic Coloring · Range Space

1 Introduction

A *range capturing hypergraph* is a geometric hypergraph $\mathcal{H}(V, \mathcal{R})$ defined by a finite point set $V \subset \mathbb{R}^2$ in the plane and a family $\mathcal{R}$ of subsets of $\mathbb{R}^2$, called *ranges*. Possible ranges are for example the family $\mathcal{R}$ of all axis-aligned rectangles, all horizontal strips, or all translates of the first (north-east) quadrant. Given the points V and ranges $\mathcal{R}$, the hypergraph $\mathcal{H}(V, \mathcal{R}) = (V, \mathcal{E})$ has V as its vertex set and a subset $E \subset V$ forms a hyperedge $E \in \mathcal{E}$ whenever there exists a range

M. A. Bekos and M. Kaufmann (Eds.): WG 2022, LNCS 13453, pp. 144–157, 2022.
https://doi.org/10.1007/978-3-031-15914-5_11

$R \in \mathcal{R}$ with $E = V \cap R$. That is, a subset of points forms a hyperedge whenever these points and no other points are captured by some range from family $\mathcal{R}$.

For a positive integer m, we are then interested in the m-uniform subhypergraph $\mathcal{H}(V, \mathcal{R}, m)$ given by all hyperedges of size exactly m. In particular, we investigate polychromatic vertex colorings $c \colon V \to [k]$ in k colors of $\mathcal{H}(V, \mathcal{R}, m)$ for different families of ranges $\mathcal{R}$ and different values of m. A vertex coloring is *polychromatic* if every hyperedge contains at least one vertex of each color. Polychromatic colorings of range capturing hypergraphs were first studied in the 1980s by Pach [17,18] in the context of cover-decomposability problems. These also relate to the planar sensor cover problem [10] and weak ε-nets [20,26]. Polychromatic colorings of geometric hypergraphs then experienced a major revival during the past decade with several breakthrough advances [2–6,11–15,19,23–25]. The interested reader is referred to the (slightly outdated) survey article [22] and the excellent website [1] which contains numerous references.

Here, we focus on polychromatic k-colorings for range capturing hypergraphs with given range family $\mathcal{R}$. In particular, we investigate the following question.

Question 1. *Given $\mathcal{R}$ and k, what is the smallest $m = m(k)$ such that for every finite point set $V \subset \mathbb{R}^2$ the hypergraph $\mathcal{H}(V, \mathcal{R}, m)$ admits a polychromatic k-coloring?*

Of course, $m(k) \geqslant k$, while $m(k) = \infty$ is also possible. It also holds that $m(k) \leqslant m(k+1)$ for any k: given a polychromatic $(k+1)$-coloring of a hypergraph, we can recolor every vertex of color $k+1$ arbitrarily, after that every hyperedge will still contain all colors in $1, \ldots, k$. For all range families considered here, we either show that $m(k) < \infty$ for every $k \geqslant 1$ or already $m(2) = \infty$ holds. Note that in the latter case, there are range capturing hypergraphs that are not properly[1] 2-colorable, even for arbitrarily large uniformity m. So although we do not consider proper colorings explicitly in this work, our results imply that the chromatic number of certain hypergraphs is larger than 2.

1.1 Related Work

There is a rich literature on range capturing hypergraphs, their polychromatic colorings, and answers to Question 1. Let us list the positive results (meaning $m(k) < \infty$ for all k) that are relevant here, whilst defining the respective ranges.

- For halfplanes $\mathcal{R} = \{\{(x,y) \in \mathbb{R}^2 \mid 1 \leqslant ax + by\} \mid a, b \in \mathbb{R}\}$ it is known that $m(k) = 2k - 1$ [25].
- For south-west quadrants $\mathcal{R} = \{\{(x,y) \in \mathbb{R}^2 \mid x \leqslant a \text{ and } y \leqslant b\} \mid a, b \in \mathbb{R}\}$ it is easy to prove that $m(k) = k$, see e.g. [12].
- For axis-aligned strips $\mathcal{R} = \{\{(x,y) \in \mathbb{R}^2 \mid a_1 \leqslant x \leqslant a_2\} \mid a_1, a_2 \in \mathbb{R}\} \cup \{\{(x,y) \in \mathbb{R}^2 \mid a_1 \leqslant y \leqslant a_2\} \mid a_1, a_2 \in \mathbb{R}\}$ it is known that $m(k) \leqslant 2k - 1$ [3].
- For bottomless rectangles $\mathcal{R} = \{\{(x,y) \in \mathbb{R}^2 \mid a_1 \leqslant x \leqslant a_2 \text{ and } y \leqslant b\} \mid a_1, a_2, b \in \mathbb{R}\}$ it is known that $1.67k \leqslant m(k) \leqslant 3k - 2$ [4].

---

[1] A vertex coloring is *proper* if every hyperedge contains two vertices of different colors.

- For axis-aligned squares $\mathcal{R} = \{\{(x,y) \in \mathbb{R}^2 \mid a \leqslant x \leqslant a+s \text{ and } b \leqslant y \leqslant b+s\} \mid a,b,s \in \mathbb{R}\}$ it is known that $m(k) \leqslant O(k^{8.75})$ [2].

On the contrary, let us also list the negative results (meaning $m(k) = \infty$ for some k) that are relevant here. In all cases, it is shown that already $m(2) = \infty$ holds. This means that there is a sequence $(\mathcal{H}_m)_{m \geqslant 1}$ of m-uniform hypergraphs such that for each $m \geqslant 1$ the hypergraph $\mathcal{H}_m$ admits no polychromatic 2-coloring and we have that $\mathcal{H}_m$ is a subhypergraph of $\mathcal{H}(V_m, \mathcal{R}, m)$ for some finite point set V_m. If the latter property holds, we say that $\mathcal{H}_m$ can be *realized* with $\mathcal{R}$. One such sequence are the *m-ary tree hypergraphs*, defined on the vertices of a complete m-ary tree of height m, where for each non-leaf vertex, its m children form a hyperedge, and for each leaf vertex, its m ancestors (including itself) form a hyperedge (introduced by Pach, Tardos, and Tóth [21]). A second such sequence is due to Pálvölgyi [23] (published in [19]), for which we do not repeat the formal definition here and simply refer to them as the *2-size hypergraphs* as their inductive construction involves hyperedges of two possibly different sizes.

- For strips $\mathcal{R} = \{\{(x,y) \in \mathbb{R}^2 \mid 1 \leqslant ax+by \leqslant c\} \mid a,b,c \in \mathbb{R}\}$ it is known that $m(2) = \infty$ as every m-ary tree hypergraph can be realized with strips [21].
- For unit disks $\mathcal{R} = \{\{(x,y) \in \mathbb{R}^2 \mid (x-a)^2 + (y-b)^2 \leqslant 1\} \mid a,b \in \mathbb{R}\}$ it is known that $m(2) = \infty$ as every 2-size hypergraph can be realized with unit disks [19].

Finally, for axis-aligned rectangles $\mathcal{R} = \{\{(x,y) \in \mathbb{R}^2 \mid a_1 \leqslant x \leqslant a_2 \text{ and } b_1 \leqslant y \leqslant b_2\} \mid a_1, a_2, b_1, b_2 \in \mathbb{R}\}$ it is also known that $m(2) = \infty$. See Theorem 2 below. However, the only known proof of Theorem 2 was a probabilistic argument and no explicit construction of a sequence $(\mathcal{H}_m)_{m \geqslant 1}$ of m-uniform hypergraphs realizable by axis-aligned rectangles that admit no polychromatic 2-coloring was known before this work.

Theorem 2 (Chen et al. [8]). *For the family $\mathcal{R}$ of all axis-aligned rectangles it holds that $m(2) = \infty$. That is, for every $m \geqslant 1$ there exists a finite point set $V \subset \mathbb{R}^2$ such that for every 2-coloring of V some axis-aligned rectangle contains m points of V, all of the same color.*

1.2 Our Results

In this paper we consider range families $\mathcal{R} = \mathcal{R}_1 \cup \mathcal{R}_2$ that are the union of two range families $\mathcal{R}_1, \mathcal{R}_2$. The corresponding hypergraph $\mathcal{H}(V, \mathcal{R}, m)$ is then the union of the hypergraphs $\mathcal{H}(V, \mathcal{R}_1, m)$ and $\mathcal{H}(V, \mathcal{R}_2, m)$ on the same vertex set $V \subset \mathbb{R}^2$. Clearly, if $\mathcal{H}(V, \mathcal{R}, m)$ is polychromatic k-colorable, then so are $\mathcal{H}(V, \mathcal{R}_1, m)$ and $\mathcal{H}(V, \mathcal{R}_2, m)$. But the converse is not necessarily true and this shall be the subject of our investigations.

Aloupis et al. [3] show that if $\mathcal{R}_1$ and $\mathcal{R}_2$ admit so-called m-hitting k-sets, then we can conclude that $m(k) \leqslant m < \infty$ for $\mathcal{R} = \mathcal{R}_1 \cup \mathcal{R}_2$; see Lemma 3. This is for example the case for all horizontal (resp. vertical) strips, but already fails for all south-west quadrants. In Sect. 3 we then consider all possible families of *unbounded* axis-aligned rectangles, such as axis-aligned strips, all four types of quadrants, or bottomless rectangles. We determine exactly for which subset of those, when taking $\mathcal{R}$ as their union, it holds that $m(k) < \infty$.

In particular, we show in Sect. 3.1 that $m(k) = \infty$ for all $k \geqslant 2$ when $\mathcal{R} = \mathcal{R}_1 \cup \mathcal{R}_2$ is the union of $\mathcal{R}_1$ all bottomless rectangles and $\mathcal{R}_2$ all horizontal strips. Our proof gives a new sequence $(\mathcal{H}_m)_{m \geqslant 1}$ of m-uniform hypergraphs that admit a geometric realization for simple ranges, but do not admit any polychromatic 2-coloring. On the positive side, we show in Sect. 3.2 that (up to symmetry) all other subsets of unbounded axis-aligned rectangles (excluding the above pair) admit polychromatic k-colorings for every k. Here, our proof relies on so-called shallow hitting sets and in particular a variant in which a subset of V hits every hyperedge defined by $\mathcal{R}_1$ at least once and every hyperedge defined by $\mathcal{R}_1 \cup \mathcal{R}_2$ at most a constant (usually 2 or 3) number of times.

Assumptions and Notation. Before we start, let us briefly mention some convenient facts that are usually assumed, and which we also assume throughout our paper: Whenever a range family $\mathcal{R}$ is given, we only consider point sets V that are in general position with respect to $\mathcal{R}$. For us, this means that the points in V have pairwise different x-coordinates, pairwise different y-coordinates, and also pairwise different sums of x- and y-coordinates. Secondly, all range families $\mathcal{R}$ that we consider here are *shrinkable*, meaning that whenever a set $X \subseteq V$ of i points is captured by a range in $\mathcal{R}$, then also some subset of $i - 1$ points of X is captured by a range in $\mathcal{R}$. This means that for every polychromatic k-coloring of $\mathcal{H}(V, \mathcal{R}, m)$, every range in $\mathcal{R}$ capturing m *or more* points of V, contains at least one point of each color. Finally, for every set $X \subseteq V$ captured by a range in $\mathcal{R}$, we implicitly associate to X one arbitrary but fixed such range $R \in \mathcal{R}$ with $V \cap R = X$. In particular, we shall sometimes consider *the* range R for a given hyperedge E of $\mathcal{H}(V, \mathcal{R}, m)$.

2   Polychromatic Colorings for Two Range Families

Let $\mathcal{R}_1, \mathcal{R}_2$ be two families of ranges, for each of which it is known that $m(k) < \infty$ for any $k \geqslant 1$. We seek to investigate whether also for $\mathcal{R} = \mathcal{R}_1 \cup \mathcal{R}_2$ we have $m(k) < \infty$. First, we identify a simple sufficient condition.

For fixed $k, m, \mathcal{R}$, we say that we have *m-hitting k-sets* with respect to $\mathcal{R}$ if the following holds. For every $V \subset \mathbb{R}^2$, there exist pairwise disjoint k-subsets of V such that every hyperedge of $\mathcal{H}(V, \mathcal{R}, m)$ fully contains at least one such k-subset. Clearly, if we have m-hitting k-sets, then $m(k) \leqslant m$ since we can simply use all colors $1, \ldots, k$ on each such k-subset (and color any remaining vertex arbitrarily). Crucially, if two range families $\mathcal{R}_1$ and $\mathcal{R}_2$ admit m-hitting k-sets, then $m(k) \leqslant m$ also carries over to their union $\mathcal{R} = \mathcal{R}_1 \cup \mathcal{R}_2$. This has already been implicitly used in [3].

Lemma 3 (Aloupis et al. [3]). *For fixed k, m, suppose that we have m-hitting k-sets with respect to $\mathcal{R}_1$ and m-hitting k-sets with respect to $\mathcal{R}_2$. Then for $\mathcal{R} = \mathcal{R}_1 \cup \mathcal{R}_2$ it holds that $m(k) \leqslant m$.*

For example, Lemma 3 gives $m(k) \leqslant 2k - 1$ when $\mathcal{R}$ consists of all axis-parallel strips [3]. In fact, for the vertical (resp. horizontal) strips it suffices to group the points into k-sets with consecutive x-coordinates (resp. y-coordinates).

Somewhat unfortunately, m-hitting k-sets appear to be very rare. Already for the range family $\mathcal{R}$ of all south-west quadrants, for which one can easily show that $m(k) = k$, we do not even have m-hitting 2-sets for any m. This will follow from the following result, which will also be useful later.

Lemma 4. *Let T be a rooted tree, and $\mathcal{H}(T)$ be the hypergraph on $V(T)$ where for each leaf vertex its ancestors (including itself) form a hyperedge. Then $\mathcal{H}(T)$ can be realized with the family $\mathcal{R}$ of all south-west quadrants.*

Moreover, the root is the bottommost and leftmost point and the children of each non-leaf vertex lie on a diagonal line of slope -1.

Proof. We do induction on the height of T, with height 1 being a trivial case of a single vertex. For height at least 2, remove the root r from T to obtain new trees $T_1, \ldots, T_p$, each of smaller height and rooted at a child of r. By induction, there are point sets in the plane for each $\mathcal{H}(T_i)$, $i = 1, \ldots, p$, with each respective root being bottommost and leftmost. We scale each of these points sets uniformly until the bounding box of each of them has width as well as height less than 1. For every $i \in [p]$, we put the point set for $\mathcal{H}(T_i)$ into the plane so that the root of T_i has the coordinate $(i, p - i)$. Finally, we place r in the origin. This gives the desired realization.

Note: in the end we can slightly perturb the point set so that it still realizes $\mathcal{H}(T)$ but it is in general position and the children of every non-leaf vertex are captured by a diagonal strip of slope -1. □

Corollary 5. *For any $k, m \geqslant 2$ and for the family $\mathcal{R}$ of all south-west quadrants, we do not have m-hitting k-sets in general.*

Proof. Take the rooted complete binary tree T_m of height m, for which $\mathcal{H}(T_m)$ is realizable with south-west quadrants by Lemma 4. By induction on m, we show that $\mathcal{H}(T_m)$ does not have m-hitting 2-sets. This is trivial for $m = 2$. Otherwise, any collection of disjoint 2-subsets either avoids the root r, or pairs r with a vertex in one of the two subtrees of T_m below r. In any case, there is a subtree T below r, none of whose vertices is paired with r and hence, there exist m-hitting 2-sets of $\mathcal{H}(T)$. Note that T is a complete binary tree of height $m - 1$, so $T = T_{m-1}$. But then $\mathcal{H}(T_{m-1})$ admits $(m - 1)$-hitting 2-sets too – a contradiction to the induction hypothesis.

Finally, if $\mathcal{H}(T_m)$ had m-hitting k-sets for some $k \geqslant 2$, then taking a 2-subset of every k-set in it, would result in m-hitting 2-sets of $\mathcal{H}(T_m)$. □

Corollary 6. *For any $k, m \geqslant 2$ and for the family $\mathcal{R}$ of all halfplanes, we do not have m-hitting k-sets in general.*

Proof. By a result of Middendorf and Pfeiffer [16], every range capturing hypergraph for south-west quadrants can also be realized by halfplanes and the result follows from Corollary 5. □

To summarize, parallel strips have m-hitting k-sets, but quadrants do not. Hence, we cannot apply Lemma 3 to conclude that $m(k) < \infty$ when we consider $\mathcal{R}$ to be the union of all quadrants of one direction and all parallel strips of one direction. In Sect. 3 we shall prove that indeed $m(k) < \infty$ for the union of all quadrants and strips, however only provided that the strips are axis-aligned. In fact, if they are not, this is not necessarily true.

Corollary 7. *Let $\mathcal{R}_1$ be the family of all south-west quadrants and $\mathcal{R}_2 = \{\{(x,y) \in \mathbb{R}^2 \mid a \leqslant x + y \leqslant b\} \mid a, b \in \mathbb{R}\}$ be the family of all diagonal strips of slope -1.*

Then for $\mathcal{R} = \mathcal{R}_1 \cup \mathcal{R}_2$ we have $m(2) = \infty$.

Proof. Given m, consider the rooted complete m-ary tree T_m of height m. By Lemma 4, $V(T_m)$ can be placed in the plane such that for each leaf vertex, its m ancestors (including itself) are captured by a south-west quadrant, and for each non-leaf vertex, its m children are captured by a diagonal strip of slope -1. Hence, every m-ary tree hypergraph $\mathcal{H}_m$ can be realized with $\mathcal{R}$. By [21] $\mathcal{H}_m$ admits no polychromatic 2-coloring for any m, which gives the result.   □

3   Families of Unbounded Rectangles

In this section we consider the following range families of unbounded rectangles:

- all (axis-aligned) south-west quadrants $\mathcal{R}_{SW}$,
- similarly all south-east $\mathcal{R}_{SE}$, north-east $\mathcal{R}_{NE}$, north-west $\mathcal{R}_{NW}$ quadrants,
- all horizontal $\mathcal{R}_{HS}$, vertical $\mathcal{R}_{VS}$, diagonal $\mathcal{R}_{DS}$ strips of slope -1,
- all bottomless rectangles $\mathcal{R}_{BL}$, and finally all topless rectangles $\mathcal{R}_{TL} = \{\{(x,y) \in \mathbb{R}^2 \mid a_1 \leqslant x \leqslant a_2, y \geqslant b\} \mid a_1, a_2, b \in \mathbb{R}\}$.

Observe that if a point set is captured by a south-east quadrant Q, then it is also captured by a bottomless rectangle having the same top and left sides as Q and whose right side lies to the right of every point in the vertex set. Analogous statements hold for other quadrants and vertical strips. Further, note that each of the above range families, except the diagonal strips $\mathcal{R}_{DS}$, is a special case of the family of all axis-aligned rectangles. Recall that for the family of *all* axis-aligned rectangles, it is known that $m(2) = \infty$ [8]. Here we are interested in the maximal subsets of $\{\mathcal{R}_{SW}, \mathcal{R}_{SE}, \mathcal{R}_{NE}, \mathcal{R}_{NW}, \mathcal{R}_{HS}, \mathcal{R}_{VS}, \mathcal{R}_{BL}, \mathcal{R}_{TL}\}$ so that for the union $\mathcal{R}$ of all these ranges, it still holds that $m(k) < \infty$ for all k. In fact, we shall show that for $\mathcal{R} = \mathcal{R}_{BL} \cup \mathcal{R}_{HS}$ we have $m(2) = \infty$, strengthening the result for axis-aligned rectangles [8]. On the other hand, for $\mathcal{R} = \mathcal{R}_{SW} \cup \mathcal{R}_{SE} \cup \mathcal{R}_{NE} \cup \mathcal{R}_{NW} \cup \mathcal{R}_{HS} \cup \mathcal{R}_{VS}$, i.e., the union of all quadrants and axis-aligned strips, we have $m(k) < \infty$ for all k, strengthening the results for south-west quadrants [12] and axis-aligned strips [3]. Secondly, for $\mathcal{R} = \mathcal{R}_{BL} \cup \mathcal{R}_{TL}$, i.e., the union of bottomless and topless rectangles (which also contains all quadrants and all vertical strips), we again have $m(k) < \infty$ for all k, thus strengthening the result for bottomless rectangles [4]. Using symmetries, this covers all cases of the considered unbounded axis-aligned rectangles. We complement our results by also considering the diagonal strips $\mathcal{R}_{DS}$ and recall that we already know by Corollary 7 that for $\mathcal{R} = \mathcal{R}_{DS} \cup \mathcal{R}_{SW}$ we have $m(2) = \infty$.

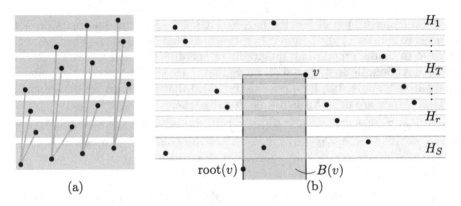

Fig. 1. (a) The forest F_2 and the desired embedding of $\mathcal{H}_2$. (b) Sketch of the embedding of $\mathcal{H}_m$ for the proof of Theorem 9. (Color figure online)

3.1 The Case with No Polychromatic Coloring: Bottomless Rectangles and Horizontal Strips

For every $m \in \mathbb{N}$, we will define a rooted forest F_m consisting of m^m trees whose vertices are partitioned into a set of the so-called *stages* (the forest F_2 is illustrated in Fig. 1(a)). The vertices of a stage S will be totally ordered and we denote this ordering by $<_S$. All vertices of a stage S will have the same distance to the root of the corresponding tree, we refer to this distance as the *level* of S. Every stage on level $j \in \{0, \ldots, m-1\}$ will consist of m^{m-j} vertices.

We start with m^m roots, one for each tree in F_m. They build the unique stage on level 0 and they are ordered in an arbitrary but fixed way. After that, for $j = 1, \ldots, m-1$, every stage S on level $j-1$, and every subset $S' \in \binom{S}{m^{m-j}}$, we add a new stage $T(S')$ on level j consisting of m^{m-j} new vertices so that every vertex in S' gets exactly one child from $T(S')$ and the vertices of $T(S')$ are ordered by $<_{T(S')}$ as their parents by $<_S$. As a result, every vertex in S gets a child for every (m^{m-j})-subset of S in which it occurs.

Now we can define the hypergraph $\mathcal{H}_m = (V, \mathcal{E})$. The vertex set V is exactly the vertex set $V(F_m)$ of the forest F_m. There are two types of hyperedges. First, stage-hyperedges $\mathcal{E}_s$: for every stage S, each m consecutive vertices in $<_S$ constitute a stage-hyperedge. Second, the path-hyperedges $\mathcal{E}_p$: every root-to-leaf path in F_m forms a path-hyperedge. Then, the set of hyperedges is defined as $\mathcal{E} = \mathcal{E}_s \cup \mathcal{E}_p$. Note that $\mathcal{H}_m = (V, \mathcal{E})$ is indeed m-uniform. For a vertex v, let root(v) denote the root of the tree in F_m containing v, and path$(v) \subset V$ denote the set of vertices on the path from v to root(v) in F_m.

Theorem 8. *For every $m \in \mathbb{N}$ the m-uniform hypergraph $\mathcal{H}_m = (V, \mathcal{E} = \mathcal{E}_s \cup \mathcal{E}_p)$ admits no polychromatic coloring with 2 colors.*

Proof. We show that every 2-coloring of V that makes all stage-hyperedges polychromatic produces a monochromatic path-hyperedge. Let $\phi : V \to \{$red, blue$\}$

be such a coloring. The key observation is that a stage S on level j (i.e., one that contains m^{m-j} vertices) can be partitioned into $m^{m-j}/m = m^{m-j-1}$ disjoint stage-hyperedges and hence, it contains at least m^{m-j-1} red vertices.

We prove for $j \in \{0, \ldots, m-1\}$ that there is a stage S_j on level j and a subset $B_j \subset S_j$ such that $|B_j| = m^{m-j-1}$ and for every $v \in B_j$, the vertices in path(v) are all red. For $j = 0$, the stage consisting of roots contains at least m^{m-1} red roots and these vertices have the desired property. Assuming the statement for some j, consider the stage $S_{j+1} = T(B_j)$ and a set B_{j+1} of m^{m-j-2} red points in it. By definition, each of these points v has its parent in B_j and hence, all vertices in path(v) are red, proving the statement for $j + 1$. By induction, it holds for $j = m - 1$ and hence, there is a vertex on level $m - 1$ (i.e., a leaf) whose root-to-leaf path is all red. So $\mathcal{H}_m$ admits no polychromatic 2-coloring.   □

Theorem 9. *For every $m \in \mathbb{N}$ the m-uniform hypergraph $\mathcal{H}_m = (V, \mathcal{E} = \mathcal{E}_s \cup \mathcal{E}_p)$ admits a realization with bottomless rectangles and horizontal strips.*

Proof. For a point $p \in \mathbb{R}^2$, let $x(p)$ and $y(p)$ denote its x- resp. y-coordinate. A sequence of points $p_1, \ldots, p_t$ is *ascending* (resp. *descending*) if $x(p_1) < \cdots < x(p_t)$ and $y(p_1) < \cdots < y(p_t)$ (resp. $x(p_1) < \cdots < x(p_t)$ and $y(p_1) > \cdots > y(p_t)$). Writing about the vertices of a stage S, we always refer to their ordering in $<_S$. We shall embed each stage S of $\mathcal{H}_m$ into a closed horizontal strip, denoted H_S, in such a way that $H_S \cap H_{S'} = \emptyset$ whenever $S \neq S'$. Note that this way, the embedded stages are vertically ordered with some available space between any two consecutive ones. For illustration see Fig. 1.

First, we embed the roots of F_m, i.e., the unique stage on level 0, as an ascending sequence in a horizontal strip for this stage. After that, until all stages are embedded, we choose some stage S that has already been embedded but the stages $T_1, \ldots, T_r$ containing its children not yet and in one step we embed $T_1, \ldots, T_r$ as follows. We pick a thin horizontal strip H between H_S and the strip above (if it exists) and within H identify disjoint horizontal strips $H_1, \ldots, H_r$. Then, every T_i is embedded inside H_i so that every vertex gets initially the same x-coordinate as its parent and the vertices of T_i build an ascending sequence in H_i. After that, for every $v \in S$ we slightly shift all children of v to the right so that they build a descending sequence but the ordering of x-coordinates relative to all other vertices remains unchanged.

The arising embedding ensures the following two properties. First, every stage-hyperedge is captured by a horizontal strip. Second, for every vertex v, the bottomless rectangle $B(v)$ with top-right corner v and root(v) on the left side captures exactly path(v), in particular every path-hyperedge is then captured by a bottomless rectangle. In the full version of the paper [7], we prove that these two properties indeed hold and this concludes the proof.   □

3.2   The Cases with Polychromatic Colorings

First, recall the result of Ackerman et al. [2] that for the range family $\mathcal{R}_{SQ}$ of all axis-aligned squares, we have $m(k) = O(k^{8.75})$. This already seals the deal for bottomless and topless rectangles.

Theorem 10. *For the family* $\mathcal{R} = \mathcal{R}_{\mathrm{BL}} \cup \mathcal{R}_{\mathrm{TL}}$ *of all bottomless and topless rectangles, we have* $m(k) = O(k^{8.75})$ *for all* k.

Proof. Let V be a finite point set and let m be arbitrary. For every bottomless (resp. topless) rectangle capturing a hyperedge of $\mathcal{H} = \mathcal{H}(V, \mathcal{R}, m)$, we introduce a bottom (resp. top) side below the bottommost (resp. above the topmost) point in V so that these rectangles are bounded now. After that, we stretch the plane horizontally until the width of every aforementioned rectangle becomes larger than its height and obtain the point set V'. This stretching preserves the ordering of x- and y-coordinates of the points so that the set of hyperedges captured by $\mathcal{R}$ remains the same. Finally, we pick every (now bounded) bottomless (resp. topless) rectangle capturing a hyperedge of $\mathcal{H}$ and shift its bottom (resp. top) side down (resp. up) until it becomes a square. Now for every hyperedge in $\mathcal{H}$, there is an axis-aligned square capturing it and hence, a hyperedge in $\mathcal{H}' = (V', \mathcal{R}_{\mathrm{SQ}}, m)$. Thus, each polychromatic coloring of $\mathcal{H}'$ yields a polychromatic coloring of $\mathcal{H}$ and this concludes the proof.   □

For the remaining cases, we utilize so-called shallow hitting sets. For a positive integer t, a subset X of vertices of a hypergraph $\mathcal{H}$ is a *t-shallow hitting set* if every hyperedge of $\mathcal{H}$ contains at least one and at most t points from X. It is known for example that for $\mathcal{R}$ being the family of all halfplanes, every range capturing hypergraph $\mathcal{H}(V, \mathcal{R}, m)$ admits a 2-shallow hitting set [25], which implies that $m(k) \leqslant 2k - 1$ in this case. In general, we have the following.

Lemma 11 (Keszegh and Pálvölgyi [13]). *Suppose that for a shrinkable range family* $\mathcal{R}$, *every hypergraph* $\mathcal{H}(V, \mathcal{R}, m)$ *admits a t-shallow hitting set. Then* $m(k) \leqslant (k - 1)t + 1$.

Remark 1. Lemma 11 states that if t-shallow hitting sets exist (for a global constant t), then $m(k) = O(k)$. However, it is not clear whether the converse is also true, for example when $\mathcal{R}$ is the family of all bottomless rectangles. Keszegh and Pálvölgyi [13] construct for this family range capturing hypergraphs without shallow hitting sets, but their constructed hypergraphs are not uniform. In fact, one can extract 3-shallow hitting sets for axis-aligned strips from the m-hitting k-sets for horizontal and vertical strips for $m = 2k - 1$: since every hyperedge of $\mathcal{H}(V, \mathcal{R}_{\mathrm{HS}} \cup \mathcal{R}_{\mathrm{VS}})$ of size $2k - 1$ or $2k$ is hit by at most three of the m-hitting k-sets, each color of the resulting k-coloring is a 3-shallow hitting set. To the best of our knowledge, it is open whether all $\mathcal{H}(V, \mathcal{R}, m)$ admit shallow hitting sets for the bottomless rectangles $\mathcal{R} = \mathcal{R}_{\mathrm{BL}}$.

Recall that for the family $\mathcal{R}_{\mathrm{NW}}$ of all north-west quadrants we have $m(k) = k$. In such a polychromatic coloring, every color class is a 1-shallow hitting set. Besides $\mathcal{R}_{\mathrm{NW}}$, we want to consider other range families, and thus are interested in t-shallow hitting sets for $\mathcal{R}_{\mathrm{NW}}$ that additionally do not hit other ranges, such as axis-parallel strips or other quadrants, too often. Let $E_t(V, m)$ (resp. $E_b(V, m)$) denote the set of m topmost (resp. bottommost) points in V.

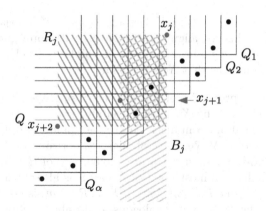

Fig. 2. Sketch for the proof of Lemma 12 for $m = 3$. The vertices in X are red.

Lemma 12. *For the family $\mathcal{R}_{\mathrm{NW}}$ of all north-west quadrants, every hypergraph $\mathcal{H}(V, \mathcal{R}_{\mathrm{NW}}, m)$ admits a 2-shallow hitting set X such that the points $x_1, \ldots, x_n$ in X have decreasing x-coordinates and decreasing y-coordinates, and*

(i) *x_1 is the leftmost point of $E_t(V, m)$,*

(ii) *the hyperedge $E_t(V, m)$ is hit by X exactly once,*

(iii) *for any two consecutive points x_j, x_{j+1} in X, the bottomless rectangle B_j with top-right corner x_j and x_{j+1} on the left side satisfies $|B_j \cap V| \geq m+1$, and*

(iv) *for any three consecutive points x_j, x_{j+1}, x_{j+2} in X, the axis-aligned rectangle R_j with top-right corner x_j and bottom-left corner x_{j+2} satisfies $|R_j \cap V| \geq m + 2$.*

Proof. For each hyperedge of $\mathcal{H}(V, \mathcal{R}_{\mathrm{NW}}, m)$ consider a fixed north-west quadrant capturing these m points of V. These quadrants can be indexed $Q_1, \ldots, Q_\alpha$ along their apices with decreasing x-coordinates (and hence also y-coordinates). I.e., Q_1 contains the topmost m points of V, while Q_α contains the leftmost m points of V. See Fig. 2 for an illustrative example.

Starting with $X = \emptyset$, we go through the north-west quadrants from Q_1 to Q_α, and whenever Q_i does not contain any point of X, we add the leftmost point of $Q_i \cap V$ to X. Label the points in X by $x_1, \ldots, x_n$ in the order of their addition to X. Along this order, the points have decreasing x-coordinates and decreasing y-coordinates. Clearly, X is a hitting set of $\mathcal{H}(V, \mathcal{R}_{\mathrm{NW}}, m)$ and satisfies (i).

Since x_1 is the leftmost point of Q_1, x_j does not belong to Q_1 for every $j \in \{2, \ldots, n\}$. Since Q_1 contains exactly vertices of $E_t(V, m)$, the corresponding hyperedge is hit by X exactly once. This proves (ii).

For any two consecutive points x_j, x_{j+1} in X, consider the bottomless rectangle B_j with top-right corner x_j and x_{j+1} on the left side. Then $|B_j \cap V| \geq m+1$ as B_j contains x_j and all points of the north-west quadrant Q for which we added x_{j+1} to X. This proves (iii).

Moreover, every point in $Q \cap V$ lies above x_{j+2} (if it exists), as x_{j+1} is leftmost in Q. Thus, the axis-aligned rectangle R_j with top-right corner x_j and bottom-left corner x_{j+2} contains x_j, all the m points in $Q \cap V$, and x_{j+2}. This proves (iv) and also implies that X is 2-shallow.   □

Let us explain the properties of this lemma. Suppose we have a south-west quadrant hit at least twice by X. Then it contains two consecutive points from X and by (iii) this quadrant contains at least $m+1$ points and hence it does not capture a hyperedge of $\mathcal{H}(V, \mathcal{R}_{\mathrm{SW}}, m)$. Similarly, a horizontal strip containing at least three points from X does not capture a hyperedge of $\mathcal{H}(V, \mathcal{R}_{\mathrm{HS}}, m)$ by (iv). Finally, since x_1 is the rightmost point of X and it is also the leftmost point of $E_t(V, m)$, every hyperedge $E \neq E_t(V, m)$ of $\mathcal{H}(V, \mathcal{R}_{\mathrm{NE}}, m)$ is not hit by X at all. For symmetry reasons, statements analogous to the above lemma hold for other types of quadrants and ranges as well. We provide a full description of these properties in the full version of the paper [7].

Intuitively speaking, Lemma 12 allows us to color some points in V, in such a way that every north-west quadrant already contains all colors, while other ranges, such as bottomless rectangles or diagonal strips, have most of their points still uncolored. The following lemma (proven in the full version of the paper [7]) provides a framework which can then be applied to color various range families.

Lemma 13. *Let $\mathcal{R}_1, \mathcal{R}_2$ be shrinkable range families, $f : \mathbb{N} \to \mathbb{N}$ be a function, and $s, t \in \mathbb{N}$ be such that:*

(i) *For every $k \in \mathbb{N}$, it holds that $m_{\mathcal{R}_2}(k) \leqslant f(k)$.*
(ii) *And for every point set V and every $m \in \mathbb{N}$, the hypergraph $\mathcal{H}(V, \mathcal{R}_1, m)$ admits a t-shallow hitting set $S \subseteq V$ such that every hyperedge of $\mathcal{H}(V, \mathcal{R}_2, m)$ is hit at most s times by S.*

Then for every $k \in \mathbb{N}$, we have $m_{\mathcal{R}_1 \cup \mathcal{R}_2}(k) \leqslant f(k) + k \max(s, t)$.

With Lemma 13 in place, we obtain the upper bounds for several range families:

Theorem 14.

(i) *For the range family $\mathcal{R}_{\mathrm{BL}} \cup \mathcal{R}_{\mathrm{NW}} \cup \mathcal{R}_{\mathrm{NE}}$, we have $m(k) \leqslant 5k - 2$ for all k.*
(ii) *For the range family $\mathcal{R}_{\mathrm{HS}} \cup \mathcal{R}_{\mathrm{VS}} \cup \mathcal{R}_{\mathrm{NW}} \cup \mathcal{R}_{\mathrm{NE}} \cup \mathcal{R}_{\mathrm{SW}} \cup \mathcal{R}_{\mathrm{SE}}$, we have $m(k) \leqslant 10k - 1$ for all k.*
(iii) *For the range family $\mathcal{R}_{\mathrm{HS}} \cup \mathcal{R}_{\mathrm{VS}} \cup \mathcal{R}_{\mathrm{DS}} \cup \mathcal{R}_{\mathrm{NW}} \cup \mathcal{R}_{\mathrm{SE}}$, we have $m(k) \leqslant \lceil 4k \ln k + k \ln 3 \rceil + 4k$ for all k.*

Proof. In all cases, we combine Lemmas 12 and 13 with some known results.

Now we prove (i). By Lemma 12, for every point set V and every $m \in \mathbb{N}$, there exist subsets $S_{\mathrm{NW}}, S_{\mathrm{NE}} \subseteq V$ such that $S_{\mathrm{NW}} \cup S_{\mathrm{NE}}$ is a 2-shallow hitting set of $\mathcal{H}(V, \mathcal{R}_{\mathrm{NW}} \cup \mathcal{R}_{\mathrm{NE}}, m)$ and every hyperedge of $\mathcal{H}(V, \mathcal{R}_{\mathrm{BL}}, m)$ is hit at most $1+1 = 2$ times by this set (see [7] for more details). So we use $s = t = 2$. Further, we set $\mathcal{R}_1 = \mathcal{R}_{\mathrm{NW}} \cup \mathcal{R}_{\mathrm{NE}}$, $\mathcal{R}_2 = \mathcal{R}_{\mathrm{BL}}$ and we use $f(k) = 3k - 2$ for all k. By [4], we know that for every k, we have $m_{\mathcal{R}_2}(k) \leqslant f(k)$. So by Lemma 13 for every k we have $m(k) \leqslant (3k-2) + k \max(2,2) = 5k - 2$.

The remaining two claims are proven similarly in the full version of the paper [7]. There we use that for axis-aligned strips we have $m(k) \leqslant 2k-1$ and for strips in three directions, we have $m(k) \leqslant \lceil 4k \ln k + k \ln 3 \rceil$ for all k [3].   □

4   Concluding Remarks

We have considered the range families $\mathcal{R}_{SW}, \mathcal{R}_{SE}, \mathcal{R}_{NE}, \mathcal{R}_{NW}$ of all south-west, south-east, north-east, and north-west quadrants, the range families $\mathcal{R}_{HS}, \mathcal{R}_{VS}$ of all horizontal and vertical strips, and the range families $\mathcal{R}_{BL}, \mathcal{R}_{TL}$ of all bottomless and topless rectangles, each being a special case of axis-aligned rectangles.

For every single family $\mathcal{R}$ in this list it is known that $m(k) = O(k)$, meaning that for every finite point set V the m-uniform hypergraph $\mathcal{H}(V, \mathcal{R}, m)$ admits a polychromatic k-coloring as long as $m = \Omega(k)$.

By Theorems 8 and 9 range capturing hypergraphs with respect to $\mathcal{R} = \mathcal{R}_{BL} \cup \mathcal{R}_{HS}$, i.e., bottomless rectangles and horizontal strips, do not even admit polychromatic 2-colorings. In other words $m(k) = \infty$ for all $k \geqslant 2$ in that case. On the other hand, by Theorems 10 and 14 such polychromatic k-colorings exist for every k (i.e., $m(k) < \infty$) whenever $\mathcal{R}$ is the union of any subset of $\{\mathcal{R}_{SW}, \mathcal{R}_{SE}, \mathcal{R}_{NE}, \mathcal{R}_{NW}, \mathcal{R}_{HS}, \mathcal{R}_{VS}, \mathcal{R}_{BL}, \mathcal{R}_{TL}\}$ that does not include both $\mathcal{R}_{BL}$ and $\mathcal{R}_{HS}$, nor any rotation of that pair. (As horizontal strips form a special case of both 90-degree rotations of bottomless rectangles, our results also cover these left-unbounded and right-unbounded axis-aligned rectangles.)

In general, we observe the same behavior as for other range families in the literature: Either $m(k) < \infty$ holds for every k or already $m(2) = \infty$. It remains an interesting open problem to determine whether in general $m(2) < \infty$ always implies $m(k) < \infty$ for all k. In the positive cases, our upper bounds on $m(k)$ are linear in k, except when $\mathcal{R}$ contains strips of three different directions or bottomless and topless rectangles. It is worth noting that no range family $\mathcal{R}$ is known for which $m(2) < \infty$ but $m(k) \in \omega(k)$. Such a candidate could be $\mathcal{R} = \mathcal{R}_{HS} \cup \mathcal{R}_{VS} \cup \mathcal{R}_{DS}$ or $\mathcal{R} = \mathcal{R}_{BL} \cup \mathcal{R}_{TL}$.

We suggest further investigations of shallow hitting sets in range capturing hypergraphs. To the best of our knowledge, their existence might be equivalent to $m(k)$ being linear in k. In particular, do bottomless rectangles (for which it is known that $m(k) \in O(k)$ [4]) allow for shallow hitting sets? And do octants in 3D (for which shallow hitting sets are known not to exist [6]) have $m(k) \in O(k)$?

Finally, let us remark that the probabilistic construction of Chen et al. [8] for the range family $\mathcal{R}$ of all axis-aligned rectangles shows that the hypergraphs $\mathcal{H}(V, \mathcal{R}, m)$ even have arbitrary large chromatic number for any fixed m, while our explicit construction for the sub-family $\mathcal{R}_{BL} \cup \mathcal{R}_{HS}$ only shows that the chromatic number is at least 3. In fact, the Union Lemma of Damasdi and Pálvölgyi [9] states that if $\mathcal{H}$ is the union of any $k-1$ hypergraphs, each of which admits a polychromatic k-coloring, then $\mathcal{H}$ has a proper k-coloring. In particular, every hypergraph $\mathcal{H}(V, \mathcal{R}_{BL} \cup \mathcal{R}_{HS}, m)$ has chromatic number at most

3 for $m \geqslant 4$, and every hypergraph $\mathcal{H}(V, \mathcal{R}, m)$ has chromatic number at most 5 for $m \geqslant 10$ when $\mathcal{R}$ is the union of all unbounded axis-aligned rectangles.

References

1. The geometric hypergraph zoo. https://coge.elte.hu/cogezoo.html
2. Ackerman, E., Keszegh, B., Vizer, M.: Coloring points with respect to squares. Discrete Comput. Geom. **58**(4), 757–784 (2017). https://doi.org/10.1007/s00454-017-9902-y
3. Aloupis, G., et al.: Colorful strips. Graphs Comb. **27**(3), 327–339 (2011). https://doi.org/10.1007/s00373-011-1014-5
4. Asinowski, A., et al.: Coloring hypergraphs induced by dynamic point sets and bottomless rectangles. In: Dehne, F., Solis-Oba, R., Sack, J.-R. (eds.) WADS 2013. LNCS, vol. 8037, pp. 73–84. Springer, Heidelberg (2013). https://doi.org/10.1007/978-3-642-40104-6_7
5. Cardinal, J., Knauer, K., Micek, P., Ueckerdt, T.: Making octants colorful and related covering decomposition problems. SIAM J. Discret. Math. **28**(4), 1948–1959 (2014). https://doi.org/10.1137/140955975
6. Cardinal, J., Knauer, K., Micek, P., Pálvölgyi, D., Ueckerdt, T., Varadarajan, N.: Colouring bottomless rectangles and arborescences. arXiv preprint arXiv:1912.05251 (2020)
7. Chekan, V., Ueckerdt, T.: Polychromatic colorings of unions of geometric hypergraphs (2021). https://arxiv.org/abs/2112.02894
8. Chen, X., Pach, J., Szegedy, M., Tardos, G.: Delaunay graphs of point sets in the plane with respect to axis-parallel rectangles. Random Struct. Algorithms **34**(1), 11–23 (2009). https://doi.org/10.1002/rsa.20246
9. Damásdi, G., Pálvölgyi, D.: Three-chromatic geometric hypergraphs. arXiv preprint arXiv:2112.01820v1 (2021)
10. Gibson, M., Varadarajan, K.: Decomposing coverings and the planar sensor cover problem. In: 2009 50th Annual IEEE Symposium on Foundations of Computer Science, pp. 159–168 (2009)
11. Keszegh, B.: Coloring half-planes and bottomless rectangles. Comput. Geom. **45**(9), 495–507 (2012). https://doi.org/10.1016/j.comgeo.2011.09.004. ISSN 0925-7721,The 19th Canadian Conference on Computational Geometry (CCCG2007) held in Ottawa, Canada on August 20–22 2007
12. Keszegh, B., Pálvölgyi, D.: More on decomposing coverings by octants. J. Computat. Geom. **6**(1), 300–315 (2015). https://doi.org/10.20382/jocg.v6i1a13
13. Keszegh, B., Pálvölgyi, D.: An abstract approach to polychromatic coloring: shallow hitting sets in ABA-free hypergraphs and pseudohalfplanes. J. Comput. Geom. **10**(1), 1–26 (2019). https://doi.org/10.20382/jocg.v10i1a1
14. Keszegh, B., Lemons, N., Pálvölgyi, D.: Online and quasi-online colorings of wedges and intervals. Order **33**(3), 389–409 (2015). https://doi.org/10.1007/s11083-015-9374-8
15. Kovács, I.: Indecomposable coverings with homothetic polygons. Discrete Comput. Geom. **53**(4), 817–824 (2015). https://doi.org/10.1007/s00454-015-9687-9
16. Middendorf, M., Pfeiffer, F.: The max clique problem in classes of string-graphs. Discrete Math. **108**(1), 365–372 (1992). https://doi.org/10.1016/0012-365X(92)90688-C. ISSN 0012-365X

17. Pach, J.: Decomposition of multiple packing and covering. In: 2. Kolloquium über Diskrete Geometrie, number CONF. Institut für Mathematik der Universität Salzburg, pp. 169–178 (1980). URL https://infoscience.epfl.ch/record/129388

18. Pach, J.: Covering the plane with convex polygons. Discrete Comput. Geom. 1(1), 73–81 (1986). https://doi.org/10.1007/BF02187684

19. Pach, J., Pálvölgyi, D.: Unsplittable coverings in the plane. Adv. Math. 302, 433–457 (2016). https://doi.org/10.1016/j.aim.2016.07.011. ISSN 0001-8708

20. Pach, J., Tardos, G.: Tight lower bounds for the size of epsilon-nets. J. Am. Math. Soc. 26(3), 645–658 (2013). https://doi.org/10.1090/S0894-0347-2012-00759-0

21. Pach, J., Tardos, G., Tóth, G.: Indecomposable coverings. In: Akiyama, J., Chen, W.Y.C., Kano, M., Li, X., Yu, Q. (eds.) CJCDGCGT 2005. LNCS, vol. 4381, pp. 135–148. Springer, Heidelberg (2007). https://doi.org/10.1007/978-3-540-70666-3_15

22. Pach, J., Pálvölgyi, D., Tóth, G.: Survey on decomposition of multiple coverings. In: Bárány, I., Böröczky, K.J., Tóth, G.F., Pach, J. (eds.) Geometry—Intuitive, Discrete, and Convex. BSMS, vol. 24, pp. 219–257. Springer, Heidelberg (2013). https://doi.org/10.1007/978-3-642-41498-5_9

23. Pálvölgyi, D.: Indecomposable coverings with unit discs. arXiv preprint arXiv:1310.6900v1 (2013)

24. Pálvölgyi, D., Tóth, G.: Convex polygons are cover-decomposable. Discrete Comput. Geom. 43(3), 483–496 (2009). https://doi.org/10.1007/s00454-009-9133-y

25. Smorodinsky, S., Yuditsky, Y.: Polychromatic coloring for half-planes. J. Comb. Theor. Ser. A 119(1), 146–154 (2012)

26. Varadarajan, K.: Weighted geometric set cover via quasi-uniform sampling. In: Proceedings of the Forty-Second ACM Symposium on Theory of Computing, STOC 2010, pp. 641–648. Association for Computing Machinery, New York (2010). https://doi.org/10.1145/1806689.1806777. ISBN 9781450300506

Kernelization for Feedback Vertex Set via Elimination Distance to a Forest

David Dekker and Bart M. P. Jansen$^{(\boxtimes)}$

Eindhoven University of Technology, P.O. Box 513,
5600 MB Eindhoven, The Netherlands
`b.m.p.jansen@tue.nl`

Abstract. We study efficient preprocessing for the undirected FEED-BACK VERTEX SET problem, a fundamental problem in graph theory which asks for a minimum-sized vertex set whose removal yields an acyclic graph. More precisely, we aim to determine for which parameterizations this problem admits a polynomial kernel. While a characterization is known for the related VERTEX COVER problem based on the recently introduced notion of bridge-depth, it remained an open problem whether this could be generalized to FEEDBACK VERTEX SET. The answer turns out to be negative; the existence of polynomial kernels for structural parameterizations for FEEDBACK VERTEX SET is governed by the elimination distance to a forest. Under the standard assumption NP $\not\subseteq$ coNP/poly, we prove that for any minor-closed graph class $\mathcal{G}$, FEEDBACK VERTEX SET parameterized by the size of a modulator to $\mathcal{G}$ has a polynomial kernel if and only if $\mathcal{G}$ has bounded elimination distance to a forest. This captures and generalizes all existing kernels for structural parameterizations of the FEEDBACK VERTEX SET problem.

Keywords: Feedback Vertex Set · Kernelization · Elimination distance

1 Introduction

For NP-complete problems, a polynomial time algorithm solving any problem instance exactly is unlikely to exist. However, as one is often interested in solving specific instances, one can try to exploit characteristics of problem instances and develop algorithms that are fast when the input has certain properties. We therefore associate a parameter with each problem instance. In our context, a problem instance is a graph for which we ask for the existence of a vertex set of size at most ℓ having certain properties. Such a parameterized instance can be denoted with a triple (G, ℓ, k), where we are asking for the existence of a solution of size at most ℓ for a graph G with parameter k. We say that an algorithm is *fixed parameter tractable* (FPT) for such a parameterization if it

B. M. P. Jansen—This project has received funding from the European Research Council (ERC) under the European Union's Horizon 2020 research and innovation programme (grant agreement No 803421, ReduceSearch).

M. A. Bekos and M. Kaufmann (Eds.): WG 2022, LNCS 13453, pp. 158–172, 2022.
https://doi.org/10.1007/978-3-031-15914-5_12

solves any instance (G, ℓ, k) of size n, as described above, in time bounded by $f(k)n^{\mathcal{O}(1)}$ for some computable function $f \colon \mathbb{N} \to \mathbb{N}$.

A strongly related field is that of *kernelization*. This field focuses on reducing a parameterized instance (G, ℓ, k) in polynomial time to an equivalent instance (G', ℓ', k') whose size is bounded by a computable function of the parameter. We speak of a polynomial kernel when this function is a polynomial. It is known that a decidable parameterized problem is fixed parameter tractable if and only if it admits a kernelization (cf. [12, Proposition 4.7.1]). In our quest for determining which parameterizations enable efficient algorithms, it is therefore interesting to determine those that allow a polynomial kernel.

This paper focuses on polynomial kernels for the undirected FEEDBACK VER-TEX SET problem, which is an NP-complete problem in graph theory as originally identified by Karp [21]. For an undirected graph G, a vertex set $X \subseteq V(G)$ is a *feedback vertex set* if the graph is acyclic after removal of X. We call a vertex set whose removal yields a graph in some graph class $\mathcal{G}$ a $\mathcal{G}$-modulator and define the deletion distance to $\mathcal{G}$ as its minimum size. The FEEDBACK VERTEX SET problem then asks for the minimum size of such a feedback vertex set, or equivalently, the deletion distance to a forest. For a graph G, we let $\text{FVS}(G)$ (the *feedback vertex number* of G) denote that minimum size. Our main question is for which parameterizations the FEEDBACK VERTEX SET problem admits a polynomial kernel.

Before exploring the FEEDBACK VERTEX SET problem further, we should mention the related VERTEX COVER problem. It asks for a minimum set of vertices hitting all edges in a graph. While a kernel in the solution size with a linear number of vertices can be obtained using various techniques [1, 7–9, 24], a polynomial kernel in a structurally smaller parameter was only discovered in 2011, when Jansen and Bodlaender developed a polynomial kernel in the feedback vertex number of a graph [18]. From there, many polynomial kernels for VERTEX COVER were discovered in modulators to even larger graph classes [4, 13, 15, 23]. In 2020, Bougeret, Jansen and Sau proved the following characterization under common hardness assumptions: VERTEX COVER admits a polynomial kernel in the size of a modulator to a minor-closed graph family $\mathcal{G}$ if and only if $\mathcal{G}$ has bounded *bridge-depth* [3]. With this result, they generalized all existing work on kernels in the size of modulators to minor-closed graph families, and they proved that their results cannot be improved further under common hardness assumptions.

For FEEDBACK VERTEX SET, the first polynomial kernel with size bound $\mathcal{O}(k^{11})$ was obtained in 2006 and it was subsequently improved to a quadratic kernel [2, 6, 26]. After the improvements for VERTEX COVER, researchers also tried to develop polynomial kernels in smaller parameters for FEEDBACK VER-TEX SET [17, 20, 22]. It remained an open problem whether these results could be generalized further or whether there exists some parameter that characterizes FEEDBACK VERTEX SET similarly to how bridge-depth characterizes VERTEX COVER. In particular, Bougeret, Jansen and Sau suggested in their paper on VERTEX COVER that the deletion distance to constant bridge-depth might also

be an interesting parameter to consider for problems such as FEEDBACK VER-
TEX SET. We therefore aim to answer the question for which graph families $\mathcal{G}$
the FEEDBACK VERTEX SET problem admits a polynomial kernel when parameterized by the size of a $\mathcal{G}$-modulator.

Our Results. To our initial surprise, the results for VERTEX COVER cannot be
generalized to FEEDBACK VERTEX SET. It turns out that a minor-closed graph
family $\mathcal{G}$ must have bounded *elimination distance to a forest* (Definition 1), in
order to allow a polynomial kernel in a $\mathcal{G}$-modulator. This concept was introduced
by Bulian and Dawar [5] and is another generalization of the more common
parameter treedepth [25]. The elimination distance to a forest is the minimum
number of rounds needed to transform the graph into a forest when removing one
vertex from each connected component in each round. Our result is described in
Theorem 1. Proofs of statements marked ($\star$) are deferred to the full version [10].

Theorem 1 ($\star$). *Assume NP $\not\subseteq$ coNP/poly and let $\mathcal{G}$ be a minor-closed graph
family. Then FEEDBACK VERTEX SET admits a polynomial kernel in the size of
a $\mathcal{G}$-modulator if and only if $\mathcal{G}$ has bounded elimination distance to a forest.*

The minor-closed and hardness assumptions are only needed for the lower
bound. To the best of our knowledge, our kernel generalizes all known polynomial kernels for the FEEDBACK VERTEX SET problem. Both the kernel and its
correctness proof follow the structure of the kernel for $\mathcal{F}$-MINOR FREE DELE-
TION in the deletion distance to a graph of constant treedepth by Jansen and
Pieterse [20]. The correctness proof of their kernel crucially relies on their Lemma
3 whose technical proof spans thirty pages. We require a variation of this lemma.
On the one hand, our variation is more involved since it deals with elimination
distance to a forest rather than treedepth; on the other hand, it is simpler since
it concerns only FEEDBACK VERTEX SET rather than $\mathcal{F}$-MINOR FREE DELE-
TION. As a result of this simplification, we can formulate the lemma without
the use of minors. Roughly speaking, the lemma says that in a graph G of
bounded elimination distance to a forest, if no minimum feedback vertex set
exists which simultaneously hits a prescribed set of *partial* cycles (single vertices
in a set S or paths between two terminals in a set T), then the same holds for
some sets $S^* \subseteq S$ and $T^* \subseteq T$ of constant size. As shown in previous work,
this limited sensitivity with respect to whether optimal solutions can break all
partial forbidden structures is crucial for the kernelization complexity. As one
of our main contributions, we prove this lemma using a strategy that differs
significantly from the one followed in earlier work [20].

Lemma 1 ($\star$). *Let G be a connected graph with disjoint vertex sets $S, T \subseteq
V(G)$. Suppose that any minimum feedback vertex set X of G misses some vertex
from S or leaves two vertices from T connected in $G - X$. Then there exist sets
$S^* \subseteq S$ and $T^* \subseteq T$ whose sizes only depend on the elimination distance to a
forest of G, such that any minimum feedback vertex set X of G misses some
vertex from S^* or leaves two vertices from T^* connected in $G - X$.*

Once Lemma 1 is proven, the kernelization upper bounds follow similarly to earlier work [20]. As for the lower bound in Theorem 1, we are also able to generalize our proof for other $\mathcal{F}$-MINOR FREE DELETION problems as described in Theorem 2.

Theorem 2 ($\star$). *Let $\mathcal{G}$ be a minor-closed family of graphs and let $\mathcal{F}$ be a finite set of biconnected planar graphs on at least three vertices. If $\mathcal{G}$ has unbounded elimination distance to an $\mathcal{F}$-minor free graph, then $\mathcal{F}$-MINOR FREE DELETION does not admit a polynomial kernel in the size of a $\mathcal{G}$-modulator, unless* NP $\subseteq$ coNP/poly.

Organization. Section 2 introduces all relevant terminology. Section 3 presents our kernel and thereby proves the 'if' direction of Theorem 1. Then Sect. 4 contains the proof of Theorem 2, thereby also proving the 'only if' direction of Theorem 1. Lastly, Sect. 5 contains our conclusions and discusses future work.

2 Preliminaries

For a positive integer n, we use the shorthand $[n]$ for the set of all natural numbers i with $1 \leq i \leq n$. All graphs we consider are finite, undirected and simple. When G is a graph, we let $V(G)$ denote the vertex set of G and $E(G)$ the edge set. For $S \subseteq V(G)$, the graph $G - S$ is the graph where all vertices in S and all incident edges are removed, and the graph $G[S]$ is the subgraph of G induced by the vertices in S. When an edge exists between two vertices in G, we say that these vertices are *adjacent*. The *neighbors* of v in G, denoted with $N_G(v)$, are the vertices adjacent to a vertex $v \in V(G)$ in G. For $S \subseteq V(G)$, we say that $v \in V(G - S)$ is adjacent to S if there exists some edge between v and a vertex in S. The set $N_G(S)$ contains all vertices $v \in V(G - S)$ for which this holds. We will sometimes slightly abuse notation and speak of a vertex being adjacent to some subgraph, rather than to the vertices in that subgraph. We say that two vertices are *connected* in G when they are in the same connected component. The set $\mathrm{CC}(G)$ denotes the set of connected components (or shortly components) of G. For sets $S, T \subseteq V(G)$, we say that S *separates* T if each component of $G - S$ contains at most one vertex from T. Notice that we do not require S and T to be disjoint. Such a set S is a *vertex multiway cut* of T in G. We use the notation $\mathcal{O}_\eta(f(n))$ to describe the functions in η and n which can be bounded by $g(\eta) \cdot f(n)$ for some computable function g.

A concept that will be used extensively is the concept of graph minors. This uses the notion of *edge contraction*. When uv is an edge in a graph G, contracting this edge replaces vertices u and v by a new vertex whose set of neighbors is $N_G(\{u, v\})$. Now H is a *minor* of G if H can be obtained from G by removing vertices, removing edges and contracting edges. Alternatively, one can define H to be a minor of G if there exists a *minor model* $\phi \colon V(H) \to 2^{V(G)}$, such that for any $v \in V(H)$ the graph $G[\phi(u)]$ is connected, for any distinct $u, v \in V(H)$ we have $\phi(u) \cap \phi(v) = \emptyset$, and for any edge uv in H there exists an edge between a vertex in $\phi(u)$ and a vertex in $\phi(v)$ in G.

A graph G has an $\mathcal{H}$-*minor* for a set of graphs $\mathcal{H}$ if G contains some graph $H \in \mathcal{H}$ as a minor. For a minor model ϕ of H in G and a set $S \subseteq V(H)$, we use the shorthand notation $\phi(S) := \bigcup_{v \in S} \phi(v)$. We say that a minor model ϕ of H in G is *minimal*, if there does not exist a minor model ϕ' of H in G with $\phi'(V(H)) \subsetneq \phi(V(H))$. A minor model ϕ of H in G and a minor model ϕ' of H' in G *intersect* if $\phi(V(H)) \cap \phi'(V(H')) \neq \emptyset$.

2.1   Elimination Distance

Our work relies crucially on the concept of elimination distance as introduced by Bulian and Dawar [5].

Definition 1 (Elimination distance). *Let G be a graph and $\mathcal{G}$ a graph family. Then the elimination distance of G to $\mathcal{G}$ is*

$$\mathrm{ED}_{\mathcal{G}}(G) = \begin{cases} 0 & \text{if } G \in \mathcal{G}, \\ \max_{G' \in \mathrm{CC}(G)} \mathrm{ED}_{\mathcal{G}}(G') & \text{if } |\mathrm{CC}(G)| > 1, \\ \min_{v \in V(G)} \mathrm{ED}_{\mathcal{G}}(G - \{v\}) + 1 & \text{otherwise.} \end{cases}$$

We only consider graph families $\mathcal{G}$ that are minor-closed. We use the shorthand $\mathcal{G}_F$ for the graph family containing precisely all forests. We say that a graph family $\mathcal{G}$ has bounded elimination distance to some graph class $\mathcal{H}$ if there is a constant $c \in \mathbb{N}$ such that all graphs $G \in \mathcal{G}$ satisfy $\mathrm{ED}_{\mathcal{H}}(G) \leq c$. The elimination distance of a graph G to the empty graph is called the *treedepth* of G and is denoted with $\mathrm{TD}(G)$. More intuitively, the elimination distance to a graph class $\mathcal{G}$ can be interpreted as the minimum number of 'elimination iterations' that are necessary to obtain a graph where every connected component is in $\mathcal{G}$. In such an iteration, one is allowed to remove one vertex from each connected component. This interpretation leads to the notion of an elimination forest.

Definition 2 ($\mathcal{G}$-elimination forest). *Let G be a graph and $\mathcal{G}$ a graph family. A $\mathcal{G}$-elimination forest of G is a tuple $\left(F, (B_u)_{u \in V(F)}\right)$ where F is a rooted forest and where each vertex $v \in V(F)$ has a bag $B_v \subseteq V(G)$ such that:*

- *The bags define a partition of $V(G)$, i.e. for any vertex $v \in V(G)$ there is a unique vertex $u \in V(F)$ with $v \in B_u$.*
- *For any non-leaf u of F, its bag B_u contains precisely one vertex.*
- *For any leaf u of F, the graph $G[B_u]$ is connected and $G[B_u] \in \mathcal{G}$.*
- *For any edge uv in G, let $s, t \in V(F)$ be the vertices such that $u \in B_s$ and $v \in B_t$. Then s is an ancestor of t, or t is an ancestor of s in F.*

We define the height of an elimination forest F to be the maximum number of edges on a path from the root to a leaf in F. One can prove with induction that this height is equal to the elimination distance we defined earlier.

We will use these elimination forests extensively for our kernel and therefore introduce some shorthand notation. Let $(F, (B_u)_{u \in V(F)})$ be a $\mathcal{G}$-elimination forest. Let v be a vertex in F. The *tail* of v, denoted with $\mathrm{TAIL}(v)$, is defined as the union of B_u over all proper ancestors u of v. The closed tail $\text{TAIL}[v]$ also includes B_v. Similarly, $\text{TREE}(v)$ denotes the union of B_u over all proper descendants u of v and $\text{TREE}[v]$ also includes B_v. The subgraph of G induced by all vertices in $\text{TREE}[v]$ is denoted with G_v. We will sometimes slightly abuse notation and use G_v as a vertex set. We use the shorthand G_v^+ to describe the induced subgraph on the vertices in $\text{TREE}[v] \cup \text{TAIL}[v]$.

We will also introduce the notion of *bridge-depth* as introduced by Bougeret, Jansen and Sau [3]. A *bridge* in a graph G is an edge whose removal increases the number of connected components of G. The concept of bridge-depth now allows us to delete a set of vertices S as long as $G[S]$ is connected and each edge in $G[S]$ is a bridge in G. Such a structure $G[S]$ is called a *tree of bridges*. Observe that a single vertex is always a tree of bridges.

Definition 3 (Bridge-depth). *Let G be a graph. The bridge-depth of G is defined as*

$$\text{BD}(G) = \begin{cases} 0 & \text{if } G \text{ is the empty graph,} \\ \max\limits_{G' \in \text{CC}(G)} \text{BD}(G') & \text{if } |\text{CC}(G)| > 1, \\ \min\limits_{\substack{S \subseteq V(G): \\ G[S] \text{ is a tree of bridges}}} \text{BD}(G - S) + 1 & \text{otherwise.} \end{cases}$$

Cf. [3] for equivalent definitions. Lastly, we sometimes use the more common concept of *treewidth*. The treewidth of a graph G is denoted with $\text{TW}(G)$. We mention some useful properties of these concepts in Proposition 1.

Proposition 1 ($\star$). *Let G be a graph with $\mathcal{G}_F$-elimination forest $(F, (B_u)_{u \in V(F)})$ and let η be an integer such that $\text{ED}_{\mathcal{G}_F}(G) \leq \eta$. Let X be a minimum feedback vertex set in G and let v be a leaf in F. Then the following claims hold.*

1. $\text{TW}(G) \leq \text{BD}(G) \leq \text{ED}_{\mathcal{G}_F}(G) + 1$.
2. *X contains at most η vertices from B_v.*
3. *If there exists a path in G from a vertex in B_v to a vertex outside B_v, then this path contains a vertex in $\text{TAIL}(v)$.*
4. *Let $u \in V(F)$, then u has at most η children c where $X \cap G_c$ is not a minimum feedback vertex set in G_c.*

3   Kernelization Upper Bounds

Our kernel follows the structure of the polynomial kernel for $\mathcal{F}$-MINOR FREE DELETION when parameterized by a treedepth-η modulator for some integer η [20]. Our kernel relies crucially on the reduction rule specified in Lemma 2.

Lemma 2 ($\star$ Cf. [20], Lemma 6). *There is a polynomial-time algorithm that, given a graph G with modulator $X \subseteq V(G)$ such that $\text{ED}_{\mathcal{G}_F}(G - X) \leq \eta$ for a constant η, outputs an induced subgraph G' of G together with an integer Δ such that $\text{FVS}(G) = \text{FVS}(G') + \Delta$ and $G' - X$ has at most $|X|^{\mathcal{O}_\eta(1)}$ components.*

We can use this reduction rule to obtain a graph G' where $G' - X$ has a bounded number of connected components. We can then identify a set of vertices $Y \subseteq V(G' - X)$ with $|Y| \leq |X|^{\mathcal{O}_{\eta}(1)}$ such that $\text{ED}_{\mathcal{G}_F}(G' - X - Y) < \eta$. By definition of elimination distance, every connected component C of $G' - X$ contains a vertex whose removal decreases $\text{ED}_{\mathcal{G}_F}(C)$. As we limited the number of connected components by applying Lemma 2, these vertices constitute a suitable set Y. Now observe that $X \cup Y$ is a modulator to a graph with elimination distance to a forest $\eta - 1$ and that $|X \cup Y|$ is bounded by a polynomial in $|X|$. One can therefore provide an inductive argument which repeatedly applies Lemma 2 and increases the modulator such that the elimination distance to a forest of the remaining graph decreases every iteration. Once we obtain a modulator to a graph with elimination distance to a forest 1 (which is a forest), we can apply a known polynomial kernel in the size of a feedback vertex set [16].

The reduction rule of Lemma 2 follows from the following key lemma using known techniques. We therefore focus our attention on the proof of Lemma 1.

Lemma 1 ($\star$). *Let G be a connected graph with disjoint vertex sets $S, T \subseteq V(G)$. Suppose that any minimum feedback vertex set X of G misses some vertex from S or leaves two vertices from T connected in $G - X$. Then there exist sets $S^* \subseteq S$ and $T^* \subseteq T$ whose sizes only depend on the elimination distance to a forest of G, such that any minimum feedback vertex set X of G misses some vertex from S^* or leaves two vertices from T^* connected in $G - X$.*

We can split up Lemma 1 into two parts. Lemma 3 will bound the number of vertices in the $\mathcal{G}_F$-elimination tree that contain a vertex in S or T. This part corresponds to the original treedepth formulation in [20, Lemma 3], but is significantly simplified for our restricted setting. Lemma 4 bounds the number of vertices in S and T in a bag of the elimination tree. This covers the generalization to elimination distance to a forest and concludes the proof of Lemma 1. In the full version [10], we show how these lemmas imply Lemma 1.

Lemma 3. *Let G be a connected graph with disjoint vertex sets $S, T \subseteq V(G)$. Let $(R, (B_u)_{u \in V(R)})$ be a $\mathcal{G}_F$-elimination tree of G of height η. Suppose that any minimum feedback vertex set X of G misses a vertex from S or leaves two vertices from T connected. Then this also holds for some subsets $S^* \subseteq S$ and $T^* \subseteq T$, such that any vertex in the elimination tree has at most $3\eta \cdot 2^{\eta}$ children u for which G_u contains a vertex from S^* or T^*.*

Proof. In analogy to the original formulation in [20], a *labeled vertex* is a vertex in S or T. When we remove a label from a vertex, we remove the vertex from S and T while the vertex remains in the graph. Suppose that we pick S^* and T^* such that no minimum feedback vertex set contains S^* and separates T^*, while the latter property does not hold for any other pair of sets S', T' with $S' \subseteq S^*$ and $T' \subseteq T^*$. We claim that for such sets S^* and T^*, any vertex in the elimination tree R has at most $3\eta \cdot 2^{\eta}$ children u for which G_u contains a labeled vertex. We will also refer to the set T^* as the set of terminals.

Assume towards a contradiction that vertex v has more child subtrees with labels. Let these children be $c_1, \ldots, c_\ell$. For each of these children c_i, there exists a minimum feedback vertex set X_i in G that witnesses the fact that the labels cannot be removed from G_{c_i}. This set X_i will therefore miss a vertex in $S^* \cap G_{c_i}$ or leave a vertex in $T^* \cap G_{c_i}$ connected to some other vertex in T^*, while X_i contains all vertices in $S^* \backslash G_{c_i}$ and separates all vertices in $T^* \backslash G_{c_i}$. Define $Z_i :=$ TAIL$[v] \backslash X_i$.

Now fix a set $Z \subseteq$ TAIL$[v]$. We will bound the number of children c_i for which $Z_i = Z$ by 3η. Suppose towards a contradiction that there are $3\eta + 1$ of these children. Let C be the set containing these vertices. Pick some child $c_j \in C$ and observe the following.

- By Proposition 1.4, there are at most η children $c_i \in C$ where $X_j \cap G_{c_i}$ is not a minimum feedback vertex set in G_{c_i}.
- There are at most η children $c_i \in C$ with $i \neq j$ such that a terminal in G_{c_i} is connected to a vertex in Z in $G_{c_i}^+ - X_j$, i.e. (recall the notation in Sect. 2.1) in the induced subgraph on the remaining vertices in the subtree and tail of c_i. Otherwise, two children other than c_j have a terminal connected to the same vertex in Z, while X_j separates all terminals outside G_{c_j}.
- There are at most $\eta - 1$ children $c_i \in C$ such that in $G_{c_i}^+ - X_j$, there exists a path between distinct vertices in Z that uses some vertex in G_{c_i}. Otherwise, we claim that we can directly construct a cycle in $G - X_j$. Consider for example the auxiliary (multi)graph on vertex set Z which contains, for each child $c_i \in C$ for which $G_{c_i}^+ - X_j$ contains such a path, say between $z_1, z_2 \in Z$, one edge $z_1 z_2$. This auxiliary graph contains a cycle since it has too many edges to be acyclic, which implies that there exists a cycle in $G - X_j$.

Pick a child $c_k \in C$ which is neither c_j nor in the list of $3\eta - 1$ children above. As $|C| > 3\eta$, such a vertex exists. By the first item above, we can deduce that $X_j \cap G_{c_k}$ is a minimum feedback vertex set in G_{c_k}. Besides, this set contains $S^* \cap G_{c_k}$, it separates all terminals in $T^* \cap G_{c_k}$, and it separates $T^* \cap G_{c_k}$ from Z. Furthermore, no path exists in $G_{c_k}^+ - X_j$ that connects two vertices in Z and also contains some vertex in G_{c_k}.

Claim 1 ($\star$). The set $X' := (X_k \backslash G_{c_k}) \cup (X_j \cap G_{c_k})$ is a minimum feedback vertex set in G which contains S^* and separates T^*.

Claim 1 contradicts that any minimum feedback vertex set in G misses a vertex in S^* or leaves two vertices in T^* connected. We conclude that there are at most 3η children c_i of v for which a witnessing minimum feedback vertex set has $Z_i = Z$. As there are at most 2^η subsets of TAIL$[v]$ for any non-leaf v, this leads to the bound of at most $3\eta \cdot 2^\eta$ children for which the labels cannot be removed. □

Lemma 4. *Let G be a connected graph with disjoint vertex sets $S, T \subseteq V(G)$. Let $\big(R, (B_u)_{u \in V(R)}\big)$ be a $\mathcal{G}_F$-elimination tree of G of height η. Suppose that any minimum feedback vertex set X of G misses a vertex from S or leaves two* *vertices from T connected. Then this also holds for some subsets $S^* \subseteq S$ and $T^* \subseteq T$, such that for any leaf u in the elimination tree, the set B_u contains at most η vertices from S^* and at most $\mathcal{O}(\eta^2)$ from T^*.*

Proof. Pick some leaf v of elimination tree R, for which we want to ensure that there are $\mathcal{O}(\eta^2)$ vertices with labels among vertices in $Y := B_v$. Define $S_Y := S \cap Y$ and $T_Y := T \cap Y$. Our goal is to obtain subsets $S_Y^* \subseteq S_Y$ and $T_Y^* \subseteq T_Y$ whose sizes are $\mathcal{O}(\eta^2)$, such that every minimum feedback vertex set misses a vertex from $S_Y^* \cup (S \backslash Y)$ or leaves a pair of terminals in $T_Y^* \cup (T \backslash Y)$ connected. By applying this operation to all leaves of the elimination tree, we obtain the sets promised by Lemma 4.

The construction of S_Y^* is straightforward. If $|S_Y| > \eta + 1$, we let S_Y^* be an arbitrary subset of S_Y of size $\eta + 1$. Otherwise, $S_Y^* = S_Y$.

Claim 2. Let X be a minimum feedback vertex set in G. If X misses a vertex in S, then it also misses a vertex in $S_Y^* \cup (S \backslash Y)$.

Proof. If X misses a vertex in $S \backslash Y$, then the implication is trivial. Therefore assume X misses a vertex in S_Y. If this vertex is not in S_Y^*, then $|S_Y^*| = \eta + 1$ by construction. By Proposition 1.2, we know that $|X \cap S_Y^*| \le \eta$ so X misses a vertex in S_Y^*.   ∎

For the construction of T_Y^* we distinguish two cases. First, we assume that T_Y cannot be separated with η vertices in $G[Y]$ and make the following observation.

Proposition 2 ($\star$). *Let G be a tree and $T \subseteq V(G)$. If T cannot be separated with η vertices, then there exist $\eta + 1$ vertex-disjoint paths whose endpoints are distinct vertices in T.*

We define T_Y^* by taking the $2\eta + 2$ endpoints of the paths guaranteed by Proposition 2. Observe that these vertices are all in T_Y.

Claim 3. Suppose that T_Y cannot be separated with η vertices in $G[Y]$. Let X be a minimum feedback vertex set in G. Then X leaves two vertices in T_Y^* connected.

Proof. By Proposition 1.2, X can only intersect η of the $\eta + 1$ vertex-disjoint paths that were obtained through Proposition 2. Therefore, at least one path is disjoint from X, so its endpoints in T_Y^* are connected in $G - X$.   ∎

It remains to consider the case where T_Y can be separated with η vertices. Let Z be a vertex multiway cut of T_Y in $G[Y]$ with $|Z| \le \eta$ and let $\mathcal{C} := \mathrm{cc}(G[Y] - Z)$. Observe that each of these connected components is a tree with at most one vertex in T_Y. Let $\mathcal{C}_T \subseteq \mathcal{C}$ be the set of components that contain a vertex in T_Y. We are now going to mark components. For each $z \in Z$, mark $\eta + 2$ components in $\mathcal{C}_T$ that are adjacent to z in $G[Y]$, or all if there are fewer. Similarly, for each $u \in \mathrm{TAIL}(v)$ we mark up to $\eta + 2$ components in $\mathcal{C}_T$ that are adjacent to v in G_v^+. Then we define T_Y^* to be the union of all vertices in T_Y in the marked components, together with $Z \cap T_Y$. These are at most $\eta(\eta + 2) + \eta(\eta + 2) + \eta = \mathcal{O}(\eta^2)$ vertices.

Claim 4. Suppose that T_Y can be separated with η vertices in $G[Y]$. Let X be a minimum feedback vertex set in G and suppose that X leaves two vertices in T connected. Then X also leaves two vertices in $T_Y^* \cup (T \setminus Y)$ connected.

Proof. Let Z be the vertex multiway cut used in the construction of T_Y^* and let $t_1, t_2 \in T$ be two terminals that are connected in $G - X$. If they are both in $T_Y^* \cup (T \setminus Y)$, then the implication is trivial, so assume that $t_1 \in T_Y$ but not in T_Y^*. Observe that therefore $t_1 \notin Z$. Let P be a path from t_1 to t_2 in $G - X$ and let z be the first vertex on this path that is not in $G[Y] - Z$. We now distinguish two cases. If $z \in Z$, then observe that t_1 was in a component in $\mathcal{C}_T$ that was not marked. Then there are $\eta + 2$ marked components in $\mathcal{C}_T$ adjacent to z in $G[Y]$ of which the terminals are in T_Y^*. Only η of these components can be intersected by X by Proposition 1.2, so there exists a path between two terminals in T_Y^* in $G[Y]$. If $z \notin Z$, then we obtain that $z \in \mathrm{TAIL}(v)$ by Proposition 1.3 and the case follows analogously.   ∎

This concludes the construction of the sets S_Y^* and T_Y^*. If any minimum feedback vertex set in G misses a vertex in S or leaves a pair of terminals in T connected, then it also misses a vertex in $S_Y^* \cup (S \setminus Y)$ or leaves a pair of terminals in $T_Y^* \cup (T \setminus Y)$ connected. By applying this operation to all leaves of the elimination tree, we obtain the promised sets S^* and T^* which concludes the proof of Lemma 4.   □

With Lemma 3 and Lemma 4 proven, we conclude the proof of Lemma 1: if any minimum feedback vertex set in a graph G misses a vertex from a set $S \subseteq V(G)$ or leaves two terminals in a set $T \subseteq V(G)$ connected, then this property also holds for sets $S^* \subseteq S$ and $T^* \subseteq T$ whose sizes only depend on $\mathrm{ED}_{\mathcal{G}_F}(G)$. This is the key ingredient for the proof of Lemma 2, which leads to the kernel upper bound.

4   Kernelization Lower Bounds

In this section we summarize the main ideas behind the lower bound. We first introduce the notion of a necklace, which turns out to be a crucial structure.

Definition 4. *Let G be a graph and let $\mathcal{F}$ be a collection of connected graphs. G is an $\mathcal{F}$-necklace of length t if there exists a partition of $V(G)$ into $S_1, \ldots, S_t$ such that*

- *$G[S_i] \in \mathcal{F}$ for each $i \in [t]$ (these subgraphs are the* beads *of the necklace),*
- *G has precisely one edge between S_i and S_{i+1} for each $i \in [t-1]$,*
- *G has no edges between any other pair of sets S_i and S_j.*

When the length of the necklace is not relevant, we simply speak of an $\mathcal{F}$-necklace. The following definition specifies a special type of necklace.

Definition 5. *Let $\mathcal{F}$ be a collection of connected graphs. Let G be an $\mathcal{F}$-necklace of length t. We say that G is a* uniform necklace *if it satisfies two additional conditions.*

- *There exists a graph $H \in \mathcal{F}$ such that each bead $G[S_i]$ is isomorphic to H.*
- *There exist $x, y \in V(H)$ and graph isomorphisms $f_i \colon V(H) \to V(G[S_i])$ for each bead $G[S_i]$, such that for each $i \in [t-1]$, the edge between $G[S_i]$ and $G[S_{i+1}]$ has precisely the endpoints $f_i(x)$ and $f_{i+1}(y)$.*

These concepts are used to derive the following characterization. We say that a set contains arbitrarily long necklaces if there does not exist a constant c such that each necklace in the set has length at most c.

Lemma 5 ($\star$). *Let $\mathcal{F}$ be a finite collection of connected planar graphs. Any minor-closed graph family $\mathcal{G}$ with unbounded elimination distance to an $\mathcal{F}$-minor free graph contains arbitrarily long uniform $\mathcal{F}$-necklaces.*

Then we will prove the following lemma by giving a reduction from CNF SATISFIABILITY parameterized by the number of variables [11].

Lemma 6 ($\star$). *Let $\mathcal{F}$ be a finite set of biconnected planar graphs on at least three vertices and let $\mathcal{G}$ be a minor-closed graph family. If $\mathcal{G}$ contains arbitrarily long uniform $\mathcal{F}$-necklaces, then $\mathcal{F}$-MINOR FREE DELETION does not admit a polynomial kernel in the size of a $\mathcal{G}$-modulator, unless $\mathrm{NP} \subseteq \mathrm{coNP/poly}$.*

Lemma 5 and Lemma 6 together directly imply Theorem 2. We will explain the main ideas of the proof of Lemma 5 here. Our proof follows the proof by Bougeret et al. when they characterize graph families with unbounded bridge-depth [3]. Similar to their work, we define $\mathrm{NM}_{\mathcal{F}}(G)$ to be the length of the longest $\mathcal{F}$-necklace that a graph G contains as a minor for a family of connected graphs $\mathcal{F}$. Our goal is now to prove the existence of a small set X such that $\mathrm{NM}_{\mathcal{F}}(G - X) < \mathrm{NM}_{\mathcal{F}}(G)$ as described in Lemma 7.

Lemma 7 ($\star$). *Let $\mathcal{F}$ be a collection of connected planar graphs. Then there exists a polynomial function $f_{\mathcal{F}} \colon \mathbb{N} \to \mathbb{N}$ such that for any connected graph G with $\mathrm{NM}_{\mathcal{F}}(G) = t$, there exists a set $X \subseteq V(G)$ with $|X| \leq f_{\mathcal{F}}(t)$ such that $\mathrm{NM}_{\mathcal{F}}(G - X) < t$.*

Bougeret et al. showed that one can derive a bounding function when the considered structures satisfy the Erdős-Pósa property [3]. This also is the case for $\mathcal{F}$-necklaces when the graphs in $\mathcal{F}$ are connected and planar, so this approach would be suitable for our purposes as well. To derive a polynomial bound on the size of X, we use a different argument that uses treewidth and grid minors. We start with the following property of planar graphs.

Proposition 3 ($\star$). *Any planar graph G on n vertices is a minor of the $4n \times 4n$ grid.*

Together with the Excluded Grid Theorem, this leads to the following treewidth bound.

Lemma 8 ($\star$). *Let $\mathcal{F}$ be a collection of connected planar graphs of at most n vertices each. There exists a polynomial $f \colon \mathbb{N} \to \mathbb{N}$ with $f(g) = \mathcal{O}(g^{19} \mathrm{poly} \log g)$ such that for any graph G with $\mathrm{NM}_{\mathcal{F}}(G) = t$, it holds that $\mathrm{TW}(G) < f(4n(t+1))$.*

To use this treewidth bound, we need a property similar to [3, Lemma 4.6].

Proposition 4 ($\star$). *For any family of connected graphs $\mathcal{F}$ and connected graph G with $\mathrm{NM}_{\mathcal{F}}(G) > 0$, any pair of minor models of $\mathcal{F}$-necklaces of length $\mathrm{NM}_{\mathcal{F}}(G)$ in G must intersect.*

Proposition 4 is a generalization of the idea that in any connected graph, two paths of maximum length must intersect at a vertex. Given a graph G with a tree decomposition, we can use this property to identify a vertex in the tree decomposition such that the removal of all vertices in its bag decreases $\mathrm{NM}_{\mathcal{F}}(G)$. This result is described in Lemma 9.

Lemma 9 ($\star$). *Let $\mathcal{F}$ be a collection of connected graphs. Let G be a connected graph with $\mathrm{TW}(G) = w$ and $\mathrm{NM}_{\mathcal{F}}(G) = t$. Then there exists a set $Z \subseteq V(G)$ with $|Z| \leq w + 1$ such that $\mathrm{NM}_{\mathcal{F}}(G - Z) < t$.*

The proof of Lemma 7 follows directly by combining Lemma 8 and Lemma 9. An inductive argument, analogous to [3, Theorem 4.8], remains to use this result to prove Lemma 5.

5   Conclusion and Discussion

We conclude that the elimination distance to a forest characterizes the FEED-BACK VERTEX SET problem in terms of polynomial kernelization. For a minor-closed graph family $\mathcal{G}$, the problem admits a polynomial kernel in the size of a $\mathcal{G}$-modulator if and only if $\mathcal{G}$ has bounded elimination distance to a forest, assuming NP $\not\subseteq$ coNP/poly. In particular, this implies that FEEDBACK VERTEX SET does not admit a polynomial kernel in the deletion distance to a graph of constant bridge-depth under the mentioned hardness assumption. We also generalize the lower bound to other $\mathcal{F}$-MINOR FREE DELETION problems where $\mathcal{F}$ contains only biconnected planar graphs on at least three vertices. It remains unknown whether such a lower bound also generalizes to collections of graphs $\mathcal{F}$ that contain non-planar graphs.

An interesting open problem is whether similar polynomial kernels can be obtained for other $\mathcal{F}$-MINOR FREE DELETION problems. Regarding the field of fixed parameter tractable algorithms, it was recently shown [19] that for any set $\mathcal{F}$ of connected graphs, $\mathcal{F}$-MINOR FREE DELETION admits an FPT algorithm when parameterized by the elimination distance to an $\mathcal{F}$-minor free graph (or even $\mathcal{H}$-treewidth when $\mathcal{H}$ is the class of $\mathcal{F}$-minor free graphs). This generalizes known FPT algorithms for the natural parameterization by solution size. Regarding polynomial kernels, $\mathcal{F}$-MINOR FREE DELETION problems admit a polynomial kernel in the solution size when $\mathcal{F}$ contains a planar graph [14]. Do polynomial kernels exist when the problem is parameterized by a modulator to a graph of constant elimination distance to being $\mathcal{F}$-minor free?

References

1. Abu-Khzam, F.N., Fellows, M.R., Langston, M.A., Suters, W.H.: Crown structures for vertex cover kernelization. Theor. Comput. Syst. **41**(3), 411–430 (2007). https://doi.org/10.1007/s00224-007-1328-0

2. Bodlaender, H.L., van Dijk, T.C.: A cubic kernel for feedback vertex set and loop cutset. Theor. Comput. Syst. **46**(3), 566–597 (2010). https://doi.org/10.1007/s00224-009-9234-2

3. Bougeret, M., Jansen, B.M.P., Sau, I.: Bridge-depth characterizes which structural parameterizations of vertex cover admit a polynomial kernel. In: Czumaj, A., Dawar, A., Merelli, E. (eds.) 47th International Colloquium on Automata, Languages, and Programming, ICALP 2020, 8–11 July 2020, Saarbrücken, Germany (Virtual Conference). LIPIcs, vol. 168, pp. 1–19. Schloss Dagstuhl - Leibniz-Zentrum für Informatik (2020). https://doi.org/10.4230/LIPIcs.ICALP.2020.16

4. Bougeret, M., Sau, I.: How much does a treedepth modulator help to obtain polynomial kernels beyond sparse graphs? Algorithmica **81**(10), 4043–4068 (2018). https://doi.org/10.1007/s00453-018-0468-8

5. Bulian, J., Dawar, A.: Fixed-parameter tractable distances to sparse graph classes. Algorithmica **79**(1), 139–158 (2016). https://doi.org/10.1007/s00453-016-0235-7

6. Burrage, K., Estivill-Castro, V., Fellows, M., Langston, M., Mac, S., Rosamond, F.: The undirected feedback vertex set problem has a poly(k) kernel. In: Bodlaender, H.L., Langston, M.A. (eds.) IWPEC 2006. LNCS, vol. 4169, pp. 192–202. Springer, Heidelberg (2006). https://doi.org/10.1007/11847250_18

7. Chen, J., Kanj, I.A., Jia, W.: Vertex cover: further observations and further improvements. J. Algorithms **41**(2), 280–301 (2001). https://doi.org/10.1006/jagm.2001.1186

8. Chlebík, M., Chlebíková, J.: Crown reductions for the minimum weighted vertex cover problem. Discret. Appl. Math. **156**(3), 292–312 (2008). https://doi.org/10.1016/j.dam.2007.03.026

9. Chor, B., Fellows, M., Juedes, D.: Linear kernels in linear time, or how to save k colors in $O(n^2)$ steps. In: Hromkovič, J., Nagl, M., Westfechtel, B. (eds.) WG 2004. LNCS, vol. 3353, pp. 257–269. Springer, Heidelberg (2004). https://doi.org/10.1007/978-3-540-30559-0_22

10. Dekker, D., Jansen, B.M.P.: Kernelization for feedback vertex set via elimination distance to a forest. CoRR abs/2206.04387 (2022). https://doi.org/10.48550/arXiv.2206.04387

11. Dell, H., van Melkebeek, D.: Satisfiability allows no nontrivial sparsification unless the polynomial-time hierarchy collapses. J. ACM **61**(4), 1–27 (2014). https://doi.org/10.1145/2629620

12. Downey, R.G., Fellows, M.R.: Fundamentals of Parameterized Complexity. TCS, Springer, London (2013). https://doi.org/10.1007/978-1-4471-5559-1

13. Fellows, M.R., Lokshtanov, D., Misra, N., Mnich, M., Rosamond, F.A., Saurabh, S.: The complexity ecology of parameters: an illustration using bounded max leaf number. Theor. Comput. Syst. **45**(4), 822–848 (2009). https://doi.org/10.1007/s00224-009-9167-9

14. Fomin, F.V., Lokshtanov, D., Misra, N., Saurabh, S.: Planar $\mathcal{F}$-Deletion: approximation, kernelization and optimal FPT algorithms. In: 53rd Annual IEEE Symposium on Foundations of Computer Science, FOCS 2012, New Brunswick, NJ, USA, 20–23 October 2012, pp. 470–479. IEEE Computer Society (2012). https://doi.org/10.1109/FOCS.2012.62

15. Hols, E.C., Kratsch, S.: Smaller parameters for vertex cover kernelization. In: Lokshtanov, D., Nishimura, N. (eds.) 12th International Symposium on Parameterized and Exact Computation, IPEC 2017, 6–8 September 2017, Vienna, Austria. LIPIcs, vol. 89, pp. 1–12. Schloss Dagstuhl - Leibniz-Zentrum für Informatik (2017). https://doi.org/10.4230/LIPIcs.IPEC.2017.20

16. Iwata, Y.: Linear-time kernelization for feedback vertex set. In: Chatzigiannakis, I., Indyk, P., Kuhn, F., Muscholl, A. (eds.) 44th International Colloquium on Automata, Languages, and Programming, ICALP 2017, 10–14 July 2017, Warsaw, Poland. LIPIcs, vol. 80, pp. 1–14. Schloss Dagstuhl - Leibniz-Zentrum für Informatik (2017). https://doi.org/10.4230/LIPIcs.ICALP.2017.68

17. Jansen, B., Raman, V., Vatshelle, M.: Parameter ecology for feedback vertex set. Tsinghua Sci.Technol. **19**(4), 387–409 (2014). https://doi.org/10.1109/TST.2014.6867520

18. Jansen, B.M.P., Bodlaender, H.L.: Vertex cover kernelization revisited. Theor. Comput. Syst. **53**(2), 263–299 (2012). https://doi.org/10.1007/s00224-012-9393-4

19. Jansen, B.M.P., de Kroon, J.J.H., Włodarczyk, M.: Vertex deletion parameterized by elimination distance and even less. In: Khuller, S., Williams, V.V. (eds.) STOC '21: 53rd Annual ACM SIGACT Symposium on Theory of Computing, Virtual Event, Italy, 21–25 June 2021, pp. 1757–1769. ACM (2021). https://doi.org/10.1145/3406325.3451068

20. Jansen, B.M.P., Pieterse, A.: Polynomial kernels for hitting forbidden minors under structural parameterizations. Theor. Comput. Sci. **841**, 124–166 (2020). https://doi.org/10.1016/j.tcs.2020.07.009

21. Karp, R.M.: Reducibility among combinatorial problems. In: Miller, R.E., Thatcher, J.W. (eds.) Proceedings of a symposium on the Complexity of Computer Computations, Held 20–22 March 1972, at the IBM Thomas J. Watson Research Center, Yorktown Heights, New York, USA. The IBM Research Symposia Series, pp. 85–103. Plenum Press, New York (1972). https://doi.org/10.1007/978-1-4684-2001-2_9

22. Majumdar, D., Raman, V.: Structural parameterizations of undirected feedback vertex set: FPT algorithms and kernelization. Algorithmica **80**(9), 2683–2724 (2018). https://doi.org/10.1007/s00453-018-0419-4

23. Majumdar, D., Raman, V., Saurabh, S.: Polynomial kernels for vertex cover parameterized by small degree modulators. Theor. Comput. Syst. **62**(8), 1910–1951 (2018). https://doi.org/10.1007/s00224-018-9858-1

24. Nemhauser, G.L., Jr., Trotter, L.E.: Vertex packings: structural properties and algorithms. Math. Program. **8**(1), 232–248 (1975). https://doi.org/10.1007/BF01580444

25. Nešetřil, J., Ossona de Mendez, P.: Sparsity - Graphs, Structures, and Algorithms, Algorithms and Combinatorics, vol. 28. Springer, Heidelberg (2012). https://doi.org/10.1007/978-3-642-27875-4

26. Thomassé, S.: A quadratic kernel for feedback vertex set. In: Mathieu, C. (ed.) Proceedings of the Twentieth Annual ACM-SIAM Symposium on Discrete Algorithms, SODA 2009, New York, NY, USA, 4–6 January 2009, pp. 115–119. SIAM (2009). http://dl.acm.org/citation.cfm?id=1496770.1496783

Finding k-Secluded Trees Faster

Huib Donkers[ID], Bart M. P. Jansen[(✉)][ID], and Jari J. H. de Kroon[ID]

Eindhoven University of Technology, Eindhoven, The Netherlands
{h.t.donkers,b.m.p.jansen,j.j.h.d.kroon}@tue.nl

Abstract. We revisit the k-SECLUDED TREE problem. Given a vertex-weighted undirected graph G, its objective is to find a maximum-weight induced subtree T whose open neighborhood has size at most k. We present a fixed-parameter tractable algorithm that solves the problem in time $2^{\mathcal{O}(k \log k)} \cdot n^{\mathcal{O}(1)}$, improving on a double-exponential running time from earlier work by Golovach, Heggernes, Lima, and Montealegre. Starting from a single vertex, our algorithm grows a k-secluded tree by branching on vertices in the open neighborhood of the current tree T. To bound the branching depth, we prove a structural result that can be used to identify a vertex that belongs to the neighborhood of any k-secluded supertree $T' \supseteq T$ once the open neighborhood of T becomes sufficiently large. We extend the algorithm to enumerate compact descriptions of all maximum-weight k-secluded trees, which allows us to count the number of such trees containing a specified vertex in the same running time.

Keywords: Secluded tree · FPT · Enumeration algorithm

1 Introduction

Background. We revisit a problem from the field of parameterized complexity: Given a graph G with positive weights on the vertices, find a connected induced acyclic subgraph H of maximum weight such that the open neighborhood of H in G has size at most k.

A parameterized problem is fixed parameter tractable (FPT) [4,6] if there is an algorithm that, given an instance I with parameter k, solves the problem in time $f(k) \cdot |I|^{\mathcal{O}(1)}$ for some computable function f. For problems that are FPT, such algorithms allow NP-hard problems to be solved efficiently on instances whose parameter is small. It is therefore desirable for the function f to grow slowly in terms of k, both out of theoretical interest as well as improving the practical relevance of these algorithms.

We say that a vertex set $S \subseteq V(G)$ is k-secluded in G if the open neighborhood of S in G has size at most k. An induced subgraph H of G is k-secluded in G if $V(H)$ is. If H is also a tree, we say that H is a k-secluded tree in G. Formally, the problem we study in this work is defined as follows.

B. M. P. Jansen—Supported by NWO Gravitation grant "Networks".
J. J. H. de Kroon—Supported by ERC Starting grant 803421, "ReduceSearch".

M. A. Bekos and M. Kaufmann (Eds.): WG 2022, LNCS 13453, pp. 173–186, 2022.
https://doi.org/10.1007/978-3-031-15914-5_13

LARGE SECLUDED TREE (LST)	**Parameter:** k
Input: An undirected graph G, a non-negative integer k, and a weight function $w \colon V(G) \to \mathbb{N}^+$.	
Task: Find a k-secluded tree H of G of maximum weight, or report that no such H exists.	

Golovach et al. [11] consider the more general CONNECTED SECLUDED Π-SUBGRAPH, where the k-secluded induced subgraph of G should belong to some target graph class Π. They mention that (LARGE) SECLUDED TREE is FPT and can be solved in time $2^{2^{\mathcal{O}(k \log k)}} \cdot n^{\mathcal{O}(1)}$ using the recursive understanding technique, the details of which can be found in the arXiv version [10]. For the case where Π is characterized by a finite set of forbidden induced subgraphs $\mathcal{F}$, they show that the problem is FPT with a triple-exponential dependency. They pose the question whether it is possible to avoid these double- and triple-exponential dependencies on the parameter. They give some examples of Π for which this is the case, namely for Π being a clique, a star, a d-regular graph, or an induced path.

Results. Our main result is an algorithm for LARGE SECLUDED TREE that takes $2^{\mathcal{O}(k \log k)} \cdot n^4$ time. This answers the question of Golovach et al. [11] affirmatively for the case of trees. We solve a more general version of the problem, where a set of vertices is given that should be part of the k-secluded tree. Our algorithm goes one step further by allowing us to find all maximum-weight solutions. As we will later argue, it is not possible to output all such solutions directly in the promised running time. Instead, the output consists of a bounded number of solution descriptions such that each maximum-weight solution can be constructed from one such description. This is similar in spirit to the work of Guo et al. [12], who enumerate all minimal solutions to the FEEDBACK VERTEX SET problem in $\mathcal{O}(c^k \cdot m)$ time. They do so by giving a list of *compact representations*, a set $\mathcal{C}$ of pairwise disjoint vertex subsets such that choosing exactly one vertex from every set results in a minimal feedback vertex set. Our descriptions are *non-redundant* (no two descriptions describe the same secluded tree), which allows us to *count* the number of maximum-weight k-secluded trees containing a specified vertex in the same running time.

Techniques. Rather than using recursive understanding, our algorithm is based on bounded-depth branching with a non-trivial progress measure. Similarly to existing algorithms to compute spanning trees with many leaves [13], our algorithm iteratively grows the vertex set of a k-secluded tree T. If we select a vertex v in the neighborhood of the current tree T, then for any k-secluded supertree T' of T there are two possibilities: either v belongs to the neighborhood of T', or it is contained in T'; the latter case can only happen if v has exactly one neighbor in T. Solutions of the first kind can be found by deleting v from the graph and searching for a $(k-1)$-secluded supertree of T. To find solutions of the second kind we can include v in T, but since the parameter does not decrease in this case we have to be careful that the recursion depth stays bounded. Using a reduction rule to deal with degree-1 vertices, we can essentially ensure that v has at least three neighbors (exactly one of which belongs to T), so that adding v to T strictly increases the open neighborhood size $|N(T)|$. Our main insight to obtain an FPT algorithm is a structural lemma showing that, whenever $|N(T)|$ becomes sufficiently large in terms of k, we can identify a vertex u that belongs to the open neighborhood of any k-secluded supertree $T' \supseteq T$. At that point, we can remove u and decrease k to make progress.

Related Work. Secluded versions of several classic optimization problems have been studied intensively in recent years [1–3, 8, 14], many of which are discussed in Till Fluschnik's PhD thesis [7]. Marx [15] considers a related problem CUTTING k (CONNECTED) VERTICES, where the aim is to find a (connected) set S of size exactly k with at most ℓ neighbors. Without the connectivity requirement, the problem is W[1]-hard by $k + \ell$. The problem becomes FPT when S is required to be connected, but remains W[1]-hard by k and ℓ seperately. Fomin et al. [9] consider the variant where $|S| \leq k$ and show that it is FPT parameterized by ℓ.

Organization. We introduce our enumeration framework in Sect. 2. We present our algorithm that enumerates maximum-weight k-secluded trees in Sect. 3 and present its correctness and running time analyses. We give some conclusions in Sect. 4. Proofs marked with ★ are deferred to the full version [5].

2   Framework for Enumerating Secluded Trees

We consider simple undirected graphs with vertex set $V(G)$ and edge set $E(G)$. We use standard notation pertaining to graph algorithms, such as presented by Cygan et al. [4]. When the graph G is clear from context, we denote $|V(G)|$ and $|E(G)|$ by n and m respectively. The open neighborhood of a vertex set X in a graph G is denoted by $N_G(X)$, where the subscript may be omitted if G is clear from context. For a subgraph H of G we may write $N(H)$ to denote $N(V(H))$. If $w: V(G) \to \mathbb{N}^+$ is a weight function, then for any $S \subseteq V(G)$ let $w(S) := \sum_{v \in S} w(s)$ and for any subgraph H of G we may denote $w(V(H))$ by $w(H)$.

It is not possible to enumerate all maximum-weight k-secluded trees in FPT time; consider the graph with n vertices of weight 1 and two vertices of weight n which are connected by $k + 1$ vertex-disjoint paths on $n/(k + 1)$ vertices each, then there are $\mathcal{O}(k \cdot (n/k)^k)$ maximum-weight k-secluded trees which consist of all vertices except one vertex out of exactly k paths. However, it is possible to give one short description for such an exponential number of k-secluded trees.

Definition 1. *For a graph G, a description is a pair $(r, \mathcal{X})$ consisting of a vertex $r \in V(G)$ and a set $\mathcal{X}$ of pairwise disjoint subsets of $V(G - r)$ such that for any set S consisting of exactly one vertex from each set $X \in \mathcal{X}$, the connected component H of $G - S$ containing r is acyclic and $N(H) = S$, i.e., H is a $|\mathcal{X}|$-secluded tree in G. The order of a description is equal to $|\mathcal{X}|$. We say that a k-secluded tree H is described by a description $(r, \mathcal{X})$ if $N(H)$ consists of exactly one vertex of each $X \in \mathcal{X}$ and $r \in V(H)$.*

Note that a single k-secluded tree H can be described by multiple descriptions. For example, for a path on $v_1, \dots, v_4$ the 1-secluded tree induced by $\{v_1, v_2\}$ is described by $(v_1, \{\{v_3, v_4\}\})$, $(v_1, \{\{v_3\}\})$, and $(v_2, \{\{v_3\}\})$. We define the concept of redundancy in a set of descriptions.

Definition 2. *For a graph G, a set of descriptions $\mathfrak{X}$ of maximum order k is called* redundant *for G if there is a k-secluded tree H in G such that H is described by two distinct descriptions in $\mathfrak{X}$. We say $\mathfrak{X}$ is* non-redundant *for G otherwise.*

Definition 3. *For a graph G and a set of descriptions $\mathfrak{X}$ of maximum order k, let $\mathcal{T}_G(\mathfrak{X})$ denote the set of all k-secluded trees in G described by a description in $\mathfrak{X}$.*

Note 1. For a graph G we have $\mathcal{T}_G(\mathfrak{X}_1) \cup \mathcal{T}_G(\mathfrak{X}_2) = \mathcal{T}_G(\mathfrak{X}_1 \cup \mathfrak{X}_2)$ for any two sets of descriptions $\mathfrak{X}_1, \mathfrak{X}_2$.

Note 2. For a graph G, a set of descriptions $\mathfrak{X}$, and non-empty vertex sets X_1, X_2 disjoint from $\bigcup_{(r, \mathcal{X}) \in \mathfrak{X}} (\{r\} \cup \bigcup_{X \in \mathcal{X}} X)$, the set $\mathcal{T}_G(\{(r, \mathcal{X} \cup \{X_1 \cup X_2\}) \mid (r, \mathcal{X}) \in \mathfrak{X}\})$ equals $\mathcal{T}_G(\{(r, \mathcal{X} \cup \{X_1\}) \mid (r, \mathcal{X}) \in \mathfrak{X}\}) \cup (r, \mathcal{X} \cup \{X_2\}) \mid (r, \mathcal{X}) \in \mathfrak{X}\})$.

For an induced subgraph H of G and a set $F \subseteq V(G)$, we say that H is a supertree of F if H induces a tree and $F \subseteq V(H)$. Let $\mathcal{S}_G^k(F)$ be the set of all k-secluded supertrees of F in G. For a set X of subgraphs of G let $\text{maxset}_w(X) := \{H \in X \mid w(H) \geq w(H') \text{ for all } H' \in X\}$. We focus our attention to the following version of the problem, where some parts of the tree are already given.

ENUMERATE LARGE SECLUDED SUPERTREES (ELSS)   **Parameter:** k
Input: A graph G, a non-negative integer k, non-empty vertex sets $T \subseteq F \subseteq V(G)$ such that $G[T]$ is connected, and a weight function $w: V(G) \to \mathbb{N}^+$.
Output: A non-redundant set $\mathfrak{X}$ of descriptions such that $\mathcal{T}_G(\mathfrak{X}) = \text{maxset}_w(\mathcal{S}_G^k(F))$.

Note that if $G[T]$, or even $G[F]$, contains a cycle, then the answer is trivially the empty set. In the end we solve the general enumeration problem by solving ELSS with $F = T = \{v\}$ for each $v \in V(G)$ and reporting only those k-secluded trees of maximum weight. Intuitively, our algorithm for ELSS finds k-secluded trees that "grow" out of T. In order to derive some properties of the types of descriptions we compute, we may at certain points demand that certain vertices non-adjacent to T need to end up in the k-secluded tree. For this reason the input additionally has a set F, rather than just T.

Our algorithm solves smaller instances recursively. We use the following abuse of notation: in an instance with graph G and weight function $w: V(G) \to \mathbb{N}^+$, when solving the problem recursively for an instance with induced subgraph G' of G, we keep the weight function w instead of restricting the domain of w to $V(G')$.

Note 3. For a graph G, a vertex $v \in V(G)$, and an integer $k \geq 1$, if H is a $(k-1)$-secluded tree in $G-v$, then H is a k-secluded tree in G. Consequently, $\mathcal{S}_{G-v}^{k-1}(F) \subseteq \mathcal{S}_G^k(F)$ for any $F \subseteq V(G)$.

Note 4. For a graph G, a vertex $v \in V(G)$, and an integer $k \geq 1$, if H is a k-secluded tree in G with $v \in N_G(H)$, then H is a $(k-1)$-secluded tree in $G-v$. Consequently, $\{H \in \mathcal{S}_G^k(F) \mid v \in N_G(H)\} \subseteq \mathcal{S}_{G-v}^{k-1}(F)$ for any $F \subseteq V(G)$.

3 Enumerate Large Secluded Supertrees

Section 3.1 proves the correctness of a few subroutines used by the algorithm. Section 3.2 describes the algorithm to solve ELSS. In Sect. 3.3 we prove its correctness and in Sect. 3.4 we analyze its time complexity. In Sect. 3.5 we show how the algorithm for ELSS can be used to count and enumerate maximum-weight k-secluded trees containing a specified vertex.

3.1 Subroutines for the Algorithm

Similar to the FEEDBACK VERTEX SET algorithm given by Guo et al. [12], we aim to get rid of degree-1 vertices. In our setting there is one edge case however. The reduction rule is formalized as follows.

Reduction Rule 1. *For an ELSS instance (G, k, F, T, w) with a degree-1 vertex v in G such that $F \neq \{v\}$, contracting v into its neighbor u yields the ELSS instance $(G - v, k, F', T', w')$ where the weight of u is increased by $w(v)$ and:*

$$F' = \begin{cases} (F \backslash \{v\}) \cup \{u\} & \textit{if } v \in F \\ F & \textit{otherwise} \end{cases} \qquad T' = \begin{cases} (T \backslash \{v\}) \cup \{u\} & \textit{if } v \in T \\ T & \textit{otherwise.} \end{cases}$$

We prove the correctness of the reduction rule, that is, the descriptions of the reduced instance form the desired output for the original instance.

Lemma 1 (★). *Let $I = (G, k, F, T, w)$ be an ELSS instance. Suppose G contains a degree-1 vertex v such that $\{v\} \neq F$. Let $I' = (G - v, k, F', T', w')$ be the instance obtained by contracting v into its neighbor u. If $\mathfrak{X}$ is a non-redundant set of descriptions for $G - v$ such that $T_{G-v}(\mathfrak{X}) = maxset_{w'}(\mathcal{S}_{G-v}^k(F'))$, then $\mathfrak{X}$ is a non-redundant set of descriptions for G such that $T_G(\mathfrak{X}) = maxset_w(\mathcal{S}_G^k(F))$.*

We say an instance is *almost leafless* if the lemma above cannot be applied, that is, if G contains a vertex v of degree 1, then $F = \{v\}$.

Lemma 2. *There is an algorithm that, given an almost leafless ELSS instance (G, k, F, T, w) such that $k > 0$ and $|N_G(T)| > k(k + 1)$, runs in time $\mathcal{O}(k \cdot n^3)$ and either:*

1. *finds a vertex $v \in V(G) \backslash F$ such that any k-secluded supertree H of F in G satisfies $v \in N_G(H)$, or*
2. *concludes that G does not contain a k-secluded supertree of F.*

Proof. We aim to find a vertex $v \in V(G) \backslash F$ with $k+2$ distinct paths $P_1, \ldots, P_{k+2}$ from $N_G(T)$ to v that intersect only in v and do not contain vertices from T. We first argue that such a vertex v satisfies the first condition, if it exists. Consider some k-secluded supertree H of F. Since the paths $P_1, \ldots, P_{k+2}$ are disjoint apart from their common endpoint v while $|N_G(H)| \leq k$, there are two paths P_i, P_j with $i \neq j \in [k+2]$ for which $P_i \backslash \{v\}$ and $P_j \backslash \{v\}$ do not intersect $N_G(H)$. These paths are contained in H since they contain a neighbor of $T \subseteq F \subseteq H$. As P_i and P_j form a cycle together with a path through the connected set T, which cannot be contained in the acyclic graph H, this implies $v \in N_G(H)$.

Next we argue that if G has a k-secluded supertree H of $F \supseteq T$, then there exists such a vertex v. Consider an arbitrary such H and root it at a vertex $t \in T$. For each vertex $u \in N_G(T)$, we construct a path P_u disjoint from T that starts in u and ends in $N_G(H)$, as follows.

- If $u \notin H$, then $u \in N_G(H)$ and we take $P_u = (u)$.
- If $u \in H$, then let ℓ_u be an arbitrary leaf in the subtree of H rooted at u; possibly $u = \ell_u$. Since T is connected and $H \supseteq T$ is acyclic and rooted in $t \in T$, the subtree rooted at $u \in N_G(T) \cap H$ is disjoint from T. Hence $\ell_u \notin T$, so that $F \neq \{\ell_u\}$. As the instance is almost leafless we therefore have $\deg_G(\ell_u) > 1$. Because ℓ_u is a leaf of H this implies that $N_G(\ell_u)$ contains a vertex y other than the parent of ℓ_u in H, so that $y \in N_G(H)$. We let P_u be the path from u to ℓ_u through H, followed by the vertex $y \in N_G(H)$.

The paths we construct are distinct since their startpoints are. Two constructed paths cannot intersect in any vertex other than their endpoints, since they were extracted from different subtrees of H. Since we construct $|N_G(T)| > k(k+1)$ paths, each of which ends in $N_G(H)$ which has size at most k, some vertex $v \in N_G(H)$ is the endpoint of $k + 2$ of the constructed paths. As shown in the beginning the proof, this establishes that v belongs to the neighborhood of any k-secluded supertree of F. Since $F \subseteq V(H)$ we have $v \notin F$.

All that is left to show is that we can find such a vertex v in the promised time bound. After contracting T into a source vertex s, for each $v \in V(G) \backslash F$, do $k + 2$ iterations of the Ford-Fulkerson algorithm in order to check if there are $k + 2$ internally vertex-disjoint sv-paths. If so, then return v. If for none of the choices of v this holds, then output that there is no k-secluded supertree of F in G. In order to see that this satisfies the claimed running time bound, note that there are $\mathcal{O}(n)$ choices for v, and $k + 2$ iterations of Ford-Fulkerson runs can be implemented to run in $\mathcal{O}(k \cdot (n + m))$ time. □

3.2   The Algorithm

Consider an input instance (G, k, F, T, w) of ELSS. If $G[F]$ contains a cycle, return $\emptyset$. Otherwise we remove all connected components of G that do not contain a vertex of F. If more than one connected component remains, return $\emptyset$. Then, while there is a degree-1 vertex v such that $F \neq \{v\}$, contract v into its neighbor as per Rule 1. While $N_G(T)$ contains a vertex $v \in F$, add v to T. Finally, if $N_G(F) = \emptyset$, return $\{(r, \emptyset)\}$ for some $r \in F$. Otherwise if $k = 0$, return $\emptyset$.

We proceed by considering the neighborhood of T as follows:

1. If any vertex $v \in N_G(T)$ has two neighbors in T, then recursively run this algorithm to obtain a set of descriptions $\mathfrak{X}'$ for $(G - v, k - 1, F, T, w)$ and return $\{(r, \mathcal{X} \cup \{\{v\}\}) \mid (r, \mathcal{X}) \in \mathfrak{X}'\}$.

2. If $|N_G(T)| > k(k + 1)$, apply Lemma 2. If it concludes that G does not contain a k-secluded supertree of F, return $\emptyset$. Otherwise let $v \in V(G)\backslash F$ be the vertex it finds, obtain a set of descriptions $\mathfrak{X}'$ for $(G - v, k - 1, F, T, w)$ and return $\{(r, \mathcal{X} \cup \{\{v\}\}) \mid (r, \mathcal{X}) \in \mathfrak{X}'\}$.

3. Pick some $v \in N_G(T)$ and let $P = (v = v_1, v_2, \ldots, v_\ell)$ be the unique[1] maximal path disjoint from T satisfying $\deg_G(v_i) = 2$ for each $1 \leq i < \ell$ and $(v_\ell \in N_G(T)$ or $\deg_G(v_\ell) > 2)$.

   (a) If $v_\ell \notin F$, obtain a set of descriptions $\mathfrak{X}_1$ by recursively solving $(G - v_\ell, k - 1, F, T, w)$. Otherwise take $\mathfrak{X}_1 = \emptyset$. (We find the k-secluded trees avoiding v_ℓ but containing $P - v_\ell$.)

   (b) If $P - F - v_\ell \neq \emptyset$, obtain a set of descriptions $\mathfrak{X}_2$ by recursively solving $(G - V(P - v_\ell), k - 1, (F\backslash V(P)) \cup \{v_\ell\}, T, w)$. Otherwise take $\mathfrak{X}_2 = \emptyset$. (We find the k-secluded trees containing both endpoints of P which have one vertex in P as a neighbor.)

   (c) If $G[F \cup V(P)]$ is acyclic, obtain a set of descriptions $\mathfrak{X}_3$ by recursively solving $(G, k, F \cup V(P), T \cup V(P), w)$. Otherwise take $\mathfrak{X}_3 = \emptyset$. (We find the k-secluded trees containing the entire path P.)

   Let M be the set of minimum weight vertices in $P - F - v_\ell$ and define:

$$\mathfrak{X}_1' := \{(r, \mathcal{X} \cup \{\{v_\ell\}\}) \mid (r, \mathcal{X}) \in \mathfrak{X}_1\}$$
$$\mathfrak{X}_2' := \{(r, \mathcal{X} \cup \{M\}) \mid (r, \mathcal{X}) \in \mathfrak{X}_2\}$$
$$\mathfrak{X}_3' := \mathfrak{X}_3.$$

For each $i \in [3]$ let w_i be the weight of an arbitrary $H \in \mathcal{T}_G(\mathfrak{X}_i')$, or 0 if $\mathfrak{X}_i' = \emptyset$. Return the set $\mathfrak{X}'$ defined as $\bigcup_{\{i \in [3] \mid w_i = \max\{w_1, w_2, w_3\}\}} \mathfrak{X}_i'$.

3.3 Proof of Correctness

In this section we argue that the algorithm described in Sect. 3.2 solves the ELSS problem. In various steps we identify a vertex v such that the neighborhood of any (maximum-weight) k-secluded supertree must include v. We argue that for these steps, the descriptions of the current instance can be found by adding $\{v\}$ to every description of the supertrees of T in $G - v$ if some preconditions are satisfied.

Lemma 3 ($\bigstar$). *Let (G, k, F, T, w) be an ELSS instance and let $v \in V(G)\backslash F$. Let $\mathfrak{X}$ be a set of descriptions for $G - v$ such that $\mathcal{T}_{G-v}(\mathfrak{X}) = \text{maxset}_w(\mathcal{S}_{G-v}^{k-1}(F))$ and $v \in N_G(H)$ for all $H \in \mathcal{T}_{G-v}(\mathfrak{X})$. Then we have:*

$$\mathcal{T}_G\left(\{(r, \mathcal{X} \cup \{\{v\}\}) \mid (r, \mathcal{X}) \in \mathfrak{X}\}\right) = \text{maxset}_w\{H \in \mathcal{S}_G^k(F) \mid v \in N_G(H)\}.$$

---
[1] To construct P, initialize $P := (v = v_1)$; then while $\deg_G(v_{|V(P)|}) = 2$ and $N_G(v_{|V(P)|})\backslash(V(P) \cup T)$ consists of a single vertex, append that vertex to P.

The following lemma is used to argue that the branches of Step 3 are disjoint.

Lemma 4. *Let (G, k, F, T, w) be an almost leafless ELSS instance such that G is connected and $N_G(F) \neq \emptyset$. Fix some $v \in N_G(T)$ and let $P = (v = v_1, v_2, \ldots, v_\ell)$ be the unique maximal path disjoint from T satisfying $\deg_G(v_i) = 2$ for each $1 \leq i < \ell$ and $(v_\ell \in N_G(T)$ or $\deg_G(v_\ell) > 2)$. Then for any maximum-weight k-secluded supertree H of F, exactly one of the following holds:*

1. $v_\ell \in N(H)$ (so $v_\ell \notin F$),
2. $|N(H) \cap V(P - F - v_\ell)| = 1$ and $v_\ell \in V(H)$, or
3. $V(P) \subseteq V(H)$.

Proof. First note that such a vertex v exists since $N_G(F) \neq \emptyset$ and G is connected, so $N_G(T) \neq \emptyset$. Furthermore since the instance is almost leafless, the path P is well defined. If there is no k-secluded supertree of F, then there is nothing to show. So suppose H is a maximum-weight k-secluded supertree of F. We have $v \in V(P)$ is a neighbor of $T \subseteq F \subseteq V(H)$, so either $V(P) \subseteq V(H)$ or $V(P)$ contains a vertex from $N(H)$. In the first case Item 3 holds, in the second case we have $|N(H) \cap V(P)| \geq 1$. First suppose that $|N(H) \cap V(P)| \geq 2$. Let $i \in [\ell]$ be the smallest index such that $v_i \in N(H) \cap V(P)$. Similarly let $j \in [\ell]$ be the largest such index. We show that in this case we can contradict the fact that H is a maximum-weight k-secluded supertree of F. Observe that $H' = V(H) \cup \{v_i, \ldots, v_{j-1}\}$ induces a tree since $(v_i, \ldots, v_{j-1})$ forms a path of degree-2 vertices and the neighbor v_j of v_{j-1} is not in H. Furthermore H' has a strictly smaller neighborhood than H and it has larger weight as vertices have positive weight. Since $F \subseteq V(H')$, this contradicts that H is a maximum-weight k-secluded supertree of F.

We conclude that $|N(H) \cap V(P)| = 1$. Let $i \in [\ell]$ be the unique index such that $N(H) \cap V(P) = \{v_i\}$. Clearly $v_i \notin F$. In the case that $i = \ell$, then Item 1 holds. Otherwise if $i < \ell$, the first condition of Item 2 holds. In order to argue that the second condition also holds, suppose that $v_\ell \notin V(H)$. Then $H \cup \{v_i, \ldots, v_{\ell-1}\}$ is a k-secluded supertree of F in G and it has larger weight than H as vertices have positive weight. This contradicts the fact that H has maximum weight, hence the second condition of Item 2 holds as well. □

Armed with Lemma 4 we are now ready to prove correctness of the algorithm.

Lemma 5. *The algorithm described in Sect. 3.2 is correct.*

Proof. Let $I = (G, k, F, T, w)$ be an ELSS instance. We prove correctness by induction on $|V(G)\backslash F|$. Assume the algorithm is correct for any input $(\hat{G}, \hat{k}, \hat{F}, \hat{T}, \hat{w})$ with $|V(\hat{G})\backslash\hat{F}| < |V(G)\backslash F|$. We prove correctness of the algorithm up to Step 3. The correctness of Step 3 is proven in the full version [5].

Before Step 1. We first prove correctness when the algorithm terminates before Step 1, which includes the base case of the induction. Note that if $G[F]$ contains a cycle, then no induced subgraph H of G with $F \subseteq V(H)$ can be acyclic. Therefore the set of maximum-weight k-secluded trees containing F is the empty set, so we correctly return $\emptyset$. Otherwise $G[F]$ is acyclic. Clearly any connected component of G that has no vertices of F can be removed. If there are two connected components of G containing vertices of F, then no induced subgraph of G containing all of F can be connected, again we correctly return the empty set. In the remainder we have that G is connected.

By iteratively applying Lemma 1 we conclude that a solution to the instance obtained after iteratively contracting (most) degree-1 vertices is also a solution to the original instance. Hence we can proceed to solve the new instance, which we know is almost leafless. In addition, observe that the contraction of degree-1 vertices maintains the property that G is connected and $G[F]$ is acyclic.

After exhaustively adding vertices $v \in N_G(T) \cap F$ to T we have that $G[T]$ is a connected component of $G[F]$. In the case that $N_G(F) = \emptyset$, then since G is connected it follows that $F = T = V(G)$ and therefore T is the only maximum-weight k-secluded tree. For any $r \in V(G)$, the description $(r, \emptyset)$ describes this k-secluded tree, so we return $\{(r, \emptyset)\}$. In the remainder we have $N_G(F) \neq \emptyset$.

Since $N_G(F) \neq \emptyset$ and G is almost leafless, we argue that there is no 0-secluded supertree of F. Suppose G contains a 0-secluded supertree H of F, so $|N_G(H)| = 0$ and since $H \supseteq F$ is non-empty and G is connected we must have $H = G$, hence G is a tree with at least two vertices (since F and $N_G(F)$ are both non-empty) so G contains at least two vertices of degree-1, contradicting that G is almost leafless. So there is no k-secluded supertree of F in G and the algorithm correctly returns $\emptyset$ if $k = 0$.

Observe that the value $|V(G) \backslash F|$ cannot have increased since the start of the algorithm since we never add vertices to G and any time we remove a vertex from F it is also removed from G. Hence we can still assume in the remainder of the proof that the algorithm is correct for any input $(\hat{G}, \hat{k}, \hat{F}, \hat{T}, \hat{w})$ with $|V(\hat{G}) \backslash \hat{F}| < |V(G) \backslash F|$. To conclude this part of the proof, we have established that if the algorithm terminates before reaching Step 1, then its output is correct. On the other hand, if the algorithm continues we can make use of the following properties of the instance just before reaching Step 1:

Property 1. If the algorithm does not terminate before reaching Step 1 then (i) the ELSS instance (G, k, F, T, w) is almost leafless, (ii) $G[F]$ is acyclic, (iii) $G[T]$ is a connected component of $G[F]$, (iv) G is connected, (v) $k > 0$, and (vi) $N_G(F) \neq \emptyset$.

Step 1. Before arguing that the return value in Step 1 is correct, we observe the following.

Claim 1. If H is an induced subtree of G that contains T and $v \in N_G(T)$ has at least two neighbors in T, then $v \in N_G(H)$.

Proof. Suppose $v \notin N_G(H)$, then since $v \in N_G(T)$ and $T \subseteq V(H)$ we have that $v \in V(H)$. But then since T is connected, subgraph H contains a cycle. This contradicts that H is a tree and confirms that $v \in N_G(H)$. ⌐

Now consider the case that in Step 1 we find a vertex $v \in N_G(T)$ with two neighbors in T, and let $\mathfrak{X}'$ be the set of descriptions as obtained by the algorithm

through recursively solving the instance $(G - v, k - 1, F, T, w)$. Since $|V(G -$
$v)\backslash F| < |V(G)\backslash F|$ (as $v \notin F$) we know by induction that $\mathcal{T}_{G-v}(\mathfrak{X}')$ is the set of
all maximum-weight $(k-1)$-secluded supertrees of F in $G-v$. Any $H \in \mathcal{T}_{G-v}(\mathfrak{X}')$
is an induced subtree of G with $T \subseteq V(H)$, so by Claim 1 we have $v \in N_G(H)$
for all $H \in \mathcal{T}_{G-v}(\mathfrak{X}')$. We can now apply Lemma 3 to conclude that $\mathcal{T}_G(\{(r, \mathcal{X} \cup$
$\{\{v\}\}) \mid (r, \mathcal{X}) \in \mathfrak{X}'\})$ is the set of all maximum-weight k-secluded supertrees H
of F in G for which $v \in N_G(H)$. Again by Claim 1 we have that $v \in N_G(H)$
for all such k-secluded supertrees of F, hence $\mathcal{T}_G(\{(r, \mathcal{X} \cup \{\{v\}\}) \mid (r, \mathcal{X}) \in \mathfrak{X}'\})$
is the set of all maximum-weight k-secluded supertrees of F in G. We argue
non-redundancy of the output. Suppose that two descriptions $(r, \mathcal{X} \cup \{\{v\}\})$ and
$(r', \mathcal{X}' \cup \{\{v\}\})$ describe the same supertree H of F in G. Note that then $(r, \mathcal{X})$
and $(r', \mathcal{X})$ describe the same supertree H of F in $G - v$, which contradicts
the induction hypothesis that the output of the recursive call was correct and
therefore non-redundant.

Concluding this part of the proof, we showed that if the algorithm terminates
during Step 1, then its output is correct. On the other hand, if the algorithm
continues after Step 1 we can make use of the following in addition to Property 1.

Property 2. If the algorithm does not terminate before reaching Step 2 then no
vertex $v \in N_G(T)$ has two neighbors in T.

Step 2. In Step 2 we use Lemma 2 if $|N_G(T)| > k(k + 1)$. The preconditions of
the lemma are satisfied since $k > 0$ and the instance is almost leafless by Property 1. If it concludes that G does not contain a k-secluded supertree of F, then
the algorithm correctly outputs $\emptyset$. Otherwise it finds a vertex $v \in V(G)\backslash F$ such
that any k-secluded supertree H of F in G satisfies $v \in N_G(H)$. We argue
that the algorithm's output is correct. Let $\mathfrak{X}'$ be the set of descriptions as
obtained through recursively solving $(G - v, k - 1, F, T, w)$. Since $v \notin F$ we
have $|(V(G - v)\backslash F| < |V(G)\backslash F|$, so by induction we have that $\mathcal{T}_{G-v}(\mathfrak{X}')$ is
the set of all maximum-weight $(k - 1)$-secluded supertrees of F in $G - v$. Fur-
thermore by Note 3 for any $H \in \mathcal{T}_{G-v}(\mathfrak{X}') = \mathcal{S}_{G-v}^{k-1}(F)$ we have $H \in \mathcal{S}_G^k(F)$,
and therefore $v \in N_G(H)$. It follows that Lemma 3 applies to $\mathfrak{X}'$ so we can
conclude that $\mathcal{T}_G(\{(r, \mathcal{X} \cup \{\{v\}\}) \mid (r, \mathcal{X}) \in \mathfrak{X}'\})$ is the set of maximum-weight
k-secluded supertrees H of F in G for which $v \in N_G(H)$. Since we know there
are no k-secluded supertrees H of F in G for which $v \notin N_G(H)$, it follows
that $\mathcal{T}_G(\{(r, \mathcal{X} \cup \{\{v\}\}) \mid (r, \mathcal{X}) \in \mathfrak{X}\})$ is the set of maximum-weight k-secluded
supertrees of F in G as required. Non-redundancy of the output follows as in
Step 1.

To summarize the progress so far, we have shown that if the algorithm terminates before it reaches Step 3, then its output is correct. Alternatively, if we
proceed to Step 3 we can make use of the following property, in addition to
Properties 1 and 2, which we will use later in the running time analysis.

Property 3. If the algorithm does not terminate before reaching Step 3, then
$|N_G(T)| \leq k(k + 1)$.

Step 3 (★). In the full version [5] we show using Properties 1 to 3 that if the algorithm reaches Step 3, then its output is correct. For this we use Lemma 4 to argue that the k-secluded supertrees of F in G can be partitioned into three sets $\mathcal{T}_1, \mathcal{T}_2, \mathcal{T}_3$. The three recursive calls in Step 3 correspond to the subproblems of finding the secluded trees in $\mathcal{T}_1, \mathcal{T}_2$, and $\mathcal{T}_3$. Each call finds maximum-weight k-secluded trees of one particular type. Since the latter restriction may cause the tree to have smaller weight than maximum k-secluded trees in general, the postprocessing step of the algorithm restricts the output to describe only those types providing the maximum global weight. □

3.4 Runtime Analysis

If all recursive calls in the algorithm would decrease k then, since for $k = 0$ it does not make any further recursive calls, the maximum recursion depth is k. However in Step 3(c) the recursive call does not decrease k. In order to bound the recursion depth, we show the algorithm cannot make more than $k(k + 1)$ consecutive recursive calls in Step 3(c), that is, the recursion depth cannot increase by more than $k(k+1)$ since the last time k decreased. This follows from Lemma 6 together with the fact that if $N_G(T) > k(k + 1)$ then the algorithm executes Step 2, decreases k when it goes into recursion, and does not proceed to Step 3.

Lemma 6 (★). *If the recursion tree generated by the algorithm contains a path of $i \geq 1$ consecutive recursive calls in Step 3(c), and (G, k, F, T, w) is the instance considered in Step 3 where the i-th of these recursive calls is made, then $|N_G(T)| \geq i$.*

Using this bound on the number of consecutive recursive calls in Step 3(c), we obtain a maximum recursion depth of $\mathcal{O}(k^3)$. We argue that each recursive call takes $\mathcal{O}(kn^3)$ time and since we branch at most three ways, we obtain a running time of $3^{\mathcal{O}(k^3)} \cdot kn^3 = 3^{\mathcal{O}(k^3)} \cdot n^3$. However, with a more careful analysis we can give a better bound on the number of nodes in the recursion tree. For this, label each edge in the recursion tree with a label from the set $\{1, 2, 3a, 3b, 3c\}$ indicating where in the algorithm the recursive call took place. Now observe that each node in the recursion tree can be uniquely identified by a sequence of edge-labels corresponding to the path from the root of the tree to the relevant node. We call such a sequence of labels a *trace*. To bound the number of nodes in the recursion tree we give a bound on the number of valid traces. Since recursive calls corresponding to labels 1, 2, 3a, and 3b each decrease k, they can occur at most k times in a valid trace. All remaining labels in the trace are 3c. So the total number of traces of length ℓ is $\binom{\ell}{k} \cdot 4^k \leq \ell^k \cdot 4^k = (4\ell)^k$. Considering valid traces have a length of at most $k^2(k + 1)$ we derive the following bound on the total number of valid traces using the fact that $(k^c)^k = (2^{\log(k^c)})^k = 2^{\mathcal{O}(k \log k)}$:

$$\sum_{1 \leq \ell \leq k^2(k+1)} (4\ell)^k \leq k^2(k + 1) \cdot (4k^2(k + 1))^k = 2^{\mathcal{O}(k \log k)}.$$

We can conclude that the total number of nodes in the recursion tree is at most $2^{\mathcal{O}(k \log k)}$ which leads to the following lemma.

Lemma 7 (★). *The algorithm described in Sect. 3.2 can be implemented to run in time $2^{\mathcal{O}(k \log k)} \cdot n^3$.*

3.5 Finding, Enumerating, and Counting Large Secluded Trees

With the algorithm of Sect. 3.2 at hand we argue that we are able to enumerate k-secluded trees, count such trees containing a specified vertex, and solve LST.

Theorem 1 (★). *There is an algorithm that, given a graph G, weight function w, and integer k, runs in time $2^{\mathcal{O}(k \log k)} \cdot n^4$ and outputs a set of descriptions $\mathfrak{X}$ such that $T_G(\mathfrak{X})$ is exactly the set of maximum-weight k-secluded trees in G. Each such tree H is described by $|V(H)|$ distinct descriptions in $\mathfrak{X}$.*

By returning an arbitrary maximum-weight k-secluded tree described by any description in the output of Theorem 1, we have the following consequence.

Corollary 1. *There is an algorithm that, given a graph G, weight function w, and integer k, runs in time $2^{\mathcal{O}(k \log k)} \cdot n^4$ and outputs a maximum-weight k-secluded tree in G if one exists.*

The following theorem captures the consequences for counting.

Theorem 2 (★). *There is an algorithm that, given a graph G, vertex $v \in V(G)$, weight function w, and integer k, runs in time $2^{\mathcal{O}(k \log k)} \cdot n^3$ and counts the number of k-secluded trees in G that contain v and have maximum weight out of all k-secluded trees containing v.*

4   Conclusion

We revisited the k-SECLUDED TREE problem first studied by Golovach et al. [11], leading to improved FPT algorithms with the additional ability to count and enumerate solutions. The non-trivial progress measure of our branching algorithm is based on a structural insight that allows a vertex that belongs to the *neighborhood* of every solution subtree to be identified, once the solution under construction has a sufficiently large open neighborhood. As stated, the correctness of this step crucially relies on the requirement that solution subgraphs are acyclic. It would be interesting to determine whether similar branching strategies can be developed to solve the more general k-SECLUDED CONNECTED $\mathcal{F}$-MINOR-FREE SUBGRAPH problem; the setting studied here corresponds to $\mathcal{F} = \{K_3\}$. While any $\mathcal{F}$-minor-free graph is known to be sparse, it may still contain large numbers of internally vertex-disjoint paths between specific pairs of vertices, which stands in the way of a direct extension of our techniques.

A second open problem concerns the optimal parameter dependence for k-SECLUDED TREE. The parameter dependence of our algorithm is $2^{\mathcal{O}(k \log k)}$. Can it be improved to single-exponential, or shown to be optimal under the Exponential Time Hypothesis?

References

1. van Bevern, R., Fluschnik, T., Mertzios, G.B., Molter, H., Sorge, M., Suchý, O.: The parameterized complexity of finding secluded solutions to some classical optimization problems on graphs. Discret. Optim. **30**, 20–50 (2018). https://doi.org/10.1016/j.disopt.2018.05.002
2. van Bevern, R., Fluschnik, T., Tsidulko, Y.O.: Parameterized algorithms and data reduction for the short secluded s-t-path problem. Networks **75**(1), 34–63 (2020). https://doi.org/10.1002/net.21904
3. Chechik, S., Johnson, M.P., Parter, M., Peleg, D.: Secluded connectivity problems. Algorithmica **79**(3), 708–741 (2016). https://doi.org/10.1007/s00453-016-0222-z
4. Cygan, M., et al.: Parameterized Algorithms. Springer, Cham (2015). https://doi.org/10.1007/978-3-319-21275-3
5. Donkers, H., Jansen, B.M.P., de Kroon, J.J.H.: Finding k-secluded trees faster (2022). https://doi.org/10.48550/ARXIV.2206.09884
6. Downey, R.G., Fellows, M.R.: Parameterized Complexity. Monographs in Computer Science. Springer, Heidelberg (1999). https://doi.org/10.1007/978-1-4612-0515-9
7. Fluschnik, T.: Elements of efficient data reduction: fractals, diminishers, weights and neighborhoods. Ph.D. thesis, Technische Universität Berlin (2020). https://doi.org/10.14279/depositonce-10134
8. Fomin, F.V., Golovach, P.A., Karpov, N., Kulikov, A.S.: Parameterized complexity of secluded connectivity problems. Theory Comput. Syst. **61**(3), 795–819 (2016). https://doi.org/10.1007/s00224-016-9717-x
9. Fomin, F.V., Golovach, P.A., Korhonen, J.H.: On the parameterized complexity of cutting a few vertices from a graph. In: Chatterjee, K., Sgall, J. (eds.) MFCS 2013. LNCS, vol. 8087, pp. 421–432. Springer, Heidelberg (2013). https://doi.org/10.1007/978-3-642-40313-2_38
10. Golovach, P.A., Heggernes, P., Lima, P.T., Montealegre, P.: Finding connected secluded subgraphs. CoRR, abs/1710.10979 (2017). arXiv:1710.10979
11. Golovach, P.A., Heggernes, P., Lima, P.T., Montealegre, P.: Finding connected secluded subgraphs. J. Comput. Syst. Sci. **113**, 101–124 (2020). https://doi.org/10.1016/j.jcss.2020.05.006
12. Guo, J., Gramm, J., Hüffner, F., Niedermeier, R., Wernicke, S.: Compression-based fixed-parameter algorithms for feedback vertex set and edge bipartization. J. Comput. Syst. Sci. **72**(8), 1386–1396 (2006). https://doi.org/10.1016/j.jcss.2006.02.001
13. Kneis, J., Langer, A., Rossmanith, P.: A new algorithm for finding trees with many leaves. Algorithmica **61**(4), 882–897 (2010). https://doi.org/10.1007/s00453-010-9454-5
14. Luckow, M.-J., Fluschnik, T.: On the computational complexity of length - and neighborhood-constrained path problems. Inf. Process. Lett. **156**, 105913 (2020). https://doi.org/10.1016/j.ipl.2019.105913
15. Marx, D.: Parameterized graph separation problems. Theor. Comput. Sci. **351**(3), 394–406 (2006). https://doi.org/10.1016/j.tcs.2005.10.007

On the Minimum Cycle Cover Problem on Graphs with Bounded Co-degeneracy

Gabriel L. Duarte[2] and Uéverton S. Souza[1,2(✉)]

[1] Institute of Informatics, University of Warsaw, Warsaw, Poland
ueverton@ic.uff.br
[2] Instituto de Computação, Universidade Federal Fluminense, Niterói, Brazil
gabrield@id.uff.br

Abstract. In 2017, Knop, Koutecký, Masařík, and Toufar [WG 2017] asked about the complexity of deciding graph problems Π on the complement of G considering a parameter p of G, especially for sparse graph parameters such as treewidth. In 2021, Duarte, Oliveira, and Souza [MFCS 2021] showed some problems that are FPT when parameterized by the treewidth of the complement graph (called co-treewidth). Since the degeneracy of a graph is at most its treewidth, they also introduced the study of co-degeneracy (the degeneracy of the complement graph) as a parameter. In 1976, Bondy and Chvátal [DM 1976] introduced the notion of *closure* of a graph: let ℓ be an integer; the $(n + \ell)$-closure, $\mathrm{cl}_{n+\ell}(G)$, of a graph G with n vertices is obtained from G by recursively adding an edge between pairs of nonadjacent vertices whose degree sum is at least $n+\ell$ until no such pair remains. A graph property Υ defined on all graphs of order n is said to be $(n+\ell)$-stable if for any graph G of order n that does not satisfy Υ, the fact that uv is not an edge of G and that $G + uv$ satisfies Υ implies $d(u) + d(v) < n + \ell$. Duarte et al. [MFCS 2021] developed an algorithmic framework for co-degeneracy parameterization based on the notion of closures for solving problems that are $(n + \ell)$-stable for some ℓ bounded by a function of the co-degeneracy. In 2019, Jansen, Kozma, and Nederlof [WG 2019] relax the conditions of Dirac's theorem and consider input graphs G in which at least $n - k$ vertices have degree at least $\frac{n}{2}$, and present an FPT algorithm concerning to k, to decide whether such graphs G are Hamiltonian. In this paper, we first determine the stability of the property of having a bounded cycle cover. After that, combining the framework of Duarte et al. [MFCS 2021] with some results of Jansen et al. [WG 2019], we obtain a $2^{\mathcal{O}(k)} \cdot n^{\mathcal{O}(1)}$-time algorithm for MINIMUM CYCLE COVER on graphs with co-degeneracy at most k, which generalizes Duarte et al. [MFCS 2021] and Jansen et al. [WG 2019] results concerning the HAMILTONIAN CYCLE problem.

Keywords: Degeneracy · Complement graph · Cycle cover · Closure · FPT · Kernel

This research has received funding from Rio de Janeiro Research Support Foundation (FAPERJ) under grant agreement E-26/201.344/2021, National Council for Scientific and Technological Development (CNPq) under grant agreement 309832/2020-9, and the European Research Council (ERC) under the

European Union's Horizon 2020 research and innovation programme under grant agreement CUTACOMBS (No. 714704).

© Springer Nature Switzerland AG 2022
M. A. Bekos and M. Kaufmann (Eds.): WG 2022, LNCS 13453, pp. 187–200, 2022.
https://doi.org/10.1007/978-3-031-15914-5_14

1 Introduction

Graph width parameters are useful tools for identifying tractable classes of instances for NP-hard problems and designing efficient algorithms for such problems on these instances. Treewidth and clique-width are two of the most popular graph width parameters. An algorithmic meta-theorem due to Courcelle, Makowsky, and Rotics [7] states that any problem expressible in the monadic second-order logic on graphs (MSO_1) can be solved in FPT time when parameterized by the clique-width of the input graph.[1] In addition, Courcelle [5] states that any problem expressible in the monadic second-order logic of graphs with edge set quantifications (MSO_2) can be solved in FPT time when parameterized by the treewidth of the input graph. Although the class of graphs with bounded treewidth is a subclass of the class of graphs with bounded clique-width [4], the MSO_2 logic on graphs extends the MSO_1 logic, and there are MSO_2 properties like "G has a Hamiltonian cycle" that are not MSO_1 expressible [6]. In addition, there are problems that are fixed-parameter tractable when parameterized by treewidth, such as MaxCut, Largest Bond, Longest Cycle, Longest Path, Edge Dominating Set, Graph Coloring, Clique Cover, Minimum Path Cover, and Minimum Cycle Cover that cannot be FPT when parameterized by clique-width [12,15–18], unless FPT = W[1].

For problems that are fixed-parameter tractable concerning treewidth, but intractable when parameterized by clique-width, the identification of tractable classes of instances of bounded clique-width and unbounded treewidth becomes a fundamental quest [11]. In 2016, Dvořák, Knop, and Masařík [13] showed that k-Path Cover is FPT when parameterized by the treewidth of the complement of the input graph. This implies that Hamiltonian Path is FPT when parameterized by the treewidth of the complement graph. In 2017, Knop, Koutecký, Masařík, and Toufar (WG 2017, [21]) asked about the complexity of deciding graph problems Π on the complement of G considering a parameter p of G (i.e., with respect to $p(G)$), especially for sparse graph parameters such as treewidth. In fact, the treewidth of the complement of the input graph, proposed be called *co-treewidth* in [11], seems a nice width parameter to deal with dense instances of problems that are hard concerning clique-width. MaxCut, Clique Cover, and Graph Coloring are example of problems W[1]-hard concerning clique-width but FPT-time solvable when parameterized by co-treewidth (see [11]).

The *degeneracy* of a graph G is the least k such that every induced subgraph of G contains a vertex with degree at most k. Equivalently, the degeneracy of G is the least k such that its vertices can be arranged into a sequence so that each vertex is adjacent to at most k vertices preceding it in the sequence. It is well-known that the degeneracy of a graph is upper bounded by its treewidth; thus, the class of graphs with bounded treewidth is also a subclass of the class of graphs with bounded degeneracy. In [11], Duarte, Oliveira, and Souza presented an algorithmic framework to deal with the degeneracy of the complement graph, called *co-degeneracy*, as a parameter.

---

[1] Originally this required a clique-width expression as part of the input.

Although the notion of co-parameters is as natural as their complementary versions, just a few studies have ventured into the world of dense instances with respect to sparse parameters of their complements. Also, note that would be natural to consider *"co-clique-width"* parameterization, but Courcelle and Olariu [8] proved that for every graph G its clique-width is at most twice the clique-width of $\overline{G}$. Thus, the co-clique-width notion is redundant from the point of view of parameterized complexity. Therefore, in the sense of being a useful parameter for many NP-hard problems in identifying a large and new class of (dense) instances that can be efficiently handled, the co-degeneracy seems interesting because it is incomparable with clique-width and stronger[2] than co-treewidth.

In [11], Duarte, Oliveira, and Souza developed an algorithmic framework for co-degeneracy parameterization based on the notion of Bondy-Chvátal closure for solving problems that have a "bounded" stability concerning some closure. More precisely, for a graph G with n vertices, and two distinct nonadjacent vertices u and v of G such that $d(u) + d(v) \geq n$, Ore's theorem states that G is hamiltonian if and only if $G + uv$ is hamiltonian. In 1976, Bondy and Chvátal [2] generalized Ore's theorem and defined the *closure* of a graph:

– let ℓ be an integer; the $(n + \ell)$-closure, $\text{cl}_{n+\ell}(G)$, of a graph G is obtained from G by recursively adding an edge between pairs of nonadjacent vertices whose degree sum is at least $n + \ell$ until no such pair remains.

Bondy and Chvátal showed that $\text{cl}_{n+\ell}(G)$ is uniquely determined from G and that G is hamiltonian if and only if $\text{cl}_n(G)$ is hamiltonian.

A property Υ defined on all graphs of order n is said to be $(n + \ell)$-stable if for any graph G of order n that does not satisfy Υ, the fact that uv is not an edge of G and that $G + uv$ satisfies Υ implies $d(u) + d(v) < n + \ell$. In other words, if $uv \notin E(G)$, $d(u) + d(v) \geq n + \ell$ and $G + uv$ has property Υ, then G itself has property Υ (c.f. [3]). The smallest integer $n + \ell$ such that Υ is $(n + \ell)$-stable is the *stability* of Υ, denoted by $s(\Upsilon)$. Note that Bondy and Chvátal showed that Hamiltonicity is n-stable. A survey on the stability of graph properties can be found in [3].

In [11], based on the fact that the class of graphs with co-degeneracy at most k is closed under completion (edge addition), it was proposed the following framework for determining whether a graph G satisfies a property Υ in FPT time regarding the co-degeneracy of G, denoted by k:

1. determine an upper bound for $s(\Upsilon)$ - the stability of Υ;
2. If $s(\Upsilon) \leq n + \ell$ where $\ell \leq f(k)$ (for some computable function f) then
   (a) set $G = \text{cl}_{n+\ell}(G)$;
   (b) since $G = \text{cl}_{n+\ell}(G)$ and G has co-degeneracy k then G has co-vertex cover number (distance to clique) at most $2k + \ell + 1$ (see [11]);
   (c) at this point, it is enough to solve the problem in FPT-time concerning co-vertex cover parameterization.

---

[2] A parameter y is stronger than x, if the set of instances where x is bounded is a subset of those where y is bounded.

In [11], using such a framework, it was shown that HAMILTONIAN PATH, HAMILTONIAN CYCLE, LONGEST PATH, LONGEST CYCLE, and MINIMUM PATH COVER are all fixed-parameter tractable when parameterized by co-degeneracy. Note that LONGEST PATH and MINIMUM PATH COVER are two distinct ways to generalize the HAMILTONIAN PATH problem just as LONGEST CYCLE and MINIMUM CYCLE COVER generalize the HAMILTONIAN CYCLE problem. However, the MINIMUM CYCLE COVER problem seems to be more challenging than the others concerning co-degeneracy parameterization, even because the stability of having a cycle cover of size at most r, to the best of our knowledge, is unknown.

In the MINIMUM CYCLE COVER problem, we are given a simple graph G and asked to find a minimum set S of vertex-disjoint cycles of G such that each vertex of G is contained in one cycle of S, where single vertices are considered trivial cycles. Note that each nontrivial cycle has size at least three. In this paper, our focus is on MINIMUM CYCLE COVER parameterized by co-degeneracy.

The Dirac's theorem from 1952 (see [10]) states that a graph G with n vertices ($n \geq 3$) is Hamiltonian if every vertex of G has degree at least $\frac{n}{2}$. In [20], Jansen, Kozma, and Nederlof relax the conditions of Dirac's theorem and consider input graphs G in which at least $n - k$ vertices have degree at least $\frac{n}{2}$, and present a $2^{\mathcal{O}(k)} \cdot n^{\mathcal{O}(1)}$-time algorithm to decide whether G has a Hamiltonian cycle. In 2022, F. Fomin, P. Golovach, D. Sagunov, and K. Simonov [19] presented the following algorithmic generalization of Dirac's theorem: if all but k vertices of a 2-connected graph G are of degree at least δ, then deciding whether G has a cycle of length at least $\min\{2\delta + k, n\}$ can be done in time $2^k \cdot n^{\mathcal{O}(1)}$. Besides, in 2020, F. Fomin, P. Golovach, D. Lokshtanov, F. Panolan, S. Saurabh, and M. Zehavi [14] proved that deciding whether a 2-connected d-degenerate n-vertex G contains a cycle of length at least $d + k$ can be done in time $2^{\mathcal{O}(k)} \cdot n^{\mathcal{O}(1)}$.

In this paper, we first determine the stability of the property of having a cycle cover of size at most r. After that, using the closure framework proposed in [11] together with some results and techniques presented in [20], we show that MINIMUM CYCLE COVER admits a kernel with linear number of vertices when parameterized by co-degeneracy. After that, by designing an exact single-exponential time algorithm for solving MINIMUM CYCLE COVER, we obtain as a corollary a $2^{\mathcal{O}(k)} \cdot n^{\mathcal{O}(1)}$-time algorithm for the MINIMUM CYCLE COVER problem on graphs with co-degeneracy at most k. These results also implies a $2^{\mathcal{O}(k)} \cdot n^{\mathcal{O}(1)}$-time algorithm for solving MINIMUM CYCLE COVER on graphs G in which at least $n - k$ vertices have degree at least $\frac{n}{2}$, generalizing the Jansen, Kozma, and Nederlof's result presented in [20] (WG 2019) for the HAMILTONIAN CYCLE problem. Also, the single-exponential FPT algorithm for MINIMUM CYCLE COVER parameterized by co-degeneracy implies that HAMILTONIAN CYCLE can be solved with the same running time, improving the current state of the art for solving the HAMILTONIAN CYCLE problem parameterized by co-degeneracy since the algorithm presented in [11] runs in $2^{\mathcal{O}(k \log k)} \cdot n^{\mathcal{O}(1)}$ time, where k is the co-degeneracy. Note that our results also imply that MINIMUM CYCLE COVER on *co-planar* graphs can be solved in polynomial time, which seemed to be unknown in the literature.

2   On the Stability of Having a Bounded Cycle Cover

Although the stability of several properties has already been studied (c.f. [3]), the stability of the property of having a cycle cover of size at most r, to the best of our knowledge, is unknown. Therefore, we show that $s(\Upsilon) \leq n$, where r is any positive integer, and Υ is the property of having a cycle cover of size at most r.

Lemma 1. *Let r be a positive integer. A simple graph G with n vertices has a cycle cover of size at most r if and only if its n-closure, $\mathrm{cl}_n(G)$, has also a cycle cover of size at most r.*

Proof. Let G be a simple graph with n vertices, r be a positive integer, and Υ be the graph property of having a cycle cover of size at most r. Since the claim trivially holds when $r = 0$ or $r \geq n$, we assume that $1 \leq r \leq n - 1$.

First, note that if G has a cycle cover S of size r then the set S is also a cycle cover of $\mathrm{cl}_n(G)$, because G is a spanning subgraph of $\mathrm{cl}_n(G)$.

Now, suppose that G does not have a cycle cover of size at most r but $\mathrm{cl}_n(G)$ has a cycle cover of size at most r.

Given that $\mathrm{cl}_n(G)$ is uniquely determined from G [2], the construction of $\mathrm{cl}_{n+\ell}(G)$ can be seen as an iterative process of adding edges, starting from G, where a single edge is added at each step i, until no more edges can be added. Let $E_0 = E(G)$. We call by E_i the resulting set of edges after adding i edges during such a process. Therefore, $G_0 = G$, $G_1 = (V, E_1)$, $G_2 = (V, E_2), \ldots, G_t = (V, E_t)$, where $G_t = \mathrm{cl}_n(G)$ is the finite sequence of graphs generated during a construction of the n-closure of G.

Since G does not have a cycle cover of size at most r but $\mathrm{cl}_n(G)$ has a cycle cover of size at most r, by the construction of $\mathrm{cl}_n(G)$, there is a single i $(1 \leq i \leq t)$ such that G_{i-1} does not has a cycle cover of size at most r but G_i has a cycle cover of size at most r. Let $\{uw\} = E_i \setminus E_{i-1}$.

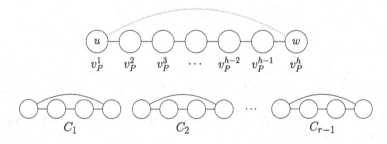

Fig. 1. Representation of a graph with $r - 1$ cycles, a path of size h and the edge vw that will be added, creating a graph with r cycles.

Suppose that G_i has a cycle cover S_i of size at most r. For simplicity, we assume that $|S_i| = r$. Therefore, the vertices of G_{i-1} can be covered by a set formed by $r - 1$ cycles $C_1, C_2, \ldots, C_{r-1}$ and a path P (the cycle of G_i that contains the edge uw). Assume that each cycle C_j is defined by the sequence $v_{C_j}^1, v_{C_j}^2, \ldots, v_{C_j}^{x_j}$ of vertices, where x_j is the number of vertices of C_j. Let $C = \{C_1, C_2, \ldots, C_{r-1}\}$, and $P = v_P^1, v_P^2, \ldots, v_P^h$, where $u = v_P^1$, $w = v_P^h$ and h is the number of vertices of P. Note that $h \geq 3$; otherwise $u = w$, implying that P is a trivial cycle and G_{i-1} has a cycle cover of size r. Figure 1 illustrates C and P.

We partition some vertices of G_{i-1} into four sets:

$$X_P = \{v_P^q \mid (v_P^{q-1}, v_P^h) \in E_{i-1} \text{ and } 2 < q < h\},$$

$$X_C = \{v_{C_j}^q \mid (v_{C_j}^{(q \bmod x_j)+1}, v_P^h) \in E_{i-1}, \ 1 \leq q \leq x_j, \text{ and } C_j \in C\},$$

$$Y_P = \{v_P^q \mid (v_P^1, v_P^q) \in E_{i-1} \text{ and } 2 < q < h\},$$

and

$$Y_C = \{v_{C_j}^q \mid (v_{C_j}^q, v_P^1) \in E_{i-1}, 1 \leq q \leq x_j, \text{ and } C_j \in C\}.$$

Note that $v_{C_j}^1 = v_{C_j}^{(1 \bmod 1)+1}$ for trivial cycles C_j. Thus, X_C it is well defined. Let $X = X_P \cup X_C$ and $Y = Y_P \cup Y_C$.

The set X, is the set of vertices (with the exception of v_P^h) in which its predecessor in the path or its successor in the cycle is adjacent to v_P^h. Also, the set Y, is the set of vertices adjacent to v_P^1 (with the exception of v_P^2). Note that the size of both X and Y are bounded by $n - 3$, since they exclude the vertices v_P^1, v_P^2 and v_P^h of P. Besides that, we can observe that

$$|X| = d(v_P^h) - 1 \text{ and } |Y| = d(v_P^1) - 1,$$

where $d(v)$ is the degree of the vertex v. Therefore, the following holds:

$$|X| + |Y| = d(v_P^h) + d(v_P^1) - 2$$

that is,

$$|X| + |Y| \geq n - 2$$

since $d(u) + d(w) \geq n$ where $u = v_P^1$, $w = v_P^h$, and $\{uw\} = E_i \setminus E_{i-1}$.

However, $|X \cup Y| \leq n - 3$ because both X and Y exclude v_P^1, v_P^2 and v_P^r. Therefore, there is at least one vertex that belong to both X and Y. Note that $(X_P \cup Y_P) \cap (X_C \cup Y_C) = \emptyset$, since, by definition, the elements of the covering are vertex disjoint.

Therefore, there are two possibilities:

1. There is a vertex v_P^q belonging to the path P such that $v_P^q \in X_P \cap Y_P$. This implies that G_{i-1} already had a cycle covering exactly the vertices of P before the addition of the edge $uw = v_P^1 v_P^h$, which could be formed as follows (see Fig. 2):

$$v_P^1, v_P^2, \ldots, v_P^{q-1}, v_P^h, v_P^{h-1}, v_P^{h-2}, \ldots, v_P^{q+1}, v_P^q, v_P^1;$$

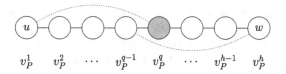

Fig. 2. Representation of case 1, where the vertex v_P^q, highlighted in gray, belongs to $X_P \cap Y_P$.

2. There is a vertex $v_{C_j}^q$ belonging to a cycle $C_j \in C$ such that $v_{C_j}^q \in X_C \cap Y_C$. In this case, G_{i-1} has a larger cycle that can be obtained by merging C_j with the path P as follows (see Fig. 3):

$$v_{C_j}^{(q \bmod x_j)+1}, v_{C_j}^{(q \bmod x_j)+2}, \ldots, v_{C_j}^q, v_P^1, v_P^2, \ldots, v_P^h, v_{C_j}^{(q \bmod x_j)+1}.$$

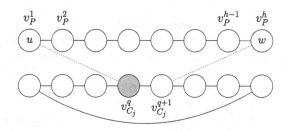

Fig. 3. Representation of case 2, where the vertex $v_{C_j}^q$, highlighted in gray, belongs to $X_C \cap Y_C$.

In the first case G_{i-1} has a cycle cover of size r, while in the second case G_{i-1} has a cycle cover of size $r - 1$. Both cases contradicts the hypothesis that G_{i-1} does not have a cycle cover of size at most r.

Therefore, there is no $1 \leq i \leq t$ such that G_{i-1} does not have a cycle cover of size at most r and G_i has such a cycle cover. Thus, if $G_t = \mathrm{cl}_n(G)$ has a cycle cover of size at most r then $G_0 = G$ also has a cycle cover of size at most r. □

Lemma 1 states that for any positive integer r, the graph property Υ of having a cycle cover of size at most r satisfies that $s(\Upsilon) \leq n$. We remark that such a bound is tight since whenever $r = 1$, the target Υ is the Hamiltonicity property, which is well known to have stability (exactly) equal to n (c.f. [2]).

Now, observe that the class of graphs with co-degeneracy at most k is closed under completion (edge addition), in the same way as the class of graphs with degeneracy at most k is closed under edge removals. Recall that $\mathrm{cl}_n(G)$ is uniquely determined from a n-vertex graph G and it can be constructed in polynomial time. Therefore, by Lemma 1, we may assume that $G = \mathrm{cl}_n(G)$ whenever G is an instance of MINIMUM CYCLE COVER parameterized by co-degeneracy.

We call by *co-vertex cover* any set of vertices whose removal makes the resulting graph complete, i.e., a vertex cover in the complement graph. The *co-vertex cover number* of a graph G, *co-vc(G)*, is the size of its minimum co-vertex cover.[3]

The following theorem is a key tool for this work.

Theorem 1 ([11]). *Let $\ell \geq 0$ be an integer. If a graph G has co-degeneracy k and $G = cl_{n+\ell}(G)$ then G has co-vertex cover number bounded by $2k + \ell + 1$. In addition, a co-vertex cover of G with size at most $2k + \ell + 1$ can be found in polynomial time.*

From Lemma 1 and Theorem 1, the problem of solving MINIMUM CYCLE COVER on instances G with co-degeneracy k can be reduced in polynomial time to the problem of solving MINIMUM CYCLE COVER on instances $G' = cl_n(G)$ with co-vertex cover number at most $2k + 1$. Therefore, in the next section we will focus on parameterization by the co-vertex cover number.

3   Polynomial Kernelization

In [20], Jansen, Kozma, and Nederlof showed that given a graph G with n vertices such that at least $n - k$ vertices of G have degree at least $\frac{n}{2}$, there is a deterministic algorithm that constructs in polynomial time a graph G' with at most $3k$ vertices, such that G is Hamiltonian if and only if G' is Hamiltonian. In other words, they showed that the HAMILTONIAN CYCLE problem parameterized by such a k has a kernel with a linear number of vertices.

First, we remark that such a parameterization that aims to explore a "distance measure" (k) of a given graph G from satisfying the Dirac property, when applied to problems that are n-stable (such as HAMILTONIAN CYCLE and MINIMUM CYCLE COVER) can be polynomial-time reduced to the case where the co-degeneracy is bounded by k. Since for such problems one can consider only instances G' such that $G' = cl_n(G')$, from a graph G with n vertices such that at least $n-k$ vertices of G have degree at least $\frac{n}{2}$, we obtain an instance $G' = cl_n(G)$ having a clique of size at least $n - k$.

Therefore, in the following, we extend the "relaxed" Dirac result from [20] by considering co-degeneracy and the MINIMUM CYCLE COVER problem.

Theorem 2. *There is a polynomial-time algorithm that, given a graph G and a nonempty set $S \subseteq V(G)$ such that $G - S$ is a clique, outputs an induced subgraph G' of G on at most $3|S|$ vertices such that G has a cycle cover of size at most r if and only if G' has a cycle cover of size at most r.*

Proof. Let $G = (V, E)$ be a graph having a co-vertex cover S. Let $C = V(G)\backslash S$. If $|C| \leq 2|S|$ then by setting $G' = G$ the claim holds. Now, assume that $|C| > 2|S|$.

As in [20], let $S' = \{v_1, v_2 : v \in S\}$ be a set containing two representatives for each vertex of S. We construct a bipartite graph H on vertex set $C \cup S'$,

---

[3] *co-vc(G) is also called the *distance to clique* of G, and a co-vertex cover set is also called a *clique modulator*.*

where for each edge $cv \in E(G)$ with $c \in C$ and $v \in S$, we add the edges cv_1, cv_2 to $E(H)$.

Now, we compute a maximum matching $M \subseteq E(H)$ of H. Let C^* be the subset of vertices of C saturated (matched) by M. If $|C^*| \geq |S| + 1$ then set $C' = C^*$; otherwise, let $C' \subseteq C$ be a superset of C^* with size $|S| + 1$. Finally, set $G' = G[C' \cup S]$.

Note that G' has at most $3|S|$ vertices, because C' has at most $2|S|$ vertices.

First, suppose that G' has cycle cover Q' of size at most r. Since G' is a subgraph of G, the set Q' is a set of vertex disjoint cycles of G covering $S \cup C' \subseteq V(G)$. Thus, only vertices of $C \setminus C'$ are not covered by Q'. However, since the size of C' is greater than the size of S, there is at least one cycle $Q_j \in Q'$ that either is a single vertex of C or contains an edge between vertices of C. If $|Q_j| = 1$ then we can replace it by a cycle containing all the vertices of $(C \setminus C') \cup Q_j$. If Q_j has an edge uv such that $u, v \in C$, then we can replace this edge by a uv-path containing the vertices of $C \setminus C'$ as internal vertices. In both cases we obtain a cycle cover of size at most $|Q'|$ in the graph G.

At this point, it remains to show that if G has a cycle cover of size r then G' has a cycle cover of size at most r.

Using a strategy similar to that in [20], we first present a structure that implies cycle covers of size at most r in G'. For a vertex set S^* in a graph G^*, we define a *cycle-path cover* of S^* in G^* as a set L of pairwise vertex-disjoint simple paths or cycles such that each vertex of S^* belongs to exactly one element of L, i.e., L can be seen as a subgraph with maximum degree two which contains every vertex of S^*. For a vertex set C^* in G^*, we say that a cycle-path cover L has C^*-endpoints if the endpoints of each path $P \in L$ belong to C^*.

Claim 1. *If G' has a cycle-path cover of S having C'-endpoints and containing at most $r - 1$ cycles, then G' has a cycle cover of size at most r.*

Proof. We have two cases to analyse: if the cycle-path cover of S contains only cycles, as the number of cycles is at most $r - 1$, then we can add a new cycle formed by the vertices not yet covered; if the cycle-path cover contains some paths, by vertex disjointness, all the paths have different endpoints in C', and, since C' is a clique, we can connect such endpoints in such a way as to form a single cycle containing these paths as subgraphs, after that, an edge uv of such a cycle having $u, v \in C'$ can replaced by a uv-path containing as internal vertices the vertices of G' that are not in such a cycle-path cover of S. In both cases, we conclude that G' has a cycle cover of size at most r.   ⌟

Now, considering the bipartite graph H and its maximum matching M, let U_C be the set of vertices of C that are not saturated by M, and let R be the vertices of H that are reachable from U_C by an M-alternating path in H (which starts with a non-matching edge). Set $R_C = R \cap C$ and $R_{S'} = R \cap S'$.

By Claim 1, it is enough to show that if G has a cycle cover of size r then G' has a cycle-path cover of S having C'-endpoints and containing at most $r - 1$ cycles. For that, we consider Claim 2 presented in [20].

Claim 2 ([20]). *The sets R, R_C, $R_{S'}$ satisfy the following.*

1. *Each M-alternating path in H from U_C to a vertex in $R_{S'}$ (resp. R_C) ends with a non-matching (resp. matching) edge.*
2. *Each vertex of $R_{S'}$ is matched by M to a vertex in R_C.*
3. *For each vertex $x \in R_C$ we have $N_H(x) \subseteq R_{S'}$.*
4. *For each vertex $v \in S$ we have $v_1 \in R_{S'}$ if and only if $v_2 \in R_{S'}$.*
5. *For each vertex $v \in S' \setminus R_{S'}$, we have $N_H(v) \cap R_C = \emptyset$ and each vertex of $N_H(v)$ is saturated by M.*

Lemma 2. *If G has a cycle cover of size at most r, then G' has a cycle-path cover of S having C'-endpoints and containing at most $r - 1$ cycles.*

Proof. Let F be a cycle cover of size at most r of G. Consider F as a 2-regular subgraph of G. Let $F_1 = F[S]$ be the subgraph of F induced by S. Since F is a spanning subgraph of G, and $S \subset V(G')$, it follows that F_1 is a cycle-path cover of S in G'. At this point, we need to extend it to have C'-endpoints. As in [20], we do that by inserting edges into F_1 to turn it into a subgraph F_2 of G' in which each vertex of S has degree exactly two. This structure F_2 must be a cycle-path cover of S in G' with C'-endpoints, since the degree-two vertices S cannot be endpoints of the paths.

Setting $F_2 = F_1, R_S = \{v \in S : v_1 \in R_{S'} \text{ or } v_2 \in R_{S'}\}$, we proceed as follows.

1. For each vertex $v \in R_S$, we have $v_1, v_2 \in R_{S'}$ by Claim 2(4), which implies by Claim 2(2) that both v_1 and v_2 are matched to distinct vertices x_1, x_2 in R_C. If v has degree zero in subgraph F_1, then add the edges vx_1, vx_2 to F_2. If v has degree one in F_2 then only add the edge vx_1. (we do not add edges if v already has degree two in F_1)
2. For each vertex $v \in S \setminus R_S$, it holds that $N_G(v) \cap R_C = \emptyset$. This follows from the fact that $N_G(v) = N_H(v_1) = N_H(v_2)$ and Claim 2(5). Note that $v \notin R_S$ implies $v_1, v_2 \notin R_{S'}$. Hence the (up to two) neighbors that $v \in S \setminus R_S$ has in C on the cycle cover F do not belong to R_C (see also Claim 2(3)), In addition, Claim 2(5) ensures that all vertices of $N_G(v)$ are saturated by H and hence belong to C'. Thus, for each vertex $v \in S \setminus R_S$, for each edge from v to $C \cap C'$ incident on v in F, we insert the corresponding edge into F_2.

It is clear that the above procedure produces a subgraph F_2 in which all vertices of S have degree exactly two. By Claim 2(5), we have that a vertex $c \in C$ does not have edges added in F_2 by both previous steps, thus each vertex $c \in C$ added in F_2 has degree at most two in it because c has at most one edge in the matching M (see Step 1), while c has two edges in the cycle cover F (see Step 2).

At this point, we know that F_2 is a cycle-path cover of S having C'-endpoints. It remains to show that it contains at most $r - 1$ cycles.

Claim 3. *Every cycle of F_2 is a cycle of F.*

Proof. Suppose that F_2 has a cycle Q that is not in F. As F_2 is formed from F_1, the edges in Q between the vertices of S are also edges of F. Furthermore, by construction, the added edges from F_1 to obtain F_2 are the edges incident to the vertices of S. Therefore, there is no edge between the vertices of the clique C in Q. By Claim 2(5), we have that a vertex $c \in C$ cannot be incident to two edges of F_2 being one added by Step 1 and the other by Step 2 of the construction. Since these steps are mutually exclusive with respect to a vertex $c \in C$, and given that c has degree two in Q (since Q is a cycle), we have that the edges of each vertex $c \in C \cap Q$ were added by Step 2 of the construction (Step 1 adds only one edge of the matching). However, by construction, the edges in Q incident to a vertex $c \in C$ are the edges in F. Therefore, every edge of Q is contained in F, contradicting the hypothesis that Q is not contained in F. ⌐

By hypothesis, F has at most r cycles. Since $|C| > |S|$, it holds that at least one cycle of F must have an edge between vertices of C. Thus, at least one cycle of F is not completely contained in F_2, which implies, by Claim 3, that F_2 has at most $r - 1$ cycles. Therefore, F_2 is a cycle-path cover of S having C'-endpoints which contains at most $r - 1$ cycles. This concludes the proof of Lemma 2. ⌐

By Lemma 2 and Claim 1, it holds that if G has a cycle cover of size at most r then G' has a cycle cover of size at most r. Since the reduction can be performed in polynomial time, and $|V(G')| \leq 3|S|$, we conclude the proof of Theorem 2. □

Corollary 1. MINIMUM CYCLE COVER *parameterized by co-degeneracy admits a kernel with at most $6k + 3$ vertices, where $k = $ co-deg.*

4   An Exact Single-Exponential Time Algorithm

By Corollary 1, it holds that an exact and deterministic single-exponential time algorithm for MINIMUM CYCLE COVER is enough to obtain an FPT algorithm for MINIMUM CYCLE COVER with single-exponential dependency concerning the co-degeneracy of the input graph. In [9], using the Cut&Count technique, M. Cygan, J. Nederlof, Ma. Pilipczuk, Mi. Pilipczuk, J. Rooij and J. Wojtaszczyk produces a $2^{\mathcal{O}(tw)} \cdot |V|^{\mathcal{O}(1)}$ time Monte Carlo algorithm for MINIMUM CYCLE COVER (UNDIRECTED MIN CYCLE COVER in [9]), where tw is the treewidth of the input graph. In [1], H. Bodlaender, M. Cygan, S. Kratsch, J. Nederlof presented two approaches to design deterministic $2^{\mathcal{O}(tw)} \cdot |V|^{\mathcal{O}(1)}$-time algorithms for some connectivity problems, and claimed that such approaches can be apply to all problems studied in [9].

Although such approaches can be used to solve MINIMUM CYCLE COVER by a single-exponential time algorithm, in order to present a simpler deterministic procedure, below we present a simple and deterministic dynamic programming based on modifying the Bellman-Held-Karp algorithm.

Theorem 3. MINIMUM CYCLE COVER *can be solved in $\mathcal{O}(2^n \cdot n^3)$ time.*

Proof. Given a graph $G = (V, E)$ with an isolated vertex w, a vertex subset $X \subseteq V$, $s, t \in X$, and a Boolean variable $P2$, we denote by $M[X, s, t, P2]$ the size of a minimum set S of vertex-disjoint cycles but one nonempty vertex-disjoint st-path of $G[X]$ such that

- every vertex of X is in an element of S;
- the st-path is not a P_2 if the variable $P2 = 0$;
- the st-path is a P_2 if the variable $P2 = 1$.

Note that $M[V, w, w, 0]$ represents the size of a minimum cycle cover of G.

In essence, the st-path represents the open cycle that is still being built. The variable $P2$ is a control variable to avoid P_2 as cycles of size two. At each step, we can interpret that the algorithm either lengthens the path by adding a new endpoint or closes a cycle and opens a new trivial path. As we can reduce the MINIMUM CYCLE COVER problem to the case where the graph has an isolated vertex w, we assume that this is the case and consider $w \in X$ just when $X = V$.

Our recurrence is as follows.

If $X = \{v\}$ then $M[X, s, t, P2] = 1$ for $s = t = v$ and $P2 = 0$; otherwise, it is ∞.

If $|X| \geq 2$ then $M[X, s, t, P2]$ is equal to

$$
\begin{cases}
\infty & \text{if } s = t, P2 = 1 \\
\min_{s', t' \in X \setminus \{t\} \; : \; s't' \in E \text{ or } s'=t'} (M[X \setminus \{t\}, s', t', 0]) + 1 & \text{if } s = t, P2 = 0 \\
\infty & \text{if } s \neq t, P2 = 1, st \notin E \\
M[X \setminus \{t\}, s, s, 0] & \text{if } s \neq t, P2 = 1, st \in E \\
\min_{t' \in X \setminus \{t\} \; s.t. \; tt' \in E, \; P_2' \in \{0,1\}} (M[X \setminus \{t\}, s, t', P2']) & \text{if } s \neq t, P2 = 0
\end{cases}
$$

The size of the table is bounded by $(2^n - 1) \cdot n^2 \cdot 2$ where n is the number of vertices of the graph. Regarding time complexity, we have three cases: when $P2 = 1$ the recurrences can be computed in $\mathcal{O}(1)$ time; when $s = t$ and $P2 = 0$ the recurrence can be computed in $\mathcal{O}(n^2)$ time, and since there are at most $(2^n - 1) \cdot n + 1$ cells in this case, the total amount of time taken to compute those cells is $\mathcal{O}(2^n \cdot n^3)$; finally, when $s \neq t$ and $P2 = 0$ the recurrence can be computed in $\mathcal{O}(n)$ time, but there are $\mathcal{O}(2^n \cdot n^2)$ cells in this case, implying into a total amount of $\mathcal{O}(2^n \cdot n^3)$ time to compute all these cells. Therefore, the dynamic programming algorithm can be performed in $\mathcal{O}(2^n \cdot n^3)$ time. Note that, in addition to determining the size of a minimum cycle cover, one can find it with the same running time. Also, the correctness of the algorithm is straightforward. $\square$

Corollary 2. MINIMUM CYCLE COVER *can be solved in* $2^{\mathcal{O}(co\text{-}deg)} \cdot n^{\mathcal{O}(1)}$ *time.*

By Corollary 2, it follows that MINIMUM CYCLE COVER on co-planar graphs can be solved in polynomial time, which seems to be unknown in the literature.

Corollary 3. MINIMUM CYCLE COVER *on graphs G in which at least $n - k$ vertices have degree at least $\frac{n}{2}$ can be solved in* $2^{\mathcal{O}(k)} \cdot n^{\mathcal{O}(1)}$ *time.*

References

1. Bodlaender, H.L., Cygan, M., Kratsch, S., Nederlof, J.: Deterministic single exponential time algorithms for connectivity problems parameterized by treewidth. Inf. Comput. **243**, 86–111 (2015). 40th International Colloquium on Automata, Languages and Programming (ICALP 2013)
2. Bondy, J.A., Chvátal, V.: A method in graph theory. Discret. Math. **15**(2), 111–135 (1976)
3. Broersma, H., Ryjáček, Z., Schiermeyer, I.: Closure concepts: a survey. Graphs Comb. **16**(1), 17–48 (2000)
4. Corneil, D.G., Rotics, U.: On the relationship between clique-width and treewidth. SIAM J. Comput. **34**(4), 825–847 (2005)
5. Courcelle, B.: The monadic second-order logic of graphs. I. recognizable sets of finite graphs. Inf. Comput. **85**(1), 12–75 (1990)
6. Courcelle, B.: The monadic second order logic of graphs VI: on several representations of graphs by relational structures. Discret. Appl. Math. **54**(2–3), 117–149 (1994)
7. Courcelle, B., Makowsky, J.A., Rotics, U.: Linear time solvable optimization problems on graphs of bounded clique-width. Theory Comput. Syst. **33**(2), 125–150 (2000)
8. Courcelle, B., Olariu, S.: Upper bounds to the clique width of graphs. Discret. Appl. Math. **101**(1–3), 77–114 (2000)
9. Cygan, M., Nederlof, J., Pilipczuk, M., Pilipczuk, M., Van Rooij, J.M.M., Wojtaszczyk, J.O.: Solving connectivity problems parameterized by treewidth in single exponential time. ACM Trans. Algorithms **18**(2), 1–31 (2022)
10. Dirac, G.A.: Some theorems on abstract graphs. Proc. Lond. Math. Soc. **3**(1), 69–81 (1952)
11. Duarte, G.L., de Oliveira Oliveira, M., Souza, U.S.: Co-degeneracy and co-treewidth: using the complement to solve dense instances. In: Bonchi, F., Puglisi, S.J. (eds.) 46th International Symposium on Mathematical Foundations of Computer Science (MFCS 2021). Leibniz International Proceedings in Informatics (LIPIcs), Dagstuhl, Germany, vol. 202, pp. 42:1–42:17. Schloss Dagstuhl - Leibniz-Zentrum für Informatik (2021)
12. Duarte, G.L., et al.: Computing the largest bond and the maximum connected cut of a graph. Algorithmica **83**(5), 1421–1458 (2021)
13. Dvořák, P., Knop, D., Masařík, T.: Anti-path cover on sparse graph classes. In: Bouda, J., Holík, L., Kofron, J., Strejcek, J., Rambousek, A. (eds.) Proceedings 11th Doctoral Workshop on Mathematical and Engineering Methods in Computer Science, MEMICS 2016, Telč, Czech Republic, 21–23 October 2016. EPTCS, vol. 233, pp. 82–86 (2016)
14. Fomin, F.V., Golovach, P.A., Lokshtanov, D., Panolan, F., Saurabh, S., Zehavi, M.: Going far from degeneracy. SIAM J. Discret. Math. **34**(3), 1587–1601 (2020)
15. Fomin, F.V., Golovach, P.A., Lokshtanov, D., Saurabh, S.: Clique-width: on the price of generality. In: Proceedings of the Twentieth Annual ACM-SIAM Symposium on Discrete Algorithms, pp. 825–834. SIAM (2009)
16. Fomin, F.V., Golovach, P.A., Lokshtanov, D., Saurabh, S.: Algorithmic lower bounds for problems parameterized by clique-width. In: Proceedings of the Twenty-First Annual ACM-SIAM Symposium on Discrete Algorithms, pp. 493–502. SIAM (2010)

17. Fomin, F.V., Golovach, P.A., Lokshtanov, D., Saurabh, S.: Intractability of clique-width parameterizations. SIAM J. Comput. **39**(5), 1941–1956 (2010)
18. Fomin, F.V., Golovach, P.A., Lokshtanov, D., Saurabh, S.: Almost optimal lower bounds for problems parameterized by clique-width. SIAM J. Comput. **43**(5), 1541–1563 (2014)
19. Fomin, F.V., Golovach, P.A., Sagunov, D., Simonov, K.: Algorithmic extensions of Dirac's theorem. In: Naor, J.S., Buchbinder, N. (eds.) Proceedings of the 2022 ACM-SIAM Symposium on Discrete Algorithms, SODA 2022, Virtual Conference, Alexandria, VA, USA, 9–12 January 2022, pp. 406–416. SIAM (2022)
20. Jansen, B.M.P., Kozma, L., Nederlof, J.: Hamiltonicity below Dirac's condition. In: Sau, I., Thilikos, D.M. (eds.) WG 2019. LNCS, vol. 11789, pp. 27–39. Springer, Cham (2019). https://doi.org/10.1007/978-3-030-30786-8_3
21. Knop, D., Koutecký, M., Masařík, T., Toufar, T.: Simplified algorithmic metatheorems beyond MSO: treewidth and neighborhood diversity. In: Bodlaender, H.L., Woeginger, G.J. (eds.) WG 2017. LNCS, vol. 10520, pp. 344–357. Springer, Cham (2017). https://doi.org/10.1007/978-3-319-68705-6_26

On the Lossy Kernelization for Connected Treedepth Deletion Set

Eduard Eiben[1] , Diptapriyo Majumdar[2]([⊠]) , and M. S. Ramanujan[3]

[1] Royal Holloway, University of London, Egham, UK
`eduard.eiben@rhul.ac.uk`
[2] Indraprastha Institute of Information Technology Delhi, New Delhi, India
`diptapriyo@iiitd.ac.in`
[3] University of Warwick, Coventry, UK
`R.Maadapuzhi-Sridharan@warwick.ac.uk`

Abstract. We study the CONNECTED η-TREEDEPTH DELETION problem, where the input instance is an undirected graph G, and an integer k and the objective is to decide whether there is a vertex set $S \subseteq V(G)$ such that $|S| \leq k$, every connected component of $G - S$ has treedepth at most η and $G[S]$ is a connected graph. As this problem naturally generalizes the well-studied CONNECTED VERTEX COVER problem, when parameterized by the solution size k, CONNECTED η-TREEDEPTH DELETION is known to not admit a polynomial kernel unless NP $\subseteq$ coNP/poly. This motivates the question of designing *approximate polynomial kernels* for this problem.

In this paper, we show that for every fixed $0 < \varepsilon \leq 1$, CONNECTED η-TREEDEPTH DELETION admits a time-efficient $(1+\varepsilon)$-approximate kernel of size $k^{2^{\mathcal{O}(\eta+1/\varepsilon)}}$ (i.e., a Polynomial-size Approximate Kernelization Scheme).

Keywords: Treedepth · Kernelization · Connected Treedepth Deletion Set · Lossy Kernelization

1 Introduction

Parameterized complexity is a popular approach to cope with NP-Completeness and the related area of kernelization studies mathematical formulations of preprocessing algorithms for (typically) NP-complete decision problems. Kernelization is an important step that preprocesses the input instance (I, k) into a smaller, equivalent instance (I', k') in polynomial-time such that $|I'| + k'$ is bounded by $g(k)$. It is desired that $g(k)$ is polynomial in k, in which case we have a *polynomial kernelization*. Over the past few decades, the design of (polynomial) kernelization for numerous problems has been explored [4–6,22] and a rich variety of algorithm design techniques have been introduced. There are, however, problems that

M. S. Ramanujan is supported by Engineering and Physical Sciences Research Council (EPSRC) grants EP/V007793/1 and EP/V044621/1.

M. A. Bekos and M. Kaufmann (Eds.): WG 2022, LNCS 13453, pp. 201–214, 2022.
https://doi.org/10.1007/978-3-031-15914-5_15

provably do not admit polynomial kernels unless $\mathsf{NP} \subseteq \mathsf{coNP/poly}$ [1,8,18,19], in which case one requires an alternate rigorous notion of preprocessing. Moreover, the notion of kernelization is defined with respect to decision problems, implying that when a suboptimal solution to the reduced instance is provided, one may not be able to get a feasible solution to the original input instance. To address both of the aforementioned issues with kernelization, Lokshtanov et al. [21] introduced the framework of Approximate Kernelization. Roughly speaking, an α-approximate kernelization is a polynomial-time preprocessing algorithm for a parameterized optimization problem with the promise that if a c-approximate solution to the reduced instance is given, then a $(c \cdot \alpha)$-approximate feasible solution to the original instance can be obtained in polynomial time. In this case, both $c, \alpha \geq 1$. When the reduced instance has size bounded by $g(k)$ for some polynomial function g, then we have a α-approximate polynomial-size approximate kernel (see Sect. 2 for formal definitions). In recent years, there has been a sustained search for polynomial-size approximate kernels for well-known problems in parameterized complexity that are known to exclude standard polynomial kernelizations. One such set of problems is the family of "vertex deletion" problems with a *connectivity constraint*. A classic example here is VERTEX COVER that admits a $2k$ vertex kernel, but CONNECTED VERTEX COVER provably does not admit a polynomial kernel unless $\mathsf{NP} \subseteq \mathsf{coNP/poly}$. Lokshtanov et al. [21] proved that for every $\varepsilon > 0$, CONNECTED VERTEX COVER admits a $(1+\varepsilon)$-approximate kernel of size $\mathcal{O}(k^{\lceil 1/\varepsilon \rceil})$. This is also called a *Polynomial-size Approximate Kernelization Scheme* (PSAKS). Subsequent efforts have mainly focused on studying the feasibility of approximate kernelization for problems that generalize CONNECTED VERTEX COVER. For instance, Eiben et al. [11] obtained a PSAKS for the CONNECTED $\mathcal{H}$-HITTING SET problem (where one wants to find a smallest connected vertex set that hits all occurrences of graphs from the finite set $\mathcal{H}$ as induced subgraphs) and Ramanujan obtained a PSAKS for CONNECTED FEEDBACK VERTEX SET [24] and a $(2 + \varepsilon)$-approximate polynomial *compression* [25] for the PLANAR $\mathcal{F}$-DELETION problem [14] with connectivity constraints on the solution. A compression is a weaker notion than kernelization, where the output can be an instance of a different problem.

In this paper, our focus is on the connectivity constrained version of the η-TREEDEPTH DELETION SET problem. In the (unconnected version of the) problem, one is given a graph G and an integer k and the goal is to decide whether there is a vertex set of size at most k whose deletion leaves a graph of treedepth at most η. We refer the reader to Sect. 2 for the formal definition of treedepth. Intuitively, it is a graph-width measure that expresses the least number of rounds required to obtain an edge-less graph, where, in each round we delete some vertex from each surviving connected component. Treedepth is a graph parameter that has attracted significant interest in the last decade. It allows improved algorithmic bounds over the better-known parameter of treewidth for many problems (see, for example, [17,26]) and it plays a crucial role in the study of kernelization [15]. In recent years, the optimal solution to the η-TREEDEPTH DELETION SET problem itself has been identified as a useful parameter in the kernelization of generic vertex-deletion problems [20]. The many insightful advances made by focusing on graphs of bounded treedepth motivates us to consider the CONNECTED η-TREEDEPTH DELETION problem as an ideal conduit between the well-understood CONNECTED VERTEX COVER problem and the connected versions of more general problems such as the η-TREEWIDTH DELETION SET problem, which is still largely unexplored from the point of view of approximate kernelization. We formally state our problem as follows.

CONNECTED η-TREEDEPTH DELETION (CON-η-DEPTH-TRANSVERSAL)
Input: An undirected graph G, and an integer k.
Parameter: k
Question: Does G have a set S of at most k vertices such that $G[S]$ is connected and $(G - S)$ has treedepth at most η?

A set $S \subseteq V(G)$ is called a *connected η-treedepth deletion set* if $G[S]$ is connected and every connected component of $G - S$ has treedepth at most η. As edgeless graphs have treedepth 1, it follows that CONNECTED η-TREEDEPTH DELETION generalizes CONNECTED VERTEX COVER and does not have a polynomial kernelization under standard hypotheses even for constant values of η, thus motivating its study through the lens of approximate kernelization. Here, two results in the literature are of particular consequence to us and form the starting point of our work:

➤ Graphs of treedepth at most η can be characterized by a finite set of forbidden induced subgraphs, where each obstruction has size at most $2^{2^{\eta-1}}$ [10] and hence, an invocation of the result of Eiben et al. [11] gives a $(1+\varepsilon)$-approximate kernelization of size $\mathcal{O}(k^{2^{2^{\eta-1}} \cdot 2^{\frac{1}{\varepsilon}}+1})$ for CON-η-DEPTH-TRANSVERSAL.

➤ On the other hand, using the fact that graphs of treedepth at most η also exclude a finite set of graphs as forbidden minors including at least one planar graph, we infer that the $(2 + \varepsilon)$-approximate polynomial compression for CONNECTED PLANAR $\mathcal{F}$-DELETION of Ramanujan [25] implies a $(2+\varepsilon)$-approximate compression for CONNECTED η-TREEDEPTH DELETION of size $k^{f(\eta) \cdot 2^{\mathcal{O}(1/\varepsilon)}}$ for some function f that is at least exponential.

Naturally, these two "meta-results" provide useful proofs of concept using which we can conclude the existence of an approximate kernel (or compression) for CONNECTED η-TREEDEPTH DELETION. However, the kernel-size bounds that one could hope for by taking this approach are far from optimal and in fact, the second result mentioned above only guarantees the weaker notion of compression. Thus, these two results raise the following natural question: "Could one exploit structure inherent to the bounded treedepth graphs and improve upon both results, by obtaining a $(1 + \varepsilon)$-approximate polynomial kernelization for CONNECTED η-TREEDEPTH DELETION with improved size bounds?" Our main result is a positive answer to this question.

Theorem 1. *For every fixed $0 < \varepsilon \leq 1$, CONNECTED η-TREEDEPTH DELETION has a time-efficient $(1 + \varepsilon)$-approximate kernelization of size $k^{\mathcal{O}(\lceil 2^{\eta + \lceil 10/\varepsilon \rceil} \eta/\varepsilon \rceil)}$.*

2   Preliminaries

Sets and Graphs: We use $[r]$ to denote the set $\{1,\ldots,r\}$ and $A \uplus B$ to denote the disjoint union of two sets. We use standard graph theoretic terminologies from Diestel's book [7]. Throughout the paper, we consider undirected graphs. We use P_ℓ to denote a path with ℓ vertices. A graph is said to be *connected* if there is a path between every pair of vertices. Let $G = (V, E)$ be a graph and a pair of vertices $u, v \in V(G)$. We call a set $A \subseteq V(G)$ an (u, v)-*vertex cut* if there is no path from u to v in $G - A$. For $u, v \in G$, we use $dist(u, v)$ to denote the length of a 'shortest path' from u to v. We use $diam(G) = \max_{u,v \in V(G)} dist(u, v)$ to denote the *diameter* of G. Let $R \subseteq V(G)$ be a vertex set the elements of which are called *terminals* and a weight function $w : E(G) \to \mathbb{N}$. A *Steiner tree* with terminal set R is a subgraph T of G such that T is a tree and $R \subseteq V(T)$. The *weight* of a Steiner tree T is $w(T) = \sum_{e \in E(T)} w(e)$. A *t-component* for R is a tree with at most t leaves and all these leaves coincide with a subset of R. A *t-restricted Steiner tree* for R is a collection $\mathcal{T}$ of t-components for R such that the union of the t-components in $\mathcal{T}$ induces a Steiner tree for R. We refer to Byrka et al. [3] for more detailed introduction on these terminologies.

Proposition 1 ([2]). *For every $t \geq 1$, given a graph G, a terminal set R, a cost function $w : E(G) \to \mathbb{N}$, and a Steiner tree T for R, there exists a t-restricted Steiner tree $\mathcal{T}$ for R of cost at most $(1 + \frac{1}{\lfloor \log_2 t \rfloor}) \cdot w(T)$.*

Proposition 2 ([9]). *Let G be a graph, R be a set of terminals, and a $w : E(G) \to \mathbb{N}$ be a cost function. Then, a minimum weight Steiner tree for R can be computed in $\mathcal{O}(3^{|R|}|V(G)||E(G)|)$-time.*

Note that if $|R|$ is constant then the above algorithm runs in polynomial time.

Treedepth: Given a graph G, we define $\mathsf{td}(G)$, the *treedepth* of G as follows.

$$\mathsf{td}(G) = \begin{cases} 1 & \text{if } |V(G)| = 1 \\ 1 + \min_{v \in V(G)} \mathsf{td}(G - v) & \text{if } G \text{ is connected and } |V(G)| > 1 \\ \max_{i=1}^{p} \mathsf{td}(G_i) & \text{if } G_1, \ldots, G_p \text{ are connected components of } G \end{cases} \tag{1}$$

A *treedepth decomposition* of graph $G = (V, E)$ is a rooted forest Y with vertex set V, such that for each edge $uv \in E(G)$, we have either that u is an ancestor of v or v is an ancestor of u in Y. Note that a treedepth decomposition of a connected graph G is equivalent to some depth-first search tree of G and in the context of treedepth is also sometimes referred to as *elimination tree* of G. It is clear from the definition that the treedepth of a graph G is equivalent to the minimum depth of a treedepth decomposition of G, where depth is defined as the maximum number of vertices along a path from the root of the tree to a leaf [23]. Let T be a tree rooted at a node r. The *upward closure* for a set of nodes $S \subseteq V(T)$ is denoted by $\mathsf{UClos}_T(S) = \{v \in V(T) \mid v \text{ is an ancestor}$

of $u \in S$ in T}. This notion has proved useful in the kernelization algorithm of Giannopoulou et al. [16] for η-TREEDEPTH DELETION. The following facts about the treedepth of a graph will be useful throughout the paper.

Proposition 3 ([23]). *Let G be a graph such that $\mathsf{td}(G) \leq \eta$. Then, the diameter of G is at most 2^η.*

Proposition 4 ([14]). *For every constant $\eta \in \mathbb{N}$, there exists a polynomial-time $\mathcal{O}(1)$-approximation for η-TREEDEPTH DELETION.*

Proposition 5 ([26]). *Let G be a connected graph and $\eta \in \mathbb{N}$. There exists an algorithm running in $\mathcal{O}(f(\eta)|V(G)|)$-time, for some computable function f, that either correctly concludes that $\mathsf{td}(G) > \eta$ or computes a treedepth decomposition for G of depth at most η.*

Parameterized Algorithms and Kernels: A parameterized problem Π is a subset of $\Sigma^* \times \mathbb{N}$ for a finite alphabet Σ. An instance of a parameterized problem is a pair (x, k) where $x \in \Sigma^*$ is the input and $k \in \mathbb{N}$ is the parameter. We assume without loss of generality that k is given in unary. We say that Π admits a *kernelization* if there exists a polynomial-time algorithm that, given an instance (x, k) of Π, outputs an equivalent instance (x', k') of Π such that $|x'| + k' \leq g(k)$. If $g(k)$ is $k^{\mathcal{O}(1)}$, then we say that Π admits a *polynomial kernelization*.

Parameterized Optimization Problem and Approximate Kernels

Definition 1. *A* parameterized optimization problem *is a computable function $\Pi : \Sigma^* \times \mathbb{N} \times \Sigma^* \to \mathbb{R} \cup \{\pm\infty\}$.*

The *instances* of a parameterized problem are pairs $(x, k) \in \Sigma^* \times \mathbb{N}$, and a *solution* to (x, k) is simply $s \in \Sigma^*$ such that $|s| \leq |x| + k$. The *value* of a solution s is $\Pi(x, k, s)$. Since the problems we deal with here are minimization problems, we state some of the definitions only in terms of minimization problems (for maximization problems, the definition would be analogous). As an illustrative example, we provide the definition of the parameterized optimization version of CONNECTED η-TREEDEPTH DELETION problem as follows. This is a minimization problem that is a function CON-η-TDS $: \Sigma^* \times \mathbb{N} \times \Sigma^* \to \mathbb{R} \cup \{\pm\infty\}$ as follows.

We define CON-η-TDS$(G, k, S) = \infty$ if S is not a connected η-treedepth deletion set of G. Otherwise, S is a connected η-treedepth deletion set of G and then we define CON-η-TDS$(G, k, S) = \min\{|S|, k + 1\}$.

Definition 2. *For a parameterized minimization problem Π, the* optimum value *of an instance (x, k) is $\mathrm{OPT}_\Pi(x, k) = \min_{s \in \Sigma^*, |s| \leq |x| + k} \Pi(x, k, s)$.*

For the case of CONNECTED η-TREEDEPTH DELETION, we define $\mathrm{OPT}(G, k) = \min_{S \subseteq V(G)} \{\mathsf{CON}\text{-}\eta\text{-}\mathsf{TDS}(G, k, S)\}$. We now recall the other relevant definitions regarding approximate kernels.

Definition 3. *Let* $\alpha \geq 1$ *be a real number and let* Π *be a parameterized minimization problem. An* α-approximate polynomial-time preprocessing algo-*rithm* $\mathcal{A}$ *is a pair of polynomial-time algorithms. The first one is called the* reduction algorithm *and the second one is called the* solution-lifting algorithm. *Given an input instance* (x, k) *of* Π, *the reduction algorithm is a function* $\mathcal{R}_{\mathcal{A}} : \Sigma^* \times \mathbb{N} \rightarrow \Sigma^* \times \mathbb{N}$ *that outputs an instance* (x', k') *of* Π.

The solution-lifting algorithm takes the input instance (x, k), *the reduced instance* (x', k') *and a solution* s' *to the instance* (x', k'). *The solution-lifting algorithm works in time polynomial in* $|x|, k, |x'|, k'$, *and* $|s'|$, *and outputs a solu-tion* s *to* (x, k) *such that the following holds:*

$$\frac{\Pi(x, k, s)}{\mathrm{OPT}_{\Pi}(x, k)} \leq \alpha \frac{\Pi(x', k', s')}{\mathrm{OPT}_{\Pi}(x', k')}$$

The size *of a polynomial-time preprocessing algorithm* $\mathcal{A}$ *is a function* $\mathrm{size}_{\mathcal{A}} :$ $\mathbb{N} \rightarrow \mathbb{N}$ *defined as* $\mathrm{size}_{\mathcal{A}}(k) = \sup\{|x'| + k' : (x', k') = \mathcal{R}_{\mathcal{A}}(x, k), x \in \Sigma^*\}$.

Definition 4 (Approximate Kernelization). *An* α-approximate kerneliza-*tion (or* α-approximate kernel) *for a parameterized optimization problem* Π, *and a real* $\alpha \geq 1$ *is an* α-approximate polynomial-time preprocessing algorithm $\mathcal{A}$ *for* Π *such that* $\mathrm{size}_{\mathcal{A}}$ *is upper-bounded by a computable function* $g : \mathbb{N} \rightarrow \mathbb{N}$. *If* g *is a polynomial function, we call* $\mathcal{A}$ *an* α-approximate polynomial kernelization *algorithm.*

Definition 5 (Approximate Kernelization Schemes). *A polynomial-size* approximate kernelization scheme (PSAKS) *for a parameterized problem* Π *is a family of* α-approximate polynomial kernelization algorithms, with one such *algorithm for every fixed* $\alpha > 1$.

Definition 6 (Time-efficient PSAKS). *A PSAKS is said to be* time efficient *if both the reduction algorithm and the solution lifting algorithms run in* $f(\alpha)|x|^c$ *time for some function* f *and a constant* c *independent of* $|x|, k,$ *and* α.

3   Approximate Kernel for CONNECTED η-TREEDEPTH DELETION

In this section, we describe a $(1 + \varepsilon)$-approximate kernel for CONNECTED η-TREEDEPTH DELETION. For the entire proof, let us fix a constant $\eta \in \mathbb{N}$, the instance (G, k) of CONNECTED η-TREEDEPTH DELETION, as well as $\varepsilon \in \mathbb{R}$ such that $0 < \varepsilon \leq 1$. We prove by Theorem 1 that CONNECTED η-TREEDEPTH DELETION admits a $(1 + \varepsilon)$-approximate kernel with $k^{\mathcal{O}(\lceil 2^{\eta + \lceil 10/\varepsilon \rceil} \eta / \varepsilon \rceil)}$ vertices. As η and ε are fixed constants, the hidden constants in Big-Oh notation could depend both on η and ε.

Overview of the Algorithm. Our reduction algorithm works in three phases. First observe that in order for a connected η-treedepth deletion set to exist, at most one connected component of G can have treedepth more than η. Hence, we can focus on the case when G is connected. We then show that we can decompose the graph G into three sets X, Z, and R such that X is an η-treedepth deletion set, the size of the neighborhood of every component C of $G[R]$ in Z is at most η and every η-treedepth deletion set S of size at most k hits all but at most η neighbors of C in X. This completes Phase 1 (details in Sect. 3.1) and closely follows similar decompositions in [14, 16]. At this point, we observe that if the neighborhood of C is large (at least some constant depending on η and ε), then including the whole neighborhood in the solution is not 'too suboptimal'. This is a key insight in our algorithm. So, we can force the neighborhood of C in every solution by adding a small gadget to G. Repeating this procedure allows us to identify a set of vertices $H \subseteq X$ that we can safely force into a solution without increasing the size of an optimal solution too much. Moreover, we obtain that every component of $G[R]$ has only constantly many neighbors outside of H. This completes the Phase 2 (Reduction Rule 2).

Notice now that the vertices of any solution for CONNECTED η-TREEDEPTH DELETION can be split into two parts - *the obstruction hitting vertices* the removal of which guarantees a graph of treedepth at most η, and *the connector vertices* that are only there to provide connectivity to a solution. This high level approach of identifying *obstruction hitters* and *connectors* among the vertices is a natural first step for problems with this flavor [11, 12, 24, 25]. Now all the connected components of $G[R]$ have treedepth at most η. Moreover, we are guaranteed that any solution of size at most k contains all but at most 2η neighbors of a connected component C of $G[R]$. Hence, if our goal was only hitting the obstructions in G, then we could assume that S contains at most 2η vertices of every connected component of $G[R]$. But we do not know which 2η vertices are in $N(C) \setminus S$ and which vertices of C can provide connectivity to S. This requires us to use careful 'problem-specific argumentation' and reduction rules. We observe that $N(C) \setminus H$ has already constant size and we can classify the (subsets of) vertices of C into types depending on their neighborhood in $N(C) \setminus H$. In addition we allow each connected component of $G[R]$ to have much larger but still a constant, intersection with S. We furthermore observe that if we chose this constant, denoted by λ, then we can include the whole neighborhood of every component C that intersects a solution S in more than λ vertices without increasing the size of the solution too much. Now, we finally can identify the vertices that are not necessary for any solution that intersects every component in at most λ vertices. Denote the set of these vertices $\mathcal{M}$. There is no danger in removing such vertices for hitting the obstructions, as for every component C of $G' - (X \cup Z)$ that intersects more than λ vertices of a solution in the reduced instance G' our solution-lifting algorithm adds the neighborhood of C into the solution. However, removing all of these vertices may very well destroy the connectivity of the solution. Here, we make use of Propositions 1 and 2 to find a small subset $\mathcal{N}$ of vertices in $\mathcal{M}$ such that $G - (\mathcal{M} \setminus \mathcal{N})$ actually have a connected η-treedepth deletion set of approximately optimal size. This completes Phase 3 (details in Sects. 3.3 and 3.4).

3.1 Decomposition of the Graph G

We first observe that we can remove all connected components of G that already have treedepth at most η, as we do not need to remove any vertex from such a component.

Reduction Rule 1. *Let C be a connected component of G such that $\operatorname{td}(G[C]) \leq \eta$. Then, delete C from G. The new instance is $(G - C, k)$.*

It follows that Reduction Rule 1 is an approximation preserving reduction rule. Hence, given a c-approximate connected η-treedepth deletion set of $(G - C, k)$, we can in polynomial time compute a c-approximate connected η-treedepth deletion set of (G, k). Hence, we can assume that G is a connected graph. We start by constructing a decomposition of the graph such that $V(G) = X \uplus Z \uplus R$ satisfying some crucial properties that we use in our subsequent phases of the preprocessing algorithm. The construction is inspired by the decompositions used by Fomin et al. [14] (for the PLANAR $\mathcal{F}$-DELETION problem) and Giannopoulou et al. [16].

Lemma 1. $(\star)^1$ *There exists a polynomial-time algorithm that either correctly concludes that no η-treedepth deletion set S for G of size at most k exists, or it constructs a partition $V(G) = X \uplus Z \uplus R$ such that the following properties are satisfied. (1) X is an η-treedepth deletion set of G and $|X| = \mathcal{O}(k)$, (2) $|Z| = \mathcal{O}(k^3)$, (3) For every connected component C of $G[R]$, $|N_G(C) \cap Z| \leq \eta$, and (4) Let C be a connected component of $G[R]$. Then, for any η-treedepth deletion set S of size at most k, it holds that $|(N_G(C) \cap X) \setminus S| \leq \eta$.*

We run the algorithm of Lemma 1 and we fix for the rest of the proof the sets of vertices X, Z, and R such that they satisfy the above lemma. Furthermore, let us fix a $\delta = \frac{\varepsilon}{10}$ and notice that since $\varepsilon \leq 1$, we have that $(1 + \delta)^4 \leq (1 + \varepsilon)$. Finally let us set $d = \lceil \frac{2^{\eta+3}\eta}{\delta} \rceil$. The next step of the algorithm is to find a set of vertices $H \subseteq X$ such that every component C of $G - (X \cup Z)$ has at most $d + \eta$ neighbors in $X \setminus H$. Our goal is to do it in a way that we can force H into every solution and increase the size of an optimal solution only by a small fraction.

3.2 Processing Connected Components of $G - (X \cup Z)$ with Large Neighborhoods

We initialize $H := \emptyset$ and we apply the following reduction rule exhaustively.

Reduction Rule 2. *Let C be a connected component of $G[R]$. If $|(N_G(C) \cap X) \setminus H| > d + \eta$, then for every $u \in N_G(C) \cap X$, add a new clique J with $\eta + 1$ vertices to G such that $J \cap X = \{u\}$ and $N_G(J \setminus \{u\}) = \{u\}$. Add the vertices of $N_G(C)$ to H.*

---

[1] Due to lack of space, omitted proofs or the proofs marked $\star$ can be found in the full version.

After we finish applying Reduction Rule 2 on (G, k) exhaustively, let G' be the resulting graph. We prove the following two lemmas using Lemma 1.

Lemma 2. ($\star$) *Let S be an optimal connected η-treedepth deletion set of (G, k) of size at most k. Then, Reduction Rule 2 is not applicable more than $|S|/d$ times.*

Using the above lemma, we prove the following lemma.

Lemma 3. ($\star$) *Let (G', k') be the instance obtained after exhaustively applying the Reduction Rule 2 on (G, k) such that $k' = k$. Then, the following conditions are satisfied.* **(i)** *Any connected η-treedepth deletion set of (G', k') is a connected η-treedepth deletion set of (G, k), and* **(ii)** *If $\mathrm{OPT}(G, k) \leq k$, then $\mathrm{OPT}(G', k') \leq (1 + \delta)\mathrm{OPT}(G, k)$*

3.3 Understanding the Structure of a Good Solution

From now on we assume that we have applied Reduction Rule 2 exhaustively and, for the sake of exposition, we denote by G the resulting graph. Moreover, we also fix the set H we obtained from the exhaustive application of Reduction Rule 2. It follows that every connected component of $G - (X \cup Z)$ have at most $d + 2\eta$ neighbors outside H. Furthermore, Reduction Rule 2 ensures that any feasible connected η-treedepth deletion set must contain H. This follows because for every vertex $u \in H$ there exists a clique J of size $\eta + 1$ that contains u and $N_G(J \setminus \{u\}) = \{u\}$. So every connected η-treedepth deletion set that contains a vertex in J and a vertex outside of J contains also u. Notice that if S is a connected η-treedepth deletion set for (G, k) and C is a component of $G[R]$, then $(S \setminus C) \cup N(C)$ is an η-treedepth deletion set. Moreover, we can connect each vertex from $N(C) \setminus H$ to H using at most 2^η vertices of C. Hence the only reason for a component of $G[R]$ to contain more than $2^\eta(d + 2\eta)$ vertices is if C also provides connectivity to S. Let us fix for the rest of the proof $\lambda = 2^\eta \lceil \frac{d+2\eta}{\delta} \rceil$. Let $T \subseteq (X \cup Z) \setminus H$. We denote by $\mathrm{Comp}(T)$ the set of all the components C of $G - (X \cup Z)$ such that $N(C) \setminus H = T$. Note that, by the definition of H, if $|T| \geq d + 2\eta + 1$, then $\mathrm{Comp}(T) = \emptyset$. Let S be an η-treedepth deletion set of G. Suppose that for every $T \subseteq (X \cup Z) \setminus H$ it holds that if S intersects $\mathrm{Comp}(T)$ in more than λ vertices (i.e. $|\bigcup_{C \in \mathrm{Comp}(T)}(S \cap C)| > \lambda$), then $T \subseteq S$. Then we call S is *nice* treedepth deletion set.

From now on, we focus on nice connected η-treedepth deletion sets. We first reduce the instance (G, k) to an instance (G', k) such that **(i)** G' is an induced subgraph of G, **(ii)** every nice connected η-treedepth deletion set for (G', k) is also a nice connected η-treedepth deletion set for (G, k), and **(iii)** (G', k) has a nice connected η-treedepth deletion set of size at most $(1 + \delta)^2\mathrm{OPT}(G, k)$.

Afterwards, we show that any connected η-treedepth deletion set S' for (G', k) can be transformed into a nice connected η-treedepth deletion set for (G', k) of size at most $(1 + \delta)|S'|$. To obtain our reduced instance we will heavily rely on the following lemma that helps us identify vertices that only serve as connectors in any nice connected η-treedepth deletion set.

Lemma 4. $(\star)$ *Let G' be an induced (not necessarily strict) subgraph of G and $T, C_1, C_2, \ldots, C_\ell$ be pairwise disjoint sets of vertices in G such that:* **(i)** $G'[C_i]$ *is connected,* **(ii)** $N(C_i) \setminus H = T$, *for some fixed set of vertices T, and* **(iii)** $\mathsf{td}(G'[C_i]) = \mathsf{td}(G'[C_j])$ *for all $i, j \in [\ell]$. Now let S be an η-treedepth deletion set in G' such that $H \subseteq S$ and let $\mathcal{J} = \{C_i \mid C_i \cap S = \emptyset\}$, i.e., $\mathcal{J}$ is the set of components in $C_1, C_2, \ldots, C_\ell$ that do not contain any vertex of S. If $|\mathcal{J}| \geq \eta + 1$, then $S' = S \setminus (\bigcup_{i \in [\ell]} C_i)$ is an η-treedepth deletion set in G'.*

3.4   Identifying Further Irrelevant Vertices

We now mark some vertices of $G - (X \cup Z)$ that we would like to remove, as these vertices are not important for hitting obstructions in a nice connected η-treedepth deletion set. We note that we will end up not removing all of these vertices, as some of them will be important as connectors for obstruction hitting vertices in the solution. However, this step lets us identify a relatively small subset of vertices such that any nice η-treedepth deletion set for the subgraph induced by this subset of vertices is indeed nice η-treedepth deletion set for G. We then make use of Propositions 1 and 2 to add some vertices back as possible connectors. Recall that we fixed $\lambda = 2^\eta \lceil \frac{d+2\eta}{\delta} \rceil$. Let us set $\mathcal{M} = \emptyset$. We now describe two reduction rules based on Lemma 4 that do not change G and only add vertices to $\mathcal{M}$. For $T \subseteq (X \cup Z) \setminus H$ and $i \in \mathbb{N}$, let $\mathrm{Comp}(T, i)$ denote the components $C \in \mathrm{Comp}(T)$ such that $\mathsf{td}(G[C]) = i$.

Reduction Rule 3. *Let $T \subseteq (X \cup Z) \setminus H$ and $i \in [\eta]$. If $|\mathrm{Comp}(T, i)| \geq \lambda + \eta + 2$, then add vertices of all but $\lambda + \eta + 1$ of the components in $\mathrm{Comp}(T, i)$ to $\mathcal{M}$.*

Reduction Rule 4. *Let C be a component of $G - (X \cup Z \cup \mathcal{M})$, Y_C be a treedepth decomposition of $G[C]$ of depth at most η, and $i \in [\eta]$. Moreover, let v be a vertex in C and $T \subseteq (N(C) \setminus H) \cup \mathsf{UClos}_{Y_C}(\{v\})$. Finally, let $\mathcal{C} = \{C_1, C_2, \ldots, C_\ell\}$ be all the components of $G - T$ such that for all $j \in [\ell]$ it holds that $C_j \subseteq C$, $N(C_j) \setminus H = T$ and $\mathsf{td}(G[C_j]) = i$. If $|\mathcal{C}| \geq \lambda + \eta + 2$, then add the vertices of all but $\lambda + \eta + 1$ of the components in $\mathcal{C}$ to $\mathcal{M}$.*

Once we apply Reduction Rules 3 and 4 exhaustively and obtain a vertex set $\mathcal{M}$, we use Lemma 4 in order to prove the following two lemmas that provide some interesting characteristics (the following two lemmas) of nice η-treedepth deletion sets of G.

Lemma 5. $(\star)$ *Let $\mathcal{M}$ be the set of vertices obtained by exhaustive application of Reduction Rules 3 and 4 and let S' be a nice η-treedepth deletion set for $G - \mathcal{M}$. Then S' is a nice η-treedepth deletion set for G.*

Lemma 6. $(\star)$ *Let $\mathcal{M}$ be the set of vertices obtained by exhaustive application of Reduction Rules 3 and 4. Then $|V(G) \setminus \mathcal{M}| = \mathcal{O}(k^{3d+6\eta})$.*

Now our next goal is to add some of the vertices from $\mathcal{M}$ back, in order to preserve also an approximate nice connected η-treedepth deletion set. We start by setting $\mathcal{N} = \emptyset$. Now for every set $L \subseteq V(G) \setminus \mathcal{M}$ of size at most $t = 2^{\lceil \frac{1}{\delta} \rceil}$ we compute a Steiner tree T_L for the set of terminals L in G. If T_L has at most $(1 + \delta)k$ vertices, we add all vertices on T_L to $\mathcal{N}$. It follows from Lemma 6 that $|\mathcal{N}| = \mathcal{O}(k^{(3d+6\eta)t+1})$. Since t is a constant, it follows from Proposition 2 that we can compute each of at most $\mathcal{O}(k^{(3d+6\eta)t})$ Steiner trees in polynomial time. We now let $G' = G - (\mathcal{M} \setminus \mathcal{N})$.

The following lemma will be useful to show that there is a small nice connected η-treedepth deletion set solution in G'. Moreover, it will be also useful in our solution-lifting algorithm, where we need to first transform the solution to a nice connected η-treedepth deletion set.

Lemma 7. (⋆) *Let $Y \subseteq (V(G) \setminus (X \cup Z))$ and let S be a connected η-treedepth deletion set for $G - Y$ of size at most k. There is a polynomial-time algorithm that takes on the input G, Y, and S and outputs a nice connected η-treedepth deletion set for $G - Y$ of size at most $(1 + \delta)|S|$.*

Using the above lemma, we now prove the following two lemmas that we will eventually use to prove our final theorem statement. The proofs of the following two lemmas use the correctness of Lemma 7.

Lemma 8. (⋆) *If $\mathrm{OPT}(G, k) \leq k$, then there exists a connected η-treedepth deletion set for G' of size at most $(1 + \delta)^2 \mathrm{OPT}(G, k)$.*

The proof of following lemma will also use both Lemma 7 and Lemma 8.

Lemma 9. (⋆) *Given a connected η-treedepth deletion set S' of (G', k') of size at most k, we can in polynomial time compute a connected η-treedepth deletion set S of (G, k) such that*

$$\frac{|S|}{\mathrm{OPT}(G, k)} \leq (1 + \delta)^3 \frac{|S'|}{\mathrm{OPT}(G', k')}.$$

We are now ready to prove our main result.

Theorem 2. *For every fixed $0 < \varepsilon \leq 1$, CONNECTED η-TREEDEPTH DELETION has a time-efficient $(1 + \varepsilon)$-approximate kernelization of size $k^{\mathcal{O}(\lceil 2^{\eta + \lceil 10/\varepsilon \rceil} \eta/\varepsilon \rceil)}$.*

Proof (Sketch). We choose δ, λ and t as described earlier. The approximate kernelization algorithm has two parts, i.e. reduction algorithm and solution lifting algorithm.

➤ **Reduction Algorithm:** Let (G, k) be an input instance and we can assume without loss of generality that G is connected. The reduction algorithm works as follows. If $|V(G)| \leq 2^{3\eta^2 + d\eta}(\lambda + \eta + 1)^{\eta+1}(1 + \delta)k^{(3d+6\eta)t+1}$, then we output (G, k). Otherwise, we first invoke Lemma 1 and construct a decomposition of $V(G) = X \uplus Z \uplus R$ and some conditions are satisfied. Then we apply Reduction Rule 2 to construct H and the instance (G_1, k). Afterwards, we apply Reduction Rules 3 and 4 on (G_1, k) exhaustively to compute $\mathcal{M}$. Afterwards, we compute an optimal Steiner tree T_L for every subset L of $V(G) \setminus \mathcal{M}$ of size at most $t = 2^{\lceil \frac{1}{\delta} \rceil}$

and if its size is at most $(1+\delta)k$, then we add T_L to the set $\mathcal{N}$. We delete $\mathcal{M} \setminus \mathcal{N}$ from (G_1, k) to compute the instance (G', k') with $k' = k$. This completes the reduction algorithm.

➤ **Solution Lifting Algorithm:** Let S' be a connected η-treedepth deletion set to (G', k'). If $|S'| > k'$, then we output the entire vertex set of a connected component whose treedepth is larger than η. Otherwise $|S'| \leq k'$. We invoke Lemma 9 to compute a nice connected η-treedepth deletion set S_1 of the instance (G_1, k). By construction, $H \subseteq S_1$. If $|S_1| \leq k$, we output $S = S_1$ as a connected η-treedepth deletion set of (G, k). Otherwise $|S_1| > k$, then also we output the entire vertex set of a connected component whose treedepth is larger than η. By case, analysis it can be proved that these two algorithm together constitutes a $(1 + \varepsilon)$-approximate kernel.

By construction, Lemma 6, and the size bound of $|\mathcal{N}|$, we have that $|V(G')| = \mathcal{O}(k^{(3d+6\eta)t+1})$. Recall that $d = \lceil 2^{\eta+3}\eta/\delta \rceil$, $t = 2^{\lceil 1/\delta \rceil}$, and $\delta = \frac{\varepsilon}{10}$. It can be observed that all the reduction rules can be performed in $k^{\mathcal{O}(3d+6\eta)t}n^{\mathcal{O}(1)}$-time and are executed only when $|V(G)| = n > 2^{3\eta^2+d\eta}(\lambda+\eta+1)^{\eta+1}(1+\delta)k^{(3d+6\eta)t+1}$.

Hence, we have a time-efficient PSAKS with the claimed bound. □

4   Conclusions

We obtained a polynomial-size approximate kernelization scheme (PSAKS) for CONNECTED η-TREEDEPTH DELETION, improving upon existing bounds and advancing the line of work on approximate kernels for vertex deletion problems with connectivity constraints. Towards our result, we combined known decomposition techniques with new preprocessing steps that exploit structure present in bounded treedepth graphs. Our work points to a few interesting questions for follow up research:

➤ Is there a PSAKS for η-TREEDEPTH DELETION with stronger connectivity constraints, e.g., when the solution is required to induce a biconnected graph. Recently, Einarson et al. [13] initiated this line of research in the context of studying approximate kernels for VERTEX COVER with biconnectivity constraints. It would be interesting to obtain similar results for η-TREEDEPTH DELETION?

➤ Could one get a PSAKS for CON-η-DEPTH-TRANSVERSAL at which the dependency of η in the exponent of k can be removed? Such a result is known for η-TREEDEPTH DELETION without connectivity constraints (a kernel with $\mathcal{O}(2^{\mathcal{O}(\eta^2)}k^6)$ vertices [16]). Some parts of our algorithm are based on this work of [16]. However, we incur the $k^{\mathcal{O}(2^{\mathcal{O}(\eta)} \cdot 1/\varepsilon)}$ cost in the kernel size in several places (e.g. Reduction Rules 3 and 4). We believe that one would need to formulate a significantly distinct approach in order to attain a such a bound.

➤ Could one get a PSAKS for CONNECTED η-TREEWIDTH DELETION? The current best approximate kernel result for this problem is a $(2 + \varepsilon)$-approximate polynomial compression from [25]. We believe that several parts of our algorithm can be adapted to work for η-TREEWIDTH DELETION. However, we have crucially used the fact that a connected bounded treedepth graph has bounded diameter that does not hold for bounded treewidth graphs.

References

1. Bodlaender, H.L., Jansen, B.M.P., Kratsch, S.: Kernelization lower bounds by cross-composition. SIAM J. Discret. Math. **28**(1), 277–305 (2014)
2. Borchers, A., Du, D.: The k-Steiner ratio in graphs. SIAM J. Comput. **26**(3), 857–869 (1997)
3. Byrka, J., Grandoni, F., Rothvoß, T., Sanità, L.: Steiner tree approximation via iterative randomized rounding. J. ACM **60**(1), 6:1–6:33 (2013)
4. Cygan, M.: Deterministic parameterized connected vertex cover. In: Fomin, F.V., Kaski, P. (eds.) SWAT 2012. LNCS, vol. 7357, pp. 95–106. Springer, Heidelberg (2012). https://doi.org/10.1007/978-3-642-31155-0_9
5. Cygan, M., et al.: Parameterized Algorithms. Springer, Cham (2015). https://doi.org/10.1007/978-3-319-21275-3
6. Cygan, M., Pilipczuk, M., Pilipczuk, M., Wojtaszczyk, J.O.: Subset feedback vertex set is fixed-parameter tractable. SIAM J. Discret. Math. **27**(1), 290–309 (2013)
7. Diestel, R.: Graph Theory. Graduate Texts in Mathematics, vol. 173, 4th edn. Springer, Cham (2012)
8. Dom, M., Lokshtanov, D., Saurabh, S.: Kernelization lower bounds through colors and IDs. ACM Trans. Algorithms **11**(2), 13:1–13:20 (2014)
9. Dreyfus, S.E., Wagner, R.A.: The Steiner problem in graphs. Networks **1**(3), 195–207 (1971)
10. Dvorák, Z., Giannopoulou, A.C., Thilikos, D.M.: Forbidden graphs for tree-depth. Eur. J. Comb. **33**(5), 969–979 (2012)
11. Eiben, E., Hermelin, D., Ramanujan, M.S.: On approximate preprocessing for domination and hitting subgraphs with connected deletion sets. J. Comput. Syst. Sci. **105**, 158–170 (2019)
12. Eiben, E., Kumar, M., Mouawad, A.E., Panolan, F., Siebertz, S.: Lossy kernels for connected dominating set on sparse graphs. SIAM J. Discret. Math. **33**(3), 1743–1771 (2019)
13. Einarson, C., Gutin, G.Z., Jansen, B.M.P., Majumdar, D., Wahlström, M.: p-edge/vertex-connected vertex cover: parameterized and approximation algorithms. CoRR abs/2009.08158 (2020)
14. Fomin, F.V., Lokshtanov, D., Misra, N., Saurabh, S.: Planar F-deletion: approximation, kernelization and optimal FPT algorithms. In: 53rd Annual IEEE Symposium on Foundations of Computer Science, FOCS 2012, New Brunswick, NJ, USA, 20–23 October 2012, pp. 470–479. IEEE Computer Society (2012)
15. Gajarský, J., et al.: Kernelization using structural parameters on sparse graph classes. J. Comput. Syst. Sci. **84**, 219–242 (2017)
16. Giannopoulou, A.C., Jansen, B.M.P., Lokshtanov, D., Saurabh, S.: Uniform kernelization complexity of hitting forbidden minors. ACM Trans. Algorithms **13**(3), 35:1–35:35 (2017)
17. Hegerfeld, F., Kratsch, S.: Solving connectivity problems parameterized by treedepth in single-exponential time and polynomial space. In: Paul, C., Bläser, M. (eds.) 37th International Symposium on Theoretical Aspects of Computer Science, STACS 2020, Montpellier, France, 10–13 March 2020. LIPIcs, vol. 154, pp. 29:1–29:16. Schloss Dagstuhl - Leibniz-Zentrum für Informatik (2020)
18. Hermelin, D., Kratsch, S., Soltys, K., Wahlström, M., Wu, X.: A completeness theory for polynomial (Turing) kernelization. Algorithmica **71**(3), 702–730 (2015)

19. Hermelin, D., Wu, X.: Weak compositions and their applications to polynomial lower bounds for kernelization. In: Proceedings of the Twenty-Third Annual ACM-SIAM Symposium on Discrete Algorithms, SODA 2012, Kyoto, Japan, 17–19 January 2012, pp. 104–113 (2012)
20. Jansen, B.M.P., Pieterse, A.: Polynomial kernels for hitting forbidden minors under structural parameterizations. Theor. Comput. Sci. **841**, 124–166 (2020)
21. Lokshtanov, D., Panolan, F., Ramanujan, M.S., Saurabh, S.: Lossy kernelization. In: Proceedings of the 49th Annual ACM SIGACT Symposium on Theory of Computing, STOC 2017, Montreal, QC, Canada, 19–23 June 2017, pp. 224–237 (2017)
22. Misra, N., Philip, G., Raman, V., Saurabh, S.: The kernelization complexity of connected domination in graphs with (no) small cycles. Algorithmica **68**(2), 504–530 (2014)
23. Nesetril, J., de Mendez, P.O.: Tree-depth, subgraph coloring and homomorphism bounds. Eur. J. Comb. **27**(6), 1022–1041 (2006)
24. Ramanujan, M.S.: An approximate kernel for connected feedback vertex set. In: 27th Annual European Symposium on Algorithms, ESA 2019, Munich/Garching, Germany, 9–11 September 2019, pp. 77:1–77:14 (2019)
25. Ramanujan, M.S.: On approximate compressions for connected minor-hitting sets. In: 29th Annual European Symposium on Algorithms, ESA 2021 (2021)
26. Reidl, F., Rossmanith, P., Villaamil, F.S., Sikdar, S.: A faster parameterized algorithm for treedepth. In: Esparza, J., Fraigniaud, P., Husfeldt, T., Koutsoupias, E. (eds.) ICALP 2014. LNCS, vol. 8572, pp. 931–942. Springer, Heidelberg (2014). https://doi.org/10.1007/978-3-662-43948-7_77

Generalized k-Center: Distinguishing Doubling and Highway Dimension

Andreas Emil Feldmann⬤ and Tung Anh Vu$^{(\boxtimes)}$⬤

Faculty of Mathematics and Physics, Charles University, Prague, Czech Republic
feldmann.a.e@gmail.com, tung@kam.mff.cuni.cz

Abstract. We consider generalizations of the k-CENTER problem in graphs of low doubling and highway dimension. For the CAPACITATED k-SUPPLIER WITH OUTLIERS (CKSwO) problem, we show an efficient parameterized approximation scheme (EPAS) when the parameters are k, the number of outliers and the doubling dimension of the supplier set. On the other hand, we show that for the CAPACITATED k-CENTER problem, which is a special case of CKSwO, obtaining a parameterized approximation scheme (PAS) is W[1]-hard when the parameters are k, and the highway dimension. This is the first known example of a problem for which it is hard to obtain a PAS for highway dimension, while simultaneously admitting an EPAS for doubling dimension.

Keywords: Capacitated k-Supplier with Outliers · Highway dimension · Doubling dimension · Parameterized approximation

1 Introduction

The well-known k-CENTER problem and its generalizations has plenty of applications, for example selecting suitable locations for building hospitals to serve households of a municipality (see [3] for a survey of healthcare facility location in practice). In this setting, the number of hospitals we can actually build is limited, e.g. by budgetary constraints. We want to choose the locations so that the quality of the provided service is optimal, and a societally responsible way of measuring the quality of service is to ensure some minimal availability of healthcare to every household. We can quantify this by measuring the distance of a household to its nearest hospital, and then minimize this distance over all households. This strategy, however, does not account for the reality that healthcare providers have (possibly different) limits on the number of patients they can serve, and thus we introduce capacity constraints. Furthermore, as the instances are given by transportation networks, we model them by the titular doubling dimension and highway dimension, which we define later.

Andreas Emil Feldmann was supported by the project 19-27871X of GA ČR. Tung Anh Vu was supported by the project 22-22997S of GA ČR.

M. A. Bekos and M. Kaufmann (Eds.): WG 2022, LNCS 13453, pp. 215–229, 2022.
https://doi.org/10.1007/978-3-031-15914-5_16

We formalize the problem as follows. In the CAPACITATED k-SUPPLIER (CKS) problem, the input consists of a graph $G = (V, E)$ with positive edge lengths, a set $V_S \subseteq V$ of *suppliers*, a set $V_C \subseteq V$ of *clients*, a *capacity function* $L \colon V_S \to \mathbb{N}$, and an integer $k \in \mathbb{N}$. A *feasible solution* is an *assignment function* $\phi \colon V_C \to V_S$ such that $|\phi(V_C)| \leq k$ and for every supplier $u \in \phi(V_C)$ we have $|\phi^{-1}(u)| \leq L(u)$. For a pair of vertices $u, v \in V$ we denote by $\mathrm{dist}_G(u, v)$ the *shortest-path distance* between vertices u and v with respect to edge lengths of G. For a subset of vertices $W \subseteq V$ and a vertex $u \in V$, we denote $\mathrm{dist}_G(u, W) = \min_{w \in W} \mathrm{dist}_G(u, w)$. We omit the subscript G if the graph is clear from context. The *cost* of a solution ϕ is defined as $\mathrm{cost}(\phi) = \max_{u \in V_C} \mathrm{dist}(u, \phi(u))$ and we want to find a feasible solution of minimum cost. Let us mention the following special cases of CKS. If $V = V_S = V_C$, then the problem is called CAPACITATED k-CENTER (CKC). If $L(u) = \infty$ for every supplier $u \in V_S$, then the problems are called k-SUPPLIER and k-CENTER respectively. It is known that k-CENTER is already NP-hard [26].

Two popular approaches of dealing with NP-hard problems are *approximation algorithms* [30,31] and *parameterized algorithms* [13]. Given an instance $\mathcal{I}$ of some minimization problem, a *c-approximation algorithm* computes in polynomial time a solution of cost at most $c \cdot \mathrm{OPT}(\mathcal{I})$ where $\mathrm{OPT}(\mathcal{I})$ is the optimum cost of the instance $\mathcal{I}$, and we say that c is the *approximation ratio* of the algorithm. If the instance is clear from context, we write only OPT. In a *parameterized problem*, the input $\mathcal{I}$ comes with a *parameter* $q \in \mathbb{N}$. If there exists an algorithm which computes the optimum solution in time $f(q) \cdot |\mathcal{I}|^{\mathcal{O}(1)}$ where f is some computable function, then we call such a problem *fixed parameter tractable (FPT)* and the algorithm an *FPT algorithm*. The rationale behind parameterized algorithms is to capture the "difficulty" of the instance by the parameter q and then design an algorithm which is allowed to run in time superpolynomial in q but retains a polynomial running time in the size of the input. In this work, we focus on the superpolynomial part of the running time of FPT algorithms, so we will express $f(q) \cdot |\mathcal{I}|^{\mathcal{O}(1)}$ as $\mathcal{O}^*(f(q))$; in particular the "$\mathcal{O}^*$" notation ignores the polynomial factor in the input size.

It is known that k-CENTER and k-SUPPLIER do not admit approximation algorithms with an approximation ratio better than 2 and 3 respectively unless P = NP [27]. It is shown in the same work that these results are tight by giving corresponding approximation algorithms. For CAPACITATED k-CENTER, An et al. [4] give a 9-approximation algorithm, and Cygan et al. [14] show a lower bound of $3 - \varepsilon$ for approximation assuming P $\neq$ NP. From the perspective of parameterized algorithms, Feldmann and Marx [21] show that k-CENTER is W[1]-hard in planar graphs of constant doubling dimension when the parameters are k, highway dimension and pathwidth. Under the standard assumption FPT $\subsetneq$ W[1] $\subsetneq$ W[2], this means that an FPT algorithm for k-CENTER in planar constant doubling dimension graphs with the aforementioned parameters is unlikely to exist. To overcome these hardness results, we will design *parameterized c-approximation algorithms*, which are algorithms with FPT runtime which output a solution of cost at most $c \cdot \mathrm{OPT}$. The approach of parameterized approximation algorithms has been studied before, see the survey in [20].

Let us discuss possible choices for a parameter for these problems. An immediate choice would be the size k of the desired solution. Unfortunately, Feldmann [18] has shown that approximating k-CENTER within a ratio better than 2 when the parameter is k is W[2]-hard. So to design parameterized approximation algorithms, we must explore other parameters. Guided by the introductory example, we focus on parameters which capture properties of transportation networks.

Abraham et al. [2] introduced the *highway dimension* in order to explain fast running times of various shortest-path heuristics in road networks. The definition of highway dimension is motivated by the following empirical observation of Bast et al. [5,6]. Imagine we want to travel from some point A to some sufficiently far point B along the quickest route. Then the observation is that if we travel along the quickest route, we will inevitably pass through a sparse set of "access points". Highway dimension measures the sparsity of this set of access points around any vertex of a graph. We give one of the several formal definitions of highway dimension, see [8,19]. Let (X, dist) be a metric, for a point $u \in X$ and a radius $r \in \mathbb{R}^+$ we call the set $B_u(r) = \{v \in X \mid \text{dist}(u, v) \leq r\}$ the *ball of radius r centered at u*.

Definition 1 ([19]). *The* highway dimension *of a graph G is the smallest integer h such that, for some universal constant[1] $\gamma \geq 4$, for every $r \in \mathbb{R}^+$, and every ball $B_v(\gamma r)$ of radius γr where $v \in V(G)$, there are at most h vertices in $B_v(\gamma r)$ hitting all shortest paths of length more than r that lie in $B_v(\gamma r)$.*

We show the following hardness of parameterized approximation for CKC in low highway dimension graphs. Among the definitions of highway dimensions, the one we use gives us the strongest hardness result, cf. [8,19].

Theorem 1. *Consider any universal constant γ in Definition 1. For any $\varepsilon > 0$, there is no parameterized $((1 + \frac{1}{\gamma}) - \varepsilon)$-approximation algorithm for CKC with parameters k, treewidth[2], and highway dimension unless FPT = W[1].*

Another parameter we consider is *doubling dimension*, defined as follows.

Definition 2. *The* doubling constant *of a metric space (X, dist) is the smallest value λ such for every $x \in X$ and every radius $r \in \mathbb{R}^+$, there exist at most λ points $y_1, \ldots, y_\lambda \in X$ such that $B_u(r) \subseteq \cup_{i=1}^{\lambda} B_{y_i}(\frac{r}{2})$. We say that the ball $B_x(r)$ is covered by balls $B_{y_1}(\frac{r}{2}), \ldots, B_{y_\lambda}(\frac{r}{2})$. The* doubling dimension *$\Delta(X)$ of X is defined as $\log_2(\lambda)$. The doubling dimension of a graph is the doubling dimension of its shortest path metric.*

Folklore results show that every metric for which the distance function is given by the ℓ_q-norm in D-dimensional space $\mathbb{R}^D$ has doubling dimension $\mathcal{O}(D)$. As a transportation network is embedded on a large sphere (namely the Earth), a reasonable model is to assume that the shortest-path metric abides to the Euclidean ℓ_2-norm. Buildings in cities form city blocks, which form a grid of

---

[1] See [19, Section 9] for a discussion. In essence, the highway dimension of a given graph can vary depending on the selection of γ.

[2] See [13] or the full version of the paper for a formal definition.

streets. Therefore it is reasonable to assume that the distances in cities are given by the Manhattan ℓ_1-norm. Road maps can be thought of as a mapping of a transportation network into $\mathbb{R}^2$. It is then sensible to assume that transportation networks have constant doubling dimension.

Prior results on problems in graphs of low doubling and highway dimension went "hand in hand" in the following sense. For the k-MEDIAN problem parameterized by the doubling dimension, Cohen-Addad et al. [12] show an *efficient parameterized approximation scheme (EPAS)*, which is a parameterized algorithm that for some parameter q and any $\varepsilon > 0$ outputs a solution of cost at most $(1 + \varepsilon)\text{OPT}$ and runs in time $\mathcal{O}^*(f(q, \varepsilon))$ where f is a computable function. In graphs of constant highway dimension, Feldmann and Saulpic [22] follow up with a *polynomial time approximation scheme (PTAS)* for k-MEDIAN. If we allow k as a parameter as well, then Feldmann and Marx [21] show an EPAS for k-CENTER in low doubling dimension graphs, while Becker et al. [7] show an EPAS for k-CENTER in low highway dimension graphs. By using the result of Talwar [29] one can obtain *quasi-polynomial time approximation schemes (QPTAS)* for problems such as TSP, STEINER TREE, and FACILITY LOCATION in low doubling dimension graphs. Feldmann et al. [19] extend this result to low highway dimension graphs and obtain analogous QPTASs. The takeaway is that approximation schemes for low doubling dimension graphs can be extended to the setting of low highway dimension graphs. In light of Theorem 1, we would then expect that CKC is also hard in graphs of low doubling dimension.

Our main contribution lies in breaking the status quo by showing an EPAS for CKC in low doubling dimension graphs. This is the first example of a problem, for which we provably cannot extend an algorithmic result in low doubling dimension graphs to the setting of low highway dimension graphs.[3] In fact, our algorithm even works in the supplier with outliers regime, where we are allowed to ignore some clients: in the CAPACITATED k-SUPPLIER WITH OUTLIERS (CKSwO) problem, in addition to the CKS input (G, k, L), we are given an integer p. A *feasible solution* is an *assignment* $\phi: V_C \to V_S \cup \{\bot\}$ which, in addition to the conditions specified in the definition of CKS, satisfies $|\phi^{-1}(\bot)| \le p$. Vertices $\phi^{-1}(\bot)$ are called *outliers*. The goal is to find a solution of minimum cost, which is defined as $\text{cost}(\phi) = \max_{u \in V_C \setminus \phi^{-1}(\bot)} \text{dist}(u, \phi(u))$. Facility location and clustering with outliers were introduced by Charikar et al. [11]. Among other results, they showed a 3-approximation algorithm for k-CENTER WITH OUTLIERS and an approximation lower bound of $2 - \varepsilon$. Later, Harris et al. [25] and Chakrabarty et al. [10] independently closed this gap and showed a 2-approximation algorithm for the problem. For CKSwO, Cygan and Kociumaka [15] show a 25-approximation algorithm. It may be of interest that the algorithm we show requires only that the doubling dimension of the supplier set to be bounded.

---

[3] We remark that for this distinction to work, one has to be careful of the used definition of highway dimension: a stricter definition of highway dimension from [1] already implies bounded doubling dimension. On the other hand, for certain types of transportation networks, it can be argued that the doubling dimension is large, while the highway dimension is small. See [22, Appendix A] for a detailed discussion.

Theorem 2. *Let $\mathcal{I} = (G, k, p, L)$ be an instance of* CAPACITATED k-SUPPLIER WITH OUTLIERS. *Moreover, let (V_S, dist) be the shortest-path metric induced by V_S and Δ be its doubling dimension. There exists an algorithm which for any $\varepsilon > 0$ outputs a solution of cost $(1+\varepsilon)\text{OPT}(\mathcal{I})$ in time $\mathcal{O}^*((k+p)^k \cdot \varepsilon^{-\mathcal{O}(k\Delta)})$.*

In light of the following results, this algorithm is almost the best we can hope for. We have already justified the necessity of approximation by the result of Feldmann and Marx [21]. An EPAS parameterized only by Δ is unlikely to exist, as Feder and Greene [17] have shown that unless $P = NP$, approximation algorithms with ratios better than 1.822 and 2 for two-dimensional Euclidean, resp. Manhattan metrics cannot exist. Hence it is necessary to parameterize by both k and Δ. The only improvement we can hope for is a better dependence on the number of outliers in the running time, e.g. by giving an algorithm which is polynomial in p.

Given our hardness of approximation result for CKC on low highway dimension graphs in Theorem 1 and the known EPAS for k-CENTER given by Becker et al. [7], it is evident that the hardness stems from the introduction of capacities. For low doubling dimension graphs we were able to push the existence of an EPAS further than just introducing capacities, by considering suppliers and outliers. It therefore becomes an interesting question whether we can show an EPAS also for low highway dimension graphs when using suppliers and outliers, but without using capacities. The following theorem shows that this is indeed possible. We prove this theorem in the full version of the paper.

Theorem 3. *Let $\mathcal{I} = (G, k, p)$ be an instance of the k-SUPPLIER WITH OUT-LIERS problem. There exists an EPAS for this problem with parameters k, p, ε, and highway dimension of G.*

1.1 Used Techniques

To prove Theorem 1, we enhance a result of Dom et al. [16] which shows that CAPACITATED DOMINATING SET is W[1]-hard in low treewidth graphs.

We prove Theorem 2 by using the concept of a δ-net which is a sparse subset of the input metric such that every input point has a net point near it. This approach was previously used by Feldmann and Marx [21] to show an EPAS for k-CENTER in low doubling dimension graphs.

To prove Theorem 3, we generalize the EPAS for k-CENTER in low highway dimension graphs by Becker et al. [7]. A major component of this algorithm is an EPAS for KSWO in low treewidth graphs, which generalizes an EPAS for k-CENTER in low treewidth graphs by Katsikarelis et al. [28]. We show how to obtain this EPAS in the full version of the paper.

2 Inapproximability in Low Highway Dimension Graphs

In this section we are going to prove Theorem 1, i.e. we show that there is no parameterized approximation scheme for CAPACITATED k-CENTER in graphs of low highway dimension unless FPT = W[1].

We reduce from MULTICOLORED CLIQUE, which is known to be a W[1]-hard problem [13]. The input of MULTICOLORED CLIQUE consists of a graph G and an integer k. The vertex set of G is partitioned into *color classes* $V_1, \ldots, V_k$ where each color class is an independent set. The goal is to find a k-clique. Note that if a k-clique exists in G, then it has exactly one vertex in each color class.

To prove Theorem 1, we will need several settings of edge lengths. Namely for every $\lambda \geq 2\gamma \geq 8$, given an instance $\mathcal{I} = (G, k)$ of MULTICOLORED CLIQUE, we produce in polynomial time a CKC instance $\overline{\mathcal{I}}_\lambda = (\overline{G}, \overline{k}, L, d_\lambda)$ where the highway dimension and treewidth of $\overline{G}$ is $\mathcal{O}(k^4)$ and $\overline{k} = 7k(k-1) + 2k$. If we are not interested in a particular setting of λ or we speak generally about all instances for all possible settings of λ, we omit the subscript.

It follows from [13] that we can assume without loss of generality that every color class consists of N vertices and the number of edges between every two color classes is M. For two integers $m \leq n$ by $\langle m, n \rangle$ we mean the set of integers $\{m, m+1, \ldots, n\}$, and $\langle m \rangle = \langle 1, m \rangle$. For distinct $i, j \in \langle k \rangle$ we denote by $E_{i,j}$ the set of ordered pairs of vertices (u, v) such that $u \in V_i$, $v \in V_j$, and $\{u, v\}$ is an edge in G. When we add an (A, B)-*arrow* from vertex u to vertex v, we add A subdivided edges between u and v and additionally we add B unique vertices to the graph and connect them to v, see Fig. 1. When we *mark* a vertex u, we add $\overline{k} + 1$ new vertices to the graph and connect them to u. We denote the set of all marked vertices by Z.

We first describe the structure of $\overline{G}$ and we set the capacities and edge lengths of $\overline{G}$ afterwards. See Fig. 2 for an illustration of the reduction.

Color Class Gadget. For each color class V_i, we create a gadget as follows. We arbitrarily order vertices of V_i and to the jth vertex $u \in V_i$ we assign numbers $u^\uparrow = j \cdot 2N^2$ and $u^\downarrow = 2N^3 - u^\uparrow$. For each vertex $u \in V_i$ we create a vertex $\overline{u}$ and we denote $\overline{V}_i = \{\overline{u} \mid u \in V_i\}$. We add a marked vertex x_i and connect it to every vertex of $\overline{V}_i$. We add a set S_i of $\overline{k} + 1$ vertices and connect each vertex of S_i to every vertex of $\overline{V}_i$. For every $j \in \langle k \rangle \setminus \{i\}$ we add a pair of marked vertices $y_{i,j}$ and $z_{i,j}$. We denote $Y_i = \cup_{j \in \langle k \rangle \setminus \{i\}} \{y_{i,j}, z_{i,j}\}$. For every vertex $\overline{u} \in \overline{V}_i$ we add a $(u^\uparrow, u^\downarrow)$-arrow from u to each vertex of $\cup_{j \in \langle k \rangle \setminus \{i\}} y_{i,j}$ and a $(u^\downarrow, u^\uparrow)$-arrow from u to each vertex of $\cup_{j \in \langle k \rangle \setminus \{i\}} z_{i,j}$.

Edge Set Gadget. For every $i \in \langle k-1 \rangle, j \in \langle i+1, k \rangle$ we create a gadget for the edge set $E_{i,j}$ as follows. For every edge $e \in E_{i,j}$ we create a vertex $\overline{e}$ and we denote $\overline{E}_{i,j} = \{\overline{e} \mid e \in E_{i,j}\}$. We add a marked vertex $x_{i,j}$ and connect it to every vertex of $\overline{E}_{i,j}$. We add a set $S_{i,j}$ of $\overline{k} + 1$ vertices and connect each vertex of $S_{i,j}$ to every vertex of $\overline{E}_{i,j}$. We add four marked vertices $p_{i,j}, p_{j,i}, q_{i,j}, q_{j,i}$. Consider an edge $e = (u, v) \in E_{i,j}$, we connect $\overline{e}$ to $p_{i,j}$ with a $(u^\downarrow, u^\uparrow)$-arrow, to $q_{i,j}$ with a $(u^\uparrow, u^\downarrow)$-arrow, to $p_{j,i}$ with a $(v^\downarrow, v^\uparrow)$-arrow, and to $q_{j,i}$ with a $(v^\uparrow, v^\downarrow)$-arrow. We denote $\mathcal{S} = (\cup_{i \in \langle k \rangle} S_i) \cup (\cup_{i \in \langle k-1 \rangle, j \in \langle i+1, k \rangle} S_{i,j})$.

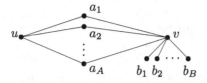

Fig. 1. The result of adding an (A, B)-arrow from u to v.

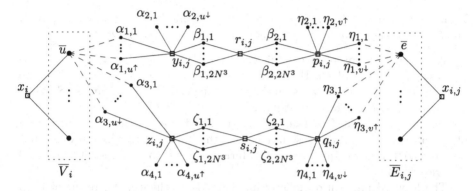

Fig. 2. Part of the reduction for color class V_i and edge set $E_{i,j}$. Vertex $\overline{e}$ represents an edge $(v, w) \in E_{i,j}$ in G. We omit the sets S_i and $S_{i,j}$. Marked vertices are drawn by boxes and we omit their $\overline{k}+1$ "private" neighbors. Edges drawn by a dashed line have length 1 and the remaining edges have length λ. We also omit the appropriate arrows from vertices of $\overline{V}_i \setminus \{\overline{u}\}$ to $y_{i,j}$ and to $z_{i,j}$, and the appropriate arrows from vertices of $\overline{E}_{i,j} \setminus \{\overline{e}\}$ to $p_{i,j}$ and to $q_{i,j}$.

Adjacency Gadget. To connect the color class gadgets and the edge set gadgets, for every distinct $i, j \in \langle k \rangle$ we add marked vertices $r_{i,j}$ and $s_{i,j}$, and we add $(2N^3, 0)$-arrows from $y_{i,j}$ to $r_{i,j}$, from $p_{i,j}$ to $r_{i,j}$, from $z_{i,j}$ to $s_{i,j}$, and from $q_{i,j}$ to $s_{i,j}$.

Capacities. We now describe the capacities $L \colon V(\overline{G}) \to \mathbb{N}$. To streamline the exposition, we will assume that each vertex of $\phi(V(\overline{G}))$ covers itself at no "cost" with respect to the capacity. For every two distinct $i, j \in \langle k \rangle$, the vertex x_i has capacity $N - 1 + \overline{k} + 1$, the vertex $x_{i,j}$ has capacity $M - 1 + \overline{k} + 1$, vertices $y_{i,j}$ and $z_{i,j}$ have capacity $2N^4 + \overline{k} + 1$, vertices $p_{i,j}$ and $q_{i,j}$ have capacity $2MN^3 + \overline{k} + 1$, vertices $r_{i,j}$ and $s_{i,j}$ have capacity $2N^3 + \overline{k} + 1$, and the remaining vertices have capacity equal to their degree.

Edge Lengths. Given $\lambda \in \mathbb{R}^+$ we set edge lengths $d_\lambda \colon E(\overline{G}) \to \mathbb{R}$ as follows. For every $i \in \langle k \rangle$ and every vertex $\overline{u} \in \overline{V}_i$, we assign length 1 to edges between $\overline{u}$ and $N(\overline{u}) \setminus (\{x_i\} \cup S_i)$, i.e. the set of vertices originating from subdivided edges of arrows between $\overline{u}$ and Y_i. Similarly for every $i \in \langle k-1 \rangle, j \in \langle i+1, k \rangle$ and every vertex $\overline{v} \in \overline{E}_{i,j}$, we assign length 1 to edges between $\overline{v}$ and $N(\overline{v}) \setminus (\{x_{i,j}\} \cup S_{i,j})$. To the remaining edges we assign length λ.

From the way we assign edge lengths d_λ, in a solution ϕ of cost $\lambda + 1$ such that $|\phi(V(\overline{G}))| \leq \overline{k}$, it must be the case that $Z \subseteq \phi(V(\overline{G}))$, since Z is the set of marked vertices with $\overline{k} + 1$ private neighbors.

Dom et al. [16, Observation 1] observe that $\mathrm{tw}(\overline{G}) = \mathcal{O}(k^4)$: The size of $\mathcal{S}$ is $\mathcal{O}(k^4)$ and the size of Z is $\frac{1}{2}(13k^2 - 11k)$. Removing $\mathcal{S} \cup Z$ leaves us with a forest and the observation follows.

We prove that $\overline{G}$ has bounded highway dimension.

Lemma 1. *For any $\lambda \geq 2\gamma$, where γ is the universal constant in Definition 1, graph $\overline{G}$ with edge lengths d_λ has highway dimension $h(\overline{G}) \in \mathcal{O}(k^4)$.*

We present the proof in the full version of the paper. Let $\mathcal{V} = (\cup_{i=1}^{k} \overline{V}_i) \cup (\cup_{1 \leq i < j \leq k} \overline{E}_{i,j})$. The idea is that in the case $r \geq 2$, we can consider $\mathcal{S} \cup Z$ as the hitting set for all shortest paths. In the remaining case $r < 2$, we can use $\mathcal{S} \cup \mathcal{V} \cup Z$ as the hitting set for all shortest paths.

Now we prove that $\mathcal{I}$ contains a k-clique if and only if $\overline{\mathcal{I}}_\lambda$ contains a solution of cost at most $\lambda + 1$. As d_λ assigns edge lengths 1 and λ, this will imply that if $\mathcal{I}$ does not contain a k-clique, then any solution of $\overline{\mathcal{I}}$ has to have cost at least $\lambda + 2$ and vice versa.

The forward implication follows implicitly from the original result of Dom et al. [16], since we can interpret a capacitated dominating set as a solution of cost λ, hence we omit the proof.

Lemma 2 ([16, Lemma 1]). *If $\mathcal{I}$ contains a k-clique, then $\overline{\mathcal{I}}_\lambda$ contains a solution of cost λ.*

To prove the backward implication, we start need to show that a solution of cost $\lambda + 1$ has to open a vertex in every $\overline{V}_i$ and every $\overline{E}_{i,j}$. In contrast to Lemma 2, the backward implication does not simply follow from the original result since we have added edge lengths to the graph.

Lemma 3. *Let ϕ be a solution of $\overline{\mathcal{I}}_\lambda$ of cost $\lambda + 1$, and $D = \phi(V(\overline{G}))$. Then for each $i \in \langle k \rangle$ we have $|\overline{V}_i \cap D| = 1$ and for each $i \in \langle k-1 \rangle, j \in \langle i+1, k \rangle$ we have and $|\overline{E}_{i,j} \cap D| = 1$.*

We defer the proof to the full version of the paper, but the idea is the following. Let $i \in \langle k \rangle$ and suppose for a contradiction that we want to cover S_i by balls of radius $\lambda + 1$ without using vertices of V_i. Then, from the construction of the graph, the size of the solution would be greater than $\overline{k}$.

Now we show that if $\overline{\mathcal{I}}_\lambda$ contains a solution of cost $\lambda + 1$, then $\mathcal{I}$ contains a k-clique.

Lemma 4. *If $\overline{\mathcal{I}}_\lambda$ has a solution ϕ of cost $\lambda + 1$, then $\mathcal{I}$ contains a k-clique.*

Proof. Let $D = \phi(V(\overline{G}))$. For $i \in \langle k \rangle$ let $\overline{u}_i$ be the vertex of $\overline{V}_i \cap D$ and for $i \in \langle k-1 \rangle, j \in \langle i+1, k \rangle$ let $\overline{e}_{i,j}$ be the vertex of $\overline{E}_{i,j} \cap D$. These vertices are well-defined by Lemma 3. To prove that these vertices encode a k-clique in G, we want to show for every $i \in \langle k-1 \rangle, j \in \langle i+1, k \rangle$ that vertices u_i and u_j, which correspond to $\overline{u}_i$ and $\overline{u}_j$ respectively, are incident to the edge $e_{i,j}$ corresponding to the vertex $\overline{e}_{i,j}$. We will only present the proof of incidence for u_i and $e_{i,j}$, for u_j we can proceed analogously. Let v be the vertex of the edge $e_{i,j}$ which belongs to V_i in G. Before we prove that u_i and $e_{i,j}$ are incident, we first argue that $u_i^\uparrow + v^\downarrow = 2N^3$.

We prove this statement by contradiction. First suppose $u_i^\uparrow + v^\downarrow < 2N^3$. Then $u_i^\uparrow + v^\downarrow \leq 2N^3 - 2N^2$ as for every two distinct vertices w_1 and w_2 of a color class we have $|w_1^\uparrow - w_2^\uparrow| \geq 2N^2$ and $|w_1^\downarrow - w_2^\downarrow| \geq 2N^2$. Consider the set

$$\mathcal{T} = (N(y_{i,j}) \cup N(r_{i,j}) \cup N(p_{i,j})) \setminus \phi^{-1}(\{x_i, u_i, e_{i,j}, x_{i,j}\}). \tag{1}$$

It follows that in a solution of cost $\lambda + 1$, vertices of $\mathcal{T}$ must be covered by $y_{i,j}$, $r_{i,j}$ or $p_{i,j}$ as edges of $\overline{G}$ have length 1 or λ. We have $L(y_{i,j}) + L(r_{i,j}) + L(p_{i,j}) = 2N^4 + 2MN^3 + 2N^3 + 3(\overline{k} + 1)$. However,

$$|\mathcal{T}| \geq 2N^4 + 2MN^3 + 4N^3 + 3(\overline{k} + 1) - ((2N^3 - 2N^2) + (N - 1) + (M - 1))$$
$$> 2N^4 + 2MN^3 + 2N^3 + 3(\overline{k} + 1), \tag{2}$$

where we used that $M \leq N^2$ and $N^2 > N$. Thus $y_{i,j}$, $r_{i,j}$, and $p_{i,j}$ cannot cover $\mathcal{T}$. This contradicts the fact that ϕ is a solution of cost $\lambda + 1$.

If $u_i^\uparrow + v^\downarrow > 2N^3$, then $u_i^\downarrow + v^\uparrow < 2N^3$ and we can apply the identical argument for vertices $z_{i,j}, s_{i,j}, q_{i,j}$.

It remains to prove that u_i is incident to $e_{i,j}$. Again, let v be the vertex of $e_{i,j}$ which lies in V_i of G. We know from the preceding argument that $u_i^\uparrow + v^\downarrow = 2N^3$. However, the only vertex $w \in V_i$ such that $w^\downarrow = 2N^3 - u_i^\uparrow$ is u_i itself; for any $w \in V_i \setminus \{u_i\}$ we would have $|(u_i^\uparrow + w^\downarrow) - 2N^3| \geq 2N^2$. Hence $v = u_i$ and so $e_{i,j}$ is incident to u_i. This concludes the proof. $\square$

We complete the proof of Theorem 1 in the full version of the paper. With Lemmas 1, 2, and 4 in hand, the main argument is the following. Suppose for a contradiction that there exists an FPT algorithm $\mathcal{A}$ which is excluded by Theorem 1. We can decide whether $\mathcal{I}$ has a k-clique by checking whether the solution $\mathcal{A}(\overline{\mathcal{I}})$ has cost at most $(1 + \frac{1}{\gamma})$. This type of argument has been previously used in both the polynomial and parameterized approximation setting, see [31], respectively [18].

3   EPAS on Graphs of Bounded Doubling Dimension

In this section we prove Theorem 2, i.e. we show the existence of an EPAS for CkSwO on instances where the supplier set has bounded doubling dimension. To be more precise, we develop a decision algorithm which, given a cost $\varrho \in \mathbb{R}^+$, and $\varepsilon > 0$, computes a solution of cost $(1 + \varepsilon)\varrho$ in FPT time with parameters k, p, doubling dimension and ε. Formally, the result is the following lemma.

Lemma 5. Let $\mathcal{I} = (G, k, p, L)$ be a CkSwO instance. Moreover, let (V_S, dist) be the shortest-path metric induced by V_S and Δ be its doubling dimension. There exists an algorithm which, given a cost $\varrho \in \mathbb{R}^+$ and $\varepsilon > 0$, either

– *computes a feasible solution of cost* $(1+\varepsilon)\varrho$ *if* $(1+\varepsilon)\varrho \geq \mathrm{OPT}(\mathcal{I})$, *or*
– *correctly decides that* $\mathcal{I}$ *has no solution of cost at most* ϱ,

running in time $\mathcal{O}^*\left((k+p)^k \varepsilon^{-\mathcal{O}(k\Delta)}\right)$.

Using Lemma 5, we can obtain the algorithm of Theorem 2 as follows. We can first assume without loss of generality, that $V_C \cup V_S = V(G)$. Suppose that we can guess the optimum cost OPT of any CKSWO instance. By using OPT as ϱ in Lemma 5, we can output a solution of cost $(1+\varepsilon)\mathrm{OPT}$. To guess the optimum cost OPT, observe that OPT must be one of the inter-vertex distances. Hence the minimum inter-vertex distance ϱ for which the algorithm outputs a solution has the property that $\varrho \leq \mathrm{OPT}$ and consequently $(1+\varepsilon)\varrho \leq (1+\varepsilon)\mathrm{OPT}$.

The main ingredient of the algorithm is the notion of a δ-net. For a metric (X, dist), a subset $Y \subseteq X$ is called a δ-*cover* if for every $u \in X$ there exists a $v \in Y$ such that $\mathrm{dist}(u,v) \leq \delta$. If a δ-cover Y has an additional property that for every two distinct $u, v \in Y$ we have $\mathrm{dist}(u,v) > \delta$, then we say that Y is a δ-*net*. Observe that a δ-net can be computed greedily in polynomial time.

Let us give the main idea behind the algorithm. Given an instance of the problem and $\varepsilon > 0$, let ϕ^* be an optimum solution of cost OPT, V_C^* be clients that are not outliers according to ϕ^*, i.e. $V_C^* = \{u \in V_C \mid \phi^*(u) \neq \bot\}$, and Y be an $(\varepsilon \cdot \mathrm{OPT})$-net of the metric (V_S, dist). Consider an assignment function ϕ constructed as follows. For each client $u \in V_C^*$ we set $\phi(u)$ to the nearest point of Y to $\phi^*(u)$, and for the remaining clients we set the value of ϕ to $\bot$. If for every selected supplier $s \in (\phi(V_C) \setminus \{\bot\})$ we have $|\phi^{-1}(s)| \leq L(s)$, then ϕ is a feasible solution. Since Y is a $(\varepsilon \cdot \mathrm{OPT})$-net, the cost of ϕ is at most $(1+\varepsilon)\mathrm{OPT}$.

The main obstacle to implementing an algorithm from this idea is that we do not know the optimum solution ϕ^*. However, by the definition of the net Y, we know that each selected supplier $\phi^*(V_C^*)$ is near some point of Y. If Y was not too large, we could guess which k of its points are near to every supplier of $\phi^*(V_C^*)$. Later, we will also show how to ensure that the solution we create respects capacities of suppliers we pick.

We now show how to bound the size of the net. Let (X, dist) be a metric of doubling dimension Δ, by the *aspect ratio* of a set $X' \subseteq X$, we mean the diameter of X' divided by the minimum distance between any two distinct points of X', that is $\frac{\max_{u,v \in X'} \mathrm{dist}(u,v)}{\min_{u,v \in X', u \neq v} \mathrm{dist}(u,v)}$. The following lemma by Gupta et al. [24] shows that the cardinality of a subset $X' \subseteq X$ can be bounded by its aspect ratio and Δ.

Lemma 6. ([24]). *Let* (X, dist) *be a metric and* Δ *its doubling dimension. Consider a subset* $X' \subseteq X$ *of aspect ratio* α *and doubling dimension* Δ'. *Then it holds that* $\Delta' = \mathcal{O}(\Delta)$ *and* $|X'| \leq 2^{\mathcal{O}(\Delta\lceil \log_2 \alpha \rceil)}$.

Using Lemma 6, we bound the size of Y.

Lemma 7. *Let* $\mathcal{I} = (G, k, p, L)$ *be an instance of the* CKSWO *problem,* $\varepsilon > 0$, *and* $\varrho \in \mathbb{R}^+$ *a cost. Moreover let* (V_S, dist) *be the shortest-path metric induced by* V_S *and* Δ *its doubling dimension. Assume that for each supplier* $s \in V_S$ *there exists a client* $c \in V_C$ *such that* $\mathrm{dist}(s,c) \leq \varrho$. *If* $\mathcal{I}$ *has a feasible solution* ϕ *with* $\mathrm{cost}(\phi) \leq \varrho$, *then an* $(\varepsilon\varrho)$-*net* Y *of* V_S *has size at most* $(k+p)\varepsilon^{-\mathcal{O}(\Delta)}$.

We present the proof in the full version of the paper. Before we give the main ideas of the proof, let us make a few comments the statement of the lemma. We do not know the cost of the optimum solution and we are merely guessing it. Hence we need to also consider the case when our guess on the cost ϱ is wrong, i.e. it is less than the cost of the optimum solution. The requirement that every supplier has a client nearby is a natural one: if we assume that our solution has cost ϱ and a supplier s has $\text{dist}(s, V_C) > \varrho$, then it will never be picked in a solution. Thus we can without loss of generality remove all such suppliers from the input. The idea behind the proof is the following. Suppose that ϕ is a solution of cost ϱ and let $D = \phi(V_C)$. Then balls of radius 2ϱ around $D \cup \phi^{-1}(\bot)$ cover the entire graph. By applying Lemma 6 in each of these balls, we get the desired bound.

When we gave the intuition behind the algorithm, we assumed that the derived solution ϕ, which replaces every optimum supplier of $\phi^*(V_C) \setminus \{\bot\}$ by its nearest net point, does not violate the capacity of any selected net point, i.e. for every $s \in \phi(V_C) \setminus \{\bot\}$ we have $|\phi^{-1}(s)| \le L(s)$. This does not have to be the case, so instead of replacing every optimum supplier by its nearest net point, we need to select the replacement net point in a more sophisticated manner, in particular to avoid violating the capacity of the replacement net point.

Let V_S^* be the optimum supplier set corresponding to the optimum assignment function ϕ^*. Suppose that we are able to guess a subset $S^* \subseteq Y$ of size k such that for every supplier $u \in V_S^*$ we have $\text{dist}(u, S^*) \le \varepsilon\varrho$. Let $A \colon V_S^* \to S^*$ map each optimum supplier to its nearest net point. As we have discussed, we cannot just replace each supplier $u \in V_S^*$ by $A(u)$ since it may happen that $|(\phi^*)^{-1}(u)| > L(A(u))$. However, there is a supplier in the ball $B_{A(u)}(\varepsilon\varrho)$ which is guaranteed to have capacity at least $L(u)$ since $u \in B_{A(u)}(\varepsilon\varrho)$ Thus we can implement the "replacement step" by replacing each optimum supplier by the supplier of highest capacity in $B_{A(u)}(\varepsilon\varrho)$ and this increases the cost of the optimum solution by at most $2\varepsilon\varrho$, i.e. the diameter of the ball.

We must also consider the case when $|A^{-1}(v)| > 1$ for some net point $v \in S^*$. Generalizing the previous idea, we replace suppliers $A^{-1}(v)$ by $|A^{-1}(v)|$ suppliers of $B_v(\varepsilon\varrho)$ with the highest capacities. As we do not know the optimum solution, we do not know $|A^{-1}(v)|$ either. Nevertheless, we know that $|S^*| \le k$, and so we can afford to guess these values after guessing the set S^*.

The final ingredient we need is the ability to verify our guesses. That is, given a set of at most k suppliers, we need to check if there exists a feasible solution of a given cost which assigns clients to a prescribed set of suppliers. This can be done by a standard reduction to network flows. We state the result formally and present the proof in the full version of the paper.

Lemma 8. *Given a* CKSwO *instance* $\mathcal{I} = (G, k, p, L)$, *a cost* $\varrho \in \mathbb{R}^+$, *and a subset* $S \subseteq V_S$, *we can determine in polynomial time whether there exists an assignment* $\phi \colon V_C \to S$ *such that* $|\phi(u)^{-1}| \le L(u)$ *for each* $u \in S$, $|\phi^{-1}(\bot)| \le p$, *and* $\text{cost}(\phi) \le \varrho$.

We now prove the correctness of the replacement strategy.

Lemma 9. *Let $\mathcal{I} = (G, k, p, L)$ be a CkSwO instance such that there exists a solution ϕ^* of cost ϱ and for each supplier there exists a client at distance at most ϱ from it. Given an $(\varepsilon\varrho)$-net Y of the shortest-path metric induced by V_S, and $\varepsilon > 0$, we can compute a solution of cost $(1 + 2\varepsilon)\varrho$ in time $\mathcal{O}^*\left(\binom{|Y|}{k}k^k\right)$.*

Proof. Let $V_S^* = \phi^*(V_C) \setminus \{\bot\}$. For an optimum supplier $u \in V_S^*$ it may happen that $|B_u(\varepsilon\varrho) \cap Y| > 1$, i.e. it is close to more than one net point. This may cause issues when we guess for each net point $v \in Y$ the size of $B_v(\varepsilon\varrho) \cap V_S^*$. To circumvent this problem, we fix a linear order $\preceq$ on the set of net points Y and we assign each optimum supplier to the first close net point. Formally, we define for a net point $v \in Y$

- $P(v) = \{v' \in Y \mid v' \prec v\}$ (note that $v \notin P(v)$),
- $M(v) = B_v(\varepsilon\varrho) \setminus (\cup_{v' \in P(v)} B_{v'}(\varepsilon\varrho))$,
- $D(v) = |M(v) \cap V_S^*|$, and
- $R(v)$ to be the set of $D(v)$ suppliers in $M(v)$ with the highest capacities.

For a net point $v \in Y$, it is easy to see that $\sum_{s \in R(v)} L(s) \geq \sum_{t \in M(v) \cap V_S^*} L(t)$. The sets $\{R(v) \mid v \in Y\}$ are disjoint by the way we defined $M(v)$.

We guess a subset $Y' \subseteq Y$ of size k such that $V_S^* \subseteq \cup_{v \in Y'} B_v(\varepsilon\varrho)$. For each $v \in Y'$ we guess $D(v)$ and select $S = \cup_{v \in Y'} R(v)$. We apply the algorithm from Lemma 8 with the set S and cost $(1 + 2\varepsilon)\varrho$. If this check passes, then the solution we obtain is in fact a solution of cost $(1 + 2\varepsilon)\varrho$ since we replaced each optimum supplier by a supplier at distance at most $2\varepsilon\varrho$ from it. Conversely, if none of our guesses pass this check, then the instance $\mathcal{I}$ has no solution of cost ϱ.

The running time of our algorithm is dominated by the time required to guess Y' and the cardinalities $D(v)$ for each $v \in Y'$. From $|Y'| \leq k$, the time required to guess the Y' is $\mathcal{O}\left(\binom{|Y|}{k}\right)$. Since $D(v) \leq k$ for every $v \in Y$, the time required to guess $D(v)$ for each $v \in Y'$ is $\mathcal{O}(k^k)$. In total, the running time of the algorithm is $\mathcal{O}^*\left(\binom{|Y|}{k}k^k\right)$. $\square$

We can prove Lemma 5 by applying Lemmas 7 and 9, and we present the argument in the full version of the paper.

4 Open Problems

We conclude with the following open problems. The algorithms given by Theorems 2 and 3 have the number of outliers in the base of the exponent. Is it possible to remove the outliers from the set of parameters? An improvement of Theorem 1 would be to show that the hardness is preserved in the case of planar graphs. It may be of interest that PLANAR CAPACITATED DOMINATING SET is W[1]-hard when parameterized by solution size [9]. Goyal and Jaiswal [23] have shown that it is possible to 2-approximate CkC when the parameter is only k, and that this result is tight. An improvement of Theorem 1 would be to show that this lower bound is tight in low highway dimension graphs. Finally, we ask whether there exists a problem which admits an EPAS in low highway dimension graphs but we cannot approximate in low doubling dimension graphs, i.e. the converse of Theorems 1 and 2.

References

1. Abraham, I., Delling, D., Fiat, A., Goldberg, A.V., Werneck, R.F.: Highway dimension and provably efficient shortest path algorithms. J. ACM (JACM) **63**(5), 1–26 (2016)
2. Abraham, I., Fiat, A., Goldberg, A.V., Werneck, R.F.: Highway dimension, shortest paths, and provably efficient algorithms. In: Proceedings of the twenty-first annual ACM-SIAM symposium on Discrete Algorithms, pp. 782–793. SIAM (2010)
3. Ahmadi-Javid, A., Seyedi, P., Syam, S.S.: A survey of healthcare facility location. Comput. Oper. Res. **79**, 223–263 (2017). https://doi.org/10.1016/j.cor.2016.05.018, https://www.sciencedirect.com/science/article/pii/S0305054816301253
4. An, H.C., Bhaskara, A., Chekuri, C., Gupta, S., Madan, V., Svensson, O.: Centrality of trees for capacitated k-center. Math. Program. **154**(1–2), 29–53 (2015). https://doi.org/10.1007/s10107-014-0857-y
5. Bast, H., Funke, S., Matijevic, D.: Transit ultrafast shortest-path queries with linear-time preprocessing. 9th DIMACS Implementation Challenge [1] (2006)
6. Bast, H., Funke, S., Matijevic, D., Sanders, P., Schultes, D.: In transit to constant time shortest-path queries in road networks. In: 2007 Proceedings of the Ninth Workshop on Algorithm Engineering and Experiments (ALENEX), pp. 46–59. SIAM (2007)
7. Becker, A., Klein, P.N., Saulpic, D.: Polynomial-time approximation schemes for k-center, k-median, and capacitated vehicle routing in bounded highway dimension. In: Azar, Y., Bast, H., Herman, G. (eds.) 26th Annual European Symposium on Algorithms (ESA 2018). Leibniz International Proceedings in Informatics (LIPIcs), vol. 112, pp. 8:1–8:15. Schloss Dagstuhl-Leibniz-Zentrum fuer Informatik, Dagstuhl, Germany (2018)
8. Blum, J.: Hierarchy of transportation network parameters and hardness results. In: Jansen, B.M.P., Telle, J.A. (eds.) 14th International Symposium on Parameterized and Exact Computation (IPEC 2019). Leibniz International Proceedings in Informatics (LIPIcs), vol. 148, pp. 4:1–4:15. Schloss Dagstuhl-Leibniz-Zentrum fuer Informatik, Dagstuhl, Germany (2019)
9. Bodlaender, H.L., Lokshtanov, D., Penninkx, E.: Planar capacitated dominating set is $W[1]$-Hard. In: Chen, J., Fomin, F.V. (eds.) IWPEC 2009. LNCS, vol. 5917, pp. 50–60. Springer, Heidelberg (2009). https://doi.org/10.1007/978-3-642-11269-0_4
10. Chakrabarty, D., Goyal, P., Krishnaswamy, R.: The non-uniform k-center problem. ACM Trans. Algorithms **16**(4) (2020). https://doi.org/10.1145/3392720
11. Charikar, M., Khuller, S., Mount, D.M., Narasimhan, G.: Algorithms for facility location problems with outliers. In: SODA, vol. 1, pp. 642–651 (2001)
12. Cohen-Addad, V., Feldmann, A.E., Saulpic, D.: Near-linear time approximation schemes for clustering in doubling metrics. J. ACM **68**(6), 1–34 (2021). https://doi.org/10.1145/3477541
13. Cygan, M., et al.: Parameterized Algorithms. Springer, Cham (2015). https://doi.org/10.1007/978-3-319-21275-3

14. Cygan, M., Hajiaghayi, M., Khuller, S.: LP rounding for k-centers with non-uniform hard capacities. In: 2012 IEEE 53rd Annual Symposium on Foundations of Computer Science, pp. 273–282. IEEE (2012)

15. Cygan, M., Kociumaka, T.: Constant factor approximation for capacitated k-center with outliers. In: Mayr, E.W., Portier, N. (eds.) 31st International Symposium on Theoretical Aspects of Computer Science (STACS 2014). Leibniz International Proceedings in Informatics (LIPIcs), vol. 25, pp. 251–262. Schloss Dagstuhl-Leibniz-Zentrum fuer Informatik, Dagstuhl, Germany (2014). https://doi.org/10.4230/LIPIcs.STACS.2014.251, https://drops.dagstuhl.de/opus/volltexte/2014/4462

16. Dom, M., Lokshtanov, D., Saurabh, S., Villanger, Y.: Capacitated domination and covering: a parameterized perspective. In: Grohe, M., Niedermeier, R. (eds.) IWPEC 2008. LNCS, vol. 5018, pp. 78–90. Springer, Heidelberg (2008). https://doi.org/10.1007/978-3-540-79723-4_9

17. Feder, T., Greene, D.: Optimal algorithms for approximate clustering. In: Proceedings of the Twentieth Annual ACM Symposium on Theory of Computing, pp. 434–444. STOC 1988, Association for Computing Machinery, New York, NY, USA (1988). https://doi.org/10.1145/62212.62255,https://doi.org/10.1145/62212.62255

18. Feldmann, A.E.: Fixed-parameter approximations for k-center problems in low highway dimension graphs. Algorithmica **81**(3), 1031–1052 (2019). https://doi.org/10.1007/s00453-018-0455-0

19. Feldmann, A.E., Fung, W.S., Konemann, J., Post, I.: A $(1+\varepsilon)$-embedding of low highway dimension graphs into bounded treewidth graphs. SIAM J. Comput. **47**(4), 1667–1704 (2018)

20. Feldmann, A.E., Karthik, C., Lee, E., Manurangsi, P.: A survey on approximation in parameterized complexity: hardness and algorithms. Algorithms **13**(6), 146 (2020)

21. Feldmann, A.E., Marx, D.: The parameterized hardness of the k-center problem in transportation networks. Algorithmica **82**(7), 1989–2005 (2020). https://doi.org/10.1007/s00453-020-00683-w

22. Feldmann, A.E., Saulpic, D.: Polynomial time approximation schemes for clustering in low highway dimension graphs. J. Comput. Syst. Sci. **122**, 72–93 (2021). https://doi.org/10.1016/j.jcss.2021.06.002, https://www.sciencedirect.com/science/article/pii/S0022000021000647

23. Goyal, D., Jaiswal, R.: Tight FPT approximation for constrained k-center and k-supplier. CoRR abs/2110.14242 (2021). https://arxiv.org/abs/2110.14242

24. Gupta, A., Krauthgamer, R., Lee, J.: Bounded geometries, fractals, and low-distortion embeddings. In: 44th Annual IEEE Symposium on Foundations of Computer Science, 2003. Proceedings, pp. 534–543. IEEE (2003)

25. Harris, D.G., Pensyl, T., Srinivasan, A., Trinh, K.: A lottery model for center-type problems with outliers. ACM Trans. Algorithms **15**(3) (2019). https://doi.org/10.1145/3311953

26. Hochbaum, D.S., Shmoys, D.B.: A best possible heuristic for the k-center problem. Math. Oper. Res. **10**(2), 180–184 (1985)

27. Hochbaum, D.S., Shmoys, D.B.: A unified approach to approximation algorithms for bottleneck problems. J. ACM **33**(3), 533–550 (1986). https://doi.org/10.1145/5925.5933

28. Katsikarelis, I., Lampis, M., Paschos, V.T.: Structural parameters, tight bounds, and approximation for (k, r)-center. Discret. Appl. Math. **264**, 90–117 (2019)

29. Talwar, K.: Bypassing the embedding: algorithms for low dimensional metrics. In: Proceedings of the Thirty-Sixth Annual ACM Symposium on Theory of Computing, pp. 281–290. STOC 2004, Association for Computing Machinery, New York, NY, USA (2004). https://doi.org/10.1145/1007352.1007399

30. Vazirani, V.V.: Approximation Algorithms. Springer, Heidelberg (2013). https://doi.org/10.1007/978-3-662-04565-7

31. Williamson, D.P., Shmoys, D.B.: The Design of Approximation Algorithmsd. Cambridge University Press, Cambridge (2011)

Extending Partial Representations
of Circular-Arc Graphs

Jiří Fiala[1] , Ignaz Rutter[2] , Peter Stumpf[2(✉)] , and Peter Zeman[3]

[1] Department of Applied Mathematics, Charles University,
Prague, Czech Republic
fiala@kam.mff.cuni.cz
[2] Faculty of Computer Science and Mathematics, University of Passau,
Passau, Germany
{rutter,stumpf}@fim.uni-passau.de
[3] Institut de mathématiques, Université de Neuchâtel, Neuchâtel, Switzerland
zeman.peter.sk@gmail.com

Abstract. The partial representation extension problem generalizes the recognition problem for classes of graphs defined in terms of geometric representations. We consider this problem for circular-arc graphs, where several arcs are predrawn and we ask whether this partial representation can be completed. We show that this problem is NP-complete for circular-arc graphs, answering a question of Klavík et al. (2014).

We complement this hardness with tractability results of the representation extension problem for various subclasses of circular-arc graphs. We give linear-time algorithms for extending normal proper Helly and proper Helly representations. For normal Helly circular-arc representations we give an $\mathcal{O}(n^3)$-time algorithm where n is the number of vertices.

Surprisingly, for Helly representations, the complexity hinges on the seemingly irrelevant detail of whether the predrawn arcs have distinct or non-distinct endpoints: In the former case the algorithm for normal Helly circular-arc representations can be extended, whereas the latter case turns out to be NP-complete. We also prove that the partial representation extension problem for unit circular-arc graphs is NP-complete.

Keywords: Partial representation extension · Circular arc graphs · Helly circular arc graphs

1 Introduction

An intersection representation $\mathcal{R}$ of a graph G is a collection of sets $\{R(v) : v \in V(G)\}$ such that $R(u) \cap R(v) \neq \emptyset$ if and only if $uv \in E(G)$. Important classes of graphs are obtained by restricting the sets $R(v)$ to some specific geometric

Funded by the grant 19-17314J of the GA ČR and by grant Ru 1903/3-1 of the German Science Foundation (DFG). Peter Zeman was also supported by the Swiss National Science Foundation project PP00P2-202667.

M. A. Bekos and M. Kaufmann (Eds.): WG 2022, LNCS 13453, pp. 230–243, 2022.
https://doi.org/10.1007/978-3-031-15914-5_17

objects. In an *interval representation* of a graph, each set $R(v)$ is a closed interval of the real line; and in a *circular-arc representation*, the sets $R(v)$ are closed arcs of a circle (but not the whole circle); see Fig. 1. A graph is an *interval graph* if it admits an interval representation and it is a *circular-arc graph* if it admits a circular-arc representation. We denote the corresponding graph classes by INT and CA, respectively.

Often, the availability of a geometric representation makes computational problems tractable that are otherwise NP-complete, e.g., maximum clique can be solved in polynomial time for both interval graphs and circular-arc graphs [15]. Another example is the coloring problem, which can be solved in polynomial time for interval graphs but remains NP-complete for circular-arc graphs [12].

A key problem in the study of geometric intersection graphs is the *recognition problem*, which asks whether a given graph has a specific type of intersection representation. It is a classic result that interval graphs can be recognized in linear time [4]. For circular-arc graphs the first polynomial-time recognition algorithm was given by Tucker [34]. McConnell gave a linear-time recognition algorithm [29].

In this paper, we are interested in a generalization of the recognition problem. For a class $\mathcal{X}$ of intersection representations, the *partial representation extension problem for $\mathcal{X}$* (REPEXT($\mathcal{X}$) for short) is defined as follows. In addition to a graph G, the input consists of a *partial representation* $\mathcal{R}'$ that is a representation of a subgraph G' of G. The question is whether there exists a representation $\mathcal{R} \in \mathcal{X}$ of G that *extends* $\mathcal{R}'$ in the sense that $R(u) = R'(u)$ for all $u \in V(G')$ where $R(u)$, $R'(u)$ are the arcs assigned to u by $\mathcal{R}$, $\mathcal{R}'$. The recognition problem is the special case where the partial representation is empty. We assume that G' is an induced subgraph of G. Otherwise, an adjacency in the subgraph induced by $V(G')$ is violated in any representation extension $\mathcal{R}$ of $\mathcal{R}'$ and we can reject. The partial representation extension problem has been recently studied for many different classes of intersection graphs, e.g., interval graphs [23], proper/unit interval graphs [20], function and permutation graphs [19], circle graphs [6], chordal graphs [22], and trapezoid graphs [24]. Related extension problems have also been considered, e.g., for planar topological [1,17] and straight-line [30] drawings, for contact representations [5], and rectangular duals [7]. In many cases, REPEXT($\mathcal{X}$) can be reduced to the corresponding simultaneous representation problem SIMREP($\mathcal{X}$), which asks for k input graphs $G_1, \ldots, G_k$ whether each G_i admits a representation $\mathcal{R}_i$ such that for any vertex shared by G_i and G_j its representations in $\mathcal{R}_i$ and $\mathcal{R}_j$ coincide. For example, REPEXT(INT) can be reduced in linear time to SIMREP(INT) with two input graphs by a result of Bläsius and Rutter [2].

In many cases, the key to solving the partial representation extension problem is to understand the structure of all possible representations. For interval representations, the basis for this is the characterization of Fulkerson and Gross [9], which establishes a bijection between the combinatorially distinct interval representations of a graph G on the one hand and the linear orders $\preceq$ of the maximal cliques of G where for each vertex v the cliques containing v appear consecutively in $\preceq$ on the other hand. This not only forms the basis for the linear-time recognition by Booth and Lueker [4], but also shows that a PQ-tree can compactly store

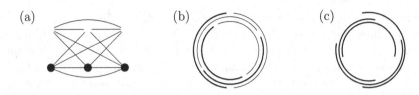

Fig. 1. (a) The graph $\overline{3K_2}$ and (b) its circular-arc representation (the arcs corresponding to the solid vertices are bold). (c) A non-Helly representation of K_4.

the set of all possible interval representations of a graph. The partial representation extension problem for interval graphs can be solved efficiently by searching this set for one that is compatible with the given partial representation.

Despite the fact that circular-arc graphs straightforwardly generalize interval graphs, the structure of their representations is much less understood. It is not clear whether there exists a way to compactly represent the structure of all representations of a circular-arc graph. There are two structural obstructions to this aim. First, in contrast to interval graphs, it may happen that two arcs have disconnected intersection, namely in the case when their union covers the entire circle. Secondly, intervals of the real line satisfy the *Helly property*: any non-empty subfamily of sets with an empty intersection contains two disjoint sets. Consequently, the maximal cliques of interval graphs can be associated to distinct points of the line and also the number of maximal cliques in an interval graph is linear in the number of its vertices. In contrast, arcs of a circle do not necessarily satisfy the Helly property and indeed the number of maximal cliques can be exponential. The complement of a perfect matching nK_2 is an example of this phenomenon, see Fig. 1b.

To capture the above properties, that may have substantial impact on explorations of circular-arc graphs, the following specific subclasses of circular-arc graphs have been defined and intensively studied [13,26,28,32]:

- *Normal circular-arc graphs* (NCA) have a circular-arc representation where the intersection of any two arcs is either empty or connected.
- *Helly circular-arc graphs* (HCA) have a circular-arc representation that satisfies the Helly property, i.e. there are no $k \geq 3$ pairwise intersecting arcs without a point in common.
- *Proper circular-arc graphs* (PCA) are circular-arc graphs that have a circular-arc representation in which no arc properly contains another.
- *Unit circular-arc graphs* (UCA) are circular-arc graphs with a circular-arc representation in which every arc has a unit length.

The above properties can be combined together in the sense that a single representation shall satisfy more properties simultaneously, e.g. *Proper Helly circular-arc graphs* (PHCA) are circular-arc graphs with a circular-arc representation that is both proper and Helly [25]. This is stronger than requiring that a graph is a proper circular-arc graph as well as a Helly circular-

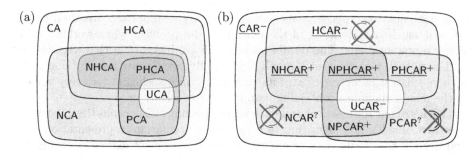

Fig. 2. (a) Relationships between classes of circular-arc graphs. (b) Relationships between classes of circular-arc representations. We study the underlined classes. REPEXT is polynomial for blue$^+$, while NP-complete for red$^-$ and open for black$^?$. (Color figure online)

arc graph (with each property guaranteed by a different representation), i.e., PHCA $\subsetneq$ PCA $\cap$ HCA.

Analogously, since C_4 has a unique representation, the wheel W_4 is a graph with a Helly representation (the degree-4 vertex intersects all four intersections) or a normal representation (it intersects three intersections) but no normal Helly representation (note that the whole circle is not an arc). Thus also NHCA $\subsetneq$ NCA $\cap$ HCA.

Moreover, Tucker [33] proved that every representation of a proper (Helly) circular-arc graph that is not normal can be transformed into a normal proper (Helly) representation. Hence, we have PCA = NPCA and PHCA = NPHCA. Figure 2a shows inclusions between the defined graph classes.

We use an analogous notation for the classes of possible representations, i.e., for $X \subseteq \{N,P,H\}$ the symbol XCAR for the class of all XCA representations, see Fig. 2b. We note that whether a graph G with a partial representation $\mathcal{R}'$ admits an extension depends crucially on the class of allowed representations, as illustrated by the example of W_4 above.

Our Results. While for many classes efficient algorithms for the representation extension problem have been found, the problem has been open for circular arc graphs for several years [21]. We prove in Sect. 3 that REPEXT(CAR) is NP-complete. With the reduction from [2] this strengthens the hardness of SIMREP(CAR) by Bok et al. [3] to two input graphs. Our hardness reduction also works for REPEXT(HCAR).

We complement this result by showing tractability for several subclasses, including all Helly variants; see Fig. 2b. In Sect. 4 we give linear-time algorithms for REPEXT(NPHCAR) and REPEXT(PHCAR). They are based on characterizations of Deng et al. [8] and Lin et al. [27], who characterize the corresponding graph classes by vertex orders. In Sect. 5 we solve REPEXT(NHCAR) and REPEXT(HCAR) in $O(n^3)$ time where the latter requires the extra condition that the prescribed arcs have pairwise distinct endpoints to circumvent the hardness result from Sect. 3. It is surprising that the complexity of the problem hinges on such degeneracies, especially since non-degeneracy assumptions are often made without much consideration of the impact on the problem when working with graph representations. The results from Sect. 5 are based on a characterization by Gavril [14], namely, that a graph G is a Helly circular-arc graph if and only if there exists a *cyclic order* $\lhd$ of its maximal cliques such that for every vertex v, the maximal cliques containing v appear consecutively in $\lhd$.

We note that the classical approach for recognizing HCA graphs [18,27], which uses McConnell's [29] algorithm to construct a circular-arc representation and transform it to a Helly circular-arc representation, does not extend to REPEXT. The reason is that such transformations cannot be exploited in the presence of a partial representation, which cannot be changed.

Finally, we show that for unit circular arcs REPEXT(UCAR) is NP-complete. We prove lemmas and theorems marked with $(\star)$ in the full version.

2   Preliminaries

Let $G = (V, E)$ be a graph. For a vertex $u \in V$ let $N[u]$ denote its *closed neighborhood* $\{v \mid uv \in E(G)\} \cup \{u\}$. If $N[u] = V$, we call u *universal*. Two adjacent vertices $u, v \in V$ form a *universal pair* if $N[u] \cup N[v] = V$.

Cyclic Orders. Let $< = v_0, \ldots, v_{n-1}$ and $<' = u_0, \ldots, u_{n-1}$ be two linear orders on a finite set S. We say that $<$ and $<'$ are *cyclically equivalent* if there is $k \in \{0, \ldots, n-1\}$ such that $v_i = u_{i+k}$, where the addition is modulo n. Clearly, this is an equivalence relation on the set of all linear orders on S. A *cyclic order* $\lhd$ on S is an equivalence class of this relation. For a linear order $<$, we denote the corresponding cyclic order by $[<]$.

Every linear order $<$ on S induces a linear order $<'$ on a subset $S' \subseteq S$ by omitting all ordered pairs in which the elements of $S \setminus S'$ occur. In this case we say that $<$ *extends* $<'$ and similarly that the cyclic order $[<]$ *extends* $[<']$.

Circular-Arc Representations. For any circular-arc representation $\mathcal{R}$ and each connected component C of a graph G the set $\bigcup_{v \in V(C)} R(v)$ is a connected subset of the circle. Therefore, if G is a disconnected circular-arc graph, then each connected component of G has to be an interval graph. These cases can be treated with the corresponding algorithms for interval graphs of [20,23]. Hence without loss of generality we restrict ourselves to connected graphs in this paper.

Let $\mathcal{R}$ be a representation of a circular-arc graph G. For a vertex v of G, we refer to the two endpoints of $R(v)$ as *tail* $R(v)_t$ and *head* $R(v)_h$. We use the convention of traversing the arc from the tail to the head in the clockwise direction along the circle. We denote such an arc as $R(v) = [R(v)_t, R(v)_h]$, and its complement $(R(v)_h, R(v)_t)$ as $R(v)^c$.

Let $\mathcal{R}$ be a Helly representation of a circular-arc graph G. Denote by $\mathcal{C}$ the set of maximal cliques of G. We assign every maximal clique $C \in \mathcal{C}$ a unique point $\mathrm{cp}(C) \in \bigcap_{v \in C} R(v)$ and call it the *clique-point* of C.

Lemma 1 (Gavril [14]). *A graph G is a Helly circular-arc graph if and only if there exists a cyclic order $\lhd$ of its maximal cliques such that for every vertex v, the maximal cliques containing v appear consecutively in $\lhd$.*

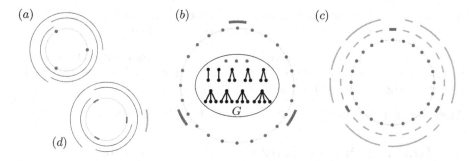

Fig. 3. (a) construction for green dots with blue universal arcs (b) the instance of REPEXT(CAR) obtained from $\{1,1,2,2,2,3,3,3,4\}$ of 3-PARTITION (predrawn universal vertices not shown). (c) the corresponding solution (d) variant without shared endpoints. (Color figure online)

Note that if we distribute clique points on the circle according to a cyclic order $\lhd$ of Lemma 1, then a representation $\mathcal{R}$ of G can be obtained by choosing for each vertex v an arc $R(v)$ that covers exactly the clique-points v belongs to.

3   Complexity

Theorem 1. *The problems* REPEXT(HCAR), REPEXT(CAR) *are* NP-*complete.* REPEXT(CAR) *is* NP-*complete even if the predrawn arcs have pairwise distinct endpoints.*

Proof (sketch). By predrawing k universal vertices as in Fig. 3a in an instance for REPEXT(HCAR) or REPEXT(CAR), we can require all other arcs to contain at least one of k points; see the *green dots* in Fig. 3a. We use this to reduce the strongly NP-complete problem 3-PARTITION [11] to REPEXT(HCAR). Let S be an instance of 3-PARTITION, i.e., a set $S = \{s_1, \ldots, s_{3n}\}$ of positive integers with a total sum of nt for some $t \in \mathbb{N}$. We may assume $t/4 < s_i < t/2$ for each $s_i \in S$. The question is whether S can be partitioned into n disjoint triples whose sum is always t. For the reduction, we generate $(t + 1)n$ green dots and predraw a *blocker vertex* that is only adjacent to the universal vertices at every $(t+1)$-th green dot. For every number $s_i \in S$ we add a star with s_i leaves to the input graph G, see Fig. 3b (each star is only adjacent to the predrawn universal vertices). Note that in every solution distinct leaf arcs must contain distinct green dots and the green dots covered by the leaves of a star are consecutive. The arc-segments between the blocker vertices thus correspond to the partition sets for S, see Fig. 3c. The restriction on the s_i ensures that every partition set contains exactly three elements.

For REPEXT(CAR), the same construction works. In fact, we can avoid shared endpoints in the partial representation with a simple modification. Namely, we slightly shorten the arc for each universal vertex; see Fig. 3d. Then we have for each former green dot a green area (on the circle) between the corresponding arc ends. Each leaf of a star must now contain a green area. Note that in every solution each leaf of a star in G contains exactly one green area and thus violates the Helly property with the two universal arcs ending there. ☐

For unit circular-arc representations we also reduce from 3-PARTITION. Here we make use of the limited space between predrawn arcs and the fact that paths of certain length cover at least a certain portion of the circle.

Theorem 2 ($\star$). *The problem* REPEXT(UCAR) *is* NP-*complete*.

4 (Normal) Proper Helly Circular-Arc Graphs

We show how to extend partial representations of NPHCA graphs in linear time. To this end, we use that the extension is essentially unique. An instance of REPEXT(NPHCA) where G is not connected can be reduced to REPEXT for proper interval graphs (PINT) by opening the circle between predrawn arcs of distinct components, or it can be reduced to REPEXT(NPHCA) with a connected input graph if only one component has predrawn vertices. REPEXT(PINT) can be solved in linear time [20]. For details see the full version.

Two vertices $u, v \in V(G)$ are called *twins* if $N[u] = N[v]$. It is possible to find the equivalence classes of twin vertices in linear time [31]. If vertices u, v are twins and if either u is not predrawn or if both u, v are predrawn with the same arc, then we remove u and in the final representation we just set $R(u) = R(v)$. We thus assume that all twins are predrawn with distinct arcs.

We use NPHCA *models* as combinatorial abstractions of NPHCA representations. Namely, an NPHCA model is a cyclic order of the heads and tails in an NPHCA representation $\mathcal{R}$ with the special case, that for any pair of touching arcs $R(u)$, $R(v)$ with $R(u)_h = R(v)_t$ we let $R(u)_h$ immediately follow $R(v)_t$. We use the following result of Lin et al. [26].

Lemma 2. *The* NPHCA *model of a connected* NPHCA *graph is unique up to permuting twins and reversal. It can be obtained in linear time.*

Theorem 3. *The problem* REPEXT(NPHCAR) *can be solved in linear time.*

Proof. We solve REPEXT(NPHCAR) as follows. Recall that we can assume that all twins are predrawn. By Lemma 2 we can compute an NPHCA model $\mathcal{M}$ of the input graph G (we just try both reversal decisions) and we reject if G is not NPHCA. Then permute twins to match their order in the partial representation $\mathcal{R}'$ (including arranging them at the corresponding end if they touch another predrawn arc).

Note that this is simple since the heads (tails) of a twin class are consecutive. Next test whether the model for $\mathcal{R}'$ is contained in $\mathcal{M}$ to ensure that $\mathcal{R}'$ extends to an NPHCA representation with model $\mathcal{M}$. Additionally, check for each pair of vertices u, v with $R(u)_h = R(v)_t$ that $R(u)_h$ and $R(v)_t$ are consecutive in $\mathcal{M}$. Otherwise, they cannot be placed on the same point and we can reject. Note that this wont occur for twin vertices. Each step can be done in linear time. ☐

For REPEXT(PHCAR), note that even though all proper Helly circular-arc graphs allow a normal proper representation, we cannot reduce

REPEXT(PHCAR) to REPEXT(NPHCAR) directly, as the given partial representation need not be normal. However, the existence of a proper Helly representation extending a pair of arcs in a non-normal position imposes strong conditions on the structure of G: after pruning all universal vertices only two disjoint cliques remain. Such instances can be solved in linear time. For details the full version.

Theorem 4 ($\star$). *The problem* REPEXT(*PHCAR*) *can be solved in linear time.*

5   (Normal) Helly Circular-Arc Graphs

Recall that in the case of interval graphs, PQ-trees can be used to capture all plausible linear orders of the maximal cliques. Klavík et al. [23] use this to solve REPEXT(INT) by determining an order that is represented by the PQ-tree and that extends a partial order that is derived from the partial representation. We generalize their approach to REPEXT(NHCAR) and REPEXT(HCAR).

By Lemma 1 a graph G is a Helly circular-arc graph if and only if there exists a *cyclic order* $\lhd$ of its maximal cliques such that for every vertex v, the maximal cliques containing v appear consecutively in $\lhd$ and Hsu and McConnell [16] use PC-trees to capture all plausible cyclic orders of the maximal cliques of a Helly circular-arc graph. However, this cannot be straightforwardly applied since extending a partial cyclic order is NP-complete, even without requiring that the order be additionally represented by some given PC-tree [10]. We overcome this problem by working with suitably linearized partial orders.

With the following lemma, we can solve instances of REPEXT(NHCAR) with universal vertices since they can be considered as instances of the interval case.

Lemma 3. *Let G be a graph with a universal vertex u. Then every NHCA representation of G has a point on the circle that is contained in no arc.*

Proof. Let $\mathcal{R}$ be an NHCA representation of G. Assume that every point of the circle is contained in some arc. Without loss of generality, we may assume that $R(u)$ is not strictly contained in any other arc of $\mathcal{R}$. We consider the complement $R^c(u)$ of $R(u)$. Let v_t, v_h be vertices whose arcs contain $R^c(u)_t$, $R^c(u)_h$ respectively, and whose arcs maximize the intersection with $R^c(u)$.

Note that, since $\mathcal{R}$ is normal, neither $R(v_t)$ nor $R(v_h)$ contain $R^c(u)$. Assume that one of $R(v_t)$, $R(v_h)$, say $R(v_t)$ contains both endpoints of $R(u)$. We then have $R(u) \subseteq R(v_t)$ which by the maximality of $R(u)$ implies $R(v_t) = R(u)$. By the choice of v_t, we find in this case a point close to $R(u)_t$ that is not contained in any arc. In the other case, each of $R(v_t)$ and $R(v_h)$ contains exactly one endpoint of $R(u)$. In particular, we have $v_t \neq v_h$. Since u is universal and since every point of $R^c(u)$ is contained in some arc, it follows that $R^c(u) \subseteq R(v_t) \cup R(v_h)$. We obtain $R(v_t) \cap R(v_h) \cap R^c(u) \neq \emptyset$ (recall that our arcs are closed sets). Due to the normal property, it follows that $R(v_t) \cap R(v_h) \cap R(u) = \emptyset$, which contradicts the Helly property. We conclude that there exists a point p that is not contained in any arc.   □

We assume for the rest of this section that G contains no universal vertices.

Lemma 4. *A graph G without a universal vertex is an NHCA graph if and only if there exists a cyclic order $\lhd$ of its maximal cliques such that*

(i) for every vertex v, the maximal cliques containing v are consecutive in $\lhd$.
(ii) for every universal pair u, w, the maximal cliques containing both u and w are consecutive in $\lhd$.

Proof. If G has an NHCA representation, then, due to the Helly property, we have for each maximal clique C a clique point $\mathrm{cp}(C)$ where the arcs of all vertices in C intersect. Let $\lhd$ be the cyclic order of the maximal cliques that corresponds to the cyclic order of their clique points on the circle. For each vertex v and for each universal pair u, w the corresponding cliques are consecutive in $\lhd$ since $R(v)$ and $R(u) \cap R(w)$ each are connected (the latter due to the normal property).

Next assume that we have a cyclic order $\lhd$ of the maximal cliques of G with properties (i) and (ii). We obtain an NHCA representation as follows. We first arrange the maximal clique points on the circle according to $\lhd$. Then, for each vertex v, we define the arc $R(v)$ as the smallest subarc of the circle that contains exactly the clique points of the maximal cliques that contain v. Note that $R(v)$ is well-defined since v is not universal. This defines a circular-arc representation $\mathcal{R}$ of G since any two intersecting arcs share a clique point. Moreover, the existence of the clique points shows the Helly property. It remains to show that $\mathcal{R}$ is normal.

Assume that there are two nodes u, w such that $R(u) \cap R(w)$ is not connected. Then u, w is a universal pair. Thus the cliques containing u and w are consecutive in $\lhd$ which contradicts $R(u) \cap R(w)$ being not connected. Hence, the representation is normal.   □

Let $\mathcal{R}'$ be a partial representation of G, let $\mathcal{C}$ be the set of all its maximal cliques and let $C \in \mathcal{C}$. We define $\mathrm{Pre}(C) = \{R'(v) : v \in C \cap V(G')\}$ to be the *predrawn arcs* corresponding to the vertices in C. We further define $\mathrm{Reg}^+(C) = \bigcap \mathrm{Pre}(C)$ to be the set of all locations covered by all predrawn arcs of vertices in C and $\mathrm{Reg}^-(C) = \bigcup(\mathcal{R}' \setminus \mathrm{Pre}(C))$ to be the set of all locations covered by any predrawn arc of a vertex not in C. The *region* of C is the set $\mathrm{Reg}(C) = \mathrm{Reg}^+(C) \setminus \mathrm{Reg}^-(C)$; see Fig. 4. Note that $\mathrm{cp}(C)$ lies in $\mathrm{Reg}(C)$ in any extension. We thus assume in the following that no region is empty.

Lemma 5. *For $C, D \in \mathcal{C}$, we have $\mathrm{Reg}(C) \cap \mathrm{Reg}(D) = \emptyset$ or $\mathrm{Reg}(C) = \mathrm{Reg}(D)$.*

Proof. If $\mathrm{Pre}(C) = \mathrm{Pre}(D)$, then clearly $\mathrm{Reg}(C) = \mathrm{Reg}(D)$. So, assume that $\mathrm{Pre}(C) \neq \mathrm{Pre}(D)$ with an arc $R' \in \mathrm{Pre}(C) \setminus \mathrm{Pre}(D)$. Then $\mathrm{Reg}(C) \subseteq \mathrm{Reg}^+(C) \subseteq R' \subseteq \mathrm{Reg}^-(D)$ which is disjoint to $\mathrm{Reg}(D)$.   □

Let C be a maximal clique. An *island* of C is a connected component of $\mathrm{Reg}(C)$ and a *gap* of C is a connected component of its complement $\mathrm{Reg}^c(C)$. We say an island and a gap are *neighboring* if they share an endpoint (where one end is open and the other is closed). Note that every island has two neighboring gaps and every gap has two neighboring islands.

Observe that if two maximal cliques $C, D \in \mathcal{C}$ satisfy $\mathrm{Pre}(C) = \mathrm{Pre}(D)$ then $\mathrm{Reg}(C) = \mathrm{Reg}(D)$ by definition. Otherwise, we obtain the following relationship:

Lemma 6. *Let $C, D \in \mathcal{C}$ with $\mathrm{Pre}(C) \neq \mathrm{Pre}(D)$. Then a single gap of D contains $\mathrm{Reg}(C)$.*

Proof. By Lemma 5 we have $\mathrm{Reg}(C) \subseteq \mathrm{Reg}^c(D)$. Assume that there are two gaps J_1, J_2 of D that contain points $j_1, j_2 \in \mathrm{Reg}(C)$. Let I_1, I_2 be the islands of D neighboring J_1. Let $i_1 \in I_1$, $i_2 \in I_2$. For $R'(v) \in \mathrm{Pre}(C)$ it holds that $j_1, j_2 \in R'(v)$ and thus $i_1 \in R'(v)$ or $i_2 \in R'(v)$ since i_1, i_2 separate j_1, j_2 on the circle. This implies $\mathrm{Pre}(C) \subseteq \{R'(v) : i_1 \in R'(v) \vee i_2 \in R'(v)\} = \{R'(v) : i_1 \in R'(v)\} \cup \{R'(v) \in V(G') : i_2 \in R'(v)\} = \mathrm{Pre}(D) \cup \mathrm{Pre}(D) = \mathrm{Pre}(D)$. Likewise, we obtain $\mathrm{Pre}(D) \subseteq \mathrm{Pre}(C)$, contradicting $\mathrm{Pre}(C) \neq \mathrm{Pre}(D)$. Thus, $\mathrm{Reg}(C)$ is contained in a single gap of D. $\square$

This means that for $C, D \in \mathcal{C}$ the clique point $\mathrm{cp}(C)$ must be placed in a given gap of D. We obtain additional consecutivity constraints. For every gap J of a maximal clique of G, we define the set $S_J = \{C \in \mathcal{C} \mid \mathrm{Reg}(C) \cap J \neq \emptyset\} = \{C \in \mathcal{C} \mid \mathrm{Reg}(C) \subseteq J\}$. Recall that we assumed that no region is empty.

Lemma 7. *Let $\leq$ be the clique order derived from an NHCA extension of $\mathcal{R}'$ and let J be a gap of a maximal clique. Then S_J is consecutive in $\leq$.*

Proof. Direct consequence of Lemma 6 since all clique points of S_J must be placed in J and all other clique points must be placed in its complement J^c. $\square$

Lemma 8. *There exists a maximal clique D with a single island.*

Proof. Let D be a maximal clique with $|\mathrm{Pre}(D)|$ maximal. We claim that $\mathrm{Reg}^+(D)$ and $\mathrm{Reg}^-(D)$ are disjoint. Otherwise there would exist a point p in $\mathrm{Reg}^+(D) \cap \mathrm{Reg}^-(D)$ with $\mathrm{Pre}(D) \subsetneq \{R'(v) : p \in R'(v)\}$. This implies the existence of a (maximal) clique C in G with $\mathrm{Pre}(D) \subsetneq \mathrm{Pre}(C)$ in contradiction to the choice of D. Since $\mathcal{R}'$ is normal, this yields that $\mathrm{Reg}(D) = \mathrm{Reg}^+(D)$ is connected. In other words, D has a single island. $\square$

For the rest of the section, let D be a maximal clique with a single island. Let p_D be a point in the interior of $\mathrm{Reg}(D)$ (or the only point in $\mathrm{Reg}(D)$, if it is a single point) and let $\leq$ be the linear order of points on the circle obtained by starting at p_D. We consider a point p to be *on the left* of a point q if we have $p \leq q$. For two sets of points P, Q, we write $P < Q$ if all points in P are on the left of all points in Q.

We define a partial order $\prec$ on $\mathcal{C}$ with minimum D (i.e., $\forall C \in \mathcal{C} \backslash \{D\}: D \prec C$) and with $C \prec C'$ for any two other cliques C, C' with $\mathrm{Reg}(C) < \mathrm{Reg}(C')$. Note that every linearization of a clique order of an extension of $\mathcal{R}'$ starting with D extends $\prec$. For any vertex v let M_v denote the set of maximal cliques containing v. Recall that for any linear order $<$ the corresponding cyclic order is denoted by $[<]$. Note that a set $S \subseteq \mathcal{C} \setminus \{D\}$ is consecutive in $[<]$ if and only if it is consecutive in $<$. On the other hand, a set $D \in S \subseteq \mathcal{C}$ is consecutive in $[<]$ if and only if its complement is consecutive in $<$.

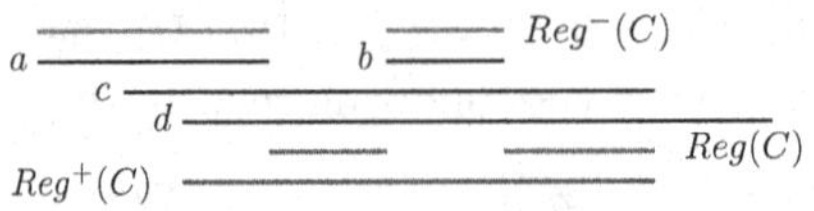

Fig. 4. $\mathrm{Reg}(C), \mathrm{Reg}^+(C)$ and $\mathrm{Reg}^-(C)$ for $c, d \in C$, $a, b \notin C$.

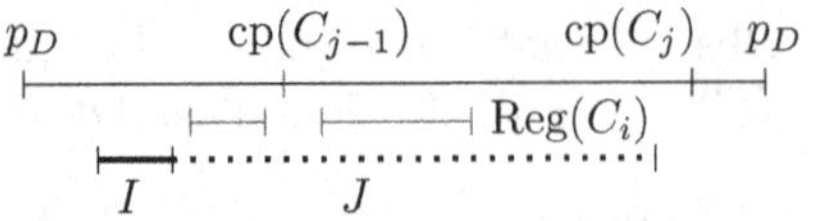

Fig. 5. Argument for the existence of a placement for $\mathrm{cp}(C_i)$.

Theorem 5. *Let G be a graph without universal vertices and let $\mathcal{R}'$ be a partial NHCA representation of G. There exists an NHCA representation of G that extends $\mathcal{R}'$ if and only if there exists a linear extension $<$ of $\prec$ such that*

1. *For any pair of distinct maximal cliques $C \neq C'$ with $\mathrm{Reg}(C) = \mathrm{Reg}(C')$, the region $\mathrm{Reg}(C)$ is not a single point.*
2. *For every vertex v, the set M_v is consecutive in $[<]$.*
3. *For every universal pair u, w, the set $M_u \cap M_w$ is consecutive in $[<]$.*
4. *For every gap J of some $C \in \mathcal{C}$, the set S_J is consecutive in $[<]$.*

Proof. We first show that, if there is an NHCA-extension of $\mathcal{R}'$, then these properties are satisfied. We obtain $<$ as the linearization of a clique order of an extension of $\mathcal{R}'$ starting with D where the clique point of D is p_D. By the construction of $\prec$, the order $<$ is a linear extension of $\prec$. By Lemmas 4 and 7, we obtain Properties 2, 3, 4. Property 1 is necessary, since no two clique points can be placed at the same point.

For the converse, let $<=C_1, \dots, C_k$ be a linear extension of $\prec$ such that Properties 1, 2, 3, 4 are satisfied. We show that each $C_i \in \mathcal{C}$ can be assigned its clique point $\mathrm{cp}(C_i) \in \mathrm{Reg}(C_i)$ such that $\mathrm{cp}(C_j) < \mathrm{cp}(C_i)$ whenever $j < i$. With Properties 2, 3, such a placement of the clique points yields an NHCA-extension of $\mathcal{R}'$ by representing every non-predrawn vertex $u \notin V(G')$ by the minimal arc $R(u)$ containing exactly the clique points of the maximal cliques from M_u as in the proof of Lemma 4.

Let $\varepsilon > 0$ be the $\frac{1}{2n+1}$-fraction of the length of the shortest island that is not a point. This choice of ε allows to place all clique points at distance at least ε but still within any island or within any side of p_D in island $\mathrm{Reg}(D)$. For $C_1 = D$ we place $\mathrm{cp}(C_1)$ on p_D. In a greedy way, when the location of the clique points $\mathrm{cp}(C_1), \dots, \mathrm{cp}(C_{i-1})$ is settled, we determine the set P of feasible points for $\mathrm{cp}(C_i)$ that is $P = \mathrm{Reg}(C_i) \cap \{p : p > \mathrm{cp}(C_{i-1}) + \varepsilon\}$. If P has minimum, we place $\mathrm{cp}(C_i)$ there, otherwise we put $\mathrm{cp}(C_i)$ at $\inf(P) + \varepsilon$. We argue that such choice always exists.

Assume for a contradiction that C_i is the first maximal clique in the order $<$ that cannot be placed, i.e., with $P = \emptyset$. Note that a clique $C \neq C_i$ can only have an island I_C consisting of a single point if we have $I_C = \mathrm{Reg}(C)$, since Reg^+ is a closed arc and all islands separated by gaps within Reg^+ have an open end. Hence, by the choice of ε, we only place a clique C at the maximum of $\mathrm{Reg}(C)$, if $\mathrm{Reg}(C)$ consists of a single point. With Property 1, this cannot happen if $\mathrm{Reg}(C) = \mathrm{Reg}(C_i)$. Therefore, one clique point must be placed to the right of $\mathrm{Reg}(C_i)$ before placing $\mathrm{cp}(C_i)$. We identify the first maximal clique $C_j, j < i$

that is placed to the right of all points in $\text{Reg}(C_i)$. Since $\text{cp}(C_j) \notin \text{Reg}(C_i)$, we have that $\text{Pre}(C_i) \neq \text{Pre}(C_j)$. By Lemma 6, the maximal clique C_j has a gap J with $\text{Reg}(C_i) \subseteq J$, see Fig. 5.

By definition of $<$, clique C_j has an island I that neighbors J on the left. Since $\text{cp}(C_j)$ was not placed on I, the clique point $\text{cp}(C_{j-1})$ has been placed to the right of I. By the choice of C_j, we have that $\text{cp}(C_{j-1})$ is not placed to the right of J and thus $\text{cp}(C_{j-1}) \in J$. Since $\text{cp}(C_j)$ has been placed to the right of $\text{Reg}(C_i)$ and thus to the right of J, we have $p_D \notin J$ and thus $D \notin S_J$. With $D < C_{j-1} < C_j < C_i$, where $C_{j-1}, C_i \in S_J$ and $D, C_j \notin S_J$ we get a contradiction with the Property 4, since S_J is not consecutive in $[<]$. $\square$

By a result of Klavík et al. [23] we can efficiently find a linear order that extends a given partial order and satisfies a set of consecutivity constraints. This yields the following theorem; see the full version for details.

Theorem 6. *The problem* REPEXT(NHCAR) *can be solved in* $\mathcal{O}(n^3)$ *time.*

In light of the hardness result from Theorem 1, it is unlikely that this can be generalized to REPEXT(HCAR). However, the hardness proof crucially relies on predrawn arcs sharing endpoints. If all predrawn arcs have distinct endpoints, the problem can be solved in a similar fashion. For details see the full version.

Theorem 7 ($\star$). *The problem* REPEXT(HCAR) *can be solved in* $\mathcal{O}(n^3)$ *time if the partial representation consists of arcs with pairwise distinct endpoints.*

Proof (sketch). We characterize extendable HCA instances similar as in Theorem 5 with Lemma 1 instead of Lemma 4. Since the predrawn arcs have pairwise distinct endpoints, we have no islands consisting of a single point. Note that Lemma 8 no longer applies and thus the placement of p_D cannot be chosen freely. Instead, we choose an arbitrary clique as C_1 and test for each island I of C_1, whether we find a representation extension by choosing $\text{cp}(C_1) \in I$. In contrast to our method for REPEXT(NHCAR) we have no special procedure for universal vertices. $\square$

6 Conclusions and Open Problems

Our study of the REPEXT problem has been restricted in two ways: First, we have considered mostly representations satisfying the Helly property as this allows us to consider the clique points of maximal cliques. For representations without this property one would use a completely different approach.

Secondly, for the recognition problem it is irrelevant whether the arcs are closed or open, but this is not the case for the representation extension. Observe that touching intervals in REPEXT(NPHCAR) imply constraints on the order. For the sake of completeness it might be worth to check whether use of open or semi-open intervals would significantly impact the computational complexity.

Acknowledgement. We thank Bartosz Walczak for inspiring comments, in particular for his hint to extend Theorem 1 to the case of CAR with distinct endpoints.

References

1. Angelini, P., et al.: Testing planarity of partially embedded graphs. ACM Trans. Algorithms **11**(4), 32:1–32:42 (2015)
2. Bläsius, T., Rutter, I.: Simultaneous PQ-ordering with applications to constrained embedding problems. ACM Trans. Algorithms (TALG) **12**(2), 1–46 (2015)
3. Bok, J., Jedličková, N.: A note on simultaneous representation problem for interval and circular-arc graphs. arXiv preprint arXiv:1811.04062 (2018)
4. Booth, K.S., Lueker, G.S.: Testing for the consecutive ones property, interval graphs, and graph planarity using PQ-tree algorithms. J. Comput. Syst. Sci. **13**(3), 335–379 (1976)
5. Chaplick, S., Dorbec, P., Kratochvíl, J., Montassier, M., Stacho, J.: Contact representations of planar graphs: extending a partial representation is hard. In: Kratsch, D., Todinca, I. (eds.) WG 2014. LNCS, vol. 8747, pp. 139–151. Springer, Cham (2014). https://doi.org/10.1007/978-3-319-12340-0_12
6. Chaplick, S., Fulek, R., Klavík, P.: Extending partial representations of circle graphs. J. Graph Theory **91**(4), 365–394 (2019)
7. Chaplick, S., Kindermann, P., Klawitter, J., Rutter, I., Wolff, A.: Extending partial representations of rectangular duals with given contact orientations. arXiv preprint arXiv:2102.02013 (2021)
8. Deng, X., Hell, P., Huang, J.: Linear-time representation algorithms for proper circular-arc graphs and proper interval graphs. SIAM J. Comput. **25**(2), 390–403 (1996)
9. Fulkerson, D.R., Gross, O.A.: Incidence matrices and interval graphs. Pac. J. Math. **15**, 835–855 (1965)
10. Galil, Z., Meggido, N.: Cyclic ordering is NP-complete. Theoret. Comput. Sci. **5**, 179–182 (1977)
11. Garey, M.R., Johnson, D.S.: Complexity results for multiprocessor scheduling under resource constraints. SIAM J. Comput. **4**(4), 397–411 (1975)
12. Garey, M.R., Johnson, D.S., Miller, G.L., Papadimitriou, C.H.: The complexity of coloring circular arcs and chords. SIAM J. Algebraic Discret. Methods **1**(2), 216–227 (1980)
13. Gavril, F.: Algorithms on circular-arc graphs. Networks **4**(4), 357–369 (1974)
14. Gavril, F.: The intersection graphs of subtrees in trees are exactly the chordal graphs. J. Comb. Theory Ser. B **16**(1), 47–56 (1974)
15. Hsu, W.L.: Maximum weight clique algorithms for circular-arc graphs and circle graphs. SIAM J. Comput. **14**(1), 224–231 (1985)
16. Hsu, W.L., McConnell, R.M.: PC trees and circular-ones arrangements. Theoret. Comput. Sci. **296**(1), 99–116 (2003)
17. Jelínek, V., Kratochvíl, J., Rutter, I.: A Kuratowski-type theorem for planarity of partially embedded graphs. Comput. Geom. **46**(4), 466–492 (2013)
18. Joeris, B.L., Lin, M.C., McConnell, R.M., Spinrad, J.P., Szwarcfiter, J.L.: Linear-time recognition of Helly circular-arc models and graphs. Algorithmica **59**(2), 215–239 (2011)
19. Klavík, P., Kratochvíl, J., Krawczyk, T., Walczak, B.: Extending partial representations of function graphs and permutation graphs. In: Epstein, L., Ferragina, P. (eds.) ESA 2012. LNCS, vol. 7501, pp. 671–682. Springer, Heidelberg (2012). https://doi.org/10.1007/978-3-642-33090-2_58
20. Klavík, P., Kratochvíl, J., Otachi, Y., Rutter, I., Saitoh, T., Saumell, M., Vyskočil, T.: Extending partial representations of proper and unit interval graphs. Algorithmica **77**(4), 1071–1104 (2017)

21. Klavík, P., et al.: Extending partial representations of proper and unit interval graphs. CoRR abs/1207.6960 (2012). https://arxiv.org/abs/1207.6960

22. Klavík, P., Kratochvíl, J., Otachi, Y., Saitoh, T.: Extending partial representations of subclasses of chordal graphs. Theoret. Comput. Sci. **576**, 85–101 (2015)

23. Klavík, P., Kratochvíl, J., Otachi, Y., Saitoh, T., Vyskočil, T.: Extending partial representations of interval graphs. Algorithmica **78**(3), 945–967 (2017)

24. Krawczyk, T., Walczak, B.: Extending partial representations of trapezoid graphs. In: Bodlaender, H.L., Woeginger, G.J. (eds.) WG 2017. LNCS, vol. 10520, pp. 358–371. Springer, Cham (2017). https://doi.org/10.1007/978-3-319-68705-6_27

25. Lin, M.C., Soulignac, F.J., Szwarcfiter, J.L.: Proper Helly circular-arc graphs. In: Brandstädt, A., Kratsch, D., Müller, H. (eds.) WG 2007. LNCS, vol. 4769, pp. 248–257. Springer, Heidelberg (2007). https://doi.org/10.1007/978-3-540-74839-7_24

26. Lin, M.C., Soulignac, F.J., Szwarcfiter, J.L.: Normal Helly circular-arc graphs and its subclasses. Discret. Appl. Math. **161**(7–8), 1037–1059 (2013)

27. Lin, M.C., Szwarcfiter, J.L.: Characterizations and linear time recognition of Helly circular-arc graphs. In: Chen, D.Z., Lee, D.T. (eds.) COCOON 2006. LNCS, vol. 4112, pp. 73–82. Springer, Heidelberg (2006). https://doi.org/10.1007/11809678_10

28. Lin, M.C., Szwarcfiter, J.L.: Characterizations and recognition of circular-arc graphs and subclasses: a survey. Discret. Math. **309**(18), 5618–5635 (2009). Combinatorics 2006, A Meeting in Celebration of Pavol Hell's 60th Birthday (1–5 May 2006)

29. McConnell, R.M.: Linear-time recognition of circular-arc graphs. Algorithmica **37**(2), 93–147 (2003)

30. Patrignani, M.: On extending a partial straight-line drawing. Int. J. Found. Comput. Sci. **17**(5), 1061–1070 (2006)

31. Rose, D.J., Tarjan, R.E., Lueker, G.S.: Algorithmic aspects of vertex elimination on graphs. SIAM J. Comput. **5**(2), 266–283 (1976)

32. Tucker, A.: Matrix characterizations of circular-arc graphs. Pac. J. Math. **39**, 535–545 (1971)

33. Tucker, A.: Structure theorems for some circular-arc graphs. Discret. Math. **7**(1–2), 167–195 (1974)

34. Tucker, A.: An efficient test for circular-arc graphs. SIAM J. Comput. **9**(1), 1–24 (1980)

Bounding Threshold Dimension: Realizing Graphic Boolean Functions as the AND of Majority Gates

Mathew C. Francis[1] [iD], Atrayee Majumder[2(✉)] [iD], and Rogers Mathew[3] [iD]

[1] Indian Statistical Institute, Chennai, India
mathew@isichennai.res.in
[2] Department of Computer Science and Engineering, Indian Institute of Technology, Kharagpur, India
atrayee.majumder@iitkgp.ac.in
[3] Department of Computer Science and Engineering, Indian Institute of Technology, Hyderabad, India
rogers@cse.iith.ac.in

Abstract. A graph G on n vertices is a *threshold graph* if there exist real numbers $a_1, a_2, \ldots, a_n$ and b such that the zero-one solutions of the linear inequality $\sum_{i=1}^{n} a_i x_i \leq b$ are the characteristic vectors of the cliques of G. Introduced in [Aggregation of inequalities in integer programming. Chvátal and Hammer, Annals of Discrete Mathematics, 1977], the *threshold dimension* of a graph G, denoted by $\dim_{\mathrm{TH}}(G)$, is the minimum number of threshold graphs whose intersection yields G. Given a graph G on n vertices, in line with Chvátal and Hammer, $f_G \colon \{0,1\}^n \to \{0,1\}$ is the Boolean function that has the property that $f_G(x) = 1$ if and only if x is the characteristic vector of a clique in G. A Boolean function f for which there exists a graph G such that $f = f_G$ is called a *graphic* Boolean function. It follows that for a graph G, $\dim_{\mathrm{TH}}(G)$ is precisely the minimum number of *majority* gates whose AND (or conjunction) realizes the graphic Boolean function f_G. The fact that there exist Boolean functions which can be realized as the AND of only exponentially many majority gates motivates us to study threshold dimension of graphs. We give tight or nearly tight upper bounds for the threshold dimension of a graph in terms of its treewidth, maximum degree, degeneracy, number of vertices, size of a minimum vertex cover, etc. We also study threshold dimension of random graphs and graphs with high girth.

Keywords: Intersection dimension · Threshold dimension · Boxicity · Threshold graphs · Graphic Boolean function · Majority gates · Depth-2 circuits · Treewidth

1 Introduction

All the graphs that are mentioned in this paper are finite, simple, and undirected. Given a graph $G = (V, E)$, we shall use $V(G)$ and $E(G)$ to denote the vertex

© Springer Nature Switzerland AG 2022
M. A. Bekos and M. Kaufmann (Eds.): WG 2022, LNCS 13453, pp. 244–256, 2022.
https://doi.org/10.1007/978-3-031-15914-5_18

set and edge set of G, respectively. For any $v \in V(G)$, we use $N_G(v)$ to denote the neighborhood of v in G, i.e., $N_G(v) = \{u \in V(G): vu \in E(G)\}$. We use $N_G[v]$ to denote $N_G(v) \cup \{v\}$. For any $S \subseteq V(G)$, we shall use $G[S]$ to denote the subgraph induced by the vertex set S in G. We use $G - S$ to denote the graph $G[V(G) \setminus S]$. A subset of vertices in a graph forms a *clique* if each pair of vertices in this subset has an edge between them; if no pair of vertices have an edge between them, then the subset is called an *independent set*.

Graphic Boolean Functions. Given a graph G on n vertices, we define the Boolean function $f_G: \{0,1\}^n \to \{0,1\}$ as follows: $\forall x \in \{0,1\}^n$, $f_G(x) = 1$ if and only if x is the characteristic vector of a clique in G. A Boolean function f such that there exists a graph G for which $f_G = f$ is called a *graphic* Boolean function. Graphic Boolean functions were defined by Chvátal and Hammer [7] (they defined the Boolean function corresponding to a graph G to be the function whose solutions are exactly the characteristic vectors of the independent sets of G; it is easy to see that this is the function $f_{\overline{G}}$ and hence this definition and the one that we gave above for graphic Boolean functions are equivalent). Below, we give a characterization of graphic Boolean functions due to Hammer and Mahadev [13].

Proposition 1 (Hammer and Mahadev [13] ($\star^1$)). *A Boolean function on n variables $x_1, x_2, \ldots, x_n$ is graphic if and only if it can be written in conjunctive normal form where each clause is of the form $(\overline{x_i} \vee \overline{x_j})$, for some distinct $i, j \in [n]$.*

Majority Gates and LTFs. A *majority gate* is a logic gate that produces an output of 1 if and only if at least half of its input bits are 1. It can be easily seen that an AND or OR gate can be realized using a majority gate by the addition of a suitable number of hardcoded input bits. A Boolean function $f: \{0,1\}^n \to \{0,1\}$ is called a *Linear Threshold Function* (LTF) if there exists a linear inequality $I: \sum_{i=1}^{n} a_i x_i \leq b$ on variables $x_1, x_2, \ldots, x_n$ such that $\forall x = (x_1, x_2, \ldots, x_n) \in \{0,1\}^n$, $f(x) = 1$ if and only if x satisfies I. We say that the linear inequality I "represents" f. It is well known that every LTF can be represented by a linear inequality in which the coefficients $a_1, a_2, \ldots, a_n, b$ are integers (from here onward, a linear inequality representing an LTF shall be implicitly assumed to have integer coefficients). This implies the well known fact that every LTF can be realized using a majority gate by wire duplication. Conversely, it is easy to see that any Boolean function that can be realized using a majority gate is an LTF.

Threshold Graphs. A graph G on n vertices is a *threshold graph* if there exist real numbers $a_1, a_2, \ldots, a_n$ and b such that the zero-one solutions of the linear inequality $\sum_{i=1}^{n} a_i x_i \leq b$ are the characteristic vectors of the cliques of G.

---

[1] The proofs of the statements marked with ($\star$) are not included in the paper due to space constraints. Please refer to [11] for these proofs.

This implies that G is a threshold graph if and only if f_G is an LTF. Since LTFs are exactly the Boolean functions that can be realized using a majority gate, we can equivalently say that a graph G is a threshold graph if and only if f_G can be realized using a majority gate. Chvátal and Hammer [7] showed that threshold graphs are exactly the graphs that contain no induced subgraph isomorphic to $2K_2$, P_4 or C_4 (the graph with four vertices and two disjoint edges, the path on four vertices and the cycle on four vertices respectively). Thus, the complement of a threshold graph is also a threshold graph, implying that one can replace 'cliques' with 'independent sets' in the definition of a threshold graph. The complete graph on n vertices is a threshold graph with the corresponding linear inequality being $\sum_{i=1}^{n} x_i \leq n$. Similarly, the star graph $K_{1,n-1}$ is a threshold graph, as shown by the linear inequality $x_1 + \sum_{i=2}^{n}(n-1)x_i \leq n$. For a graph G, the characteristic vectors of the subsets of $V(G)$ correspond to the corners of the n-dimensional hypercube. Thus, a graph G is threshold if and only if there is a hyperplane in $\mathbb{R}^n$ that separates the corners of the n-dimensional hypercube that correspond to the cliques of G from the other corners of the hypercube. Threshold graphs, which find applications in integer programming and set packing problems, were introduced by Chvátal and Hammer [7]. Refer to the book [12] by Golumbic to know more about the different properties of threshold graphs. A more comprehensive study of threshold graphs can be found in the book [17] by Mahadev and Peled.

The following equivalent characterization of threshold graphs (Corollary 1B in [7]) will be useful for us.

Proposition 2 (Chvátal and Hammer [7]). *G is a threshold graph if and only if there is a partition of $V(G)$ into an independent set A and a clique B, and an ordering $u_1, u_2, \ldots, u_k$ of A such that $N_G(u_k) \subseteq N_G(u_{k-1}) \subseteq \cdots \subseteq N_G(u_1)$.*

Threshold Dimension. If $G_1, G_2, \ldots, G_k$ are graphs on the same vertex set as G such that $E(G) = E(G_1) \cap E(G_2) \cap \cdots \cap E(G_k)$, then we say that $G = G_1 \cap G_2 \cap \cdots \cap G_k$. In a similar way, if $E(G) = E(G_1) \cup E(G_2) \cup \cdots \cup E(G_k)$, then we say that $G = G_1 \cup G_2 \cup \cdots \cup G_k$. Given a class $\mathcal{A}$ of graphs, Kratochvíl and Tuza [16] defined the $\mathcal{A}$-*dimension* of a graph G, denoted as $\dim_{\mathcal{A}}(G)$, to be the minimum integer k such that there exist k graphs in $\mathcal{A}$ whose intersection is G. Let TH denote the class of threshold graphs. Chacko and Francis [4] studied the parameter $\dim_{\mathrm{TH}}(G)$ of a graph G, which in the language of [16], can be called the *threshold dimension* of G.

Definition 1 (Threshold dimension). *The threshold dimension of a graph G, denoted by $\dim_{\mathrm{TH}}(G)$, is the smallest integer k for which there exist threshold graphs $G_1, G_2, \ldots, G_k$ such that $G = G_1 \cap G_2 \cap \cdots \cap G_k$.*

Let $f \colon \{0,1\}^n \to \{0,1\}$ be a Boolean function. Let $\gamma(f)$ denote the minimum number of LTFs whose AND (or conjunction) realizes f, or equivalently, the minimum number of majority gates in a depth-2 circuit realizing f whose first layer consists of only majority gates and second layer consists of a single output AND gate. Chvátal and Hammer proved the following theorem connecting the parameters $\gamma(f_G)$ and $\dim_{TH}(G)$ for a graph G.

Theorem 1 (Chvátal and Hammer [7] ($\star$)). *For every graph G, $\gamma(f_G) = \dim_{TH}(G)$.*

For any Boolean function f on n variables, $\gamma(f) \leq 2^n$ (since any Boolean function on n variables can be realized using a depth-2 circuit in which the first layer contains at most 2^n OR gates and the second layer contains an AND gate—which is just another way of saying that f can be written in conjunctive normal form), and there are families of Boolean functions $\{f^1, f^2, \ldots\}$, where f^i is a Boolean function on i variables, for which $\gamma(f^n)$ is exponential in n [18]. For a Boolean function f on n variables that can be expressed as a 2-CNF formula, the number of clauses in it is at most $\binom{2n}{2}$, which means that f can be realized using a depth-2 circuit containing at most $\binom{2n}{2}$ majority gates. If further, f is a graphic Boolean function, then the number of clauses when written in 2-CNF form is at most $\binom{n}{2}$ (by Proposition 1), implying that f can be realized using a depth-2 circuit containing at most $\binom{n}{2}$ majority gates. As for any graph G, we have $\gamma(f_G) = \dim_{TH}(G) \leq n$ (this can be seen as follows: for every vertex $u \in V(G)$, we define the graph G_u on vertex set $V(G)$ and having edge set $\{xy: x, y \in V(G) \setminus \{u\}$ and $x \neq y\} \cup \{uv: v \in N_G(u)\}$; then $G = \bigcap_{u \in V(G)} G_u$ and each G_u is a threshold graph), every graphic Boolean function on n variables can be realized using a depth-2 circuit whose first layer contains at most n majority gates. This can be improved further by deriving better upper bounds for threshold dimension (see for example, Corollary 4). Further, when the graphs corresponding to the graphic Boolean functions have some nice properties, we can show even better bounds on the number of majority gates required in a depth-2 circuit realizing the function.

Note that Chvátal and Hammer [7] use the term "threshold dimension" of a graph G with a slightly different meaning: they define it to be the minimum integer k for which there exist threshold graphs $G_1, G_2, \ldots, G_k$ such that $G = G_1 \cup G_2 \cup \cdots \cup G_k$. We call this the *threshold cover number* of G and denote it by $\mathrm{cov}_{TH}(G)$. Since the complement of a threshold graph is also a threshold graph, we have the following.

Observation 1. *For every graph G, $\mathrm{cov}_{TH}(G) = \dim_{TH}(\overline{G})$.*

For a graph G, let $\alpha(G)$, $\omega(G)$, and $\chi(G)$ denote the size of a maximum independent set, the size of a maximum clique, and the chromatic number of G, respectively. It was shown in [7] that for every graph G on n vertices, $\mathrm{cov}_{TH}(G) \leq n - \alpha(G)$. In the same paper, the authors also showed that for every positive ϵ, there is a graph G on n vertices such that $\mathrm{cov}_{TH}(G) > (1 - \epsilon)n$. Yannakakis [23] showed that it is NP-complete to recognize graphs having threshold cover number at most k, for all fixed $k \geq 3$. Raschle and Simon [20] showed that there is a polynomial time algorithm that recognizes graphs having threshold cover number at most 2. Combining Observation 1 with the results due in [7, 20, 23] mentioned above directly yields the following.

Corollary 1.

(a) *For every graph G on n vertices, $\dim_{TH}(G) \leq n - \omega(G)$, where $\omega(G)$ denotes the size of a largest clique in G.*

(b) *For every positive ϵ, there is a graph G on n vertices such that $\dim_{TH}(G) > (1 - \epsilon)n$.*

(c) *It is NP-complete to recognize graphs having threshold dimension at most k, for all fixed $k \geq 3$.*

(d) *There is a polynomial time algorithm that recognizes graphs having threshold dimension at most 2.*

We now give a lower bound on the threshold dimension of a graph.

Proposition 3 ($\star$). *For every graph G, $\dim_{TH}(G) \geq \min\{\chi(G - C): C$ is a clique of $G\}$.*

Note that the above proposition actually gives a lower bound on $\dim_{SPLIT}(G)$, where SPLIT is the class of "split graphs"—the graphs whose vertex set can be partitioned into an independent set and a clique—of which the class of threshold graphs is a subclass.

A graph is an *interval graph* if there is a mapping from the set of vertices of the graph to the set of closed intervals on the real line such that two vertices in the graph are adjacent to each other if and only if the intervals they are mapped to have a non-empty intersection. Let INT denote the class of interval graphs. The parameter $\dim_{INT}(G)$ is more commonly known as the *boxicity* of the graph G and denoted as $\text{box}(G)$. It is known that threshold graphs form a subclass of the class of interval graphs. This implies the following.

Observation 2. *For every graph G, $\text{box}(G) \leq \dim_{TH}(G)$.*

The graph parameter 'boxicity' was introduced by Roberts [21] in 1969 and, since then, it has been extensively studied (see [1,2,5,6,10,15]). We will see how Observation 2 helps us get tight examples to various bounds we prove for threshold dimension in this paper. Chacko and Francis [4] gave the following upper bound for the threshold dimension of a graph G in terms of its boxicity and chromatic number.

Theorem 2 (Theorem 19 in [4]). *For every graph G, $\dim_{TH}(G) \leq \text{box}(G) \cdot \chi(G)$.*

We note here that the above upper bound is tight, as shown by the following observation, which also shows that the threshold dimension of a graph cannot be bounded by any function of its boxicity.

Proposition 4 ($\star$). *There is an interval graph G for which $\dim_{TH}(G) = \chi(G) = |V(G)|/2$.*

In this paper, we prove tighter upper bounds for the threshold dimension of a graph that cannot be obtained from Theorem 2 by plugging in known upper bounds for boxicity.

1.1   Our Results

Let G be a graph with n vertices. Let Δ denote the maximum degree of a vertex in G and let tw(G) denote the treewidth of G. Let $\alpha(G)$ and $\omega(G)$ denote the sizes of a maximum independent set and a maximum clique, respectively, in G. We prove the following results.

1. Chandran and Sivadasan [6] showed that for any graph G, box$(G) \leq$ tw$(G) + 2$. Chacko and Francis [4] note that for any graph G, $\dim_{TH}(G) \leq$ $(\text{tw}(G) + 1)(\text{tw}(G) + 2)$ and ask if the threshold dimension of every graph can be bounded by a linear function of its treewidth. In Sect. 2, we answer this question in the affirmative by showing that $\dim_{TH}(G) \leq 2(\text{tw}(G) + 1)$. We show that this bound is tight up to a multiplicative factor of 2. Co-comparability graphs, AT-free graphs, and chordal graphs are known to have $O(\Delta)$ upper bounds on their treewidth. We thus get an $O(\Delta)$ upper bound to the threshold dimension of such graphs.

2. Let $\dim_{TH}(\Delta) := \max\{\dim_{TH}(G) \colon G \text{ is a graph having maximum degree } \Delta\}$. In Sect. 3, we show that $\dim_{TH}(\Delta) = O(\Delta \ln^{2+o(1)} \Delta)$. It was shown by Erdős, Kierstead, and Trotter in [9] that there exist graphs G having boxicity $\Omega(\Delta \ln \Delta)$. Using Observation 2, we get $\dim_{TH}(\Delta) = \Omega(\Delta \ln \Delta)$. Bridging the gap between the upper and lower bounds for $\dim_{TH}(\Delta)$ would be interesting. Since, by Theorem 1, $\dim_{TH}(G) = \gamma(f_G)$, it may be worthwhile to see if techniques from complexity theory could be used to bridge this gap.

3. Let G be k-degenerate. We show in Sect. 4 that $\dim_{TH}(G) \leq 10k \ln n$. It was shown in Sect. 3.1 in [2] that there exist k-degenerate graphs on n vertices with boxicity in $\Omega(k \ln n)$. Together with Observation 2, this implies that the upper bound for $\dim_{TH}(G)$ we prove in Sect. 4 is tight up to constants. This bound gives some interesting corollaries.
   (a) Let $G \in \mathcal{G}(n, m)$, where $m \geq n/2$. Then, asymptotically almost surely $\dim_{TH}(G) \in O(d_{av} \log n)$, where $d_{av} = \frac{2m}{n}$ denotes the average degree of G.
   (b) If G has a girth greater than $g + 1$, then $\dim_{TH}(G) = O(n^{\frac{1}{\lfloor g/2 \rfloor}} \ln n)$.

4. In Sect. 5, we show that the threshold dimension of any graph is upper bounded by its minimum vertex cover number, which implies that for any graph G, $\dim_{TH}(G) \leq n - \max\{\alpha(G), \omega(G)\}$. We show that this bound is tight. As a corollary we show that if n is sufficiently large, then $\dim_{TH}(G) \leq n - 0.72 \ln n$.

1.2   Preliminaries

Definition 2. *Given a graph G, an independent set $A = \{u_1, u_2, \ldots, u_t\}$ in G, and a total ordering $\sigma \colon u_1, u_2, \ldots, u_t$ of the vertices of A, we define the threshold supergraph $\tau(G, A, \sigma)$ of G as below. Let $B = V(G) \setminus A$ and for $v \in B$, let $s(v) = \max\{i \colon u_i \in N_G(v)\}$ if $N(v) \cap B \neq \emptyset$ and $s(v) = 0$ otherwise. In $\tau(G, A, \sigma)$, the vertices of A form an independent set and those of B form a clique and each vertex $v \in B$ is adjacent to exactly the vertices $u_1, u_2, \ldots, u_{s(v)}$. Formally,*

$$V(\tau(G, A, \sigma)) = V(G)$$

$$E(\tau(G, A, \sigma)) = E(G) \cup \{xy \colon x, y \in B \text{ and } x \neq y\} \cup \bigcup_{v \in B} \{vu_1, vu_2, \ldots, vu_{s(v)}\}$$

The following proposition follows directly from the above definition and Proposition 2.

Proposition 5. *Given a graph G, an independent set A of G, and an ordering σ of A, the graph $\tau(G, A, \sigma)$ is a threshold graph and G is its subgraph.*

2 Threshold Dimension and Treewidth

In this section, we show that for every graph G, we have $\dim_{\mathrm{TH}}(G) \leq 2(\mathrm{tw}(G) + 1)$, where $\mathrm{tw}(G)$ denotes the treewidth of G. We set up some notations and discuss some necessary existing results before going into the proof of the main result. The notion of treewidth was first introduced by Robertson and Seymour in [22].

Definition 3 (Tree decomposition). *A tree decomposition of a graph $G = (V, E)$ is a pair $(T, \{X_i \colon i \in V(T)\})$ where T is a tree and for each $i \in V(T)$, X_i is a subset of $V(G)$ (sometimes called a bag), such that the following conditions are satisfied:*

- $\bigcup_{i \in V(T)} X_i = V(G)$.
- $\forall uv \in E(G), \exists i \in V(T)$, *such that $u, v \in X_i$.*
- $\forall i, j, k \in V(T)$: *if j is on the path in T from i to k, then $X_i \cap X_k \subseteq X_j$.*

The width of a tree-decomposition $(T, \{X_i \colon i \in V(T)\})$ is $\max_{i \in V(T)} |X_i| - 1$.

Definition 4 (Treewidth). *The treewidth of a graph G, denoted by $\mathrm{tw}(G)$, is the minimum width over all possible tree decompositions of G.*

A tree decomposition $(T, \{X_i \colon i \in V(T)\})$ of a graph G is said to be a *path decomposition* of G if T is a path. The *pathwidth* of G, denoted by $\mathrm{pw}(G)$, is defined as the minimum width over all possible path decompositions of G. The following result by Chacko and Francis connects threshold dimension of a graph with its pathwidth.

Theorem 3 (Theorem 7 in [4]). *For every graph G, $\dim_{\mathrm{TH}}(G) \leq \mathrm{pw}(G) + 1$.*

Since path decompositions are special cases of tree decompositions, it can be seen that $\mathrm{tw}(G) \leq \mathrm{pw}(G)$. Korach and Solel showed that $\mathrm{pw}(G) = O(\log n \cdot \mathrm{tw}(G))$, where $n = |V(G)|$ (Theorem 6 in [14]). We thus have $\dim_{\mathrm{TH}}(G) = O(\log n \cdot \mathrm{tw}(G))$. Chacko and Francis note that for any graph G, $\dim_{\mathrm{TH}}(G) \leq (\mathrm{tw}(G) + 1)(\mathrm{tw}(G) + 2)$ and ask if there is a linear bound on the threshold dimension of a graph in terms of its treewidth. We give an affirmative answer to this question.

Given an ordering σ of the vertices of a graph G and $u, v \in V(G)$, we denote by $u <_\sigma v$ the fact that u appears before v in the ordering.

Let T be a rooted tree. For any $u, v \in V(T)$, u is an *ancestor* of v, and v a *descendant* of u, if u lies on the path from v to the root of T. It follows from this definition that every vertex of T is both an ancestor and descendant of itself. For a rooted tree T, a *preorder traversal* of T is an ordering of $V(T)$ in the order in which a depth-first search algorithm starting from the root may visits the vertices of T. The following is not difficult to see.

Proposition 6. *If π is a preorder traversal of a rooted tree T, then:*

(i) for $u, v \in V(T)$ such that v is a descendant of u, we have $u <_\pi v$, and
(ii) for $u, v, w \in V(T)$ such that $u <_\pi v <_\pi w$, if w is a descendant of u, then v is also a descendant of u.

Let G be a graph and $\mathcal{T} = (T, \{X_i : i \in V(T)\})$ be a tree decomposition of G having width k. We choose an arbitrary vertex r to be the root of T and henceforth consider T to be a rooted tree. Then a function $b : V(G) \to V(T)$ is defined as follows: for a vertex $v \in V(G)$, $b(v)$ is the bag containing v in the tree decomposition that is closest to r. Formally, $b(v)$ is the vertex of T such that $v \in X_{b(v)}$ and $v \notin X_i$ for any $i \in V(T)$ that is an ancestor of $b(v)$.

Lemma 1 (Lemma 10 in [6]). *If $uv \in E(G)$, then $b(u)$ is either an ancestor or descendant of $b(v)$ in T.*

Lemma 2 (Lemma 8 in [6]). *There exists a function $\theta : V(G) \to \{0, 1, \ldots, k\}$, such that for any $i \in V(T)$ and for any two distinct nodes $u, v \in X_i$, $\theta(u) \neq \theta(v)$.*

Remark. The function θ is a proper vertex colouring of the chordal graph G' that one obtains from G by adding edges between every pair of vertices that appear together in some bag of the tree decomposition. Clearly, $\mathcal{T}$ is a tree decomposition of G' as well. From the fact that every clique in G' has to be contained in some bag of $\mathcal{T}$, and the fact that chordal graphs are perfect, it follows that θ needs to use only $\max\{|X_i| : i \in V(T)\}$ different colours.

The following lemmas from [6] describe some properties of the functions θ and b that we will use later. These are direct corollaries of the definition of θ and that of tree decompositions.

Lemma 3 (Lemma 9 in [6]). *If $uv \in E(G)$ then $\theta(u) \neq \theta(v)$.*

Lemma 4 Lemma 11 in [6]). *Let $uv \in E(G)$ and let $b(u)$ be an ancestor of $b(v)$. For any vertex $w \in V(G) \setminus \{u\}$, $\theta(w) \neq \theta(u)$ if $b(w)$ is in the path from $b(v)$ to $b(u)$ in T.*

Let π be a preorder traversal of T. Let σ be an ordering of $V(G)$ such that for any two vertices $u, v \in V(G)$, $u <_\sigma v$ in σ if $b(u) <_\pi b(v)$. (In σ, we let the ordering between two vertices $u, v \in V(G)$ such that $b(u) = b(v)$ to be arbitrary. Thus, if $u <_\sigma v$, then $b(u) \leq_\pi b(v)$.) Let σ^{-1} denote the ordering of $V(G)$ obtained by reversing the ordering σ. Given a set $A \subseteq V(G)$, we denote by $\sigma|_A$ the ordering of vertices of A in the order in which they appear in σ.

For $i \in \{0, 1, \ldots, k\}$, we define $C_i = \{v \in V(G) : \theta(v) = i\}$. From Lemma 3, we know that θ is a proper colouring of G, which implies that C_i is an independent set of G. For each class C_i, where $0 \leq i \leq k$, we define two graphs $G_i^1 = \tau(G, C_i, \sigma|_{C_i})$ and $G_i^2 = \tau(G, C_i, \sigma^{-1}|_{C_i})$.

Lemma 5. *Let u, v be distinct vertices in G. Then there do not exist $x_u, y_u \in N_G(u)$ and $x_v, y_v \in N_G(v)$ such that $x_u <_\sigma v <_\sigma y_u$, $x_v <_\sigma u <_\sigma y_v$, $\theta(u) = \theta(x_v)$, and $\theta(v) = \theta(x_u)$.*

Proof. Clearly, we have either $u <_\sigma v$ or $v <_\sigma u$. Let us assume without loss of generality that $u <_\sigma v$. Then we have $u <_\sigma v <_\sigma y_u$, which implies that $b(u) \leq_\pi b(v) \leq_\pi b(y_u)$. Since $u y_u \in E(G)$, we have from Lemma 1 that $b(u)$ is either an ancestor or descendant of $b(y_u)$. As π is a preorder traversal of T, Proposition 6(i) implies that $b(u)$ is an ancestor of $b(y_u)$ in T. As $b(u) \leq_\pi b(v) \leq_\pi b(y_u)$, it now follows from Proposition 6(ii) that $b(v)$ is a descendant of $b(u)$. Similarly, $x_v <_\sigma u <_\sigma v$ implies that $b(x_v) \leq_\pi b(u) \leq_\pi b(v)$, and $v x_v \in E(G)$ then implies by Lemma 1, Proposition 6(i) and (ii) that $b(u)$ is a descendant of $b(x_v)$. Now applying Lemma 4 to x_v, u and v, we have that $\theta(x_v) \neq \theta(u)$, which is a contradiction. $\qquad\square$

Lemma 6. $G = \bigcap_{0 \leq i \leq k} (G_i^1 \cap G_i^2)$

Proof. Consider any two distinct vertices u and v of G. Since G_i^1 and G_i^2, for $1 \leq i \leq k$, are both supergraphs of G by definition, we have that if $uv \in E(G)$, then uv is an edge of both G_i^1 and G_i^2. So in order to prove the lemma, we only need to prove that whenever $uv \notin E(G)$, there exists $i \in \{0, 1, \dots, k\}$ and $j \in \{1, 2\}$ such that $uv \notin E(G_i^j)$.

Suppose $\theta(u) = \theta(v) = i$. Since the class C_i is an independent set in G_i^1 and G_i^2, uv is an edge in neither G_i^1 nor G_i^2, and we are done. So let us assume that $\theta(u) \neq \theta(v)$. Let $\theta(u) = i$ and $\theta(v) = j$. We claim that uv is not an edge in one of the graphs G_i^1, G_i^2, G_j^1, or G_j^2. Suppose for the sake of contradiction that $uv \in E(G_i^1) \cap E(G_i^2) \cap E(G_j^1) \cap E(G_j^2)$. Then uv is an edge in each of the graphs $\tau(G, C_i, \sigma|_{C_i})$, $\tau(G, C_i, \sigma^{-1}|_{C_i})$, $\tau(G, C_j, \sigma|_{C_j})$, $\tau(G, C_j, \sigma^{-1}|_{C_j})$. Since $uv \in E(\tau(G, C_i, \sigma|_{C_i}))$, by Definition 2, we have that there exists $y_v \in C_i \cap N_G(v)$ such that $u <_\sigma y_v$. Further, since $uv \in E(\tau(G, C_i, \sigma^{-1}|_{C_i}))$, there exists $x_v \in C_i \cap N_G(v)$ such that $u <_{\sigma^{-1}} x_v$, or in other words, $x_v <_\sigma u$. As $uv \in E(\tau(G, C_j, \sigma|_{C_j}))$ and $uv \in E(\tau(G, C_j, \sigma^{-1}|_{C_j}))$, we can similarly conclude that there exist $x_u, y_u \in C_j \cap N_G(u)$ such that $x_u <_\sigma v <_\sigma y_u$. Since $\theta(x_u) = \theta(v) = j$ and $\theta(x_v) = \theta(u) = i$, we now have a contradiction to Lemma 5. $\qquad\square$

From Proposition 5 and Definition 2, it follows that G_i^1 and G_i^2 are both threshold graphs for each $i \in \{0, 1, 2, \dots, k\}$. Thus by Lemma 6, we get that $\dim_{\mathrm{TH}}(G) \leq 2(k + 1)$, which leads to the following theorem.

Theorem 4. *For any graph G, $\dim_{\mathrm{TH}}(G) \leq 2(\mathrm{tw}(G) + 1)$.*

Tightness of the Bound. Note that from Proposition 4, we know that the graph $2K_n$ has threshold dimension n and it is easy to see that the treewidth of this graph is $n - 1$. Thus the upper bound on threshold dimension given by Theorem 4 is tight up to a multiplicative factor of 2. Please see Example 1 in [11] for the construction of another graph that shows the same tightness bound.

3   Threshold Dimension and Maximum Degree

Let $\dim_{TH}(\Delta) := \max\{\dim_{TH}(G)\colon G$ is a graph having maximum degree $\Delta\}$. In this section, we show that $\dim_{TH}(\Delta) = O(\Delta \ln^{2+o(1)} \Delta)$.

Given a graph G and an $S \subseteq V(G)$, recall that we use $G[S]$ to denote the subgraph induced by the vertex set S in G. For any disjoint pair of sets $S, T \subseteq V(G)$, we use $G[S, T]$ to denote the bipartite subgraph of G where $V(G[S, T]) = S \cup T$ and $E(G[S, T]) = \{uv\colon u \in S,\ v \in T,\ uv \in E(G)\}$. Let $G^*[S, T]$ denote the graph constructed from $G[S, T]$ by making T a clique. That is, $V(G^*[S, T]) = S \cup T$ and $E(G^*[S, T]) = E(G[S, T]) \cup \{uv\colon u, v \in T\}$.

Lemma 7 ($\star$). *Let G be a bipartite graph with bipartition $\{A, B\}$, where vertices in A have degree at most Δ and vertices in B have degree at most d, for some $2 \le d \le \Delta$. Then,*

$$\dim_{TH}(G^*[A, B]) \le (81 + o(1))d \ln (d\Delta) \ln \ln \Delta (2e)^{\sqrt{\ln d}},$$

when $d \to \infty$.

Theorem 5 ($\star$). *For a graph G with maximum degree Δ,*

$$\dim_{TH}(G) \le (24300 + o(1))\Delta \ln^2 \Delta \ln \ln \Delta (2e)^{\sqrt{(1+o(1)) \ln \ln \Delta}},$$

when $\Delta \to \infty$.

Since $(2e)^{\sqrt{(1+o(1)) \ln \ln \Delta}} \ln \ln \Delta = (\ln \Delta)^{\frac{\ln(2e)\sqrt{(1+o(1))\ln \ln \Delta}}{\ln \ln \Delta} + \frac{\ln \ln \ln \Delta}{\ln \ln \Delta}} = \ln^{o(1)} \Delta$ we get the following corollary.

Corollary 2.

$$\dim_{TH}(\Delta) \in O(\Delta \ln^{2+o(1)} \Delta).$$

4   Threshold Dimension and Degeneracy

Given a graph G and a positive integer k, an ordering of the vertices of G such that no vertex has more than k neighbors after it is called a k-*degenerate ordering* of G. We say a graph is k-*degenerate* if it has a k-degenerate ordering. The minimum k such that G is k-degenerate is called the *degeneracy* of G. From its definition, it is clear that the degeneracy of a graph is at most its maximum degree. In this section, we derive upper bounds on the threshold dimension of a graph in terms of its degeneracy. The techniques we adopt are mostly inspired by those in [2].

Throughout this section, we shall assume that G is a k-degenerate graph on n vertices with vertex set $\{v_1, v_2, \ldots, v_n\}$ and that $v_1, v_2, \ldots, v_n$ is a k-degenerate ordering of G. Thus, for each $i \in \{1, 2, \ldots, n\}$, $|N_G(v_i) \cap \{v_{i+1}, v_{i+2}, \ldots, v_n\}| \le k$. The vertices in $N_G(v_i) \cap \{v_{i+1}, v_{i+2}, \ldots, v_n\}$ are called the *forward neighbors* of v_i. Let $i < j$ and $v_i v_j \notin E(G)$. A coloring f of the vertices of G is *desirable* for the non-adjacent pair (v_i, v_j) if (i) f is a proper coloring, and (ii) $f(v_j) \ne f(v_t)$, for all neighbors v_t of v_i such that $t > j$.

Lemma 8 ($\star$). *Let G be a k-degenerate graph on n vertices and let $v_1, v_2, \ldots, v_n$ be a k-degenerate ordering of G. Let $r = \lceil \ln n \rceil$. Then there is a collection $\{f_1, \ldots, f_r\}$, where each $f_i \colon V(G) \to [10k]$ is a proper coloring of the vertices of G, such that for every non-adjacent pair (v_i, v_j), where $i < j$, there exists an $\ell \in [r]$ such that f_ℓ is a desirable coloring for the pair (v_i, v_j).*

Theorem 6 ($\star$). *For every k-degenerate graph G on n vertices, $\dim_{\text{TH}}(G) \leq 10k \ln n$.*

4.1   Random Graphs

The following lemma was proved in [2].

Lemma 9 (Lemma 12 in [2]). *For a random graph $G \in \mathcal{G}(n, p)$, where $p = \frac{c}{n-1}$ and $1 \leq c \leq n-1$, $Pr[G$ is 4ec-degenerate$] \geq 1 - \frac{1}{\Omega(n^2)}$.*

Applying Lemma 9 and Theorem 6, we get the following lemma.

Lemma 10. *For a random graph $G \in \mathcal{G}(n, p)$, where $p = \frac{c}{n-1}$ and $1 \leq c \leq n-1$, $Pr[\dim_{\text{TH}}(G) \in O(c \ln n)] \geq 1 - \frac{1}{\Omega(n^2)}$.*

It is known that (see page 35 of [3])

$$P_m(Q) \leq 3\sqrt{m} P_p(Q) \tag{1}$$

where (i) Q is a property of graphs of order n, (ii) $P_m(Q)$ is the probability that Property Q is satisfied by a graph $G \in \mathcal{G}(n, m)$, and (iii) $P_p(Q)$ is the probability that Property Q is satisfied by a graph $G \in \mathcal{G}(n, p)$ with $p = \frac{m}{\binom{n}{2}} = \frac{2m/n}{n-1}$. Assume $m \geq n/2$. Then, $p = \frac{2m/n}{n-1} \geq \frac{1}{n-1}$ and by Lemma 10, $Pr[\dim_{\text{TH}}(G) \notin O(\frac{2m}{n} \ln n)] \leq \frac{1}{\Omega(n^2)}$. Applying Eq. 1, for a random graph $G \in \mathcal{G}(n, m)$, $m \geq n/2$, $Pr[\dim_{\text{TH}}(G) \notin O(\frac{2m}{n} \ln n)] \leq \frac{3\sqrt{m}}{\Omega(n^2)} \leq \frac{1}{\Omega(n)}$. We thus have the following theorem.

Theorem 7. *For a random graph $G \in \mathcal{G}(n, m)$, $m \geq n/2$, $Pr[\dim_{\text{TH}}(G) \in O(\frac{2m}{n} \ln n)] \geq 1 - \frac{1}{\Omega(n)}$. In other words, $Pr[\dim_{\text{TH}}(G) \in O(d_{av} \ln n)] \geq 1 - \frac{1}{\Omega(n)}$, where d_{av} denotes the average degree of G.*

4.2   Graphs of High Girth

The *girth* of a graph is the length of a smallest cycle in it. We assume that if the graph is acyclic, then its girth is ∞. We apply Theorem 6 to prove an upper bound for the threshold dimension of a graph in terms of its girth and the number of vertices. The following lemma was proved in [19].

Lemma 11 (Lemma 23 in [19]). *Let G be a graph on n vertices having girth greater than $g + 1$. Then, G is k-degenerate, where $k = \lceil n^{\frac{1}{\lfloor g/2 \rfloor}} \rceil$.*

Applying the above lemma, we get the following corollary to Theorem 6.

Corollary 3. *Let G be a graph on n vertices with girth greater than $g+1$. Then, $\dim_{TH}(G) \leq 10\lceil n^{\frac{1}{\lfloor g/2 \rfloor}} \rceil \ln n$.*

The bipartite graph G obtained by removing a perfect matching from the complete bipartite graph $K_{n,n}$ is known to have a boxicity of $\frac{n}{2}$. From Observation 2 and by applying Corollary 3 with $g = 2$, we have $\frac{n}{2} \leq \dim_{TH}(G) = O(n \ln n)$. Thus, we cannot expect to get an upper bound of $O(n^{\alpha/g})$, with $\alpha < 2$, for the threshold dimension of a graph with girth greater than $g + 1$.

5   Threshold Dimension and Minimum Vertex Cover

A *vertex cover* of G is a set of vertices $S \subseteq V(G)$ such that $\forall e \in E(G)$, at least one endpoint of e is in S. A *minimum vertex cover* of G is a vertex cover of G of the smallest cardinality. We use $\beta(G)$ to denote the cardinality of a minimum vertex cover. In this section, we prove a tight upper bound for the threshold dimension of a graph in terms of the size of its minimum vertex cover.

Proposition 7 ($\star$). *For every graph G, $\dim_{TH}(G) \leq \beta(G)$.*

Since $\alpha(G) = |V(G)| - \beta(G)$, by combining Corollary 1(a) with Proposition 7, we get the following theorem.

Theorem 8. *For every graph G on n vertices, $\dim_{TH}(G) \leq n - \max\{\omega(G), \alpha(G)\}$.*

In Ramsey theory, $R(k, k)$ denotes the smallest positive integer n such that every graph on n vertices has either an independent set of size k or a clique of size k. It is known due to [8] that $R(k, k) \leq k^{\frac{-c \ln k}{\ln \ln k}} 4^k$, where c is a constant. This implies that for sufficiently large n, every graph on n vertices has either an independent set or a clique (or both) of size $0.72 \ln n$. This gives us the following corollary.

Corollary 4. *When n is sufficiently large, a graph G on n vertices satisfies $\dim_{TH}(G) \leq n - 0.72 \ln n$.*

Tightness of the bound in Theorem 8 It can be verified that the graph H on $2n$ vertices having threshold dimension n constructed in Example 1 in [11] satisfies $\alpha(H) = \omega(H) = \beta(H) = n$. Hence, the bounds in Theorem 8 and Proposition 7 are tight.

Acknowledgment. We thank Karteek Sreenivasaiah for helpful discussions and the anonymous reviewers for their valuable suggestions.

References

1. Adiga, A., Bhowmick, D., Chandran, L.S.: The hardness of approximating the boxicity, cubicity and threshold dimension of a graph. Discrete Appl. Math. **158**(16), 1719–1726 (2010)
2. Adiga, A., Chandran, L.S., Mathew, R.: Cubicity, degeneracy, and crossing number. Eur. J. Comb. **35**, 2–12 (2014)
3. Bollobás, B.: Random graphs, vol. 73. Cambridge University Press, Cambridge (2001)
4. Chacko, D., Francis, M.C.: Representing graphs as the intersection of cographs and threshold graphs. Electron. J. Comb. **28**(3), P3.11 (2021)
5. Chandran, L.S., Francis, M.C., Sivadasan, N.: Geometric representation of graphs in low dimension using axis parallel boxes. Algorithmica **56**(2), 129–140 (2010). https://doi.org/10.1007/s00453-008-9163-5
6. Chandran, L.S., Sivadasan, N.: Boxicity and treewidth. J. Combinatorial Theory Ser. B **97**(5), 733–744 (2007)
7. Chvátal, V., Hammer, P.L.: Aggregation of inequalities in integer programming. Ann. Discrete Math. **1**, 145–162 (1977)
8. Conlon, D.: A new upper bound for diagonal Ramsey numbers. Ann. Math. 941–960, (2009)
9. Erdős, P., Kierstead, H.A., Trotter, W.T.: The dimension of random ordered sets. Random Struct. Algorithms **2**(3), 253–275 (1991)
10. Esperet, L., Wiechert, V.: Boxicity, poset dimension, and excluded minors. Electron. J. Comb. **25**(4), P4.51 (2018)
11. Francis, M.C., Majumder, A., Mathew, R.: Bounding threshold dimension: realizing graphic Boolean functions as the AND of majority gates (2022). https://arxiv.org/abs/2202.12325
12. Golumbic, M.C.: Algorithmic Graph Theory and Perfect Graphs. Elsevier, Amsterdam (2004)
13. Hammer, P.L., Mahadev, N.V.: Bithreshold graphs. SIAM J. Algebraic Discrete Methods **6**(3), 497–506 (1985)
14. Korach, E., Solel, N.: Tree-width, path-width, and cutwidth. Discret. Appl. Math. **43**(1), 97–101 (1993)
15. Kratochvíl, J.: A special planar satisfiability problem and a consequence of its NP-completeness. Discret. Appl. Math. **52**, 233–252 (1994)
16. Kratochvíl, J., Tuza, Z.: Intersection dimensions of graph classes. Graphs Comb. **10**(2–4), 159–168 (1994). https://doi.org/10.1007/BF02986660
17. Mahadev, N.V., Peled, U.N.: Threshold Graphs and Related Topics. Elsevier, Amsterdam (1995)
18. Mahajan, M.: Depth-2 threshold circuits. Resonance **24**(3), 371–380 (2019)
19. Majumder, A., Mathew, R.: Local boxicity and maximum degree (2021). https://arxiv.org/abs/1810.02963
20. Raschle, T., Simon, K.: Recognition of graphs with threshold dimension two. In Proceedings of the Twenty-seventh Annual ACM Symposium on Theory of computing, pp. 650–661 (1995)
21. Roberts, F.S.: Recent Progresses in Combinatorics, chapter On the boxicity and cubicity of a graph, pp. 301–310. Academic Press, New York (1969)
22. Robertson, N., Seymour, P.D.: Graph minors. II. algorithmic aspects of tree-width. J. Algorithms **7**(3), 309–322 (1986)
23. Yannakakis, M.: The complexity of the partial order dimension problem. SIAM J. Algebraic Discrete Methods **3**(3), 351–358 (1982)

Parameterized Complexity of Weighted Multicut in Trees

Esther Galby[1], Dániel Marx[1] , Philipp Schepper[1(✉)] , Roohani Sharma[2] ,
and Prafullkumar Tale[1]

[1] CISPA Helmholtz Center for Information Security, Saarbrucken, Germany
{esther.galby,marx,philipp.schepper,prafullkumar.tale}@cispa.de
[2] Max Planck Institute for Informatics, Saarland Informatics Campus,
Saarbrucken, Germany
rsharma@mpi-inf.mpg.de

Abstract. The EDGE MULTICUT problem is a classical cut problem where given an undirected graph G, a set of pairs of vertices $\mathcal{P}$, and a budget k, the goal is to determine if there is a set S of at most k edges such that for each $(s,t) \in \mathcal{P}$, $G - S$ has no path from s to t. EDGE MULTICUT has been relatively recently shown to be fixed-parameter tractable (**FPT**), parameterized by k, by Marx and Razgon [SICOMP 2014], and independently by Bousquet et al. [SICOMP 2018]. In the weighted version of the problem, called WEIGHTED EDGE MULTICUT one is additionally given a weight function $\mathsf{wt} : E(G) \to \mathbb{N}$ and a weight bound $\mathbf{w}$, and the goal is to determine if there is a solution of size at most k and weight at most $\mathbf{w}$. Both the **FPT** algorithms for EDGE MULTICUT by Marx et al. and Bousquet et al. fail to generalize to the weighted setting. In fact, the weighted problem is non-trivial even on trees and determining whether WEIGHTED EDGE MULTICUT on trees is **FPT** was explicitly posed as an open problem by Bousquet et al. [STACS 2009]. In this article, we answer this question positively by designing an algorithm which uses a very recent result by Kim et al. [STOC 2022] about directed flow augmentation as subroutine.

We also study a variant of this problem where there is no bound on the size of the solution, but the parameter is a structural property of the input, for example, the number of leaves of the tree. We strengthen our results by stating them for the more general vertex deletion version.

Keywords: Weighted multicut in trees · Directed flow augmentation · Weighted digraph pair cut

1 Introduction

EDGE MULTICUT is a generalization of the classical (s, t)-CUT problem where given a graph G, a set of terminal pairs $\mathcal{P} = \{(s_1, t_1), \ldots, (s_p, t_p)\}$, and an integer k, the goal is to determine if there exists a set of at most k edges whose deletion disconnects s_i from t_i, for each $i \in [p]$. Such a set is called a $\mathcal{P}$-*multicut*

© Springer Nature Switzerland AG 2022
M. A. Bekos and M. Kaufmann (Eds.): WG 2022, LNCS 13453, pp. 257–270, 2022.
https://doi.org/10.1007/978-3-031-15914-5_19

in G. The case $p = 1$ corresponds to the classical (s, t)-CUT problem. EDGE MULTICUT is polynomial time solvable for $p \leq 2$ [22] and is NP-hard even for $p = 3$ [8]. From the parameterized complexity point of view, it was a long-standing open question to determine if the problem is fixed-parameter tractable (FPT) parameterized by the solution size. This question was resolved independently by Marx and Razgon [21] and Bousquet et al. [1], proving that the problem is FPT. Both algorithms extensively use the notion of important separators, a technique introduced earlier by Marx [20]. Bousquet et al. [1] additionally use several problem-specific observations and arguments about the structure of multicut instances, while Marx and Razgon [21] formulated the technique of random sampling of important separators, which found further applications for many other problems [4–6, 16, 18, 19].

Weighted Multicut. One drawback of the algorithms using important separators is that they are essentially based on a replacement argument: if a subset X of the solution satisfies some property, then this technique allows us to find a set X' such that X can be replaced with X', thereby making progress towards fully identifying a solution. This local replacement argument inherently fails if the overall solution is also required to satisfy additional properties, such as minimizing the overall weight, since replacing X with X' may violate these additional constraints. Thus, the ideas from the algorithms of Marx and Razgon [21] and Bousquet et al. [1] fail to generalize to the edge deletion version of WEIGHTED MULTICUT (wMC) where we are, additionally, given a weight function $\mathsf{wt} : E(G) \to \mathbb{N}$ and an integer $\mathbf{w}$, and the goal is to determine if there exists a $\mathcal{P}$-multicut in G of size at most k and weight at most $\mathbf{w}$.

(Weighted) Multicut on Trees. EDGE MULTICUT remains NP-hard on trees [10] but can be solved in $\mathcal{O}(2^k \cdot n)$-time, where n is the number of vertices in the input tree, using an easy branching algorithm [11]: for the "deepest" (s_i, t_i)-path branch on the deletion of the two edges on this path which are incident to the lowest common ancestor of s_i and t_i. A series of work shows improvement over this simple running time [3, 13], and also the problem admits a polynomial kernel [2, 3]. Since the algorithmic approaches for EDGE MULTICUT on trees are based on greedily finding partial solutions, they do not generalize to the weighted setting. In fact, the question whether wMC on trees is FPT (parameterized by the solution size), was explicitly posed as an open problem by Bousquet et al. [2]. In this article, we answer this question in the positive.

Flow Augmentation. As mentioned earlier, most of the available techniques used to design FPT algorithms, especially for cut problems, do not work in the weighted setting. Kim et al. [15] recently developed the technique of flow augmentation in directed graphs. This technique offers a new perspective to design FPT algorithms for cut problems and positively settles the parameterized complexity of some long standing open problems, such as WEIGHTED (s, t)-CUT, WEIGHTED DIRECTED FEEDBACK VERTEX SET and WEIGHTED DIGRAPH PAIR CUT.

Our main goal is to use this technique for the underlying core difficulty in wMC on trees. More precisely, we do not use the directed flow augmentation technique as such but we crucially use the FPT algorithm for WEIGHTED DIGRAPH PAIR CUT (wDPC) which is one important example of the use of this technique. The wDPC problem is defined as follows [15,17]: given a *directed* graph G, a source vertex $\mathbf{r} \in V(G)$, terminal pairs $\mathcal{P} = \{(s_1, t_1), \ldots, (s_p, t_p)\}$, a weight function $\mathtt{wt} : E(G) \to \mathbb{N}$, a positive integer k, the goal is to determine if there exists a set S of at most k arcs of G such that $\mathtt{wt}(S)$ is minimum[1], and for each $i \in [p]$, if $G - S$ has a path from $\mathbf{r}$ to s_i, then $G - S$ has no path from $\mathbf{r}$ to t_i. Such a set is called a $\mathcal{P}$-*dpc* with respect to $\mathbf{r}$ in G. Kim et al. [15, Section 6.1ff] showed that wDPC can be solved in randomized $2^{\mathcal{O}(k^4)} \cdot n^{\mathcal{O}(1)}$-time. The randomized running time of this algorithm is an artifact of the use of the directed flow augmentation procedure which is randomized. Apart from this step, all the other steps of the algorithm are deterministic. Our basic observation is that the algorithm for wDPC can be used to solve a non-trivial base case of wMC in trees. Indeed, the following three statements are equivalent: (a) if there is a vertex $\mathbf{r} \in V(T)$ such that all the terminal pair paths of $\mathcal{P}$ pass through $\mathbf{r}$, then S is a $\mathcal{P}$-multicut of T, (b) for all $(s, t) \in \mathcal{P}$, S intersects the $(\mathbf{r}, s)$-path or the $(\mathbf{r}, t)$-path, and (c) S is a $\mathcal{P}$-dpc for T (in wDPC we interpret each edge of T to be directed away from $\mathbf{r}$).[2]

Edge Deletion vs. Vertex Deletion. In the weighted setting, the edge deletion version of wMC (on trees) reduces to its vertex deletion version (on trees), by subdividing each edge and assigning the weight of the original edge to the newly added vertex corresponding to the edge, and by setting the weights of the original vertices to ∞ (or larger than the weight budget parameter). Note that such a reduction does not work in the unweighted setting as the vertex deletion version of MULTICUT in trees is polynomial time solvable [7].

Main Result. From now on we only study the vertex deletion version of wMC on trees which, as mentioned above, is more general than the edge deletion version. It is formally defined below.

WEIGHTED MULTICUT ON TREE (wMC-TREE)
Input: A tree T, a collection of terminal pairs $\mathcal{P} \subseteq V(T) \times V(T)$, a vertex weight function $\mathtt{wt} : V(T) \to \mathbb{N}$, and positive integers $\mathbf{w}$ and k.
Question: Does there exist $S \subseteq V(T)$ such that $|S| \leq k$, $\mathtt{wt}(S) \leq \mathbf{w}$, and S intersects the unique (s, t)-path in T, for each $(s, t) \in \mathcal{P}$?

---

[1] Though the formal description of the problem in [15] asks for a solution S with $\mathtt{wt}(S) \leq \mathbf{w}$, the authors remark that the algorithm in fact finds a minimum weight solution.

[2] When dealing with undirected graphs, the flow augmentation restricted to undirected graphs given by Kim et al. [14] may suffice to solve wDPC on undirected graphs. As this problem is not mentioned explicitly in [14], we stick to the directed setting.

We set $\mathtt{wt}(S) = \sum_{v \in S} \mathtt{wt}(v)$ for the ease of notation. We use the FPT algorithm for wDPC (restricted to trees) [15, Section 6.1ff] as a subroutine to prove our main result, namely that wMC-TREE is FPT.

Theorem 1. wMC-TREE *can be solved in randomized* $2^{\mathcal{O}(k^4)} \cdot n^{\mathcal{O}(1)}$ *time.*

Structural Parameterizations. In scenarios where the size of the solution is large, it might be desired to drop the constraint on the size of the solution altogether, and seek to parameterize the problem with some structural parameter of the input. In this setting, we first consider the problem parameterized by the number of leaves of the tree and then extend this result to a more general parameter that takes into account the number of requests (terminal pair paths) passing through a vertex. Technically, we solve a different problem in this setting, where we only have a uni-objective function seeking to minimize the weight of the solution (in contrast to the bi-objective function in the case of wMC-TREE). This problem is formally defined below.

UNCONSTRAINED WEIGHTED MULTICUT ON TREE (uwMC-TREE)
Input: A tree T, a collection of terminal pairs $\mathcal{P} \subseteq V(T) \times V(T)$, a vertex weight function $\mathtt{wt} : V(T) \to \mathbb{N}$ and a positive integer $\mathbf{w}$.
Question: Does there exist $S \subseteq V(T)$ such that $\mathtt{wt}(S) \le \mathbf{w}$ and S intersects the unique (s,t)-path in T, for each $(s,t) \in \mathcal{P}$?

---

uwMC is another generalization of the vertex deletion variant of MULTICUT. The former problem has been studied on trees in the parameterized complexity setting with respect to certain structural parameters. In particular, Guo et al. [12, Theorem 9] showed that uwMC-TREE is FPT when the parameter is the maximum number of (s,t)-paths that pass through any vertex of the input. We call this parameter the *request degree d* of an instance. Guo et al. [12] gave an algorithm for uwMC-TREE that runs in time $\mathcal{O}(3^d \cdot n)$.

We first study uwMC-TREE when the parameter is the number of leaves of the tree. The problem is polynomial time solvable on paths (Lemma 5) but becomes NP-hard on (general) trees. Thus, the number of leaves appears to be a natural parameter which could explain the contrast between the above two results. Formally, we prove the following theorem.

Theorem 2. uwMC-TREE *can be solved in* $2^{\mathcal{O}(\ell^2 \log \ell)} \cdot n^{\mathcal{O}(1)}$ *time, where ℓ is the number of leaves in the input tree.*

At the core of the algorithm for Theorem 2, we again solve instances of wDPC on trees, but, in this case, these instances have a special structure: they are subdivided stars (i.e. trees with at most one vertex of degree at least 3). We show that these instances do not require the use of the flow augmentation technique. In fact, these instances correspond to the arcless instances of wDPC in [15, Section 6.2.2] defined roughly as follows: the input graph comprises of two designated vertices s, t with internally vertex-disjoint paths from s to t, and the solution picks exactly one arc from each of these internally vertex-disjoint paths. Since the arcless instances can be solved faster than the general instances of wDPC and do not require the usage of the flow augmentation technique [15, Lemma 6.12], the algorithm for uwMC-Tree is deterministic and has a better running time.

As a final result, we use the algorithm of Theorem 2 as a subroutine to give an FPT algorithm for uwMC-Tree that generalizes the result of Guo et al. [12, Theorem 9] and Theorem 2. To do so, we define a new parameter that comprises both the request degree and the number of leaves of the input instance. An instance $(T, \mathcal{P}, \mathtt{wt}, \mathbf{w})$ is (d, q)-*light* if the following hold. Let Y be the set of vertices through which at most d terminal pair paths of $\mathcal{P}$ pass. Such vertices are called d-*light vertices*. Then for each connected component C of $T - Y$, the number of leaves of $T[N[C]]$ must be at most q. We show in the full version [9] that it is crucial to consider the *neighborhood* of the component, as the problem is otherwise already NP-hard for $d = 3$ and $q = 2$. We design a dynamic programming algorithm that stores partial solutions for every d-light vertex using the algorithm of Theorem 2 as a subroutine to solve the problem on (d, q)-light instances.

Theorem 3 ($\star$). uwMC-Tree *can be solved in* $3^d \cdot 2^{dq} \cdot 2^{\mathcal{O}(q^2 \log q)} \cdot n^{\mathcal{O}(1)}$ *time on* (d, q)-*light instances.*

Observe that an instance with a tree on ℓ leaves is a $(0, \ell)$-light instance, and an instance with the request degree at most d is a $(d, 0)$-light instance. Thus, Theorem 3 implies Theorem 2 and Theorem 9 in [12], up to the polynomial factors in the running time.

Our Methods. Our algorithms for Theorems 1 and 2, are crucially based on the observation mentioned earlier: if every terminal path goes through a root $\mathbf{r}$, then the problem reduces to wDPC. In the vertex deletion version, the *vertex $\mathcal{P}$-multicut* in a tree can be found using the algorithm for wDPC, by assigning the weight of a vertex to the unique edge connecting it to its parent in T. The general idea for both our algorithms is to design a branching algorithm that effectively solves instances of the above-mentioned type to reduce the measure in each branch. Let T be a rooted tree. The goal is to identify two vertices $x, y \in V(T)$ where x is a descendant of y, and branch on the possibility of a hypothetical solution intersecting the (y, x)-path. If the solution does not intersect the (y, x)-path, then contracting the edges of the (y, x)-path and making the resulting vertex undeletable, is a safe operation. If the solution intersects the (y, x)-path, then for each vertex v on the (y, x)-path, we increase the weight of v by adding to it the minimum weight of a solution in $T_v - \{v\}$ (where T_v is the subtree of T rooted at V), and then forget about the terminal pair paths in $T_v - \{v\}$. To update the weight of v, one therefore needs to find a minimum weight solution in $T_v - \{v\}$. For this reason, we choose the vertices x, y so that the instance restricted to $T_v - \{v\}$ can be solved using the algorithm for wDPC.

If x, y are vertices of degree at least 3 (branching vertices) in T, then contracting the (y, x)-path decreases the number of branching vertices in the resulting instance. This choice of x, y allows to design a branching algorithm where the measure is the number of branching vertices, and thus the number of leaves (Theorem 2). If x, y are vertices of a minimum-size (unweighted) $\mathcal{P}$-multicut (which can be found in polynomial time), then contracting the (y, x)-path decreases the size of a $\mathcal{P}$-multicut in the resulting instance. This choice of x, y allows the design of a branching algorithm parameterized by the solution size (Theorem 1). Additionally, if we choose x to be the furthest branching vertex in T (resp. furthest vertex of X) from the root and y to be its unique closest ancestor that is a branching vertex (resp. in X), then for each vertex v on the (y, x)-path, the instance restricted on $T_v - \{v\}$ can indeed be solved using the algorithm for wDPC. The proofs of statements marked with $\star$ appear in the full version [9].

2   Basic Notation

Let T be a tree. For any $u, v \in V(T)$, $P_{u,v}$ denotes the unique (u, v)-path in T. For a set $\mathcal{P} \subseteq V(T) \times V(T)$ of terminal pairs, we interchangeably refer to a pair $(s, t) \in \mathcal{P}$ as the terminal pair (s, t) and as the path $P_{s,t}$ in T. For any subtree T' of T, $\mathcal{P}|_{T'}$ denotes the paths of $\mathcal{P}$ that are contained in T' and for any $E' \subseteq E(T)$, $\mathcal{P}/E'$ denotes the paths in $\mathcal{P}$ obtained by contracting the edges of E' in T. Also, T/E' denotes the tree obtained from T after contracting the edges in E'. The sets $V_{\geq 3}(T)$ and $V_{=1}(T)$ denote the set of vertices of degree at least 3 (*branching vertices*) and of degree equal to 1 (*leaves*), respectively. We denote by T_u the subtree of T rooted at u and $T_u^\dagger = T_u \setminus \{u\}$. For any descendant x of u, the tree denoted by $T_{u,x}$ is defined as follows. Let $\{v_1, \ldots, v_p\}$ be the children of u in T and say $x \in V(T_{v_i})$. Then $T_{u,x} = T_u \setminus (\cup_{j \in [p] \setminus \{i\}} T_{v_j})$. We define $T_{u,x}^\dagger = T_{u,x} \setminus \{u\}$. Observe that, both $T_{u,x}$ are $T_{u,x}^\dagger$ are connected.

3   wMC-Tree Parameterized by the Solution Size

In this section, we prove Theorem 1 by designing a branching algorithm. In order to reduce the measure of a given instance, our branching algorithm requires a solution for the instances where every terminal pair path passes through a single vertex. Let $\mathcal{I} = (T, \mathcal{P}, \mathbf{r}, \mathtt{wt}, k)$ be an instance such that all the terminal pair paths of $\mathcal{P}$ pass through $\mathbf{r}$, and $\mathtt{wt} : V(T) \to \mathbb{N}$ is a vertex weight function. Let $\overrightarrow{T}$ be the directed tree obtained by orienting the edges of T so that all the vertices, except for $\mathbf{r}$, have in-degree exactly one, while $\mathbf{r}$ has in-degree zero. In other words, the oriented tree $\overrightarrow{T}$ is an out-tree rooted at $\mathbf{r}$. We define an edge weight function $\mathtt{wt}' : E(\overrightarrow{T}) \to \mathbb{N}$ such that for every arc $e = (u, v) \in E(\overrightarrow{T})$, $\mathtt{wt}'(e) = \mathtt{wt}(v)$. Then it can be easily seen that $Z \subseteq E(\overrightarrow{T})$ is a $\mathcal{P}$-dpc in T with $\mathtt{wt}'(Z) = \mathbf{w}$ if and only if $S = \{v : (u, v) \in Z\} \subseteq V(T) \setminus \{\mathbf{r}\}$ (that is, S is obtained from Z by picking the heads of all the arcs in Z) is a $\mathcal{P}$-multicut in T with $\mathtt{wt}(S) = \mathbf{w}$. Let $\mathcal{A}_{\mathsf{dpc}}$ be the algorithm that takes as input an instance $\mathcal{I}$ as above, constructs the edge-weight function $\mathtt{wt}'$ and uses the wDPC algorithm of Kim et al. [15, Section 6.1ff] to solve the instance $(\overrightarrow{T}, \mathcal{P}, \mathbf{r}, \mathtt{wt}', k)$. This runs in randomized $2^{\mathcal{O}(k^4)} \cdot n^{\mathcal{O}(1)}$-time[3]. Therefore, $\mathcal{A}_{\mathsf{dpc}}$ outputs the minimum *weight* of a solution of $\mathcal{I}$ if it exists, and ∞ otherwise. In particular, if $\mathcal{P} = \emptyset$ then $\mathcal{A}_{\mathsf{dpc}}$ outputs 0.

Branching Algorithm. Let $(T, \mathcal{P}, \mathsf{wt}, \mathbf{w}, k)$ be an instance of wMC-TREE. Fix an arbitrary vertex $\mathbf{r} \in V(T)$ to be the root of T. We begin by finding a set $X \subseteq V(T)$ which is a $\mathcal{P}$-multicut in T and is closed under taking lca (least common ancestor). To find X, we first compute a *unweighted* $\mathcal{P}$-multicut $X_{\mathsf{opt}} \subseteq V(T)$ in T of minimum size. The set X_{opt} can be found in polynomial time (folklore) by the following greedy algorithm. Initialize $X_{\mathsf{opt}} = \emptyset, T' = T$, and $\mathcal{P}' = \mathcal{P}$. Let $v \in V(T')$ be a furthest vertex from $\mathbf{r}$ such that there exists $(s, t) \in \mathcal{P}'$ with $s, t \in V(T_v)$. By the choice of v, the (s, t)-path (and every terminal pair path in $\mathcal{P}|_{T_v}$) passes through v. It is easy to see that there is a minimum-size $\mathcal{P}'$-multicut containing v. Set $X_{\mathsf{opt}} = X_{\mathsf{opt}} \cup \{v\}, \mathcal{P}' = \mathcal{P}' \setminus \mathcal{P}|_{T_v}, T' = T' \setminus T_v$, and repeat the procedure until $\mathcal{P}' = \emptyset$. At the end of the procedure, X_{opt} is a minimum-size $\mathcal{P}$-multicut in T. If $|X_{\mathsf{opt}}| > k$, report No. Otherwise, let X be the lca-closure of X_{opt} in T. Hence, $|X| \leq 2k$.

A notable property of a $\mathcal{P}$-multicut X closed under taking lca is that for any $x \in X$, if $y \in X$ is the unique closest ancestor of x in T, then for each $v \in V(P_{y,x}) \setminus \{y\}$, all the terminal pair paths of $\mathcal{P}|_{T_v}$ either pass through x, or are contained in T_x. Indeed, if $T_v \setminus T_x$ contains a path of $\mathcal{P}$, then any $\mathcal{P}$-multicut intersects $V(T_v \setminus T_x)$. Then there exists a vertex $y' \in X$ such that $y' \neq x$ lies on $P_{v,x} \subseteq P^\dagger_{y,x}$, contradicting the choice of y.

We design a branching algorithm whose input is $\mathcal{I} = (T, \mathcal{P}, \mathsf{wt}, \mathbf{w}, k, X)$ where X is $\mathcal{P}$-multicut $X \subseteq V(T)$ closed under taking lca, and where the measure of an instance $\mathcal{I}$ is defined as $\mu(\mathcal{I}) = |X|$. Note that, as mentioned above, $\mu(\mathcal{I}) \leq 2k$. The base case of the branching algorithm occurs in the following scenarios.

1. If $\mu(\mathcal{I}) = 0$, then $\emptyset$ is a solution of $\mathcal{I}$. Return YES iff $k \geq 0$ and $\mathbf{w} \geq 0$.
2. If $\mu(\mathcal{I}) = 1$, let $X = \{x\}$. In this case, since all the paths of $\mathcal{P}$ pass through x, return YES if and only if the $\mathcal{A}_{\mathsf{dpc}}(T, \mathcal{P}, x, \mathsf{wt}, k) \leq \mathbf{w}$.
3. If $k < 0$, or $k \leq 0$ and $\mathcal{P} \neq \emptyset$, then return No.

If $\mu(\mathcal{I}) \geq 2$ (that is, $|X| \geq 2$), then let $x \in X$ be a furthest vertex from $\mathbf{r}$ and let $y \in X$ be its unique closest ancestor. We branch in the following two cases.

Case 1. *There exists a solution of $\mathcal{I}$ that does not intersect $V(P_{y,x})$.* In this case, we return the instance $\mathcal{I}_1 = (T_1, \mathcal{P}_1, \mathsf{wt}_1, \mathbf{w}, k, X_1)$ where $T_1 = T/E(P_{y,x})$, $\mathcal{P}_1 = \mathcal{P}/E(P_{y,x})$. Let the vertex onto which the edges of $P_{y,x}$ are contracted be y°. Then $\mathsf{wt}_1(y^\circ) = \mathbf{w} + 1$ and, for each $v \in V(T_1) \setminus \{y^\circ\}$, $\mathsf{wt}_1(v) = \mathsf{wt}(v)$. Observe that $(X \setminus \{x, y\}) \cup \{y^\circ\}$ is a $\mathcal{P}_1$-multicut in T_1 and is closed under taking lca and thus, we may set $X_1 = (X \setminus \{x, y\}) \cup \{y^\circ\}$. Clearly, $\mu(\mathcal{I}_1) < \mu(\mathcal{I})$ and $\mathcal{I}_1$ can be constructed in polynomial time.

Case 2. *There exists a solution of $\mathcal{I}$ that intersects $V(P_{y,x})$.* In this case, the idea is the following: for each vertex v on the $P_{y,x}$ path, we increase the weight

---

[3] The dependency in k is not explicit in [15] but can be easily deduced.

of v by the weight of the solution in the tree $T_{v,x}^\dagger$ (the tree strictly below v). To do so, the size of a solution in the tree $T_{v,x}^\dagger$ is first guessed. Once the weights are updated, we can forget the terminal pairs contained in the tree $T_{y,x}^\dagger$ and just remember that the solution picks a vertex from $P_{y,x}$. This is formalized below.

Let S be a solution which intersects $V(P_{y,x})$ and let $z \in V(P_{y,x})$ be the vertex in S closest to y. Then we further branch into $k+1$ branches where each branch corresponds to the guess on $|S \cap T_{z,x}^\dagger|$. More precisely, for every $i \in \{0\} \cup [k]$, we create the instance $\mathcal{I}_{2,i} = (T_2, \mathcal{P}_2, \mathtt{wt}_{2,i}, \mathtt{w}, k - i, X_2)$ where $T_2 = T \setminus T_x^\dagger$, $\mathcal{P}_2 = \mathcal{P}|_{T_2} \setminus (V(T_{y,x}^\dagger) \times V(T_{y,x}^\dagger)) \cup \{(y, x)\}$ and $\mathtt{wt}_{2,i}$ is defined below.

$$\mathtt{wt}_{2,i}(v) = \begin{cases} \mathtt{wt}(v) + \mathcal{A}_{\mathsf{dpc}}(T_{v,x}^\dagger, \mathcal{P}|_{T_{v,x}^\dagger}, x, \mathtt{wt}|_{V(T_{v,x}^\dagger)}, i) & \text{if } v \in V(P_{y,x}) \\ \mathtt{wt}(v) & \text{otherwise.} \end{cases}$$

Observe that the set $X \setminus \{x\}$ is a $\mathcal{P}_2$-multicut in T_2 with $y \in X \setminus \{x\}$. The only paths that might not be cut are the ones in $\mathcal{P}|_{T_{y,x}^\dagger}$ as they pass through x, but they are not contained in $\mathcal{P}_2$ by definition. Also $X \setminus \{x\}$ is closed under taking $\mathtt{lca}$ in T_2, thus we may set $X_2 = X \setminus \{x\}$. Clearly, $\mu(\mathcal{I}_{2,i}) < \mu(\mathcal{I})$ for each $i \in \{0\} \cup [k]$.

Lemma 4. $\mathcal{I}$ *is a* YES-*instance if and only if at least one of* $\mathcal{I}_1, \mathcal{I}_{2,0}, \ldots, \mathcal{I}_{2,k}$ *is a* YES-*instance.*

Proof. ($\Rightarrow$) Assume that $\mathcal{I}$ is a YES-instance and let S be a minimal solution of $\mathcal{I}$. Suppose first that $S \cap V(P_{y,x}) = \emptyset$ and consider a path $P_{s,t}$ of $\mathcal{P}_1$. Then $(s, t) \neq (y^\circ, y^\circ)$ for otherwise, S would not intersect the path in $\mathcal{P}$ corresponding to the pair $(s, t) \in \mathcal{P}_1$. If $y^\circ \notin \{s, t\}$ then $(s, t) \in \mathcal{P}$ and so, S intersects the path $P_{s,t}$. Otherwise, assume, without loss of generality, that $s = y^\circ$ and let $(z, t) \in \mathcal{P}$ where $z \in V(P_{y,x})$, be the terminal pair in $\mathcal{P}$ corresponding to (s, t). Then, since $P_{z,t}$ is intersected by $S \setminus V(P_{y,x})$, $P_{s,t}$ is also intersected by $S \setminus \{y^\circ\}$. Thus, S is a solution for $\mathcal{I}_1$.

Suppose next that $S \cap V(P_{y,x}) \neq \emptyset$ and let $z \in V(P_{y,x})$ be the vertex in S closest to y. Observe that since X is a $\mathcal{P}$-multicut in T and $x \in X$ is a furthest vertex in T from $\mathtt{r}$, every path of $\mathcal{P}$ contained in T_x passes through x. Similarly, if $z \neq x$ then, from the choice of x and y, each terminal pair path contained in $T_{z,x}^\dagger$ passes through x: indeed, if there exists a terminal pair path contained in $T_{z,x}^\dagger \setminus T_x$, then it is not intersected by X, a contradiction to the fact that X is a $\mathcal{P}$-multicut. Let $S^* = S \cap T_{z,x}^\dagger$ and let $i = |S^*|$. Note that if $z = x$ then $\mathcal{P}_{T_{x,x}^\dagger} = \emptyset$ by the above, and thus, $S^* = \emptyset$ by minimality of S. Since S^* is a $\mathcal{P}|_{T_{z,x}^\dagger}$-multicut, it follows that $\mathtt{wt}(S^*) \geq \mathcal{A}_{\mathsf{dpc}}(T_{z,x}^\dagger, \mathcal{P}|_{T_{z,x}^\dagger}, x, \mathtt{wt}|_{T_{z,x}^\dagger}, i)$. Now let $S' = S \setminus S^*$. Note that $z \in S'$; in fact, $S' \cap V(P_{y,x}) = \{z\}$ by the choice of z. We claim that S' is a solution for $\mathcal{I}_{2,i}$. Clearly, $|S'| = |S| - |S^*| \leq k - i$. Furthermore, $\mathtt{wt}_{2,i}(S') = \mathtt{wt}(S) - \mathtt{wt}(S^*) - \mathtt{wt}(z) + \mathtt{wt}_{2,i}(z)$ and since $z \in V(P_{y,x})$, $\mathtt{wt}_{2,i}(z) \leq \mathtt{wt}(z) + \mathtt{wt}(S^*)$. Thus, $\mathtt{wt}_{2,i}(S') \leq \mathtt{wt}(S) \leq \mathtt{w}$. We now show that S' is a $\mathcal{P}_2$-multicut. Consider a path $P_{s,t}$ of $\mathcal{P}_2$. Since by construction, $\mathcal{P}_2 \cap (V(T_{y,x}^\dagger) \times V(T_{y,x}^\dagger)) = \emptyset$, at most one of s and t belongs to $V(T_{y,x}^\dagger)$. Suppose first that $\{s, t\} \cap V(T_{y,x}^\dagger) \neq \emptyset$, say

$s \in V(T_{y,x}^{\dagger})$ without loss of generality. If $s \in V(P_{y,z}) \setminus \{y, z\}$ then, by the choice of z and because S is a $\mathcal{P}$-multicut, $P_{s,t}$ is intersected by $S \setminus V(T_{y,x}^{\dagger}) \subseteq S'$. Otherwise, $P_{s,t}$ passes through z and is therefore intersected by S'. Since it is clear that $P_{s,t}$ is intersected by S' if $\{s, t\} \cap V(T_{y,x}^{\dagger}) = \emptyset$, we conclude that S' is indeed a $\mathcal{P}_2$-multicut.

($\Leftarrow$) Suppose first that $\mathcal{I}_1$ is a YES-instance and let S_1 be a solution of $\mathcal{I}_1$. Since $\mathtt{wt}_1(y^{\circ}) = \mathbf{w} + 1$, $y^{\circ} \notin S$. This implies, in particular, that $(y^{\circ}, y^{\circ}) \notin \mathcal{P}_1$ and thus, no path of $\mathcal{P}$ is contained in $P_{y,x}$. Therefore, S_1 is a solution for $\mathcal{I}$. Suppose next that there exists $i \in \{0, \ldots, k\}$ such that $\mathcal{I}_{2,i}$ is a YES-instance and let $S_{2,i}$ be a minimal solution of $\mathcal{I}_{2,i}$. We first claim that $|S_{2,i} \cap V(P_{y,x})| = 1$. Indeed, observe that $S_{2,i} \cap V(P_{y,x}) \neq \emptyset$ since $(y, x) \in \mathcal{P}_2$. For the sake of contradiction, suppose that there exist $z, z' \in S_{2,i} \cap V(P_{y,x})$ such that $z' \neq z$, say z' is a descendant of z. Since, by construction, no path of $\mathcal{P}_2$ is contained in $T_{y,x}^{\dagger}$, each path of $\mathcal{P}_2$ that passes through z', also passes through z. Thus, $S_{2,i} \setminus \{z'\}$ is a $\mathcal{P}_2$-multicut, contradicting the minimality of $S_{2,i}$. Let $S_{2,i} \cap V(P_{y,x}) = \{z\}$. As argued above, if $z \neq x$, then, from the choice of x and y, every path of $\mathcal{P}$ contained in $T_{z,x}^{\dagger}$ passes through x. Similarly, every path of $\mathcal{P}$ contained in T_x passes through x. Let S^* be a $\mathcal{P}|_{T_{z,x}^{\dagger}}$-multicut of size at most i such that $\mathtt{wt}(S^*)$ is minimum. Then $\mathtt{wt}(S^*) = \mathcal{A}_{\mathsf{dpc}}(T_{z,x}^{\dagger}, \mathcal{P}_{T_{z,x}^{\dagger}}, x, \mathtt{wt}|_{T_{z,x}^{\dagger}}, i)$. Let $S = S_{2,i} \cup S^*$. We claim that S is a solution for $\mathcal{I}$. Indeed, first note that $|S| = |S_{2,i}| + |S^*| \leq k - i + i = k$. Furthermore, since $S_{2,i} \cap V(P_{y,x}) = \{z\}$, $\mathtt{wt}(S_{2,i}) = \mathtt{wt}_{2,i}(S_{2,i}) - \mathtt{wt}_{2,i}(z) + \mathtt{wt}(z)$ and $\mathtt{wt}_{2,i}(z) = \mathtt{wt}(z) + \mathtt{wt}(S^*)$. Thus, $\mathtt{wt}(S) = \mathtt{wt}(S_{2,i}) + \mathtt{wt}(S^*) \leq \mathtt{wt}_{2,i}(S_{2,i}) \leq \mathbf{w}$. We now show that S is a $\mathcal{P}$-multicut. Since $S_{2,i} \subseteq S$ and $S_{2,i}$ is a $\mathcal{P}_2$-multicut, any path of $\mathcal{P}$ fully contained in $V(T) \setminus V(T_{y,x}^{\dagger})$ is intersected by S. Consider now a path $P_{s,t}$ of $\mathcal{P}$ that intersects $V(T_{y,x})$ If $P_{s,t}$ is fully contained in $T_{z,x}^{\dagger}$, then it is intersected by S^*. Similarly, if $P_{s,t}$ passes through z, then it is intersected by S since $z \in S$. If $P_{s,t}$ passes through y without containing z, then $P_{s,t} \in \mathcal{P}_2$ and so, by the choice of z, $P_{s,t}$ is intersected by $S_{2,i} \setminus \{z\} \subseteq S$. Observe finally that $P_{s,t}$ is not fully contained in $V(P_{y,z}^{\dagger}) \setminus \{z\}$ for otherwise, $P_{s,t}$ is not intersected by X, a contradiction to the fact that X is a $\mathcal{P}$-multicut. Therefore, S is a solution for $\mathcal{I}$. $\square$

Proof (Theorem 1). Let $\mathcal{I} = (T, \mathcal{P}, \mathtt{wt}, \mathbf{w}, k)$ be an instance of wMC-TREE. Lemma 4 shows that the above algorithm correctly solves the problem. The described algorithm does a $(k + 2)$-way branching, where the measure of the input instance is bounded by $2k$ and drops by at least 1 in every branch. Since the branching stops when the measure is at most 1, the total number of branching nodes of the algorithm is at most $(k + 2)^{2k+1}$. Since $\mathcal{I}_1$ can be constructed in polynomial time and each instance $\mathcal{I}_{2,i}$ can be constructed by making $\mathcal{O}(n)$ calls to $\mathcal{A}_{\mathsf{dpc}}$, the final running time is $2^{\mathcal{O}(k^4)} \cdot n^{\mathcal{O}(1)}$. $\square$

4   uwMC-Tree Parameterized by the Number of Leaves

In this section, we prove Theorem 2. We first show that the problem on subdivided stars can be solved without using the flow augmentation from [15]

(Lemma 7). Towards this, we first design a simple polynomial-time algorithm for the problem on paths (Lemma 5) and use it to eliminate the terminal pair paths that do not pass through the high degree vertex of the sub-divided star. We then observe that the problem on sub-divided stars, when each terminal pair path pass through the high degree vertex, corresponds to the arcless instances of [15, Section 6.2.2], which can be solved faster [15, Lemma 6.12] (Proposition 6). We then use the algorithm of Lemma 7 as a subroutine to design a branching algorithm that proves Theorem 2.

Lemma 5 ($\star$). *Let T be a disjoint union of paths, $\mathcal{P} \subseteq V(T) \times V(T)$ and $wt : V(T) \to \mathbb{N}$. There is an algorithm $\mathcal{A}_{path}$ that outputs the weight of a $\mathcal{P}$-multicut $S \subseteq V(T)$, in T such that $wt(S)$ is minimum, in polynomial time.*

The following result follows from [15, Section 6.2.2, Lemma 6.12]. The root of a subdivided star, that is not a path, is the unique branching vertex.

Proposition 6 ($\star$, [15]). *Given an instance $(T, \mathcal{P} \subseteq V(T) \times V(T), \mathbf{r} \in V(T),$ $wt : V(T) \to \mathbb{N})$ such that T is a subdivided star with root $\mathbf{r}$ and $\ell \geq 3$ leaves. Suppose all the terminal pair paths in $\mathcal{P}$ pass through $\mathbf{r}$. Then one can find the weight of a $\mathcal{P}$-multicut $S \subseteq V(T)$ such that $wt(S)$ is minimum, in time $2^{\mathcal{O}(\ell^2 \log \ell)} \cdot n^{\mathcal{O}(1)}$.*

Lemma 7. UWMC *can be solved in $2^{\mathcal{O}(\ell^2 \log \ell)} \cdot n^{\mathcal{O}(1)}$-time on a subdivided star with ℓ leaves.*

Proof. Let the input instance be $\mathcal{I} = (T, \mathcal{P}, \mathtt{wt}, \mathbf{w})$. If $\ell = 2$, then T is a path. In this case, report YES if and only if $\mathcal{A}_{path}(T, \mathcal{P}, \mathtt{wt}) \leq \mathbf{w}$. Otherwise, let $\mathbf{r} \in V(T)$ be the root of T, that is $\mathbf{r}$ is the unique vertex of degree at least 3 in T. In the first step, the algorithm guesses whether $\mathbf{r}$ is in the solution or not. If $\mathbf{r}$ belongs to the solution, then delete $\mathbf{r}$ from T and solve the resulting instance using $\mathcal{A}_{path}$. Formally, the algorithm returns YES if and only if $\mathtt{wt}(\mathbf{r}) + \mathcal{A}_{path}(T, \mathcal{P}, \mathtt{wt}) \leq \mathbf{w}$. Henceforth, we assume that the solution does not contain $\mathbf{r}$, or equivalently, we set $\mathtt{wt}(\mathbf{r}) = \mathbf{w} + 1$. The remaining algorithm has two phases. In the first phase, it eliminates all the paths in $\mathcal{P}$ that do not pass through $\mathbf{r}$. In the second phase, it uses the algorithm of Proposition 6 to solve the problem.

Suppose that there exists a path in $\mathcal{P}$ that does not pass through $\mathbf{r}$. Let $z \in V(T)$ be a vertex that is closest to $\mathbf{r}$ such that there exists a path $P_{s,t}$ in $\mathcal{P}$ where $P_{s,t} \subseteq P^{\dagger}_{\mathbf{r},z}$. We create a new instance $\mathcal{I}' = (T', \mathcal{P}', \mathtt{wt}', \mathbf{w})$ (in polynomial time) such that $\mathcal{I}'$ is equivalent to $\mathcal{I}$. Here, $T' = T \setminus T^{\dagger}_z$ and, $\mathcal{P}' = \mathcal{P} \setminus (V(T^{\dagger}_{\mathbf{r},z}) \times V(T^{\dagger}_{\mathbf{r},z})) \cup \{(\mathbf{r}, z)\}$. Observe that the new terminal pair path $P_{\mathbf{r},z}$ in $\mathcal{P}'$ intersects $\mathbf{r}$ and thus, $\mathcal{P}'$ contains strictly fewer paths that do not pass through $\mathbf{r}$ (compared to $\mathcal{P}$). Since T is a subdivided star, for each $v \in V(T) \setminus \{\mathbf{r}\}$, $T^{\dagger}_v$ is a path. The new weight function $\mathtt{wt}'$ is defined as follows.

$$\mathtt{wt}'(v) = \begin{cases} \mathtt{wt}(v) + \mathcal{A}_{path}(T^{\dagger}_v, \mathcal{P}|_{T^{\dagger}_v}, \mathtt{wt}|_{V(T^{\dagger}_v)}) & \text{if } v \in V(P^{\dagger}_{\mathbf{r},z}) \\ \mathtt{wt}(v) & \text{otherwise.} \end{cases}$$

($\Rightarrow$) Let S be a $\mathcal{P}$-multicut of T such that $\mathtt{wt}(S) \leq \mathbf{w}$. Since $P_{\mathbf{r},z}^{\dagger}$ contains a path of $\mathcal{P}$, $S \cap V(P_{\mathbf{r},z}^{\dagger}) \neq \emptyset$. Let $y \in S \cap V(P_{\mathbf{r},z}^{\dagger})$ be the vertex that is closest to $\mathbf{r}$. Construct $S' = S \setminus V(T_y^{\dagger})$. We claim that S' is a solution for $\mathcal{I}'$. Observe that $S' \cap V(T_{\mathbf{r},z}^{\dagger}) = \{y\}$. Observe that $S \cap V(T_y^{\dagger})$ is a $\mathcal{P}|_{T_y^{\dagger}}$-multicut in $T_y^{\dagger}$. Thus, $\mathtt{wt}(S \cap V(T_y^{\dagger})) \geq \mathcal{A}_{\mathsf{path}}(T_y^{\dagger}, \mathcal{P}|_{T_y^{\dagger}}, \mathtt{wt}|_{V(T_y^{\dagger})})$. From the construction of S' and the weight function $\mathtt{wt}'$, $\mathtt{wt}'(S') = \mathtt{wt}(S) - \mathtt{wt}(S \cap V(T_y^{\dagger})) - \mathtt{wt}(y) + \mathtt{wt}'(y) \leq \mathtt{wt}(S) \leq \mathbf{w}$. We now show that S' is a $\mathcal{P}'$-multicut. Since $y \in S' \cap V(P_{\mathbf{r},z}^{\dagger})$, $T - S'$ has no $(\mathbf{r}, z)$-path. Consider any path of $\mathcal{P}'$ that intersects a vertex of $T_y^{\dagger}$. Since the paths of $\mathcal{P}'$ are not contained in $T_{\mathbf{r},z}^{\dagger}$, such a path also pass through $\mathbf{r}$ and hence y. Since $y \in S'$, S' is a $\mathcal{P}'$-multicut.

($\Leftarrow$) Let S' be a minimal $\mathcal{P}'$-multicut in T such that $\mathtt{wt}'(S') \leq \mathbf{w}$. Then $T - S'$ has no $(\mathbf{r}, z)$-path. Since $\mathtt{wt}'(\mathbf{r}) = \mathbf{w} + 1$, $S' \cap V(P_{\mathbf{r},z}^{\dagger}) \neq \emptyset$. Since S' is a minimal solution, $|S' \cap V(P_{\mathbf{r},z}^{\dagger})| = 1$ for otherwise, deleting the vertex of S on the $(\mathbf{r}, z)$-path that is furthest from $\mathbf{r}$ would result in a smaller solution. Let $S' \cap V(P_{\mathbf{r},z}^{\dagger}) = \{y\}$. Let S^* be a minimum weight $\mathcal{P}|_{T_y^{\dagger}}$-multicut. Then $\mathtt{wt}(S^*) = \mathcal{A}_{\mathsf{path}}(T_y^{\dagger}, \mathcal{P}|_{T_y^{\dagger}}, \mathtt{wt}|_{V(T_y^{\dagger})})$. Construct $S = S' \cup S^*$. We will now show that S is a solution of $\mathcal{I}$. From the construction of S and $\mathtt{wt}'$, $\mathtt{wt}(S) = \mathtt{wt}(S') + \mathtt{wt}(S^*) = \mathtt{wt}'(S') - \mathtt{wt}'(y) + \mathtt{wt}(y) + \mathtt{wt}(S^*) \leq \mathtt{wt}'(S') \leq \mathbf{w}$. Since $S' \subseteq S$, S is a $\mathcal{P}'$-multicut. Consider a path of $\mathcal{P}$ that is contained in $T_{\mathbf{r},z}^{\dagger}$. If such a path passes through y or is contained in T_y, then it is intersected by $S^* \cup \{y\}$ (and hence S). Otherwise such a path is contained in $P_{\mathbf{r},y}^{\dagger} \setminus \{y\}$. But this contradicts the choice of z. Therefore, S is indeed a $\mathcal{P}$-multicut.

We conclude that whenever there exists a path in $\mathcal{P}$ that does not pass through $\mathbf{r}$, we can apply the above procedure in polynomial time. Since every application of the above procedure decreases the number of paths of $\mathcal{P}$ that do not pass through $\mathbf{r}$ by at least one, the above procedure can be exhaustively applied in polynomial time. This ends the first phase of the algorithm. At the end of the first phase, all the paths of $\mathcal{P}$ pass through $\mathbf{r}$. Therefore, in this case, we solve the instance $(T, \mathcal{P}, \mathbf{r}, \mathtt{wt})$ using the algorithm of Proposition 6. Since the first phase of the algorithm takes polynomial time and the second phase takes $2^{\mathcal{O}(\ell^2 \log \ell)} \cdot n^{\mathcal{O}(1)}$-time, the algorithm runs in time $2^{\mathcal{O}(\ell^2 \log \ell)} \cdot n^{\mathcal{O}(1)}$. □

Observe that we can use Lemma 7 to find the minimum weight $\mathcal{P}$-multicut in a subdivided star by doing a simple binary search starting with $\mathbf{w} = 0, 1, 2, 4, 8, \ldots$ and so on. This would incur an extra $\mathcal{O}(\log \mathbf{w})$ factor in the running time. Thus, even if $\mathbf{w}$ is given as a unary input, the resulting algorithm is still polynomial in the input size. Therefore, the following corollary follows from Lemmas 5 and 7.

Corollary 8. *Let T be a subdivided star with ℓ leaves. Let $\mathcal{P} \subseteq V(T) \times V(T)$ and $\mathtt{wt} : V(T) \to \mathbb{N}$. There is an algorithm $\mathcal{A}_{\mathsf{star}}$ that finds the weight of a $\mathcal{P}$-multicut $S \subseteq V(T)$ such that $\mathtt{wt}(S)$ is minimum, in $2^{\mathcal{O}(\ell^2 \log \ell)} \cdot n^{\mathcal{O}(1)}$-time.*

We are now equipped to design the branching algorithm for Theorem 2. Let $\mathcal{I} = (T, \mathcal{P}, \mathtt{wt}, \mathbf{w})$ be an instance of uWMC-Tree. Root T at an arbitrary vertex

r. With each instance $\mathcal{I}$, we associate the measure $\mu(\mathcal{I}) = |V_{\geq 3}(T)| + |V_{=1}(T)|$. Since $|V_{=1}(T)| \leq \ell$ and $|V_{\geq 3}(T)| \leq |V_{=1}(T)| - 1$, $\mu(\mathcal{I}) \leq 2\ell$. We now design a branching algorithm such that the measure μ drops in each branch. The following cases appear as base cases: (1) If $|V_{\geq 3}(T)| \leq 1$, then return YES if and only if $\mathcal{A}_{\mathsf{star}}(T, \mathcal{P}, \mathsf{wt}) \leq \mathbf{w}$, and (2) If $\mathbf{w} < 0$ or, $\mathbf{w} \leq 0$ and $\mathcal{P} \neq \emptyset$, then return NO.

If $|V_{\geq 3}(T)| \geq 2$, let $x, y \in V_{\geq 3}(T)$ such that x is a furthest in T and, y is its unique closest ancestor. We branch into the following two cases.

Case 1. *There exists a solution of $\mathcal{I}$ that does not intersect $V(P_{y,x})$.* In this branch, we return the instance $\mathcal{I}_1 = (T_1, \mathcal{P}_1, \mathsf{wt}_1, \mathbf{w})$ where $T_1 = T/E(P_{y,x})$ and $\mathcal{P}_1 = \mathcal{P}/E(P_{y,x})$. Let the vertex onto which the edges of $P_{y,x}$ are contracted be y°. The new weight function wt_1 is defined as follows: $\mathsf{wt}_1(v) = \mathsf{wt}(v)$ for each $v \in V(T_1) \setminus \{y^\circ\}$, and $\mathsf{wt}_1(y^\circ) = \mathbf{w} + 1$. Observe that $\mathcal{I}_1$ can be constructed in polynomial time. Furthermore, since $x, y \in V_{\geq 3}(T)$ and the edges of $P_{y,x}$ are contracted in $\mathcal{I}_1$, $|V_{\geq 3}(T_1)| = |V_{\geq 3}(T)| - 1$ and thus, $\mu(\mathcal{I}_1) = \mu(\mathcal{I}) - 1$.

Case 2. *There exists a solution of $\mathcal{I}$ that intersects $V(P_{y,x})$.* In this case, let $z \in V(P_{y,x})$ be the closest vertex to y such that $P_{y,z}$ contains a path of $\mathcal{P}$. If no such vertex exists then set $z = x$. Return the instance $\mathcal{I}_2 = (T_2, \mathcal{P}_2, \mathsf{wt}_2, \mathbf{w})$ where $T_2 = T \setminus T_z^\dagger$ and $\mathcal{P}_2 = (\mathcal{P} \setminus (V(T_{y,x}) \times V(T_{y,x}))) \cup \{(y, z)\}$. Observe that, by construction, any solution of $\mathcal{I}$ intersects $V(P_{y,z})$. The new weight function wt_2 is defined as follows.

$$\mathsf{wt}_2(v) = \begin{cases} \mathsf{wt}(v) + \mathcal{A}_{\mathsf{star}}(T_{v,x}^\dagger, \mathcal{P}|_{T_{v,x}^\dagger}, \mathsf{wt}|_{V(T_{v,x}^\dagger)}) & \text{if } v \in V(P_{y,z}) \\ \mathsf{wt}(v) & \text{otherwise.} \end{cases}$$

Observe that, for each $v \in V(P_{y,x}) \setminus \{x\}$, $T_{v,x}^\dagger$ has exactly one branching vertex, namely x, since x is a furthest branching vertex in T from $\mathbf{r}$, y is the branching vertex that is the closest ancestor of x and $v \in V(P_{y,x})$. Also, $T_{x,x}^\dagger = T_x^\dagger$ is a disjoint union of paths. Since $x \in V_{\geq 3}(T)$, from the construction of T_2, $|V_{=1}(T_3)| < |V_{=1}(T)|$ and so, $\mu(\mathcal{I}_2) < \mu(\mathcal{I})$. The full proof is given in [9].

5   Future Directions

The natural question to ask is, whether the running times of our algorithms can be improved. Faster algorithms for the arcless instances in [15], directly yield faster algorithms for UWMC-TREE parameterized by the number of leaves. Another interesting question is to determine the parameterized complexity of the (bi-objective) WMC problem with respect to structural parameters such as the number of leaves. There it seems difficult to use the flow augmentation technique from [15], since such a step takes exponential time in the solution size, but the number of the leaves in the input may be much smaller than the solution size.

Another interesting follow up question is to determine if one can use the directed flow augmentation to resolve the parameterized complexity of WEIGHTED STEINER MULTICUT on trees, where given a tree T, sets $P_1, \ldots, P_r \subseteq V(T)$ each of size $p \geq 1$, a weight function $\mathsf{wt} : V(T) \to \mathbb{N}$ and positive integers $\mathbf{w}, k$, the goal is to determine if there exists a set $S \subseteq V(T)$ such that $|S| \leq k$,

$\mathtt{wt}(S) \leq \mathbf{w}$ and for each $i \in [q]$ there exists $u_i, v_i \in P_i$ such that $T - S$ has no (u_i, v_i)-path. Observe that WMC-TREE is a special case of this problem when $p = 2$.

Acknowledgements. Research supported by the European Research Council (ERC) consolidator grant No. 725978 SYSTEMATICGRAPH. Philipp Schepper is part of Saarbrücken Graduate School of Computer Science, Germany.

References

1. Bousquet, N., Daligault, J., Thomassé, S.: Multicut is FPT. SIAM J. Comput. **47**(1), 166–207 (2018). https://doi.org/10.1137/140961808
2. Bousquet, N., Daligault, J., Thomassé, S., Yeo, A.: A polynomial kernel for multicut in trees. In: Albers, S., Marion, J. (eds.) 26th International Symposium on Theoretical Aspects of Computer Science, STACS 2009, 26–28 February 2009, Freiburg, Germany, Proceedings. LIPIcs, vol. 3, pp. 183–194. Schloss Dagstuhl - Leibniz-Zentrum für Informatik, Germany (2009). https://doi.org/10.4230/LIPIcs.STACS.2009.1824
3. Chen, J., Fan, J.H., Kanj, I., Liu, Y., Zhang, F.: Multicut in trees viewed through the eyes of vertex cover. J. Comput. Syst. Sci. **78**(5), 1637–1650 (2012)
4. Chitnis, R., Egri, L., Marx, D.: List H-coloring a graph by removing few vertices. Algorithmica **78**(1), 110–146 (2016). https://doi.org/10.1007/s00453-016-0139-6
5. Chitnis, R.H., Cygan, M., Hajiaghayi, M.T., Marx, D.: Directed subset feedback vertex set is fixed-parameter tractable. ACM Trans. Algorithms **11**(4), 28:1-28:28 (2015). https://doi.org/10.1145/2700209
6. Chitnis, R.H., Hajiaghayi, M., Marx, D.: Fixed-parameter tractability of directed multiway cut parameterized by the size of the cutset. SIAM J. Comput. **42**(4), 1674–1696 (2013). https://doi.org/10.1137/12086217X
7. Cygan, M., et al.: Parameterized Algorithms. Springer, Cham (2015). https://doi.org/10.1007/978-3-319-21275-3
8. Dahlhaus, E., Johnson, D.S., Papadimitriou, C.H., Seymour, P.D., Yannakakis, M.: The complexity of multiterminal cuts. SIAM J. Comput. **23**(4), 864–894 (1994). https://doi.org/10.1137/S0097539792225297
9. Galby, E., Marx, D., Schepper, P., Sharma, R., Tale, P.: Parameterized complexity of weighted multicut in trees. CoRR abs/2205.10105 (2022). https://doi.org/10.48550/arXiv.2205.10105
10. Garg, N., Vazirani, V.V., Yannakakis, M.: Primal-dual approximation algorithms for integral flow and multicut in trees. Algorithmica **18**(1), 3–20 (1997). https://doi.org/10.1007/BF02523685
11. Guo, J., Niedermeier, R.: Fixed-parameter tractability and data reduction for multicut in trees. Networks **46**(3), 124–135 (2005). https://doi.org/10.1002/net.20081
12. Guo, J., Niedermeier, R.: Exact algorithms and applications for tree-like weighted set cover. J. Discrete Algorithms **4**(4), 608–622 (2006). https://doi.org/10.1016/j.jda.2005.07.005
13. Kanj, I., et al.: Algorithms for cut problems on trees. In: Zhang, Z., Wu, L., Xu, W., Du, D.-Z. (eds.) COCOA 2014. LNCS, vol. 8881, pp. 283–298. Springer, Cham (2014). https://doi.org/10.1007/978-3-319-12691-3_22

14. Kim, E.J., Kratsch, S., Pilipczuk, M., Wahlström, M.: Solving hard cut problems via flow-augmentation. In: Marx, D. (ed.) Proceedings of the 2021 ACM-SIAM Symposium on Discrete Algorithms, SODA 2021, Virtual Conference, 10–13 January 2021, pp. 149–168. SIAM (2021). https://doi.org/10.1137/1.9781611976465. 11

15. Kim, E.J., Kratsch, S., Pilipczuk, M., Wahlström, M.: Directed flow-augmentation. In: Leonardi, S., Gupta, A. (eds.) STOC 2022: 54th Annual ACM SIGACT Symposium on Theory of Computing, Rome, Italy, 20–24 June 2022, pp. 938–947. ACM (2022). https://doi.org/10.1145/3519935.3520018 Full version: arXiv:2111.03450

16. Kratsch, S., Pilipczuk, M., Pilipczuk, M., Wahlström, M.: Fixed-parameter tractability of multicut in directed acyclic graphs. SIAM J. Discret. Math. **29**(1), 122–144 (2015). https://doi.org/10.1137/120904202

17. Kratsch, S., Wahlström, M.: Representative sets and irrelevant vertices: new tools for kernelization. J. ACM **67**(3), 16:1-16:50 (2020). https://doi.org/10.1145/3390887

18. Lokshtanov, D., Marx, D.: Clustering with local restrictions. Inf. Comput. **222**, 278–292 (2013). https://doi.org/10.1016/j.ic.2012.10.016

19. Lokshtanov, D., Ramanujan, M.S.: Parameterized tractability of multiway cut with parity constraints. In: Czumaj, A., Mehlhorn, K., Pitts, A., Wattenhofer, R. (eds.) ICALP 2012. LNCS, vol. 7391, pp. 750–761. Springer, Heidelberg (2012). https://doi.org/10.1007/978-3-642-31594-7_63

20. Marx, D.: Parameterized graph separation problems. Theor. Comput. Sci. **351**(3), 394–406 (2006). https://doi.org/10.1016/j.tcs.2005.10.007

21. Marx, D., Razgon, I.: Fixed-parameter tractability of multicut parameterized by the size of the cutset. SIAM J. Comput. **43**(2), 355–388 (2014). https://doi.org/10.1137/110855247

22. Yannakakis, M., Kanellakis, P.C., Cosmadakis, S.S., Papadimitriou, C.H.: Cutting and partitioning a graph after a fixed pattern. In: Diaz, J. (ed.) ICALP 1983. LNCS, vol. 154, pp. 712–722. Springer, Heidelberg (1983). https://doi.org/10.1007/BFb0036950

The Segment Number: Algorithms and Universal Lower Bounds for Some Classes of Planar Graphs

Ina Goeßmann[1], Jonathan Klawitter[1], Boris Klemz[1], Felix Klesen[1],
Stephen Kobourov[2], Myroslav Kryven[2], Alexander Wolff[1],
and Johannes Zink[1(✉)]

[1] Institut für Informatik, Universität Würzburg, Würzburg, Germany
zink@informatik.uni-wuerzburg.de
[2] Department of Computer Science, University of Arizona, Tucson, USA

Abstract. The *segment number* of a planar graph G is the smallest number of line segments needed for a planar straight-line drawing of G. Dujmović, Eppstein, Suderman, and Wood [CGTA'07] introduced this measure for the *visual complexity* of graphs. There are optimal algorithms for trees and worst-case optimal algorithms for outerplanar graphs, 2-trees, and planar 3-trees. It is known that every *cubic* triconnected planar n-vertex graph (except K_4) has segment number $n/2+3$, which is the only known *universal* lower bound for a meaningful class of planar graphs.

We show that every triconnected planar 4-regular graph can be drawn using at most $n + 3$ segments. This bound is tight up to an additive constant, improves a previous upper bound of $7n/4 + 2$ implied by a more general result of Dujmović et al., and supplements the result for cubic graphs. We also give a simple optimal algorithm for cactus graphs, generalizing the above-mentioned result for trees. We prove the first linear universal lower bounds for outerpaths, maximal outerplanar graphs, 2-trees, and planar 3-trees. This shows that the existing algorithms for these graph classes are constant-factor approximations. For maximal outerpaths, our bound is best possible and can be generalized to circular arcs.

Keywords: Visual complexity · Segment number · Lower/upper bounds

1 Introduction

A drawing of a given graph can be evaluated by various quality measures depending on the concrete purpose of the drawing. Classic examples of such measures include drawing area, number of edge crossings, neighborhood preservation, and stress of the embedding. More recently, Schulz [20] proposed the *visual complexity* of a drawing, determined by the number of geometric objects (such as line segments or circular arcs) that the drawing consists of. It has been experimentally verified that people without mathematical background tend to prefer

© Springer Nature Switzerland AG 2022
M. A. Bekos and M. Kaufmann (Eds.): WG 2022, LNCS 13453, pp. 271–286, 2022.
https://doi.org/10.1007/978-3-031-15914-5_20

drawings with low visual complexity [13]. The visual complexity of a graph drawing depends on the drawing style, as well as on the underlying graph properties.
A well-studied measure of the visual complexity of a graph is its segment number, introduced by Dujmović, Eppstein, Suderman, and Wood [5]. It is defined
as follows. A *straight-line drawing* of a graph maps (i) the vertices of the graph
injectively to points in the plane and (ii) the edges of the graph to straight-line
segments that connect the corresponding points. A *segment* in such a drawing is
a maximal set of edges that together form a line segment. Given a straight-line
drawing Γ of a graph, the set of segments it induces is unique. Its cardinality is
the *segment number* of Γ. The *segment number*, seg(G), of a planar graph G is
the smallest segment number over all crossing-free straight-line drawings of G.

Previous Work. Dujmović et al. [5] pointed out two natural lower bounds for the
segment number: (i) $\eta(G)/2$, where $\eta(G)$ is the number of odd-degree vertices
of G, and (ii) the *slope number*, slope(G), of G, which is defined as follows.
The slope number slope(Γ) of a straight-line drawing Γ of G is the number of
different slopes used by any of the straight-line edges in Γ. Then slope(G) is
the minimum of slope(Γ) over all straight-line drawings Γ of G. Dujmović et al.
also showed that any tree T admits a drawing with seg(T) = $\eta(T)/2$ segments
and slope(T) = $\Delta(T)/2$ slopes, where $\Delta(T)$ is the maximum degree of a vertex
in T. These drawings, however, use exponential area. Recall that an *outerplanar
graph* is a plane graph that can be drawn such that all vertices lie on the outer
face. The *weak dual graph* of an outerplane graph is its dual graph without the
vertex corresponding to the outer face; it is known to be a tree. An outerplane
graph whose weak dual is a path is called an *outerpath*. A *maximal outerplanar
graph* is an outerplanar graph with the maximum number of edges. Dujmović et
al. showed that every maximal outerplanar graph G with n vertices admits an
outerplanar straight-line drawing with at most n segments. They showed that
this is worst-case optimal. They also gave (asymptotically) worst-case optimal
algorithms for 2-trees and plane (where the combinatorial embedding and outer
face is fixed) 3-trees. Finally, they showed that every triconnected planar graph
with n vertices can be drawn using at most $5n/2 - 3$ segments. For the special
cases of triangulations and 4-connected triangulations, Durocher and Mondal [6]
improved the upper bound of Dujmović et al. to $(7n - 10)/3$ and $(9n - 9)/4$,
respectively. The former bound implies a bound of $(16n - 3m - 28)/3$ for
arbitrary planar graphs with n vertices and m edges. Kindermann et al. [12]
observed that this implies that seg(G) $\leq (8n - 14)/3$ for any planar graph G: if
$m > (8n-14)/3$ this follows from the bound, otherwise any drawing of G is good
enough. Constructive linear-time algorithms that compute the segment number
of series-parallel graphs of maximum degree 3 and of maximal outerpaths were
given by Samee et al. [19] and by Adnan [1], respectively. Mondal et al. [17]
and Igamberdiev et al. [11] showed that every cubic triconnected planar graph
(except K_4) has segment number $n/2 + 3$. Hültenschmidt et al. [10] showed that
trees, maximal outerplanar graphs and planar 3-trees admit drawings on a grid of
polynomial size, using slightly more segments. Kindermann et al. [12] improved
some of these bounds.

Table 1. Bounds on the segment number for subclasses of planar graphs. By *existential upper bound* we mean an upper bound for the universal lower bound. Here, η is the number of odd-degree vertices and $\gamma = 3c_0 + 2c_1 + c_2$, where c_i is the number of simple cycles with exactly i cut vertices. Our results are shaded in gray.

Graph class	Universal lower bound		Existential upper bound		Existential lower bound		Universal upper bound	
planar conn.	1		1		$2n-2$	[5]	$(8n-14)/3$	[6,12]
planar 3-conn.	$\sqrt{2n}$	[5]	$O(\sqrt{n})$	[5]	$2n-6$	[5]	$5n/2-3$	[5]
planar 3-conn. 4-reg.	$\Omega(\sqrt{n})$	[7]	$O(\sqrt{n})$	[7]	n	P1	$n+3$	T2
planar 3-conn. 3-reg.	$n/2+3$	[5]	—		—		$n/2+3$	[11,17]
triangulation	$\Omega(\sqrt{n})$	[5]	$O(\sqrt{n})$	[5]	$2n-2$	[5]	$(7n-10)/3$	[6]
4-conn. triangulation	$\Omega(\sqrt{n})$	[5]	$O(\sqrt{n})$	[5]	$2n-6$	[5]	$(9n-9)/4$	[6]
planar 3-trees	$n+4$	T6	$n+7$	[7]	$3n/2$	[7]	$2n-2$	[5]
2-trees	$(n+7)/5$	T5	$(5n+24)/13$	[7]	$3n/2-2$	[5]	$3n/2$	[5]
maximal outerplanar	$(n+7)/5$	T5	$(5n+24)/13$	[7]	n	[5]	n	[5]
maximal outerpath	$\lfloor n/2 \rfloor +2$	T3	$\lfloor n/2 \rfloor +2$	P3	n	[5]	n	[5]
cactus	$\eta/2+\gamma$	T7	—		—		$\eta/2+\gamma$	T7

Other Related Work. Okamoto et al. [18] investigated variants of the segment number. For planar graphs in 2D, they allowed bends. For arbitrary graphs, they considered crossing-free straight-line drawings in 3D and straight-line drawings with crossings in 2D. They showed that all segment number variants are $\exists\mathbb{R}$-complete to compute, and they gave upper and existential lower bounds for the segment number variants of cubic graphs. The *arc number*, $\mathrm{arc}(G)$, of a graph G is the smallest number of circular arcs in any circular-arc drawings of G. It has been introduced by Schulz [20], who gave algorithms for drawing series-parallel graphs, planar 3-trees, and triconnected planar graphs with few circular arcs. For trees, he reduced the drawing area (from exponential to polynomial). Chaplick et al. [3,4] considered a different measure of the visual complexity, namely the number of lines (or planes) needed to cover crossing-free straight-line drawings of graphs in 2D (and 3D). Kryven et al. [16] considered spherical covers.

Contribution and Outline. In terms of universal upper bounds, we first show that every triconnected planar 4-regular graph with n vertices can be drawn using at most $n+3$ segments (note that there are $2n$ edges); see Sect. 2. This bound is tight up to an additive constant, improves a previous upper bound of $7n/4+2$ implied by a more general result [5, Theorem 15] of Dujmović et al., and supplements the result for cubic graphs due to Mondal et al. [17] and Igamberdiev et al. [11]. Our algorithm works even for *plane* graphs and produces drawings that are *convex*, that is, the boundary of each face corresponds to a convex polygon. We remark that triconnected planar 4-regular graphs are a rich and natural graph class that comes with a simple set of generator rules [2]. It might seem tempting to prove our result inductively by means of these rules, though we have not been able make this idea work. Instead, our algorithm relies on a decomposition of the graph along carefully chosen paths (Lemma 3), which might be of independent interest. We also give a simple optimal (cf. Table 1) algorithm for cactus graphs[1] (see Sect. 4), generalizing the result of Dujmović et al. for trees.

We prove the first linear universal lower bounds for maximal outerpaths ($\lfloor n/2 \rfloor + 2$; see Sect. 3), maximal outerplanar graphs as well as 2-trees ($(n+7)/5$; see Sect. 4), and planar 3-trees ($n + 4$; see Sect. 4). This makes the corresponding algorithms of Dujmović et al. constant-factor approximation algorithms. For Adnan's algorithm [1] that computes the segment number of maximal outerpaths, our result provides a lower bound on the size of the solution. For maximal outerpaths, our bound is best possible and can be generalized to circular arcs. For planar 3-trees, the bound is best possible up to the additive constant. Known and new results are listed in Table 1. Claims marked with "$\star$" are proved in [7].

Notation and Terminology. All graphs in this paper are simple. For any graph G, let $\mathrm{V}(G)$ be the vertex set and $\mathrm{E}(G)$ the edge set of G. Now let Γ be a planar drawing of a planar and connected graph G. The boundary ∂f of each face f of Γ can be uniquely described by a counterclockwise sequence of edges. If G is biconnected, then ∂f is a simple cycle. The collection of the boundaries of all faces of Γ is called the *combinatorial embedding* of Γ. The unique unbounded face of Γ is called its *outer* face; the remaining faces are called *internal*. Vertices (edges) belonging to the boundary of the outer face are called *outer* vertices (edges); the remaining vertices (edges) are called *internal*. A *plane* graph is a planar graph equipped with a combinatorial embedding and a distinguished outer face. A path in a plane graph is *internal* if its edges and interior vertices do not belong to its outer face. For any $k \in \mathbb{N}$, we use $[k]$ as shorthand for $\{1, 2, \ldots, k\}$.

2 Triconnected 4-Regular Planar Graphs

This section is concerned with the segment number of 3-connected 4-regular planar graphs. We establish a universal upper bound of $n + 3$ segments, which we complement with an existential lower bound of n segments, where n denotes the number of vertices.

Overview. Towards the upper bound, we will show that each graph of the considered class admits a drawing where all but three of its vertices are placed in the interior of some segment. In such a drawing, each of these vertices is the endpoint of at most two segments. The claimed bound then follows from the fact that each segment has exactly two endpoints.

To construct the desired drawings, we follow a strategy that has already been used in an algorithm by Hong and Nagamochi [9], which was sped up by Klemz [15]. Both algorithms generate convex drawings of so-called hierarchical

---

[1] A *cactus* is a connected graph where any two simple cycles share at most one vertex.

plane st-graphs, but they can also be applied to "ordinary" plane graphs. In this context, the algorithmic framework is as follows: the input is an internally (defined below, see Definition 1) 3-connected plane graph G and a convex drawing Γ^o of the boundary of its outer face. The task is to extend Γ^o to a convex drawing of G. The main idea of both algorithms is to choose a suitable internal vertex y of the given graph G and compute three disjoint (except for y) paths P_1, P_2, P_3 from y to the outer face. Each of these paths is then embedded as a straight-line segment so that Γ^o is dissected into three convex polygons, for an illustration see Fig. 1a. The graphs corresponding to the interior of these polygons can now be handled recursively. To ensure that a solution exists, the computed paths (as well as the paths corresponding to the segments of Γ^o) need to be *archfree*, meaning that they are not arched by an internal face: a path P is *arched* by a face a between $u, v \in V(\partial a) \cap V(P)$ if the subpath P_{uv} of P between u and v is interior-disjoint from ∂a, see Fig. 1a. Indeed, if a is internal, then such a path P cannot be realized as a straight-line segment in a convex drawing since the interior of the segment uv has to be disjoint from the realization of a. We follow the idea of dissecting our graphs along archfree paths. However, to ensure that each internal vertex is placed in the interior of some segment, the way in which we construct our paths is necessarily quite different. Specifically, we will show that a large subfamily of the considered graph class can be dissected along three archfree paths that are arranged in a windmill pattern as depicted in Fig. 2a.

Existence of Convex Drawings. It is well-known that a plane graph admits a convex drawing if and only if it is a subdivision of an *internally 3-connected* graph [8,9,21,22]. There are multiple ways to define this property and it will be convenient to refer to all of them. Therefore, we use the following well-known characterization; for a proof, see, e.g., [14].

Definition 1. *Let G be a plane 2-connected graph. Let o denote its outer face. Then G is called* internally 3-connected *if and only if the following equivalent statements are satisfied:*

(I1) Inserting a new vertex v in o and adding edges between v and all vertices of ∂o results in a 3-connected graph.

(I2) From each internal vertex w of G there exist three paths to o that are pairwise disjoint except for the common vertex w.

(I3) Every separation pair u, v of G is external, *i.e., u and v lie on ∂o and every connected component of the subgraph of G induced by $V(G) \setminus \{u, v\}$ contains a vertex of ∂o.*

Observation 1 ($\star$, folklore). *Let G be an internally 3-connected plane graph, and let C be a simple cycle in G. The closed interior C^- of C is an internally 3-connected plane graph.*

In the context of our recursive strategy, we face a special case of the following problem: given an internally 3-connected plane graph G and a convex drawing Γ^o

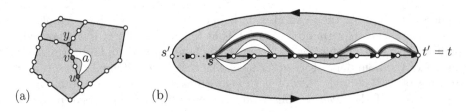

Fig. 1. (a) Splitting Γ^o along three straight-line paths. The subpolygon containing arch a cannot be extended to a convex drawing of its subgraph. (b) Left-aligned path $L_G(P)$ of $P = (s, \ldots, t)$.

of the boundary of its outer face, extend Γ^o to a convex drawing of G. It is known that such an extension exists if and only if each segment of Γ^o corresponds to an archfree path of G [8, 9, 21, 22]. Hence, we say that Γ^o is *compatible* with G if and only if it satisfies this property.

Construction of Archfree Paths. The following lemma gives rise to a strategy for transforming a given internal path into an archfree path:

Lemma 1 ([9, Lem. 1]). *Let G be an internally 3-connected plane graph, f an internal face of G. Any subpath P of ∂f with $|\mathrm{E}(P)| \leq |\mathrm{E}(\partial f)| - 2$ is archfree.*

Let G be an internally 3-connected graph. Consider the edges of the outer face ∂o of G to be directed in counterclockwise direction. Assume that there are two distinct vertices s' and t' on ∂o that are joined by a simple internal path P'. Consider P' to be directed from s' to t' and let $P = (s, \ldots, t)$ be a directed subpath of P'. Suppose that P is arched by an internal face a. Then we say a arches P *from the left* if a is interior to the cycle formed by P' and the directed $t's'$-path on ∂o; otherwise, we say that a arches P *from the right*. The *left-aligned* path $L_G(P)$ of P is obtained be exhaustively applying the following modification (for an illustration see Fig. 1b): suppose that an internal face a arches P from the left between two vertices u, v such u precedes v along P. Transform P by replacing its uv-subpath with the uv-path obtained by walking along ∂a in counterclockwise direction from u to v. The *right-aligned* path $R_G(P)$ is defined symmetrically.

Lemma 2 ([8, Lemma 5, Corollary 6]). *Let G be an internally 3-connected plane graph. Let $P = (s, \ldots, t)$ be a subpath of a simple internal directed path P' between two distinct outer vertices of G. Then:*

- *$L_G(P)$ $(R_G(P))$ is a simple internal st-path not arched from the left (right).*
- *If P is not arched from the right (left) by an internal face, then $L_G(P)$ $(R_G(P))$ is archfree.*
- *$R_G(L_G(P))$ $(L_G(R_G(P)))$ is archfree.*

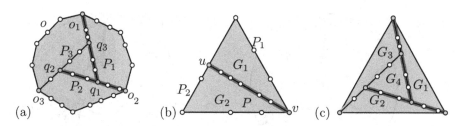

Fig. 2. (a) A windmill (P_1, P_2, P_3). (b,c) The 3-connected case in the proof of Theorem. 1.

Existence of Archfree Windmills. Recall that our plan is to dissect our given (internally) 3-connected graph along three archfree paths that form a windmill pattern; see Fig. 2a.

Definition 2. *Let G be an internally 3-connected plane graph and let o denote its outer face. For $i \in [3]$, let $P_i = (o_i, \ldots, q_i)$ be a simple path in G. We call (P_1, P_2, P_3) a* windmill *of G if and only if all of the following properties hold (all indices are considered modulo 3):*

(W1) The vertices o_1, o_2, o_3 are pairwise distinct and belong to ∂o.
(W2) For $i \in [3]$, no vertex of $V(P_i) \setminus \{o_i\}$ belongs to ∂o.
(W3) For $i \in [3]$, no interior vertex of P_i belongs to P_{i+1}.
(W4) For $i \in [3]$, the endpoint q_i is an interior vertex of P_{i+1}.

If (P_1, P_2, P_3) is a windmill of G, we call it archfree *if P_1, P_2, P_3 are archfree.*

A necessary condition for the existence of an archfree windmill is the existence of a *strictly* internal face (a face without outer vertices). For the considered graph class we show that the condition is sufficient. The following lemma is the main technical contribution of this section:

Lemma 3 ($\star$). *Let G be an internally 3-connected plane graph of maximum degree 4 with a strictly internal face f. Then G contains an archfree windmill.*

Proof (sketch). Let o be the outer face of G. By means of the internal 3-connectivity of G and Lemma 2, it can be shown that there are three pairwise disjoint archfree paths $P_i = (o_i, \ldots, f_i), i \in [3]$ between ∂o and ∂f as depicted in Fig. 3a. We now walk along ∂f in a clockwise fashion and append appropriate parts of ∂f to the paths P_1, P_2, P_3 to obtain an initial windmill $(P_1^{\mathrm{cw}}, P_2^{\mathrm{cw}}, P_3^{\mathrm{cw}})$ as illustrated in Fig. 3b. Specifically, we extend each P_i by the $f_i f_{i+1}$ subpath of ∂f that does not contain f_{i+2} (indices are considered modulo 3). This windmill is not necessarily archfree, but its paths can only be arched in a controlled way: suppose that P_i^{cw} is arched by an internal face a_i^{cw}. The subpath of P_i^{cw} that belongs to ∂f is archfree by Lemma 1. Combined with the fact that P_i is archfree, it follows that a_i^{cw} arches P_i^{cw} between some vertex $s_i^{\mathrm{cw}} \in V(P_i) \setminus \{f_i\}$ and a vertex $t_i^{\mathrm{cw}} \in V(P_i^{\mathrm{cw}}) \setminus V(P_i)$. Moreover, by planarity, a_i^{cw} has to arch P_i^{cw}

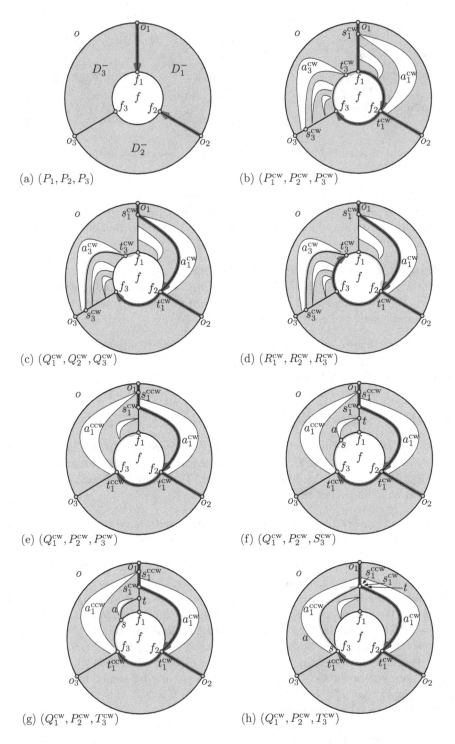

(a) (P_1, P_2, P_3)

(b) $(P_1^{\mathrm{cw}}, P_2^{\mathrm{cw}}, P_3^{\mathrm{cw}})$

(c) $(Q_1^{\mathrm{cw}}, Q_2^{\mathrm{cw}}, Q_3^{\mathrm{cw}})$

(d) $(R_1^{\mathrm{cw}}, R_2^{\mathrm{cw}}, R_3^{\mathrm{cw}})$

(e) $(Q_1^{\mathrm{cw}}, P_2^{\mathrm{cw}}, P_3^{\mathrm{cw}})$

(f) $(Q_1^{\mathrm{cw}}, P_2^{\mathrm{cw}}, S_3^{\mathrm{cw}})$

(g) $(Q_1^{\mathrm{cw}}, P_2^{\mathrm{cw}}, T_3^{\mathrm{cw}})$

(h) $(Q_1^{\mathrm{cw}}, P_2^{\mathrm{cw}}, T_3^{\mathrm{cw}})$

Fig. 3. Evolution of the three paths in the first (subfigures (a)–(d)) and second (subfigures (e)–(h)) part of the proof of Lemma 3.

from the left, as illustrated in Fig. 3b. We remark that there might be multiple "nested" faces that arch P_i^{cw}. W.l.o.g., we use a_i^{cw} to denote the "outermost" one, that is, the unique arch whose boundary replaces a part of P_i^{cw} in the left-aligned path $Q_i^{cw} = L_G(P_i^{cw})$, see Fig. 3c. The paths of $(Q_1^{cw}, Q_2^{cw}, Q_3^{cw})$ are now archfree by Lemma 2, though, (W4) from Definition 2 is satisfied only for exactly those $i \in [3]$ where the arch P_{i+1}^{cw} is archfree. For each Q_i^{cw} where (W4) is violated, we append the $f_{i+1}t_{i+1}^{cw}$-path of ∂f that does not contain f_i, see Fig. 3d. This modification maintains the archfreeness by planarity and Lemma 1. However, the resulting path triple $(R_1^{cw}, R_2^{cw}, R_3^{cw})$ might still not be a wind-mill: suppose that a path P_i^{cw} is not archfree and its arching face a_i^{cw} is *big*, that is, $t_i^{cw} = f_{i+1}$, while additionally the path P_{i+1}^{cw} is archfree (this is the case for $i = 1$ in Fig. 3b). Then (W3) from Definition 2 is violated for R_{i+1}^{cw} and (W4) is violated for R_{i+2}^{cw}. Suppose that $(R_1^{cw}, R_2^{cw}, R_3^{cw})$ is indeed not a windmill. We construct path triples $(P_1^{ccw}, P_2^{ccw}, P_3^{ccw})$, $(Q_1^{ccw}, Q_2^{ccw}, Q_3^{ccw})$, and $(R_1^{ccw}, R_2^{ccw}, R_3^{ccw})$ in a symmetric fashion by walking around ∂f in counterclock-wise direction. If $(R_1^{ccw}, R_2^{ccw}, R_3^{ccw})$ is also not a windmill, it follows that both $(R_1^{cw}, R_2^{cw}, R_3^{cw})$ and $(R_1^{ccw}, R_2^{ccw}, R_3^{ccw})$ contain a path that is arched by a big face. By planarity and the degree bounds, we can now argue that there is exactly one $i \in [3]$ such that both (P_i^{cw}) and (P_i^{ccw}) are arched by big faces while both (P_{i+1}^{cw}) and (P_{i+2}^{ccw}) are archfree, which is illustrated in Fig. 3e for $i = 1$. Assume w.l.o.g. that $i = 1$ and that s_1^{cw} is not closer to o_1 on P_1 than s_1^{ccw}. In view of the previous observations, it is now easy to argue that the paths of $(Q_1^{cw}, P_2^{cw}, P_3^{cw})$ are archfree and satisfy all windmill properties with the exception of (W4) for $i = 3$. We restore (W4) by appending the $f_1s_1^{cw}$-subpath of P_1 to P_3^{cw}, see Fig. 3f. By means of the degree bounds, it can be argued that (W2) and (W4) are main-tained for $i = 3$. The resulting path S_3^{cw} might now be arched (from the left, by planarity), which can be remedied by applying Lemma 2, see Figs. 3g and h. By means of the degree bounds and planarity arguments, it can be shown that this modification maintains all windmill properties.   □

A plane graph G is *internally 4-regular* if all of its internal vertices have degree 4 and its outer vertices have degree at most 4. In Lemma 3, we established that the existence of an internal face suffices for the existence of an archfree windmill. By means of simple counting arguments, it can be shown that this condition is satisfied if G has a triangular outer face.

Lemma 4 ($\star$). *Let G be an internally 3-connected plane graph that is internally 4-regular. Let o denote the outer face of G and assume $|\partial o| = 3$. Then G has a strictly internal face.*

Theorem 1 ($\star$). *Let G be an internally 3-connected internally 4-regular plane graph and let Γ^o be a compatible convex drawing of its outer face. There exists a convex drawing Γ of G that uses Γ^o as the realization of the outer face where each internal vertex of G is contained in the interior of some segment of Γ.*

Proof (sketch). Our goal is to (recursively) compute coordinates for the internal vertices to obtain the desired drawing of G. The base case of the recursion is that G contains no internal edges, in which case there is nothing to show. Assume that G is 3-connected – we deal with the case where G is not 3-connected in the full version. If $|\mathrm{V}(\Gamma^o)| \geq 4$, then there exist two distinct outer vertices u, v that do not belong to a common segment of Γ^o, see Fig. 2b. By 3-connectivity and Lemma 2, they are joined by an archfree internal path P. We split Γ^o into two simple convex polygons along P and handle the two corresponding subgraphs recursively. If $|\mathrm{V}(\Gamma^o)| = 3$, then G contains an archfree windmill (P, S, Q) by Lemmas 4 and 3. Since the three outer endpoints of P, S, Q do not belong to a common segment of Γ^o, we can embed them in a straight-line fashion such that Γ^o is dissected into four simple convex polygons, see Fig. 2c. We handle the corresponding four subgraphs recursively.   □

Universal Upper Bound. Recall that to establish the claimed upper bound, it suffices to create a drawing where all but three of the vertices of the graph are drawn in the interior of some segment. To achieve this goal, we can now draw the outer face of the graph as a triangle and then apply Theorem 1.

Theorem 2 ($\star$). *Every 3-connected internally 4-regular plane graph G admits a convex drawing on at most $n + 3$ segments where n is the number of vertices.*

Existential Lower Bound. For a graph G, let G^2 denote the *square* of G, that is, $\mathrm{V}(G) = \mathrm{V}(G^2)$, and two vertices in G^2 are adjacent if and only if their distance in G is at most 2. For $n \geq 6$, the square of the n-cycle, C_n^2, is 4-regular and triconnected. By removing three edges from a drawing Γ of C_n^2, we obtain a drawing of a graph whose segment number is n [5, proof of Theorem 7]. Consequently, Γ uses at least $n-3$ segments. We prove a slightly stronger bound.

Proposition 1 ($\star$). *For even $n \geq 6$, C_n^2 is planar and $\mathrm{seg}(C_n^2) \geq n$.*

3 Maximal Outerpaths

In this section, we generalize segments and arcs to pseudo-k-arcs (defined below) and give a universal lower bound for the number of pseudo-k-arcs in drawings of maximal outerpaths.

We call a sequence $v_1, v_2, \ldots, v_n$ of the vertices of a maximal outerpath G a *stacking order* of G if for each i, the graph G_i induced by the vertices $v_1, v_2, \ldots, v_i$ is a maximal outerpath. An arrangement of *pseudo-k-arcs* is a set of curves in the plane such that any two of the curves intersect at most k times. (If two curves share a tangent, this counts as two intersections.) We forbid self-intersections, but for $k \geq 2$ we allow a pseudo-k-arc to be closed.

To show the bound, we present a charging scheme that assigns internal edges to pseudo-k-arcs. Any drawing of a maximal outerpath has exactly $n-3$ internal edges. A pseudo-k-arc is *long* if it contains at least $k+1$ internal edges; otherwise it is *short*. Let arc_k denote the number of pseudo-k-arcs, and let arc_k^i denote the number of pseudo-k-arcs with i internal edges. The internal edges of a long arc α subdivide the outerpath into subgraphs $H_0, H_1, \ldots, H_\ell$ called *bays*; see Fig. 4. Given a drawing Γ of a maximal outerpath, we denote the sub-drawings

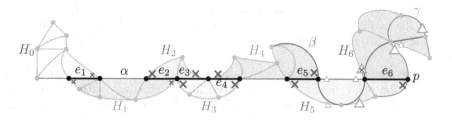

Fig. 4. An outerpath represented by a pseudo-2-arc arrangement. The internal edges $e_1, \ldots, e_6$ of arc α subdivide the outerpath into bays $H_0, \ldots, H_6$. We marked the bay crossings of α and β by red crosses and violet triangles, respectively. For the bay crossings in C that are relevant for our charging scheme we used larger symbols. (Color figure online)

of $G_3, G_4, \ldots, G_n$ within Γ by $\Gamma_3, \Gamma_4, \ldots, \Gamma_n$, respectively. A pseudo-k-arc α is *incident* to a face f if α contains an edge incident to a vertex of f. We say that α is *active* in Γ_i if α is incident to the last face that has been added.

Lemma 5 ($\star$). *For any $i \in \{3, \ldots, n\}$, a partial outerpath drawing Γ_i contains at most one active long pseudo k-arc.*

We do a 2-round assignment to assign each internal edge to a pseudo-k-arc. We start with the *round-1 assignment*. Let I denote the set of internal edges of long pseudo-k-arcs starting at the $(k + 1)$-th internal edge (as for the first k internal edges an arc is still short). We assign all $n - 3$ internal edges except for the edges in I to their own pseudo-k-arcs:

$$(n-3)-|I| = k\,\mathrm{arc}_k^{\geq k} +(k-1)\,\mathrm{arc}_k^{k-1} + \cdots +\mathrm{arc}_k^1 = k\,\mathrm{arc}_k -\textstyle\sum_{i=0}^{k}(k-i)\,\mathrm{arc}_k^i \quad (1)$$

Now we describe the *round-2 assignment*. There, we charge the internal edges of I to specific crossings, which we can charge in turn to pseudo-k-arcs. A *crossing* is a triple (α, β, p) that consists of two pseudo-k-arcs α and β and a point p at which α and β intersect. These specific crossings involve long arcs and we call them *bay crossings*. Next, we define them such that for each long pseudo-k-arc α with ℓ internal edges ($\ell > k$), there are 2ℓ bay crossings $(\alpha, *, *)$ where $*$ is a wildcard. For each bay $H \in \{H_1, \ldots, H_{\ell-1}\}$, we have two bay crossings: a crossing of α with another pseudo-k-arc at each of the two vertices of H that have degree 2 within H; see the red crosses in Fig. 4. Clearly, they exist for each H because H is an outerpath. Since these two vertices are distinct for each pair of consecutive bays, their bay crossings are distinct as well. Note that a tangential point may be shared by some H_j and H_{j+2} (for $j \in [\ell - 3]$); see, e.g., H_2 and H_4 in Fig. 4. However, we still have distinct bay crossings for H_j and H_{j+2} since a tangential point counts for two crossings. For each of H_0 and H_ℓ, there is one bay crossing defined next. In H_0 and H_ℓ, consider the two crossings of α at the internal edge e_1 and e_ℓ, respectively – one at each of the vertices of the internal edge. One of these vertices is the degree-2 vertex of H_1 ($H_{\ell-1}$) and hence may be identical with a bay crossing of H_1 ($H_{\ell-1}$). E.g., in Fig. 4, the bay crossing (α, γ, p) of H_5 occurs as one of the considered crossings of H_6. The other one of the two considered crossings cannot be a bay crossing in a neighboring bay and this is our bay crossing of H_0 (H_ℓ); see the red crosses at H_0 and H_6 in Fig. 4.

In the round-2 assignment, we charge the surplus internal edges of a long arc α to the other pseudo-k-arcs involved in bay crossings with α. For each internal edge of I, we have two distinct bay crossings of the preceding bay, e.g., in Fig. 4 H_2 provides two bay crossings for e_3. Let C be the set of these bay crossings. The bay crossings of $H_0, \ldots, H_{k-1}$, and H_ℓ are not included in C as the internal edges $e_1, \ldots, e_k$ are not contained in I and there is no $e_{\ell+1}$. Clearly, $2|I| = |C|$.

Next, we give an upper bound for $|C|$ in terms of arc_k. The main argument we exploit is that, by definition, each pseudo-k-arc can participate in at most k crossings with the (current) long arc and, hence, also in at most k bay crossings with the (current) long arc. However, we need to be a bit careful when one long pseudo-k-arc becomes inactive and a new pseudo-k-arc becomes long, i.e., we consider the transition between one long arc to a new long arc. A (not necessarily long) pseudo-k-arc γ could potentially contribute k crossings in C with each long arc. To compensate for the double counting at transitions, we introduce the *transition loss* t_k, which we define as $t_k = \sum_{\gamma \in \mathcal{A} \setminus \{\alpha_1\}} (|\{c = (*, \gamma, *) \mid c \in C\}| - k)$, where $\mathcal{A}$ is the set of all pseudo-k-arcs and α_1 is the first long arc in Γ. In other words, each pseudo-k-arc, while it is short, contributes to t_k the number of its bay crossings minus k. For example, in Fig. 4, γ contributes 1 to t_k: γ has one bay crossing in C with the long arc α (red cross at e_6) and two bay crossings in C with the long arc β (violet triangles on the top right). The arc β contributes -1 to t_k: β has one bay crossing in C with the long arc α.

Note that, while it is long, an arc does not cross other long arcs. Also, we do not count the crossings of the first k bays and the very last bay. Hence,

$$\underbrace{2|I| = |C| \leq}_{} \underbrace{k \cdot (\mathrm{arc}_k - 1)}_{} \qquad \underbrace{-(2k-1)}_{} \quad \overbrace{-1} \quad \underbrace{+ t_k}_{} \qquad (2)$$

The first long pseudo-k-arc does not provide crossings with another long pseudo-k-arc.

Each pseudo-k-arc intersects the current long arc at most k times.

Crossings of $H_0, H_1, \ldots, H_{k-1}$ of the first long arc are not in C.

The crossing of H_ℓ of the last long arc is not in C.

transition loss

Plugging Eq. (2) into Eq. (1), we obtain the following general formula, which gives a lower bound on the number of pseudo-k-arcs for any outerpath.

$$\mathrm{arc}_k \geq \left(2n - 6 + 2 \cdot \sum_{i=0}^{k} (k-i)\, \mathrm{arc}_k^i - t_k\right) / (3k) + 1 \qquad (3)$$

Since this formula still contains unresolved variables, we now resolve t_k.

Lemma 6 ($\star$). *There is a loss of at most one crossing per transition from one long pseudo-k-arc to another long pseudo-k-arc. Hence, $t_k \leq \max\{0, \mathrm{arc}_k^{>k} - 1\} \leq \mathrm{arc}_k^{>k} = \mathrm{arc}_k - \sum_{i=0}^{k} \mathrm{arc}_k^i$, where $\mathrm{arc}_k^{>k}$ is the number of long pseudo-k-arcs.*

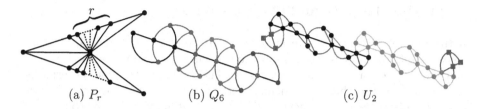

(a) P_r (b) Q_6 (c) U_2

Fig. 5. Families of maximal outerpaths with (a) $n/2 + 2$ segments (matching the lower bound in Theorem 3), (b) $n/3 + 1$ circular arcs, and (c) $(5n + 18)/16 < n/3$ pseudo 2-arcs.

By Lema 6 and Eq. (3),

$$\text{arc}_k \geq \left(2n + 3k - 6 + \sum_{i=0}^{k}(2k - 2i + 1)\,\text{arc}_k^i\right)/(3k + 1). \tag{4}$$

Into this general formula, we plug specific values of k and prove lower bounds on arc_k^i. We start with $k = 1$, i.e., outerpath drawings on pseudo segments.

Lemma 7 ($\star$). *For $k = 1$ and $n \geq 3$, in any outerpath drawing either $\text{arc}_1^0 \geq 3$ or ($\text{arc}_1^0 \geq 2$ and $\text{arc}_1^1 \geq 3$).*

Using Lema 7, we fill the gaps in Eq. 4 for $k = 1$ and obtain Theorem 3.

Theorem 3 ($\star$). *For any n-vertex maximal outerpath G, $\text{seg}(G) \geq \lfloor \frac{n}{2} \rfloor + 2$.*

For $k = 2$, i.e., for (pseudo) circular arcs, Eq. (4) leads to the following bound.

Theorem 4 ($\star$). *For any n-vertex maximal outerpath G, $\text{arc}(G) \geq \lceil \frac{2n}{7} \rceil$.*

For $k > 2$, it is not obvious how to generalize circular arcs. Still, we can make a similar statement for curve arrangements, which follows directly from Eq. (4).

Proposition 2. *Let G be an n-vertex maximal outerpath drawn on a curve arrangement in the plane s.t. curves intersect pairwise $\leq k$ times, can be closed, but do not self-intersect. Then, the number $\text{arc}_k(G)$ of curves required is $\lceil \frac{2n+3k-6}{3k+1} \rceil$.*

The infinite families of examples in Proposition 3 and Fig. 5 show that our bounds for segments and arcs are tight. This implies, somewhat surprisingly, that, at least for worst-case instances, using pseudo segments requires as many elements as using straight line segments. Whether this also holds for pseudo circular arcs and circular arcs is an open question. With circular arcs, we could not beat a bound of $n/3$, which we could do for pseudo circular arcs.

Proposition 3 ($\star$). *For every $r \in \mathbb{N}$, maximal outerpaths P_r, Q_r, U_r exist s.t.*

(i) P_r has $n = 2r + 6$ vertices and $\text{seg}(P_r) \leq r + 5 = n/2 + 2$,
(ii) Q_r has $n = 3r$ vertices and $\text{arc}(Q_r) \leq r + 1 = n/3 + 1$,
(iii) U_r has $n = 16r + 6$ vertices and $\text{arc}_2(U_r) \leq 5r + 3 = \frac{5n+18}{16} \approx 0.3125\,n$.

4   Further Results and Open Problems

In the full version [7], we give an alternative proof for Theorem 3, charging segment ends to vertices. We also give universal lower bounds on the segment numbers of 2-trees and maximal outerpaths. The key idea is to "glue" outerpaths, while adjusting the charging scheme. With a different charging scheme from segment ends to faces, we show an (almost) tight universal lower bound for planar 3-trees.

Theorem 5 ($\star$). *For a 2-tree (or a maximal outerplanar graph) G with n vertices, $\operatorname{seg}(G) \geq (n + 7)/5$.*

Theorem 6 ($\star$). *For a planar 3-tree G with $n \geq 6$ vertices, $\operatorname{seg}(G) \geq n + 4$.*

For cactus graphs, we can compute the segment number in linear time.

Theorem 7 ($\star$). *Given a cactus graph G, we can compute $\operatorname{seg}(G)$ in linear time. Within this timebound, we can draw G using $\operatorname{seg}(G)$ many segments. If G is given with an outerplanar embedding, the drawing will respect the given embedding.*

Now we turn to open problems. The most prominent one is to close the gaps in Table 1. Since circular-arc drawings are a generalization of straight-line drawings, it is natural to ask about the maximum ratio between the segment number and the arc number of a graph. We make some initial observations regarding this question in the full version [7]. Finally, what is the complexity of deciding whether the arc number of a given graph is strictly smaller than its segment number?

References

1. Adnan, M.A.: Minimum segment drawings of outerplanar graphs. Master's thesis, Department of Computer Science and Engineering, Bangladesh University of Engineering and Technology (BUET), Dhaka (2008). http://lib.buet.ac.bd:8080/xmlui/bitstream/handle/123456789/1565/Full%20%20Thesis%20.pdf?sequence=1&isAllowed=y
2. Broersma, H.J., Duijvestijn, A.J.W., Göbel, F.: Generating all 3-connected 4-regular planar graphs from the octahedron graph. J. Graph Theory **17**(5), 613–620 (1993). https://doi.org/10.1002/jgt.3190170508
3. Chaplick, Steven, Fleszar, Krzysztof, Lipp, Fabian, Ravsky, Alexander, Verbitsky, Oleg, Wolff, Alexander: The complexity of drawing graphs on few lines and few planes. In: Ellen, F., Kolokolova, A., Sack, J.-R. (eds.) WADS 2017. LNCS, vol. 10389, pp. 265–276. Springer, Cham (2017). https://doi.org/10.1007/978-3-319-62127-2_23
4. Chaplick, S., Fleszar, K., Lipp, F., Ravsky, A., Verbitsky, O., Wolff, A.: Drawinggraphs on few lines and few planes. J. Comput. Geom **11**(1), 433–475 (2020). https://doi.org/10.20382/jocg.v11i1a17

5. Dujmović, V., Eppstein, D., Suderman, M., Wood, D.R.: Drawings of planar graphs with few slopes and segments. Comput. Geom. Theory Appl. **38**(3), 194–212 (2007). https://doi.org/10.1016/j.comgeo.2006.09.002
6. Durocher, S., Mondal, D.: Drawing plane triangulations with few segments. Comput. Geom. Theory Appl. **77**, 27–39 (2019). https://doi.org/10.1016/j.comgeo.2018.02.003
7. Goeßmann, I., et al.: The segment number: Algorithms and universal lower bounds for some classes of planar graphs. arXiv preprint (2022). https://arxiv.org/abs/2202.11604
8. Hong, S., Nagamochi, H.: Convex drawings of graphs with non-convex boundary constraints. Discret. Appl. Math. **156**(12), 2368–2380 (2008). https://doi.org/10.1016/j.dam.2007.10.012
9. Hong, S., Nagamochi, H.: Convex drawings of hierarchical planar graphs and clustered planar graphs. J. Discrete Algorithms **8**(3), 282–295 (2010). https://doi.org/10.1016/j.jda.2009.05.003
10. Hültenschmidt, G., Kindermann, P., Meulemans, W., Schulz, A.: Drawing planar graphs with few geometric primitives. J. Graph Alg. Appl. **22**(2), 357–387 (2018). https://doi.org/10.7155/jgaa.00473
11. Igamberdiev, A., Meulemans, W., Schulz, A.: Drawing planar cubic 3-connected graphs with few segments: Algorithms & experiments. J. Graph Algorithms Appl. **21**(4), 561–588 (2017). https://doi.org/10.7155/jgaa.00430
12. Kindermann, P., Mchedlidze, T., Schneck, T., Symvonis, A.: Drawing planar graphs with few segments on a polynomial grid. In: Archambault, D., Tóth, C.D. (eds.) GD 2019. LNCS, vol. 11904, pp. 416–429. Springer, Cham (2019). https://doi.org/10.1007/978-3-030-35802-0_32
13. Kindermann, P., Meulemans, W., Schulz, A.: Experimental analysis of the accessibility of drawings with few segments. J. Graph Alg. Appl. **22**(3), 501–518 (2018). https://doi.org/10.7155/jgaa.00474
14. Kleist, L., Klemz, B., Lubiw, A., Schlipf, L., Staals, F., Strash, D.: Convexity-increasing morphs of planar graphs. Comput. Geom. **84**, 69–88 (2019). https://doi.org/10.1016/j.comgeo.2019.07.007
15. Klemz, B.: Convex drawings of hierarchical graphs in linear time, with applications to planar graph morphing. In: Mutzel, P., Pagh, R., Herman, G., (eds.) Proceedings of 29th Annual European Symposium on Algorithms (ESA 2021), vol. 204 of LIPIcs, pp. 57:1–57:15. Schloss Dagstuhl – Leibniz-Zentrum für Informatik (2021). https://doi.org/10.4230/LIPIcs.ESA.2021.57
16. Kryven, M., Ravsky, A., Wolff, A.: Drawing graphs on few circles and few spheres. J. Graph Alg. Appl. **23**(2), 371–391 (2019). https://doi.org/10.7155/jgaa.00495
17. Mondal, D., Nishat, R.I., Biswas, S., Rahman, M.: Minimum-segment convex drawings of 3-connected cubic plane graphs. J. Comb. Optim. **25**(3), 460–480 (2013). https://doi.org/10.1007/s10878-011-9390-6
18. Okamoto, Y., Ravsky, A., Wolff, A.: Variants of the segment number of a graph. In: Archambault, D., Tóth, C.D. (eds.) GD 2019. LNCS, vol. 11904, pp. 430–443. Springer, Cham (2019). https://doi.org/10.1007/978-3-030-35802-0_33
19. Samee, M.A.H., Alam, M.J., Adnan, M.A., Rahman, M.S.: Minimum segment drawings of series-parallel graphs with the maximum degree three. In: Tollis, I.G., Patrignani, M. (eds.) GD 2008. LNCS, vol. 5417, pp. 408–419. Springer, Heidelberg (2009). https://doi.org/10.1007/978-3-642-00219-9_40
20. Schulz, A.: Drawing graphs with few arcs. J. Graph Alg. Appl. **19**(1), 393–412 (2015). https://doi.org/10.7155/jgaa.00366

21. Thomassen, C.: Plane representations of graphs. In: Bondy, J.A., Murty, U.S.R., (eds.) Progress in Graph Theory, pp. 43–69. Academic Press (1984)
22. Tutte, W.T.: Convex representations of graphs. Proc. London Math. Soc. s3–10(1), 304–320 (1960). https://doi.org/10.1112/plms/s3-10.1.304

Bounding Twin-Width for Bounded-Treewidth Graphs, Planar Graphs, and Bipartite Graphs

Hugo Jacob[1]([⊠]) and Marcin Pilipczuk[2]

[1] ENS Paris-Saclay, Gif-sur-Yvette, France
hugo.jacob@ens-paris-saclay.fr
[2] Institute of Informatics, University of Warsaw, Warsaw, Poland
malcin@mimuw.edu.pl

Abstract. Twin-width is a newly introduced graph width parameter that aims at generalizing a wide range of "nicely structured" graph classes. In this work, we focus on obtaining good bounds on twin-width $\mathbf{tww}(G)$ for graphs G from a number of classic graph classes. We prove the following:

- $\mathbf{tww}(G) \leq 3 \cdot 2^{\mathbf{tw}(G)-1}$, where $\mathbf{tw}(G)$ is the treewidth of G,
- $\mathbf{tww}(G) \leq \max(4\mathbf{bw}(G), \frac{9}{2}\mathbf{bw}(G) - 3)$ for a planar graph G with $\mathbf{bw}(G) \geq 2$, where $\mathbf{bw}(G)$ is the branchwidth of G,
- $\mathbf{tww}(G) \leq 183$ for a planar graph G,
- the twin-width of a universal bipartite graph $(X, 2^X, E)$ with $|X| = n$ is $n - \log_2(n) + \mathcal{O}(1)$.

An important idea behind the bounds for planar graphs is to use an embedding of the graph and sphere-cut decompositions to obtain good bounds on neighbourhood complexity.

Keywords: Twin-width · Planar graphs · Treewidth

1 Introduction

Twin-width is a graph parameter recently introduced by Bonnet et al. [7], which has already proven to be very versatile and useful. It is defined via iterated contraction of vertices that are almost twins, while limiting the amount of errors that are carried on. Twin-width is known, for instance, to be bounded on classes of graphs of bounded treewidth, bounded rank-width, or excluding a fixed minor

This research is part of projects that have received funding from the European Research Council (ERC) under the European Union's Horizon 2020 research and innovation programme Grant Agreement 714704. Initial part of the reseach was done when Hugo Jacob was on an internship at University of Warsaw in Spring and Summer 2021. The authors acknowledge support from the ERC starting grant "CRACKNP" (Grant Agreement 853234) for attending the conference.

© Springer Nature Switzerland AG 2022
M. A. Bekos and M. Kaufmann (Eds.): WG 2022, LNCS 13453, pp. 287–299, 2022.
https://doi.org/10.1007/978-3-031-15914-5_21

[7]. It is also possible to design algorithms on the contraction sequences, thus providing a common framework for efficient algorithms on several graph classes [6, 7]. Twin-width is also linked to First Order logic, FO model checking is FPT for graphs of bounded twin-width, and FO transductions preserve twin-width boundedness [7] (see also [12]). However, finding good contraction sequences is hard [4]. Furthermore, no efficient approximation algorithm is known yet. This motivates looking at some simple classes and comparing twin-width to other parameters (the case of poset width has already been considered [2] for instance). Bounds on the twin-width of general graphs and random graphs have also been established [1].

Many currently known bounds on the twin-width, in particular for minor-closed graph classes such as planar graphs, rely on very general arguments and result in unreasonably large constants. Finding a better bound was explicitly mentionned as an open problem. In this paper, we present a few results we obtained while looking for an improved bound.

We first give some results on graphs of bounded treewidth: an exponential bound on the twin-width of a graph of bounded treewidth, and a linear bound on the twin-width of planar graphs of bounded treewidth. We then obtain a bound of 183 on the twin-width of planar graphs, which is, to the best of our knowledge, currently the best known bound. We were not able to prove a matching exponential lower bound for the twin-width of graphs of bounded treewidth. As a partial result in this direction, we determine the twin-width of universal bipartite graphs up to a constant additive term.

Independently of this work, Bonnet, Kwon, and Wood [8] obtained a bound of 583 on the twin-width of planar graphs, among other results on more general classes such as bounded genus graphs.

2 Preliminaries

In the following $[n]$ denotes $\{1, \ldots, n\}$. Given a set X, $|X|$ denotes its cardinality and 2^X denotes the set of subsets of X.

The subgraph induced by a vertex subset A in a graph G is denoted by $G[A]$, $G - A$ denotes $G[V \setminus A]$. For a graph $G = (V, E)$ and B a subset of E, $G - B$ denotes $(V, E \setminus B)$. The neighbourhood of vertex v in $G = (V, E)$ is $N(v) = \{w \in V | \{v, w\} \in E\}$, and we extend this notation with $N(X) = \left(\bigcup_{x \in X} N(x) \right) \setminus X$. To emphasize that the neighbourhood is taken in graph G, we use N_G instead of N.

We call neighbourhood classes with respect to Y in X the set $\Omega(X, Y) = \{N(x) \cap Y : x \in X\}$. Note that if $|Y| = k$, then $|\Omega(X, Y)| \leq 2^k$.

We call universal bipartite graph the bipartite graph $\mathcal{B}(n) = ([n], 2^{[n]}, \{(k, A \cup \{k\}) : k \in [n], A \in 2^{[n] \setminus \{k\}}\})$.

We now define formally the notion of *twin-width* of a graph. A *trigraph* is a triple $G = (V, E, R)$ where E and R are disjoint sets of edges on V, the (usual) edges and the red edges respectively. The notion of induced subgraph is extended to trigraphs in the obvious way. We denote by $R(v)$ the red neighbourhood of

v. A trigraph (V, E, R) such that (V, R) has maximum degree at most d is a d-trigraph. Any graph (V, E) can be seen as the trigraph $(V, E, \varnothing)$. Given a trigraph $G = (V, E, R)$ and two vertices u, v of V, the trigraph $G' = (V', E', R')$ obtained by the contraction[1] of u, v into a new vertex w is defined as the trigraph on vertex set $V' = V \setminus \{u, v\} \cup \{w\}$, such that $G - \{u, v\} = G' - \{w\}$, and such that $N_{G'}(w) = N_G(u) \cap N_G(v)$ and $R_{G'}(w) = R_G(u) \cup R_G(v) \cup (N_G(u) \Delta N_G(v))$, where Δ denotes the symmetric difference. A d-contraction sequence of G is a sequence of trigraph contractions starting with G and ending with the single-vertex trigraph, such that all intermediate trigraphs have maximum red degree d. The twin-width of graph G is the minimum d such that there exists a d-contraction sequence, it is denoted $\mathbf{tww}(G)$. A partial d-contraction sequence is a sequence of trigraph contractions starting with G such that intermediate trigraphs have maximum red degree d (i.e. we removed the constraint of ending with a single-vertex trigraph).

We use the notation of [9] for tree decompositions. Given a rooted tree T, N_T denotes its nodes, $\leq_T$ denotes its ancestor relation which is a partial order on N_T where the root is the maximal element, and the leaves are the minimal elements. For a fixed node u of T, we denote by $p(u)$ its parent (minimal strict ancestor), by $T_{\leq}(u)$ the set $\{w \in N_T | w \leq_T u\}$ and similarly for $T_<(u), T_{\geq}(u), T_>(u)$. A tree T is normal for graph G if $V(G) = N_T$, and for each edge of G, its endpoints are comparable under $<_T$. We denote by (T, f) a tree decomposition of G where T is a rooted tree, f maps N_T to $2^{V(G)}$ and satisfies the following conditions: every vertex of G is contained in at least one bag $f(u)$, for every edge of G there is a bag containing its two endpoints, and for every vertex v of G, the nodes u such that $f(u)$ contains v induce a connected subgraph of T. (T, f) is normal if T is normal for G, $f(u) \subseteq T_{\geq}(u)$ and $u \in f(u)$, for every $u \in N_T$. $f^*(u)$ denotes $f(u) \setminus \{u\}$. (T, f) is clean if it is normal, $f^*(u) = N_G(T_{\leq}(u)) \cap T_>(u)$ for every node u of T, and $p(u) \in f(u)$ for every node u of T except its root. The width of (T, f) is $\max_{u \in N_T} |f(u)| - 1$, and the treewidth of a graph is the minimum width over its tree decompositions. It is denoted by $\mathbf{tw}(G)$.

Let Σ be a sphere $\{(x, y, z) \in \mathbb{R}^3 | x^2 + y^2 + z^2 = 1\}$. A Σ-plane graph G is a planar graph embedded in Σ without crossing edges. To simplify notations, we do not distinguish vertices and edges from the points of Σ representing them. An O-arc is a subset of Σ homeomorphic to a circle. An O-arc in Σ is a noose if it meets G only in vertices and intersects every face at most once. The set of vertices met by a noose N is denoted $V(N)$, the length of the noose is $|V(N)|$, the number of vertices it meets. Every noose N bounds two open discs Δ_1, Δ_2 in Σ, i.e., $\Delta_1 \cap \Delta_2 = \varnothing$ and $\Delta_1 \cup \Delta_2 \cup N = \Sigma$.

A branch decomposition (T, μ) of a graph G consists of a ternary tree T (internal vertices of degree 3) and a bijection $\mu : L \rightarrow E(G)$ from the set L of leaves of T to the edge set of G. For every edge e of T, the middle set of e is a subset of $V(G)$ corresponding to the common vertices of the two graphs induced by the edges associated to the leaves of the two connected components of $T - e$. The width of the decomposition is the maximum cardinality of the

---

[1] The vertices are not required to be adjacent.

middle sets over all edges of T. An optimal decomposition is one with minimum width, which is called *branchwidth* and denoted by $\mathbf{bw}(G)$.

For a Σ-plane graph G, a *sphere-cut decomposition* (T, μ) is a branch decomposition such that for every edge e of T, there exists a noose N_e meeting G only on the vertices of the middle set of e and such that the two graphs induced by the edges associated to the leaves of the two connected components of $T - e$ are each on one side of N_e. The following result is stated in [10] as a consequence of the results of Seymour and Thomas [15], and Gu and Tamaki [13].

Lemma 1. *Let G be a connected Σ-plane graph of branchwidth at most ℓ without vertices of degree one. There exists a sphere-cut decomposition of G of width at most ℓ, and it can be computed in time $\mathcal{O}(|V(G)|^3)$.*

A sphere-cut decomposition (T, μ) can be rooted by subdividing an edge e of T into two edges e', e'' with middle vertex s, and adding a root r connected to s. The middle set of e' and e'' is the middle set of e, and $\{r, s\}$ has an empty middle set. For every edge e of T, the subtree of $T - e$ that does not contain the root is called the *lower part*, we denote by G_e the subgraph induced by the edges associated to the leaves of the lower part. For an internal node v of T, the edge incident to v on the path to r, is called the parent edge, and the other two are called children edges. There can be at most 2 vertices common to the middle sets of these three edges [10].

We slightly extend sphere-cut decompositions to cover the case of connected graphs with minimal degree one and branchwidth at least 2. Consider a connected graph G, let G' be its maximal induced subgraph with no vertex of degree one. Note that G' must be connected and that the graph H induced by the edges $E(G) \setminus E(G')$ is a forest where each tree has only a vertex in common with G', which we will consider as its root. We can first compute a sphere-cut decomposition (T', μ') of G' and then for each root r of a tree H_i in H, we can find an edge e of T' such that r is in its middle set (it exists because r has degree at least 2 in G'), and attach an optimal branch decomposition of H_i on e. This does not increase the branchwidth because r was already in the middle set of e. Once this is done for all trees H_i in H, we obtain a branch decomposition (T, μ) of G, such that there exists a noose meeting exactly the middle set of each edge of T. However, the nooses do not correspond to cycles in the radial graph anymore since we have to embed the H_i in faces of G'.

Lemma 2. *Let G be a connected Σ-plane graph of branchwidth $\ell \geq 2$. There exists a sphere-cut decomposition of G of width ℓ, and it can be computed in time $\mathcal{O}(|V(G)|^3)$.*

Proof. Computing G' and H can be done in time $\mathcal{O}(|E(G)|)$ and the optimal decompositions of the trees in H can be produced in total time $\mathcal{O}(|V(G)|)$. $\square$

3   Twin-Width of Graphs of Bounded Treewidth

The following result reuses a method to bound clique-width described in [9, Proposition 13].

Theorem 1. *For an undirected graph G, $\mathbf{tww}(G) \le 3 \cdot 2^{\mathbf{tw}(G)-1}$.*

Proof. We consider a connected graph G as the twin-width of a disconnected graph is simply the maximum twin-width of its connected components.

We consider a clean tree decomposition (T, f) of G of width $\mathbf{tw}(G)$ (this is always possible [9, Lemma 3, Lemma 5]).

We proceed by structural induction on the tree T. Consider a node v with children $u_1, \ldots, u_k$.

We assume that for each u_i, we have contracted $V(T_{\le}(u_i))$ into A_i consisting of at most $|\Omega(T_{\le}(u_i), f^*(u_i))|$ vertices such that their incident red edges have both endpoints within A_i. Equivalently, vertices of $V(T_{\le}(u_i))$ that were contracted have the same neighbourhood class with respect to $G - V(T_{\le}(u_i))$, i.e. the same neighbourhood class with respect to $f^*(u_i)$, because $f^*(u_i)$ separates $V(T_{\le}(u_i))$ from the rest of the graph.

We will contract these sets of vertices into a set C consisting of at most $|\Omega(T_{\le}(v), f^*(v))|$ vertices.

Let $B_0 = \varnothing$. We will inductively obtain for each $i \in [k]$ a vertex set B_i of

size at most $\left|\Omega\left(\bigcup_{j=1}^{i} T_{\le}(u_j), f^*(v)\right)\right|$, by contracting vertices of $\bigcup_{j=1}^{i} A_i$.

For each $i \in [k]$, we first contract vertices of A_i that have the same neighbourhood in $f^*(v)$, this produces $\widetilde{A}_i$ consisting of at most $|\Omega(T_{\le}(u_i), f^*(u_i) - \{v\})|$ vertices. Doing so will produce at most $|\widetilde{A}_i|$ red edges incident to v, which now has at most $|B_{i-1}| + |\widetilde{A}_i|$ incident red edges. We then contract vertices of $\widetilde{A}_i \cup B_{i-1}$ that have the same neighbourhood in $f^*(v)$, producing B_i consisting of at most

$\left|\Omega\left(\bigcup_{j=1}^{i} T_{\le}(u_j), f^*(v)\right)\right|$ vertices. Note that the red degree of a vertex resulting

from one of these contractions is at most $|\widetilde{A}_i| - 1 + |B_{i-1}| - 1 + |\{v\}| \le |B_{i-1}| + |\widetilde{A}_i|$. The two -1 terms are because we bound after the first contraction, and there is no selfloop. Vertex v now has $|B_i|$ incident red edges.

After this we can contract v with the vertex of B_k having the same neighbourhood in $f^*(v)$ if it exists. This produces C consisting of at most $|\Omega(T_{\le}(v), f^*(v))|$ vertices and such that their incident red edges remain within C.

In all of the described steps, the red degree of a vertex is at most $3 \cdot 2^{\mathbf{tw}(G)-1}$:

- Vertices in A_i have red degree at most $|A_i| \le |\Omega(T_{\le}(u_i), f^*(u_i))| \le 2^{\mathbf{tw}(G)}$.
- Vertices in $\widetilde{A}_i$ have red degree at most $|\widetilde{A}_i| \le |\Omega(T_{\le}(u_i), f^*(u_i) - \{v\})| \le 2^{\mathbf{tw}(G)-1}$.
- v has red degree at most

$$|B_{i-1}| + |\widetilde{A}_i| \le \left|\Omega\left(\bigcup_{j=1}^{i-1} T_{\le}(u_j), f^*(v)\right)\right| + |\Omega(T_{\le}(u_i), f^*(u_i) - \{v\})| \le 3 \cdot 2^{\mathbf{tw}(G)-1}$$

- When contracting $B_{i-1} \cup \widetilde{A}_i$, vertices have red degree at most

$$|B_{i-1}| + |\widetilde{A}_i| \le 3 \cdot 2^{\mathbf{tw}(G)-1}$$

Since the property is trivial on leaves of the tree, we conclude that

$$\mathbf{tww}(G) \le 3 \cdot 2^{\mathbf{tw}(G)-1}$$

□

Using sphere-cut decompositions, we establish the following theorem. Similar results are shown for clique-width in [9,11], but would lead to a worse constant, if we combined them with the bound on clique-width.

Theorem 2. *For an undirected connected planar graph G with $\mathbf{bw}(G) \ge 2$,*

$$\mathbf{tww}(G) \le \max\left(4\mathbf{bw}(G), \frac{9}{2}\mathbf{bw}(G) - 3\right) \le \max\left(4\mathbf{tw}(G) + 4, \frac{9}{2}\mathbf{tw}(G) + \frac{3}{2}\right)$$

For an undirected connected planar graph G with $\mathbf{bw}(G) \le 1$, $\mathbf{tww}(G) = 0$.

This mainly relies on the following result.

Lemma 3. *If N is a noose with $|V(N)| = k > 1$ that separates a plane graph G into G_1 and G_2, then $\Omega(V(G_1) \setminus V(G_2), V(G_2)) = \Omega(V(G_1) \setminus V(N), V(N))$ and $|\Omega(V(G_1) \setminus V(N), V(N))| \le 4k - 4 =: h(k)$.*

Proof. We will count the different possible neighbourhoods with respect to N by size:

- The only possibility for size 0 is the empty neighbourhood.
- The possibilities for size 1 are the singletons of $V(N)$ and there are k of them.
- For the neighbourhoods of size 2, we pick one vertex for each of them, and call A the set of picked vertices. We now consider $G_1[A \cup V(N)]$ and smooth the vertices of A in it, i.e. for each vertex a of A with incident edges ua, av, we remove vertex a and edges ua, av and replace them by edge uv, this operation preserves planarity and the resulting graph H is an outerplanar graph on vertices $V(N)$ because they were on the outerface of $G_1[A \cup V(N)]$. Since the number of edges of H is at most $2k - 3$ because it is outerplanar and is equal to $|A|$, the number of different neighbourhoods is bounded by $2k - 3$.
- For the neighbourhoods of size at least 3, we once again pick one vertex for each of them, and call B the set of picked vertices. We now consider $G_1[B \cup V(N)] - E(G[B])$ which is planar. We show $|B| \le n_3(k) \le k - 2$ by induction on $k = V(N)$, where $n_3(k)$ denotes the maximum number of vertices of B of degree more than 3 we can have in $G_1[B \cup V(N)]$. First, if $k \le 2$ then there are no such neighbourhoods, and if $k = 3$, there is exactly one. Then for $k > 3$,

$$n_3(k) = 1 + \max\left\{\sum_{i=1}^{\ell} n_3(a_i + 1) : \ell \ge 3, \forall i \in [\ell], a_i \ge 1, \sum_{i=1}^{\ell} a_i = k\right\}$$

because after placing one vertex v of degree $\ell \ge 3$, we must have subdivided our instance into ℓ smaller instances because edges incident to v will not be crossed by other edges. Note that with two consecutive edges incident to v and the part of the noose between their other endpoints x, y, we can obtain a smaller instance with only the vertices between x and y (inclusive) on the noose. Using the induction hypothesis, we have

$$n_3(k) \leq 1 + \sum_{i=1}^{\ell}(a_i - 1) \leq 1 + k - l \leq k - 2$$

By summing the previous bounds, we conclude that

$$|\Omega(V(G_1) \setminus V(N), V(N))| \leq 4k - 4$$

$\square$

Note that this bound is tight: denote the vertices in their order on the noose by $[k]$, we can place vertices with neighbourhoods $\{\varnothing\} \cup \{\{i\} : i \in [k]\} \cup \{\{i, i+1\} : i \in [k-1]\} \cup \{\{1, i, i+1\}, \{1, i+1\} : i \in [2, k-1]\}$.

Proof (of Theorem 2). Consider a connected planar graph G. If G has branchwidth at most 1, it cannot contain a path on 4 vertices as a subgraph, hence it is a star and has twin-width 0 (first contract twins and finish with the root).

We now consider the case when $\mathbf{bw}(G) \geq 2$. G admits a sphere-cut decomposition (T, μ) of width $k := \mathbf{bw}(G)$. Let $\widehat{G}_e$ denote $G_e - V(N_e)$.

We root T arbitrarily and proceed by structural induction on T. Consider a parent edge e with children edges e_1, e_2. We assume that, for $i \in \{1, 2\}$, $V(\widehat{G}_{e_i})$, has been contracted to a set A_i according to the neighbourhood in $V(N_{e_i})$. Consequently, $|A_i|$ is at most $|\Omega(V(\widehat{G}_{e_i}), V(N_{e_i}))|$, and red edges incident to A_i have both endpoints in A_i.

Let $x := |V(N_e) \cap V(N_{e_1})|$ and $y := |V(N_e) \cap V(N_{e_2})|$.

Note that $x + y - 2 \leq |V(N_e)| \leq k$.

Let $I := V(N_{e_1}) \cap V(N_{e_2}) \setminus V(N_e)$, and $z := |I|$

For $i \in \{1, 2\}$, we contract vertices of A_i that have the same neighbourhood in $V(N_{e_i}) \setminus I$, and call the resulting set of vertices $\widetilde{A}_i$. The vertices of I now have red degree at most $|\widetilde{A}_1| + |\widetilde{A}_2|$, while the vertices of $\widetilde{A}_i$ have red degree at most $|I| + |\widetilde{A}_i| - 1$.

We then contract the vertices of $I \cup \widetilde{A}_1 \cup \widetilde{A}_2$ that have the same neighbourhood in $V(N_e)$, and call A the resulting set of vertices. Contracted vertices have red degree at most $|\widetilde{A}_1| + |\widetilde{A}_2| + |I| - 2$. Using Lemma 3, we obtain the following inequalities:

$$|\widetilde{A}_1| + |\widetilde{A}_2| \leq |\Omega(V(\widehat{G}_{e_1}), V(N_{e_1}) \setminus I)| + |\Omega(V(\widehat{G}_{e_2}), V(N_{e_2}) \setminus I)| \leq (4x-4) + (4y-4) \leq 4k$$

$$
\begin{aligned}
|\widetilde{A}_1| + |\widetilde{A}_2| + |I| - 2 \leq\ & |\Omega(V(\widehat{G}_{e_1}), V(N_{e_1}) \setminus I)| + |\Omega(V(\widehat{G}_{e_2}), V(N_{e_2}) \setminus I)| + z - 2 \\
\leq\ & 4x + 4y + z - 10 = \frac{7}{2}(x+y) + \frac{1}{2}(x+z) + \frac{1}{2}(y+z) - 10
\end{aligned}
$$

We have the following constraints on x, y, z:

$$x + y \leq k + 2, \ |V(N_{e_1})| = x + z \leq k, \ |V(N_{e_2})| = y + z \leq k$$

By summing inequalities, we obtain $|\widetilde{A}_1| + |\widetilde{A}_2| + |I| - 2 \leq \frac{9}{2}k - 3$.

One can check that in the degenerate cases when we can't apply Lemma 3, the bounds still hold.

$V(\widehat{G}_e)$ has been contracted to a set A of at most $|\Omega(V(\widehat{G}_e), V(N_e))|$ vertices. We conclude that $tww(G) \leq \max(4k, \frac{9}{2}k - 3)$. $\qquad \Box$

4   Twin-Width of Planar Graphs

Theorem 3. *The twin-width of planar graphs is at most* 183.

Proof. We will make use of the argument used to decompose planar graphs in [16, Lemma 5], and produce a d-contraction sequence of a planar graph G inductively on the decomposition, $d \leq 183$. The embedding of the graph will be useful in our arguments to make use of Lemma 3. Recall that $h(k) = 4k - 4$.

We may suppose that G is connected since the twin-width of a graph is simply the maximum of the twin-width over its connected components. We denote by G^+ a triangulation containing G as a spanning subgraph. Let T be a BFS spanning tree in G^+ with root r on its outerface. Note that since G is a subgraph of G^+, the plane embedding of G^+ gives a plane embedding of G and its subgraphs. We call *vertical* a subpath of a path from a leaf to the root in T. We call *layer* a set of vertices that are at the same distance from r in T.

For a cycle C, we write $C = [P_1, \ldots, P_k]$ if the P_i are pairwise disjoint paths, and the last vertex of P_i is adjacent to the first vertex of P_{i+1} for $i \in [k]$, with $P_{k+1} = P_1$. For a path P, we write $P = [P_1, \ldots, P_k]$ if the P_i are pairwise disjoint, and the last vertex of P_i is adjacent to the first vertex of P_{i+1} for $i \in [k-1]$.

The following version of Sperner's Lemma is used to recursively decompose G^+.

Lemma 4 (Sperner's Lemma). *Let G be a near-triangulation[2] whose vertices are coloured $1, 2, 3$, with the outerface $F = [P_1, P_2, P_3]$ where each vertex in P_i is coloured i. Then G contains an internal face whose vertices are coloured $1, 2, 3$.*

We prove inductively the following:

Lemma 5. *Let $P_1, \ldots, P_k$ for some $k \in [5]$ be pairwise disjoint vertical paths of T such that $F = [P_1, \ldots, P_k]$ is a cycle in G^+, let H be the subgraph of G induced by the vertices of F and the set X of vertices in the (strict) interior of F, with $r \notin X$. Let X^j denote the set of vertices of X that are at a distance j from r in T. We can construct a partial d-contraction sequence of H to trigraph H' such that for each j, the vertices of X^j are contracted to obtain a set of vertices A^j in H', $|A^j| \leq h(3k)$, the vertices of A^j have red neighbours only in A^{j-1}, A^j, A^{j+1}, and $d \leq 183$.*

---

[2] A near-triangulation is a planar graph with only one face that is not a triangle.

Proof. If we have 3 vertices then there is no vertex in the interior of the triangle, the empty contraction sequence satisfies the properties.

Otherwise, we decompose H using the argument of [16], see Fig. 1. First, we colour the vertices of H with k colours as follows. For each vertex $v \in V(H)$, we assign colour $i \in [k]$ if the first vertex of F on the path from v to r in T is a vertex of P_i. This is well defined because r is on the outerface of G^+. Since G^+ is planar, the minor obtained by contracting each colour class to a single vertex cannot be K_5. If $k = 5$, there must be a pair of non-consecutive antiadjacent colours. Without loss of generality, we assume this is the case for $(2, 5)$.

We set up for Sperner's Lemma with the following constructions:

- If $k = 1$ then, since F is a cycle, P_1 has at least 3 vertices so we can write $P_1 = [u, R_2, v]$, and set $R_1 := u, R_3 := v$.
- If $k = 2$ then, since F is a cycle, one of P_1 and P_2 has at least 2 vertices. W.l.o.g. assume it is P_1, then we write $P_1 = [u, R_2]$, and set $R_1 := u, R_3 := P_2$.
- If $k = 3$ then set $R_1 := P_1, R_2 := P_2, R_3 := P_3$.
- If $k = 4$ then set $R_1 := P_1, R_2 := P_2, R_3 := [P_3, P_4]$
- If $k = 5$ then set $R_1 := P_1, R_2 := [P_2, P_3], R_3 := [P_4, P_5]$

Note that $F = [R_1, R_2, R_3]$. We give colour i to the vertices of H whose first vertex of F on their path to the root in T is in R_i (i.e. we merge the previous colour classes in the same way we merged the P_i to obtain the R_i).

Applying Sperner's Lemma, we obtain a triangular face of G^+, with vertices v_1, v_2, v_3 where v_i is of colour i. We denote Q'_i the path in T from v_i to r restricted to its vertices in X (it might be empty). These paths delimit at most 3 faces F_1, F_2, F_3, each of which having at most 5 vertical paths around it. For the face delimited by R_2, R_3, Q'_3, Q'_2, P_2 and P_5 can't both be on its border without contradicting the assumption that their respective colour classes were antiadjacent.

We can apply the induction hypothesis on each of the faces to obtain partial contraction sequences. We first apply them in an arbitrary order (the contents of the faces are antiadjacent to each other). We denote by A_α^j the set of contracted vertices in face F_α obtained by the partial contraction sequence.

For each α and increasing j, we contract all vertices of A_α^j that are in the same neighbourhood class with respect to $P_1, \ldots, P_k$ in G. Note that only vertices on layers $j - 1, j, j + 1$ of the P_i may be adjacent and that there are at most 3 of the P_is that are adjacent to F_α. This gives us sets $\widetilde{A}_\alpha^j$ of size at most $h(9)$ by Lemma 3, since by removing the vertices of Q'_i and keeping only vertices of layers $j - 1, j, j + 1$ we obtain a graph that is still planar and in which the cycle delimiting F_α gives a noose with at most 9 vertices (vertical paths have at most 1 vertex per layer).

Then for increasing j, we contract vertices of $\widetilde{A}_1^j \cup \widetilde{A}_2^j \cup \widetilde{A}_3^j \cup q_1^j \cup q_2^j \cup q_3^j$ that are in the same neighbourhood class with respect to $P_1, \ldots, P_k$ in G, see Fig. 2, where q_α^j is the vertex of Q'_α in layer j. This gives sets A^j of size at most $h(15)$ by Lemma 3, because we can deduce a noose from $F = [P_1, \ldots, P_k]$ and by keeping only the vertices of layers $j - 1, j, j + 1$ we have at most 15 vertices on the noose.

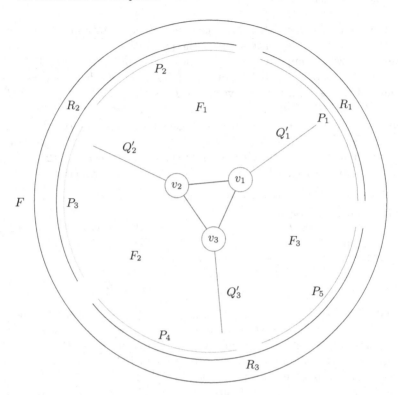

Fig. 1. Decomposition of F

We now bound the red degree that may appear in our contraction sequence. When first contracting A_α^j the red degree of its vertices is at most $|\widetilde{A}_\alpha^{j-1}| + |A_\alpha^j| + |A_\alpha^{j+1}| + 6 - 2$. The 6 term bounds the number of vertices on the Q_i' that are adjacent to vertices of A_α^j, and the -2 term is because we bound the red degree *after* the first contraction (decreasing the number of vertices by 1) and because there is no selfloop. This amounts to at most $h(9) + 2h(15) + 4 = 148$.

We then observe that the number of contractions of pairs of vertices of $\widetilde{A}_1^j \cup \widetilde{A}_2^j \cup \widetilde{A}_3^j$ that may happen when obtaining A^j is at most 5 for the following reasons. We have at most two contractions to contract the potential vertices with empty neighbourhoods coming from each F_i. Furthermore, at most 3 vertices of the P_i can have adjacent vertices in two F_α (the first vertices of F on the path from each v_i to r in T), so we may contract the two potential representatives of the neighbourhood classes consisting of a singleton of such a vertex in the two adjacent F_α. Since we know $|A^j| \le h(15)$ and each contraction may reduce the number of vertices by at most 1, we have $|\widetilde{A}_1^j| + |\widetilde{A}_2^j| + |\widetilde{A}_3^j| \le h(15) + 5$.

The red degree of a vertex of Q_i' is bounded by the sizes of the $|\widetilde{A}_\alpha^j|$ of its 3 adjacent layers on the two faces to which it is adjacent, this is because by always contracting to the same vertex in each neighbourhood class we can ensure that

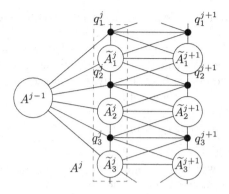

Fig. 2. Second phase of the contraction procedure

the number of red edges to this vertex is always increasing. If we add the size of the last face for each layer (positive terms), we can easily bound using the previous inequality, by $3(h(15) + 5) = 183$.

The red degree of a vertex of layer j when contracting to form A^j is at most $|A^{j-1}| + |\widetilde{A}_1^j| + |\widetilde{A}_2^j| + |\widetilde{A}_3^j| + |\widetilde{A}_1^{j+1}| + |\widetilde{A}_2^{j+1}| + |\widetilde{A}_3^{j+1}| + 6 - 2$. The 6 term bounds the number of vertices of Q_i' in the layers j and $j+1$, and the -2 term is because we bound the red degree *after* the first contraction (decreasing the number of vertices by 1) and because there is no selfloop. Combining previous inequalities, we may bound by $3h(15) + 2 \cdot 5 + 4 = 182$.   □

When the outerface is reached, we can contract arbitrarily to a single vertex layer by layer, and then contract the path. Doing so we have red degree at most $3h(9) + 1 < 183$ because there are only 3 vertices on the outerface.

We conclude that we have constructed a d-contraction sequence of G such that $d \leq 183$.   □

5   Bipartite Graph

Theorem 4. *The twin-width of the universal bipartite graph $\mathcal{B}(n)$ is $n - \log(n) + \mathcal{O}(1)$.*

Proof. We first prove an upper bound. Let $k \in [n]$. We denote by A a subset of k vertices in $X = [n]$. First, contract vertices of $Y = 2^{[n]}$ that have the same neighbourhood in A. When this is done, vertices of A have no incident red edges, while vertices of $X \setminus A$ have red edges going to all remaining vertices of Y (there are 2^k such vertices).

At this point the red degree is at most $\max(2^k, n - k)$. When contracting a neighbourhood class with respect to A in Y, we can pick an arbitrary vertex contract all others to it. This way there is only one vertex with incident red edges in the class, and its red degree is increasing. In consequence, the bound on the red degree extends to the partial contraction sequence leading to this point.

The vertices of $X \setminus A$ can then be contracted into a single vertex without creating new red edges. We can then contract all the remaining vertices of Y into a new vertex of red degree $k+1$. Finally, we contract A onto the said vertex. This establishes that for any choice of k,

$$\mathbf{tww}(\mathcal{B}(n)) \leq \max(2^k, n - k, k + 1).$$

By choosing $k = \lfloor \log(n) - 1 \rfloor$, we obtain $\mathbf{tww}(\mathcal{B}(n)) \leq n - \log(n) + \mathcal{O}(1)$.

We now prove a lower bound. Consider a $(n - k)$-contraction sequence for $\mathcal{B}(n)$. We focus on the moment before the first contraction with a vertex of X.

Note that the number of initial vertices contained in a current vertex of Y with red degree d is at most 2^d, hence at most 2^{n-k}.

A contracted vertex of Y has red degree at least 1. From the bound on the red degree of vertices of X, we know that there are at most $n(n - k)$ red edges. More precisely, if we denote by l_a for $a \in [n - k]$ the number of vertices of Y with red degree a, we have $\sum_{a=1}^{n-k} a l_a \leq n(n - k)$.

The number of vertices that were contracted in Y is therefore at most

$$\sum_{a=1}^{n-k} l_a 2^a = \sum_{a=1}^{n-k} a l_a \cdot \frac{2^a}{a} \leq n(n - k) \cdot \frac{2^{n-k}}{n - k} = n 2^{n-k}.$$

When contracting with a vertex of X for the first time, the number of red edges that become incident to it is therefore at least

$$2^{n-1} - n 2^{n-k} - 1.$$

This is bounded by $n - k$, which implies $k \leq \log_2(n) + \mathcal{O}(1)$.

We can thus conclude that $\mathbf{tww}(\mathcal{B}(n)) = n - \log_2(n) + \mathcal{O}(1)$ □

6   Conclusion

Although, we provide no lower bound matching our upper bound on the twin-width of graphs of bounded treewidth, the exponential dependency is necessary [5]. One might want to consider k-trees with heavy branching in order to find a tight lower bound.

As for the twin-width of planar graphs, it seemed reasonable that one could improve the given bound with a more careful analysis. Subsequent to this work, [3] improves the bound to 37 using the same decomposition but with a stronger invariant on the partial contraction sequence. Hliněný further improved the bound to 9 using a modified decomposition and a careful analysis [14]. Another interesting prospect would be to adapt our arguments for planar graphs to graphs of bounded genus, for which properties of the embedding might also prove useful.

References

1. Ahn, J., Hendrey, K., Kim, D., Oum, S.: Bounds for the twin-width of graphs. CoRR abs/2110.03957 (2021). https://arxiv.org/abs/2110.03957
2. Balabán, J., Hliněný, P.: Twin-width is linear in the poset width. In: Golovach, P.A., Zehavi, M. (eds.) 16th International Symposium on Parameterized and Exact Computation, IPEC 2021, 8–10 September 2021, Lisbon, Portugal. LIPIcs, vol. 214, pp. 6:1–6:13. Schloss Dagstuhl - Leibniz-Zentrum für Informatik (2021). https://doi.org/10.4230/LIPIcs.IPEC.2021.6
3. Bekos, M.A., Lozzo, G.D., Hliněný, P., Kaufmann, M.: Graph product structure for h-framed graphs. CoRR abs/2204.11495 (2022). https://doi.org/10.48550/arXiv.2204.11495
4. Bergé, P., Bonnet, É., Déprés, H.: Deciding twin-width at most 4 is NP-complete. CoRR abs/2112.08953 (2021). https://arxiv.org/abs/2112.08953
5. Bonnet, É., Déprés, H.: Twin-width can be exponential in treewidth. CoRR abs/2204.07670 (2022). https://doi.org/10.48550/arXiv.2204.07670
6. Bonnet, É., Geniet, C., Kim, E.J., Thomassé, S., Watrigant, R.: Twin-width III: max independent set, min dominating set, and coloring. In: Bansal, N., Merelli, E., Worrell, J. (eds.) 48th International Colloquium on Automata, Languages, and Programming, ICALP 2021, 12–16 July 2021, Glasgow, Scotland (Virtual Conference). LIPIcs, vol. 198, pp. 35:1–35:20. Schloss Dagstuhl - Leibniz-Zentrum für Informatik (2021). https://doi.org/10.4230/LIPIcs.ICALP.2021.35
7. Bonnet, É., Kim, E.J., Thomassé, S., Watrigant, R.: Twin-width I: tractable FO model checking. In: Irani, S. (ed.) 61st IEEE Annual Symposium on Foundations of Computer Science, FOCS 2020, Durham, NC, USA, 16–19 November 2020, pp. 601–612. IEEE (2020). https://doi.org/10.1109/FOCS46700.2020.00062
8. Bonnet, É., Kwon, O., Wood, D.R.: Reduced bandwidth: a qualitative strengthening of twin-width in minor-closed classes (and beyond). CoRR abs/2202.11858 (2022). https://arxiv.org/abs/2202.11858
9. Courcelle, B.: From tree-decompositions to clique-width terms. Discret. Appl. Math. **248**, 125–144 (2018). https://doi.org/10.1016/j.dam.2017.04.040
10. Dorn, F., Penninkx, E., Bodlaender, H.L., Fomin, F.V.: Efficient exact algorithms on planar graphs: Exploiting sphere cut decompositions. Algorithmica **58**(3), 790–810 (2010). https://doi.org/10.1007/s00453-009-9296-1
11. Fomin, F.V., Oum, S., Thilikos, D.M.: Rank-width and tree-width of H-minor-free graphs. Eur. J. Comb. **31**(7), 1617–1628 (2010). https://doi.org/10.1016/j.ejc.2010.05.003
12. Gajarský, J., Pilipczuk, M., Torunczyk, S.: Stable graphs of bounded twin-width. CoRR abs/2107.03711 (2021). https://arxiv.org/abs/2107.03711
13. Gu, Q., Tamaki, H.: Optimal branch-decomposition of planar graphs in $O(n^3)$ time. ACM Trans. Algorithms **4**(3), 30:1-30:13 (2008). https://doi.org/10.1145/1367064.1367070
14. Hliněný, P.: Twin-width of planar graphs is at most 9 (2022). https://doi.org/10.48550/ARXIV.2205.05378. https://arxiv.org/abs/2205.05378
15. Seymour, P.D., Thomas, R.: Call routing and the ratcatcher. Comb. **14**(2), 217–241 (1994). https://doi.org/10.1007/BF01215352
16. Ueckerdt, T., Wood, D.R., Yi, W.: An improved planar graph product structure theorem. CoRR abs/2108.00198 (2021). https://arxiv.org/abs/2108.00198

On Anti-stochastic Properties
of Unlabeled Graphs

Sergei Kiselev[4] , Andrey Kupavskii[1] , Oleg Verbitsky[2]([✉]) ,
and Maksim Zhukovskii[3]

[1] CNRS, Grenoble, France
kupavskii@ya.ru
[2] Humboldt-Universität zu Berlin, Berlin, Germany
verbitsky@informatik.hu-berlin.de
[3] Weizmann Institute of Science, Rehovot, Israel
[4] Grenoble, France

Abstract. We study vulnerability of a uniformly distributed random
graph to an attack by an adversary who aims for a global change of the
distribution while being able to make only a local change in the graph.
We call a graph property A *anti-stochastic* if the probability that a random graph G satisfies A is small but, with high probability, there is a small
perturbation transforming G into a graph satisfying A. While for labeled
graphs such properties are easy to obtain from binary covering codes, the
existence of anti-stochastic properties for unlabeled graphs is not so evident. If an admissible perturbation is either the addition or the deletion
of one edge, we exhibit an anti-stochastic property that is satisfied by a
random unlabeled graph of order n with probability $(2 + o(1))/n^2$, which
is as small as possible. We also express another anti-stochastic property
in terms of the degree sequence of a graph. This property has probability
$(2 + o(1))/(n \ln n)$, which is optimal up to factor of 2.

Keywords: Network resilience · Random graphs · Canonical labeling

1 Introduction

The asymptotic properties of a random graph are the subject of a rich and
comprehensive theory [2,10]. Specifically, let $\mathbf{G}_n$ be a graph chosen equiprobably
from among all graphs on the vertex set $\{1, \ldots, n\}$. Identifying a graph property
with the set of all graphs possessing this property, we say that $\mathbf{G}_n$ has a property
P *with high probability (whp)* or *asymptotically almost surely* if $\mathsf{P}[\mathbf{G}_n \in P] =
1 - o(1)$ as n increases.

If the "error probability" $o(1)$ is very small, then the definition above is
stable with respect to local perturbations of $\mathbf{G}_n$. To be specific, here and below
a perturbation means adding one edge to a graph or deleting one edge from
it. If $\mathsf{P}[\mathbf{G}_n \in P] = 1 - o(1/n^2)$, then the union bound implies that, whp, P
holds not only for $\mathbf{G}_n$ but even for each of the $\binom{n}{2}$ perturbed versions of $\mathbf{G}_n$. In

The third author is supported by DFG grant KO 1053/8–2.

other words, the property P is robust with respect to the following adversarial attack. An adversary receives a random graph $\mathbf{G}_n$ and is allowed to change the (non)adjacency of a single pair of vertices in $\mathbf{G}_n$. Whatever he does, the modified graph $\mathbf{G}'_n$ still satisfies P whp.

We here address an opposite scenario when an adversary is able to modify $\mathbf{G}_n$ so that, whp, the corrupted graph $\mathbf{G}'_n$ has a property which is unlikely for a random graph. More precisely, we call a property A *anti-stochastic* if the following two conditions are true:

– $\mathsf{P}[\mathbf{G}_n \in A] = o(1)$, and
– there is an adversary such that $\mathsf{P}[\mathbf{G}'_n \in A] = 1 - o(1)$.

More formally, by an adversary we understand an arbitrary function f which if applied to a graph G, produces a graph $f(G)$ such that G and $f(G)$ differ by at most a single edge. Thus, $\mathbf{G}'_n = f(\mathbf{G}_n)$ is a random variable which we observe instead of a uniformly distributed $\mathbf{G}_n$.

Our interest in anti-stochastic properties is motivated by the fact that they yield a conceptual formalization of the *global* damage effect on a source of random graphs that can be caused by a malicious adversary allowed to make only a *local* perturbation in a graph he accesses. Adversarial attacks on a random graph are studied in [3,4,7] focusing on the question on how many vertices can or must be deleted in order to make a dynamically evolving random graph highly disconnected. In general, resilience and vulnerability of graphs have been studied in network science in many various contexts; we refer to the recent survey [15] for overview of the concepts and results in this large research area.

Let A_n denote the set of n-vertex graphs with property A, and set $N = \binom{n}{2}$. Since a graph in A_n can be a perturbed version of at most N graphs, the second condition in the above definition implies that $|A_n|(N + 1) \geq (1 - o(1))2^N$ for any anti-stochastic property. It immediately follows that if A is anti-stochastic, then

$$\mathsf{P}[\mathbf{G}_n \in A] \geq (2 - o(1))/n^2. \tag{1}$$

This argument readily reveals a notable source of anti-stochastic properties. A set $C_N \subset \{0,1\}^N$ is a *covering code* [5] if every string in $\{0,1\}^N$ is within Hamming distance 1 of some string in C_N. For $N = \binom{n}{2}$, we can identify the graphs on the vertex set $\{1, \ldots, n\}$ with the binary strings of length N. Thus, if the density $|C_N|/2^N$ tends to 0 as N increases, then the binary code can be seen as an anti-stochastic property. The lower bound (1) turns out to be tight as there are covering codes of asymptotically optimal density $(1 + o(1))/N$; see [11] or [5, Ch. 12.4].

Being a natural combinatorial concept in the realm of strings, covering codes can hardly be considered natural graph properties. As a minimum criterion for a graph property to be natural, we require that it should be isomorphism invariant, that is, it should hold or not hold for every two isomorphic graphs simultaneously. Our first result meets this expectation, implying that the damage caused by a combinatorially optimal adversary can be, in a way, conscious.

Theorem 1. *There is an isomorphism invariant anti-stochastic property holding for a random graph of order n with probability $(2 + o(1))/n^2$.*

In this theorem and the preceding discussion, we consider *labeled* graphs (whose vertices are labeled by $1, \ldots, n$). An *unlabeled* graph can be defined formally as an isomorphism class of labeled graphs. Theorem 1 translates into virtually the same statement saying that there exists an anti-stochastic property of unlabeled graphs holding with an asymptotically optimal probability $(2 + o(1))/n^2$; cf. [8, Ch. 9].

Our proof of Theorem 1 uses the existence of covering codes with optimal density. Ensuring the invariance under graph isomorphism is, however, a subtle business. If we identify a binary code with the corresponding set of labeled graphs and just take the closure of this set under isomorphism, then we cannot exclude that this closure will have too high density, violating the first condition of an anti-stochastic property. To rectify this problem, we use the following strategy.

- For a graph G, we define a set W of vertices of G in terms of their degrees.
- We define another set of vertices W' such that W and W' are disjoint. For a vertex $w \notin W$, its membership in W' is determined by the degree of w in the subgraph of G induced on the complement of W. Moreover, W' is split into ten parts $W_1', \ldots, W_{10}'$ according to the vertex degrees in this induced subgraph.
- If G is chosen randomly, then every two vertices in W have, whp, differently many neighbors in W_i' for some $i \leq 10$, which determines a canonical labeling of W.
- Somewhat loosely speaking, the subgraphs of G induced on W and W' are almost uniformly distributed and independent from each other as well as from the adjacency pattern between W and W'. This ensures that, whp, the subgraph induced on W remains uniformly distributed also with respect to the aforementioned canonical labeling. In this way we can simulate the property of a binary word to belong to a covering code by an isomorphism invariant graph property.

The above strategy resembles the classical canonization procedure for almost all graphs due to Babai, Erdős, and Selkow [1]. Note that we cannot use canonization of a random graph directly because we need the graph to be uniformly distributed after relabeling, which makes our problem more sofisticated.

It is natural to ask which computational power must the adversary have in order to corrupt the random graph by enforcing an anti-stochastic property A. Obviously, it would be enough for him to be able to recognize whether or not a given graph satisfies A. The decision complexity of the property constructed in the proof of Theorem 1 is no more than the decision complexity of the covering code used in the construction. If $n = 2^k - 1$, we can use the Hamming code, which is perfect and can serve as a covering code. The membership in the Hamming code is efficiently recognizable. If n is not of this kind, we can expand the closest Hamming code to the length n just by appending the missing bits in all possible ways, which retains the polynomial-time complexity. However, the probability of the corresponding anti-stochastic property becomes twice higher, i.e., $(4 + o(1))/n^2$ instead of $(2 + o(1))/n^2$. If the adversary is not content with this, he has to use an asymptotically optimal covering code. Such a code is suggested by

Kabatyanskii and Panchenko [11]. Since their construction uses randomization, the recognition complexity of this code seems to be a subtle issue.

Our second construction of an isomorphism invariant anti-stochastic property is, in a certain sense, more natural as it is defined solely in terms of the degree sequence of a graph. Note that any condition on the degree sequence defines an isomorphism invariant graph property.

Theorem 2.

1. *There is an anti-stochastic property expressible in terms of the degree sequence that holds for a random graph of order n with probability $(2 + o(1))/(n \ln n)$.*
2. *On the other hand, every such anti-stochastic property has probability at least $(1 - o(1))/(n \ln n)$.*

After preliminary technical results in Sect. 2, we present the proofs of Theorems 1 and 2 in Sects. 3 and 4 respectively. Due to space limitations, some parts of the proofs are omitted or only sketched. Full proofs are available in a long version of this paper [12].

2   Preliminaries

The vertex set of a graph G is denoted by $V(G)$. If $U \subset V(G)$, then we write $G|_U$ to denote the subgraph of G induced on the set of vertices U. The degree of a vertex $v \in V(G)$ is denoted by $\deg_G(v)$.

For the uniformly distributed random graph $\mathbf{G}_n$ on n vertices, it is supposed that $V(\mathbf{G}_n) = [n]$, where $[n] = \{1, \ldots, n\}$. In some contexts, a set A of graphs on $[n]$ can be identified with the event $\mathbf{G}_n \in A$, whose probability $\mathsf{P}[\mathbf{G}_n \in A]$ will then be denoted by $\mathsf{P}[A]$.

Recall that the *characteristic function* of a random variable X is defined as $\phi(t) = \mathsf{E}e^{itX}$ where $t \in \mathbb{R}$. For an n-dimensional random vector $X = (X_1, \ldots, X_n)$, this generalizes to $\phi(t_1, \ldots, t_n) = \mathsf{E}e^{i \sum_{k=1}^{n} t_k X_k}$. The following lemma is a simple extension of its 1-dimensional analogue than can be found in [9, Theorem 1].

Lemma 3. *Let m be a positive integer. Let $X = (X_1, \ldots, X_n)$ be an integer-valued random vector with characteristic function ϕ, then*

$$\mathsf{P}[m \text{ divides } X_k \text{ for every } k = 1, \ldots, n] = \frac{1}{m^n} \sum_{j_1=0}^{m-1} \cdots \sum_{j_n=0}^{m-1} \phi\left(\frac{2\pi j_1}{m}, \ldots, \frac{2\pi j_n}{m}\right).$$

We use Lemma 3 to prove that if we take the remainders modulo m of the vertex degrees in $\mathbf{G}_n$, then the resulting sequences are evenly distributed.

Lemma 4. *Fix an odd integer $m \geq 3$ and let $r_v \in \{0, 1, \ldots, m-1\}$ for $v = 1, \ldots, n$. Then*

$$\left| \mathsf{P}[\deg_{\mathbf{G}_n}(v) \equiv r_v \pmod{m} \text{ for every } v = 1, \ldots, n] - \frac{1}{m^n} \right| = O(e^{-cn}),$$

where the constant $c > 0$ as well as the constant absorbed in the big O notation depend solely on m.

Proof. Let $Y = (Y_1, \ldots, Y_n)$ be the vector of vertex degrees of $\mathbf{G}_n$. Let $R = (r_1, \ldots, r_n)$ and $X = Y - R$. Denote the characteristic functions of X and Y by ϕ_X and ϕ_Y respectively and note that $\phi_X(t_1, \ldots, t_n) = e^{-i\sum_{v=1}^n r_v t_v} \phi_Y(t_1, \ldots, t_n)$. By Lemma 3, the probability that $Y_v \equiv r_v \pmod{m}$ for all v is equal to

$$\frac{1}{m^n} \sum_{x \in \{0,1,\ldots,m-1\}^n} \phi_X\left(\frac{2\pi x}{m}\right) = \frac{1}{m^n}\left(1 + \sum_{x \neq 0} \phi_X\left(\frac{2\pi x}{m}\right)\right)$$

where $2\pi x/m$ is an n-dimensional real vector and the equality is just due to the observation that $\phi_X(0) = 1$ for the n-dimensional zero vector. Noting that $\left|\sum_{x \neq 0} \phi_X(2\pi x/m)\right| \leq \sum_{x \neq 0} |\phi_Y(2\pi x/m)|$, we have to prove that the last sum is exponentially small. To this end, we bound the term $|\phi_Y(2\pi x/m)|$ from above for each non-zero vector $x \in \{0, 1, \ldots, m-1\}^n$.

For distinct $u, v \in [n]$, let $\xi_{u,v}$ denote the indicator random variable of the presence of the edge $\{u, v\}$ in $\mathbf{G}_n$. For a real vector $t = (t_1, \ldots, t_n)$, we have

$$\phi_Y(t) = \mathbb{E} e^{i\sum_{u=1}^n \sum_{v \neq u} t_u \xi_{u,v}} = \mathbb{E} e^{i\sum_{u<v}(t_u+t_v)\xi_{u,v}} = \prod_{u<v}\left(\frac{1}{2} + \frac{1}{2}e^{i(t_u+t_v)}\right).$$

Set $\alpha = \max_{1 \leq j < m} |1 + e^{2\pi i j/m}|$ and note that $\alpha < 2$. Let $n_0(x)$ be the number of zeros in x. Note that $|\frac{1}{2} + \frac{1}{2}e^{i(t_u+t_v)}| \leq 1$. Moreover, this number does not exceed $\alpha/2$ for $t = 2\pi x/m$ such that exactly one of the coordinates x_u and x_v is equal to 0. It follows that $|\phi_Y(2\pi x/m)| \leq (\alpha/2)^{n_0(x)(n-n_0(x))}$. Now, fix k to be the smallest integer such that $(\alpha/2)^k < 1/m$. Since $x \neq 0$, we conclude that

$$|\phi_Y(2\pi x/m)| \leq \begin{cases} (\alpha/2)^{n-1} & \text{if } n_0(x) \geq n - k \\ (\alpha/2)^{k(n-k)} & \text{if } n/2 \leq n_0(x) < n - k. \end{cases}$$

Consider the case that $n_0(x) < n/2$. Since each of the more than $n/2$ non-zero coordinates of x can take on at most $m - 1$ values, there are at least $(m-1)\binom{\lceil n/(2(m-1))\rceil}{2}$ pairs (u, v) with $u < v$ and $x_u = x_v \neq 0$. For such a pair, the assumption that m is odd implies that the sum $x_u + x_v$ is not divisible by m. It follows that $|\phi_Y(2\pi x/m)| \leq (\alpha/2)^{(m-1)\binom{\lceil n/(2(m-1))\rceil}{2}}$. Thus,

$$\sum_{x \neq 0} \left|\phi_Y\left(\frac{2\pi x}{m}\right)\right| < k\binom{n}{k}m^k\left(\frac{\alpha}{2}\right)^{n-1}$$

$$+ m^n\left(\left(\frac{\alpha}{2}\right)^{k(n-k)} + \left(\frac{\alpha}{2}\right)^{(m-1)\binom{\lceil n/(2(m-1))\rceil}{2}}\right) = O\left((m(\alpha/2)^k)^n\right),$$

completing the proof. □

3 Proof of Theorem 1

Let $k > 11$ be an odd integer non-divisible by 11. For a graph G, we define $U(G)$ to be the set of vertices of G whose degrees are divisible by k. For $r = 0, 1, \ldots, 10$, let $U_r(G)$ denote the set of those vertices in $U(G)$ whose degrees in $G|_{U(G)}$ are congruent to r modulo 11. We also set $W(G) = V(G) \setminus U(G)$ and $R(G) = U(G) \setminus U_0(G)$. In what follows, an important role will be played by the partition

$$V(G) = W(G) \cup U_0(G) \cup R(G). \tag{2}$$

For notational simplicity, we suppress the dependence of this partition on k. The value of the parameter k is supposed to be fixed until the final step of the proof.

For the random graph $\mathbf{G}_n$, the partition (2) translates in $[n] = \mathbf{W} \cup \mathbf{U}_0 \cup \mathbf{R}$, where $\mathbf{W} = W(\mathbf{G}_n)$, $\mathbf{U}_r = U_r(\mathbf{G}_n)$, and $\mathbf{R} = R(\mathbf{G}_n)$. Also, $\mathbf{U} = U(\mathbf{G}_n)$.

Distribution of Induced Subgraphs. For a set X, we write $\mathbf{G}_X$ to denote the uniformly distributed random graph on the vertex set X. Moreover, if $X \cap Y = \emptyset$, then $\mathbf{G}_{X \times Y}$ stands for the uniformly distributed random bipartite graph with vertex classes X and Y.

Our main technical tool will be a lemma about asymptotical independence and uniformity of the subgraphs $\mathbf{G}_n|_{\mathbf{R}}$, $\mathbf{G}_n|_{\mathbf{W}}$, and $\mathbf{G}_n|_{\mathbf{R} \times \mathbf{W}}$. Note that this is equivalent to asymptotical uniformity of the subgraph $\mathbf{G}_n|_{\mathbf{W} \cup \mathbf{R}}$. Somewhat loosely speaking, we show that the random graphs $\mathbf{G}_n|_{\mathbf{W} \cup \mathbf{R}}$ and $\mathbf{G}_{W \cup R}$ have almost the same distribution under the condition that $\mathbf{W} = W$ and $\mathbf{R} = R$.

Specifically, we fix a real $\varepsilon \in (0, 1)$ and suppose that $U_0, U_1, \ldots, U_{10}$ are disjoint subsets of $[n]$ such that $\frac{n}{11k}(1 - \varepsilon) < |U_i| < \frac{n}{11k}(1 + \varepsilon)$. We also set $U = \bigcup_{i=0}^{10} U_i$, $W = [n] \setminus U$, and $R = U \setminus U_0$. Moreover, $\vec{U} = (U_0, \ldots, U_{10})$. These sets, in contrast to the sets $U_r(G)$, $W(G)$ etc. defined above, are considered irrespectively of any graph G. Set $\vec{\mathbf{U}} = (\mathbf{U}_0, \ldots, \mathbf{U}_{10})$. Lemma 4 makes it intuitively clear that, conditioned on $\mathbf{U} = U$ and on $\vec{\mathbf{U}} = \vec{U}$ respectively, the graphs $\mathbf{G}_n|_U$ and $\mathbf{G}_n|_{W \cup R}$ are 'almost' uniformly distributed. We formalize this as follows.

Lemma 5. *Under the above assumption we have the following equalities.*

1. *For every property A of graphs on U,*

$$\mathsf{P}\left[\mathbf{G}_n|_U \in A \mid \mathbf{U} = U\right] = (1 + o(1))\,\mathsf{P}\left[\mathbf{G}_U \in A\right].$$

2. *For every property A of graphs on $[n] \setminus U_0$,*

$$\mathsf{P}\left[\mathbf{G}_n|_{W \cup R} \in A \mid \vec{\mathbf{U}} = \vec{U}\right] = (1 + o(1))\,\mathsf{P}\left[\mathbf{G}_{W \cup R} \in A\right].$$

The Property. For an n-vertex graph G, we will suppose that $V(G) = [n]$. Let Q be an anti-stochastic property of labeled graphs. More specifically, for each n we fix a covering code in $\{0, 1\}^{\binom{n}{2}}$ of asymptotically optimal density. An n-vertex graph G belongs to Q if the $\binom{n}{2}$-dimensional vector of adjacencies of G belongs to the code. By [11], we may assume that $\mathsf{P}[\mathbf{G}_n \in Q] \leq (2 + 1/k)/n^2$ for large enough n.

We say that $R(G) = U_1(G) \cup \ldots \cup U_{10}(G)$ *resolves* $W(G)$ if every two distinct vertices in $W(G)$ have differently many neighbors in $U_r(G)$ for some $r \in [10]$. More specifically, for $v \in W(G)$, let $\overrightarrow{d}(v) = (d_1(v), \ldots, d_{10}(v))$ where $d_r(v)$ denotes the number of neighbors of v in $U_r(G)$. Then $R(G)$ resolves $W(G)$ if $\overrightarrow{d}(v) \neq \overrightarrow{d}(u)$ for any distinct $u, v \in W(G)$. If this is the case, consider the lexicographical order on $\{0, 1, \ldots, n-1\}^{10}$ and relabel the vertices in $W(G)$ by the integers $1, \ldots, |W(G)|$ according to this order. This results in an isomorphic copy of $G|_{W(G)}$, and we will say that the subgraph $G|_{W(G)}$ is *canonically relabeled*.

Let B denote the set of graphs G such that $R(G)$ resolves $W(G)$. We define the property Q_k by setting $G \in Q_k$ if

- either $G \notin B$,
- or $G \in B$ and the canonically relabeled subgraph $G|_{W(U)}$ belongs to Q.

Note that Q_k is isomorphism invariant. Indeed, this is obvious for the property B. Now, if G satisfies the second condition above and $G' \cong G$, then $G' \in B$ too. Let f be an isomorphism from G to G'. Note that f induces an isomorphism from $G|_{W(G)}$ to $G'|_{W(G')}$ preserving the canonical labels. As a consequence, G' also satisfies the second condition in the definition of Q_k.

We split the proof of Theorem 1 in three parts.

1. We will prove that Q_k has small probability, specifically, $\mathsf{P}[\mathbf{G}_n \in Q_k] \leq \left(1 + \frac{3}{k}\right) \frac{2}{n^2}$ for large enough n.
2. Then, we will prove that Q_k is close to an anti-stochastic property in the sense that an adversary is able to transform $\mathbf{G}_n$ in $\mathbf{G}'_n$ such that $\mathsf{P}[\mathbf{G}'_n \in Q_k] > 1 - \frac{4}{k}$ for large enough n.
3. These facts will allow us to combine a sequence of properties Q_k into a single anti-stochastic property Q^*.

The Probability of Q_k is Small. Technically, this part of the proof will be accomplished by showing that

- B holds with probability $1 - o(1/n^2)$, and
- the canonically relabeled subgraph $\mathbf{G}_n|_{\mathbf{W}}$ remains almost uniformly distributed.

The following fact is the first step towards showing that $\mathbf{G}_n \in B$ whp. We call a set $U \subset [n]$ *standard* if $\frac{n}{k} - \sqrt{n} \ln n \leq |U| \leq \frac{n}{k} + \sqrt{n} \ln n$. Fix a standard $U \subset [n]$ and let $\overrightarrow{U} = (U_0, \ldots, U_{10})$ be a partition of U. We call $\overrightarrow{U}$ *standard* if $\frac{|U|}{11} - \sqrt{n} \ln n \leq |U_r| \leq \frac{|U|}{11} + \sqrt{n} \ln n$ for every $r = 0, \ldots, 10$. Recall that $\mathbf{U}$ consists of all vertices in $\mathbf{G}_n$ with degrees divisible by k.

Claim 6. 1. $\mathbf{U}$ is standard with probability $1 - o(1/n^3)$.
2. $(U_0(\mathbf{G}_U), \ldots, U_{10}(\mathbf{G}_U))$ is standard with probability $1 - o(1/n^3)$.

The proof is based on the approximation of the degree sequence of $\mathbf{G}_n$ by a vector of independent binomial random variables due to McKay and Wormald [14] and an application of Lemma 3.

Using part 1 of Lemma 5, we conclude from part 2 of Claim 6 that

$$P\left[\overrightarrow{\mathbf{U}} \text{ is standard} \mid \mathbf{U} = U\right] = 1 - o\left(\frac{1}{n^3}\right). \tag{3}$$

Assume that $\overrightarrow{U}$ is standard and consider a random graph $\mathbf{G}_{W \cup R}$ where, as usually, $W = [n] \setminus U$ and $R = U \setminus U_0$. As follows from de Moivre–Laplace limit theorem, two fixed vertices $u, v \in W$ of $\mathbf{G}_{W \cup R}$ have equally many neighbors in U_r for every $r = 1, \ldots, 10$ with probability $O(1/n^5)$. By the union bound, the partition $R = U_1 \cup \ldots \cup U_{10}$ does not resolve W in $\mathbf{G}_{W \cup R}$ with probability $O(1/n^3)$. By part 2 of Lemma 5 we conclude that, conditioned on $\overrightarrow{\mathbf{U}} = \overrightarrow{U}$, $R(\mathbf{G}_n)$ does not resolve $W(\mathbf{G}_n)$ with asymptotically the same probability, that is,

$$P\left[\overline{B} \mid \overrightarrow{\mathbf{U}} = \overrightarrow{U}\right] = O\left(\frac{1}{n^3}\right), \tag{4}$$

where $\overline{B}$ denotes the event $\mathbf{G}_n \notin B$.

We now can see that the event B holds with high probability. Indeed, taking into account Estimate (4), part 1 of Claim 6, and Estimate (3), we have

$$P\left[\overline{B}\right] \leq \sum_{U \text{ and } \overrightarrow{U} \text{ standard}} P\left[\overline{B} \mid \overrightarrow{\mathbf{U}} = \overrightarrow{U}\right] P[\overrightarrow{\mathbf{U}} = \overrightarrow{U}] + P\left[\mathbf{U} \text{ is not standard}\right]$$

$$+ \sum_{U \text{ standard}} P\left[\overrightarrow{\mathbf{U}} \text{ is not standard} \mid \mathbf{U} = U\right] P[\mathbf{U} = U] = O\left(\frac{1}{n^3}\right).$$

With this upper bound for the probability of $\overline{B}$, we are ready to estimate the probability of Q_k from above. Note that

$$P[Q_k] = P[Q_k \cap B] + P[\overline{B}] = P[Q_k \cap B] + o(1/n^2).$$

By part 1 of Claim 6 and Estimate (3),

$$P[Q_k \cap B] = \sum_{U \text{ and } \overrightarrow{U} \text{ standard}} P[Q_k \cap B \mid \overrightarrow{\mathbf{U}} = \overrightarrow{U}] P[\overrightarrow{\mathbf{U}} = \overrightarrow{U}] + o\left(\frac{1}{n^3}\right).$$

By part 2 of Lemma 5, the probability $P[Q_k \cap B \mid \overrightarrow{\mathbf{U}} = \overrightarrow{U}]$ is asymptotically the same as the probability that in the uniformly distributed random graph $\mathbf{G}_{W \cup R}$, simultaneously,

(1) $R = U_1 \cup \ldots \cup U_{10}$ resolves $W = [n] \setminus U$, and
(2) the canonically relabeled subgraph $\mathbf{G}_{W \cup R}|_W$ belongs to Q.

Assume that the former condition is fulfilled. Since the random graphs $\mathbf{G}_{W \cup R}|_W = \mathbf{G}_W$ and $\mathbf{G}_{W \times R}$ are independent, the latter condition has the same probability as the event $\mathbf{G}_{|W|} \in Q$, which does not exceed $\frac{2 + 1/k + o(1)}{(n - n/k - \sqrt{n} \ln n)^2}$. We conclude that

$$P[\mathbf{G}_n \in Q_k] \leq \frac{(2 + 1/k)(1 + o(1))}{(1 - 1/k)^2 n^2} + o\left(\frac{1}{n^2}\right) \leq \left(1 + \frac{3}{k}\right) \frac{2}{n^2}, \qquad (5)$$

where the last inequality is fulfilled for all sufficiently large n.

Q_k *is Almost Anti-stochastic.* For a graph G and two distinct vertices $u, v \in V(G)$, let $G(u, v)$ denote the graph obtained from G by changing the adjacency between u and v. If $u = v$, we set $G(u, v) = G$.

Let A denote the event $\{\exists u, v \in \mathbf{W} \quad \mathbf{G}_n(u, v) \in Q_k\}$. It is enough to prove that A has high probability.

In what follows, for a partition $\overrightarrow{U} = (U_0, \ldots, U_{10})$ of U, the event $\overrightarrow{\mathbf{U}} = \overrightarrow{U}$ will for brevity be denoted by $C_{\overrightarrow{U}}$. With some abuse of notation, we write A and $C_{\overrightarrow{U}}$ also to denote the corresponding *sets* of graphs. We have

$$P[A] \geq P[A \cap B] \geq \sum_{U \text{ and } \overrightarrow{U} \text{ standard}} P[A \cap B \mid C_{\overrightarrow{U}}] P[C_{\overrightarrow{U}}]$$

$$= \sum_{U \text{ and } \overrightarrow{U} \text{ standard}} P[A \mid B \cap C_{\overrightarrow{U}}] P[B \mid C_{\overrightarrow{U}}] P[C_{\overrightarrow{U}}].$$

We can bound $P[B \mid C_{\overrightarrow{U}}]$ from below according to Estimate (4). The probability $P[A \mid B \cap C_{\overrightarrow{U}}]$ can also be bounded according to the following claim.

Claim 7. If a set $U \subset [n]$ and its partition $\overrightarrow{U} = (U_0, \ldots, U_{10})$ are standard, then

$$P[A \mid B \cap C_{\overrightarrow{U}}] \geq \left(1 - \frac{4}{k} + \frac{4}{k^2}\right)(1 - o(1)).$$

We, therefore, obtain

$$P[A] \geq \left(1 - o\left(\frac{1}{n^2}\right)\right)\left(1 - \frac{4}{k} + \frac{4}{k^2}\right)(1 - o(1)) \sum_{U \text{ and } \overrightarrow{U} \text{ standard}} P[C_{\overrightarrow{U}}].$$

Note that

$$\sum_{U \text{ and } \overrightarrow{U} \text{ standard}} P[C_{\overrightarrow{U}}] = P[\mathbf{U} \text{ and } \overrightarrow{\mathbf{U}} \text{ are standard}]$$

$$= P[\overrightarrow{\mathbf{U}} \text{ is standard} \mid \mathbf{U} \text{ is standard}] P[\mathbf{U} \text{ is standard}] \geq 1 - o\left(\frac{1}{n^3}\right),$$

where the last inequality follows from Estimate (3) and Claim 6. We conclude that

$$P[A] \geq 1 - \frac{4}{k} \text{ for sufficiently large } n. \qquad (6)$$

Merging All Q_k's Together. It remains to convert the sequence of graph properties Q_k into a single anti-stochastic property Q^*. Based on Estimates (5) and (6), for each Q_k we define an integer N_k such that $N_k > N_{k'}$ if $k' < k$ and the inequalities $\mathsf{P}[\mathbf{G}_n \in Q_k] \le (1+3/k)\frac{2}{n^2}$ and $\mathsf{P}[\exists u, v\ \mathbf{G}_n(u, v) \in Q_k] \ge 1 - 4/k$ are true for all $n \ge N_k$. Let $k(n)$ be the maximum k such that $N_k \le n$. Define Q^* to be the event that $\mathbf{G}_n \in Q_{k(n)}$. The graph property Q^* is anti-stochastic because $k(n) \to \infty$ as $n \to \infty$. The proof of Theorem 1 is complete.

4   Proof of Theorem 2

Upper Bound. Let $[a, b]$ denote the interval of integers $a, a+1, \ldots, b$. We consider the interval of integers

$$D_n = \left\{ d \in \mathbb{Z} : |d - n/2| \le \frac{1}{2}\sqrt{n\left(\ln n - 2\sqrt{\ln n}\right)} \right\}.$$

The smallest and the largest integers in D_n are denoted by d_* and d^* respectively; thus, $D_n = [d_*, d^*]$. The integer $\lfloor n/2 \rfloor$ splits D_n in two parts $D_{n,1} = [d_*, \lfloor n/2 \rfloor - 1]$ and $D_{n,2} = [\lfloor n/2 \rfloor + 1, d^*]$, that is, $D_n = D_{n,1} \cup \{\lfloor n/2 \rfloor\} \cup D_{n,2}$. Set $\delta = |D_n|$, $\delta_1 = |D_{n,1}|$, and $\delta_2 = |D_{n,2}|$. Shifting $D_{n,1}$ and $D_{n,2}$ in 1, we obtain the intervals $D_{n,1}^+ = [d_* + 1, \lfloor n/2 \rfloor]$ and $D_{n,2}^+ = [\lfloor n/2 \rfloor + 2, d^* + 1]$.

For a non-negative integer y, we denote the number of vertices of degree y in $\mathbf{G}_n$ by N_y. For an integer $y > d_*$, we define X_y to be the total number of vertices v such that $d_* \le \deg(v) \le y$, that is, $X_y = \sum_{d=d_*}^y N_d$.

We define two integer sequences $\mathbf{X}^\downarrow = (X_y)_{y \in D_n^1}$ and $\mathbf{X}^\uparrow = (X_y)_{y \in D_n^2}$ of length δ_1 and δ_2 respectively, and make a simple observation.

Claim 8. Let u and v be vertices of degrees $\deg(u) = x$ and $\deg(v) = y$.

1. If $x \in D_{n,1}$, $y \in D_{n,2}$, and u and v are non-adjacent, then addition of an edge between u and v changes only one coordinate of $\mathbf{X}^\downarrow$, namely the one indexed by x, and only one coordinate of $\mathbf{X}^\uparrow$, namely the one indexed by y. Both coordinates decrease by 1.
2. If $x \in D_{n,1}^+$, $y \in D_{n,2}^+$, and u and v are adjacent, then deletion of the edge between u and v changes only one coordinate of $\mathbf{X}^\downarrow$, namely the one indexed by $x - 1$, and only one coordinate of $\mathbf{X}^\uparrow$, namely the one indexed by $y - 1$. Both coordinates increase by 1.

Denote by $\mathbf{Y}^\downarrow \in \{0, 1\}^{\delta_1}$ the vector of parities of $\mathbf{X}^\downarrow$, that is, $(\mathbf{Y}^\downarrow)_i = (\mathbf{X}^\downarrow)_{d_*+i-1} \bmod 2$. Similarly, $\mathbf{Y}^\uparrow \in \{0, 1\}^{\delta_2}$ is the vector of parities of $\mathbf{X}^\uparrow$. Let $\mathcal{S}^\downarrow \subset \{0, 1\}^{\delta_1}$ and $\mathcal{S}^\uparrow \subset \{0, 1\}^{\delta_2}$ be covering codes with asymptotically optimal densities. Define

$$\mathbf{Z} = \sum_{d_* \le y \le n/2} y N_y, \tag{7}$$

where the summation goes over all $y \in D_{n,1} \cup \{\lfloor n/2 \rfloor\}$. Let A be the property that $\mathbf{Y}^\downarrow \in \mathcal{S}^\downarrow$, $\mathbf{Y}^\uparrow \in \mathcal{S}^\uparrow$, and $\mathbf{Z} \bmod 4 \in \{0, 1\}$. We show that this is the desired anti-stochastic property.

The approximation of the degree sequence of $\mathbf{G}_n$ by a vector of independent binomial random variables [14] and Lemma 3 imply that the distribution of the remainders of $N_{d_*+j-1}, j \in [\delta]$, modulo 4 is almost uniform. Then, the optimality of the covering codes $\mathcal{S}^\downarrow$ and $\mathcal{S}^\uparrow$ yields the following.

Lemma 9.

1. $\mathsf{P}[\mathbf{Y}^\downarrow \in \mathcal{S}^\downarrow] = \frac{2+o(1)}{\sqrt{n \ln n}}$ and $\mathsf{P}[\mathbf{Y}^\uparrow \in \mathcal{S}^\uparrow] = \frac{2+o(1)}{\sqrt{n \ln n}}$.
2. $\mathsf{P}[\mathbf{G}_n \in A] = \frac{2+o(1)}{n \ln n}$.

As the probability of A is asymptotically determined by part 2 of Lemma 9, it remains to prove the existence of an adversary. A characteristic $\mathbf{V}$ of a random graph $\mathbf{G}_n$ is, formally, a function defined on the set of graphs with vertices $1, \ldots, n$. For a graph G on this vertex set, we therefore write $\mathbf{V}(G)$ to denote the value of $\mathbf{V}$ on G. Let G be a graph on $[n]$ such that $\mathbf{Y}^\downarrow(G) \notin \mathcal{S}^\downarrow$, $\mathbf{Y}^\uparrow(G) \notin \mathcal{S}^\uparrow$, and for every two integers $x, y \in [d_*, d^* + 1]$ there exist a pair of adjacent vertices of degrees x and y and a pair of non-adjacent vertices of degrees x and y. The first two conditions hold whp for $\mathbf{G}_n$ by part 1 of Lemma 9. The third condition also holds whp since both $\mathbf{G}_n$ and its complement whp contain no cliques of size $(\ln n)^{1.5}$ and no complete bipartite graphs with both parts of size at least $(\ln n)^{1.5}$, and because $N_y > \ln^2 n$ for all $y \in D_n$. Therefore, it suffices to show that, by adding or deleting one edge, the adversary can transform G into a graph G' possessing the property A.

The adversary is going either to add an edge between two vertices u and v as in part 1 of Claim 8 or to delete an edge between two vertices u and v as in part 2 of this claim. Note that $\mathbf{Z}(G') = \mathbf{Z}(G) + 1$ in the case of addition, and $\mathbf{Z}(G') = \mathbf{Z}(G) - 1$ in the case of deletion. This allows the adversary to choose a type of action (deletion or insertion) ensuring that $\mathbf{Z}(G') \bmod 4 \in \{0, 1\}$ whatever $\mathbf{Z}(G) \bmod 4$ is. Once the action type is fixed, the adversary chooses a codeword $S^\downarrow \in \mathcal{S}^\downarrow$ at the Hamming distance 1 from $\mathbf{Y}^\downarrow(G)$ and a codeword $S^\uparrow \in \mathcal{S}^\uparrow$ at the Hamming distance 1 from $\mathbf{Y}^\uparrow(G)$. Suppose that $S^\downarrow$ and $\mathbf{Y}^\downarrow(G)$ differ at coordinate i and that $S^\uparrow$ and $\mathbf{Y}^\uparrow(G)$ differ at coordinate j. The adversary changes the adjacency relation between vertices u and v of degree x and y respectively where $x = d_* + i - 1$ and $y = d_* + j - 1$ in the case of addition or $x = d_* + i$ and $y = d_* + j$ in the case of deletion. By Claim 8, $\mathbf{Y}^\downarrow(G') = S^\downarrow$ and $\mathbf{Y}^\uparrow(G') = S^\uparrow$ and, therefore, G' has the property A. This proves part 1 of Theorem 2.

Lower Bound. Using the coding-theoretic terminology, we say that a graph G *covers* a graph G' if G' is obtained from G by changing the adjacency relation between two vertices. All graphs considered in this section are on the vertex set $[n]$. Let $\Gamma(G)$ denote the set of all $\binom{n}{2}$ graphs covered by G. For a set of graphs $\mathcal{Q}$, let $\Gamma(\mathcal{Q}) = \bigcup_{G \in \mathcal{Q}} \Gamma(G)$. If $\mathcal{Q}$ is an anti-stochastic property, then $|\Gamma(\mathcal{Q})| = (1 - o(1))2^{\binom{n}{2}}$ and, therefore, if we also have $|\Gamma(\mathcal{Q})| \le (1 + o(1))n \ln n |\mathcal{Q}|$, this readily implies the desired lower bound $\mathsf{P}[\mathbf{G}_n \in \mathcal{Q}] \ge \frac{1+o(1)}{n \ln n}$. Moreover, for a set of graphs $\mathcal{A}$, let $\Gamma_\mathcal{A}(\mathcal{Q}) = \Gamma(\mathcal{Q}) \cap \mathcal{A}$. If $\mathbf{G}_n \in \mathcal{A}$ whp, then it suffices to have $|\Gamma_\mathcal{A}(\mathcal{Q})| \le (1 + o(1))n \ln n |\mathcal{Q}|$. We prove that the latter inequality holds for every anti-stochastic property $\mathcal{Q}$ expressible in terms of vertex degrees when

$\mathcal{A}$ is chosen to be the set of graphs with 'typical' degree sequences. Roughly speaking, the labeled degree sequence $(d_1(G), \ldots, d_n(G))$ of G is typical, if

(i) all $d_i = d_i(G)$ are not 'far' from $n/2$, namely $|d_i - n/2| < \sqrt{(n \ln n)/2}$,
(ii) the number of i such that d_i is not 'close enough' to $n/2$, namely $|d_i - n/2| > \frac{1}{2}\sqrt{n \ln n}(1 - o(1))$, is small,
(iii) for all y 'close enough' to $n/2$, there exists a sufficiently large set of i such that $d_i = y$.

Note that the number of distinct elements in a typical degree sequence is at most $\sqrt{2n \ln n}$. Moreover, almost all degrees lie within the interval $\mathcal{I} = (\frac{n}{2} - \frac{1}{2}\sqrt{n \ln n}, \frac{n}{2} + \frac{1}{2}\sqrt{n \ln n})$ of length $\sqrt{n \ln n}$. Therefore, perturbations of a graph with a typical degree sequence produce at most $2\frac{(\sqrt{2n \ln n})^2}{2}(1+o(1)) = 2n \ln n(1+o(1))$ degree sequences. The factor 2 appears due to the choice of either an insertion or a deletion of an edge. Some technical work is required to show that only perturbations inside $\mathcal{I}$ contribute to the asymptotics of the produced degree sequences, leading to the upper bound $n \ln n(1+o(1))$. Finally, the desired bound $|\Gamma_{\mathcal{A}}(\mathcal{Q})| \le (1 + o(1))n \ln n|\mathcal{Q}|$ follows from the fact that, for $G \in \mathcal{A}$ and for all $y \in \mathcal{I}$ we have many i such that $d_i(G) = y$ and from the asymptotics of the number of graphs with given degrees due to Liebenau and Wormald [13].

5   Conclusion and Further Questions

Anti-stochastic properties of graphs studied in this paper are a natural concept in the context of the research on network vulnerability [15]. Our focus on isomorphism-invariant properties (or, equivalently, on *unlabeled* graphs) is motivated by an observation that, in realistic scenarios, an adversary can only be interested in forcing somehow meaningful, structured properties. Theorem 1 determines the optimum probability of an anti-stochastic property in this setting, and Theorem 2 concerns the even more constrained scenario when an adversary aims at an anti-stochastic property expressible solely in terms of vertex degrees.

There are several further questions naturally arising in this context. We here consider the random graph model $G(n, p)$ in the case $p = 1/2$. Not all tools in our analysis can be directly applied to other edge probabilities $p = p(n)$. For example, the assumption $p = 1/2$ is essentially used in our proof of the lower bound in Theorem 2. Other random graph models, especially those designed to describe real-life networks (e.g., [3,4,7]), would also be of considerable interest.

We restrict ourselves to the case of a limited adversary who is able to change the adjacency relation just between a single pair of vertices. Consideration of other perturbation types, like changing adjacencies of multiple pairs, vertex/edge deletions or insertions, etc., would be also well motivated (corresponding to various types of errors studied in coding theory [6]).

References

1. Babai, L., Erdős, P., Selkow, S.M.: Random graph isomorphism. SIAM J. Comput. **9**(3), 628–635 (1980). https://doi.org/10.1137/0209047

2. Bollobás, B.: Random Graphs. Cambridge Studies in Advanced Mathematics, vol. 73, 2nd edn. Cambridge University Press (2001). https://doi.org/10.1017/CBO9780511814068

3. Bollobás, B., Riordan, O.: Coupling scale-free and classical random graphs. Internet Math. **1**(2), 215–225 (2003). https://doi.org/10.1080/15427951.2004.10129084

4. Bollobás, B., Riordan, O.: Robustness and vulnerability of scale-free random graphs. Internet Math. **1**(1), 1–35 (2003). https://doi.org/10.1080/15427951.2004.10129080

5. Cohen, G.D., Honkala, I.S., Litsyn, S., Lobstein, A.: Covering Codes. North-Holland Mathematical Library, vol. 54. North-Holland (2005)

6. Firer, M.: Alternative metrics. In: Concise Encyclopedia of Coding Theory, pp. 555–574. CRC Press (2021)

7. Flaxman, A.D., Frieze, A.M., Vera, J.: Adversarial deletion in a scale-free random graph process. Comb. Probab. Comput. **16**(2), 261–270 (2007). https://doi.org/10.1017/S0963548306007681

8. Harary, F., Palmer, E.M.: Graphical Enumeration. Academic Press, New York-London (1973)

9. Herschkorn, S.J.: On the modular value and fractional part of a random variable. Probab. Eng. Inf. Sci. **9**(4), 551–562 (1995). https://doi.org/10.1017/S0269964800004058

10. Janson, S., Łuczak, T., Ruciński, A.: Random Graphs. Wiley, Hoboken (2000)

11. Kabatyanskiĭ, G.A., Panchenko, V.I.: Packings and coverings of the Hamming space by balls of unit radius. Problems Inform. Transm. **24**(4), 261–272 (1988)

12. Kiselev, S., Kupavskii, A., Verbitsky, O., Zhukovskii, M.: On anti-stochastic properties of unlabeled graphs. arXiv:2112.04395 (2021)

13. Liebenau, A., Wormald, N.: Asymptotic enumeration of graphs by degree sequence, and the degree sequence of a random graph. arXiv:1702.08373 (2018)

14. McKay, B.D., Wormald, N.C.: The degree sequence of a random graph I. The models. Random Struct. Algorithms **11**(2), 97–117 (1997)

15. Schaeffer, S., Valdés, V., Figols, J., Bachmann, I., Morales, F., Bustos-Jiménez, J.: Characterization of robustness and resilience in graphs: a mini-review. J. Complex Netw. **9**(2) (2021). https://doi.org/10.1093/comnet/cnab018

Computing List Homomorphisms
in Geometric Intersection Graphs

Sándor Kisfaludi-Bak[1] , Karolina Okrasa[2,3(✉)] , and Paweł Rzążewski[2,3]

[1] Department of Computer Science, Aalto University, Espoo, Finland
`sandor.kisfaludi-bak@aalto.fi`
[2] Faculty of Mathematics and Information Science, Warsaw University
of Technology, Warsaw, Poland
`karolinaokrasa@gmail.com, pawel.rzazewski@pw.edu.pl`
[3] Institute of Informatics, Faculty of Mathematics, Informatics and Mechanics,
University of Warsaw, Warsaw, Poland

Abstract. A homomorphism from a graph G to a graph H is an edge-preserving mapping from $V(G)$ to $V(H)$. Let H be a fixed graph with possible loops. In the list homomorphism problem, denoted by LHom(H), the instance is a graph G, whose every vertex is equipped with a subset of $V(H)$, called list. We ask if there exists a homomorphism from G to H, such that every vertex from G is mapped to a vertex from its list.

We study the complexity of the LHom(H) problem in intersection graphs of various geometric objects. In particular, we are interested in answering the question for what graphs H and for what types of geometric objects, the LHom(H) problem can be solved in time subexponential in the number of vertices of the instance.

We fully resolve this question for string graphs, i.e., intersection graphs of continuous curves in the plane. Quite surprisingly, it turns out that the dichotomy coincides with the analogous dichotomy for graphs excluding a fixed path as an induced subgraph [Okrasa, Rzążewski, STACS 2021].

Then we turn our attention to intersections of fat objects. We observe that the (non) existence of subexponential-time algorithms in such classes is closely related to the size mrc(H) of a maximum reflexive clique in H, i.e., maximum number of pairwise adjacent vertices, each of which has a loop. We study the maximum value of mrc(H) that guarantees the existence of a subexponential-time algorithm for LHom(H) in intersection graphs of (i) convex fat objects, (ii) fat similarly-sized objects, and (iii) disks. In the first two cases we obtain optimal results, by giving matching algorithms and lower bounds.

Keywords: Graph homomorphisms · Geometric intersection graphs · Subexponential-time algorithms · Exponential Time Hypothesis

Karolina Okrasa–Supported by the European Research Council (ERC) under the European Union's Horizon 2020 research and innovation programme Grant Agreement no. 714704.

Paweł Rzążewski–Supported by Polish National Science Centre grant no. 2018/31/D/ST6/00062.

M. A. Bekos and M. Kaufmann (Eds.): WG 2022, LNCS 13453, pp. 313–327, 2022.
https://doi.org/10.1007/978-3-031-15914-5_23

314   S. Kisfaludi-Bak et al.

1   Introduction

For a family $\mathcal{S}$ of sets, its intersection graph is the graph whose vertex set is $\mathcal{S}$, and two sets are adjacent if and only if they have a nonempty intersection.

A prominent role is played by *geometric intersection graphs*, i.e., intersection graphs of some geometrically defined object (usually subsets of the plane). Some best studied families of this type are interval graphs [13,24] (intersection graphs of segments on a line), disk graphs [7,9] (intersection graphs of disks in the plane), segment graphs [21] (intersection graphs of segments), or string graphs [22,23] (intersection graphs of continuous curves). Geometric intersection graphs are studied not only for their elegant structural properties, but also for potential applications. Indeed, many real-life graphs have some underlying geometry [14, 17,18]. Thus the complexity of graph problems restricted to various classes of geometric intersection graphs has been an active research topic [2,10–12,25–27].

The underlying geometric structure can sometimes be exploited to obtain much faster algorithms than for general graphs. For example, for each fixed k, the k-COLORING problem is polynomial-time solvable in interval graphs, while for $k \geqslant 3$ the problem is NP-hard and thus unlikely to be solvable in polynomial time in general graphs. For disk graphs, the k-COLORING problem remains NP-hard for $k \geqslant 3$, but still it is in some sense more tractable than for general graphs. Indeed, for every fixed $k \geqslant 3$, then k-COLORING problem can be solved in *subexponential time* $2^{\mathcal{O}(\sqrt{n})}$ in n-vertex disk graphs, while assuming the Exponential-Time Hypothesis (ETH) [15,16] no such algorithm can exist for general graphs. Furthermore, the running time of the above algorithm is optimal under the ETH [20]. Biró et al. [4] studied the problem for superconstant number of colors and showed that if $k = o(n)$, then k-COLORING admits a subexponential-time algorithm in disk graphs, and proved almost tight complexity bounds conditioned on the ETH.

As a stark contrast, they showed that 6-COLORING does not admit a subexponential-time algorithm in segment graphs. This was later improved by Bonnet and Rzążewski [6] who showed that already 4-COLORING cannot be solved in subexponential time in segment graphs, but 3-COLORING admits a $2^{\mathcal{O}(n^{2/3} \log n)}$-algorithm in all string graphs. They also showed several positive and negative results concerning subexponential-time algorithms for segment and string graphs.

This line of research was continued in a more general setting by Okrasa and Rzążewski [29] who considered variants of the graph homomorphism problem in string graphs. For graphs G and H, a homomorphism from G to H is an edge-preserving mapping from $V(G)$ to $V(H)$. Note that a homomorphism to K_k is precisely a proper k-coloring, so graph homomorphisms generalize colorings. Among other results, Okrasa and Rzążewski [29] fully classified the graphs H for which a weighted variant of the homomorphism problem admits a subexponential-time algorithm in string graphs (assuming the ETH). It turns out that the substructure of H that makes the problem hard is an induced 4-cycle.

Separators in Geometric Intersection Graphs. Almost all subexponential-time algorithms for geometric intersection graphs rely on the existence of balanced separators that are small or simple in some other way. This is very convenient for a divide-&-conquer approach – due to the simplicity of the separator we can guess how the solution looks on the separator, and then recurse into connected components of the graph with the separator removed.

For example it is known that n-vertex disk graphs, where each point is contained in at most k disks, admit a balanced separator of size $\mathcal{O}(\sqrt{nk})$ [28,30]. This separator theorem was recently significantly extended by De Berg et al. [2] who introduced the notion of *clique-based separators*. Roughly speaking, a clique-based separator consists of cliques, and, instead of measuring its size (i.e., the number of vertices), we measure its *weight* defined as the sum of logarithms of sizes of the cliques. This approach shifts the focus from "small" separators to separators with "simple" structure, and proved helpful in obtaining ETH-tight algorithms for various combinatorial problems in intersection graphs of similarly sized fat or convex fat objects. The direction was followed by De Berg et al. [3] who proved that some other classes of intersection graphs admit balanced clique-based separators of small weight.

In the above approaches the size (or the weight) of the separator, as well as the balance factor, were measured in purely combinatorial terms. However, some alternative approaches, with more geometric flavor, were also used. For example Alber and Fiala [1] showed a separator theorem for intersection graphs of disks with diameter bounded from below and from above, where both the size of the separator and the size of each component of the remaining part of the graph is measured in terms of the *area* occupied by the geometric representation.

Our Contribution. In this paper we study the complexity of the list variant of the graph homomorphism problem in intersection graphs of geometric objects. For a fixed graph H (with possible loops), by $\text{LHom}(H)$ we denote the computational problem, where every vertex of the input graph G is equipped with the subset of $V(H)$ called *list*, and we need to determine whether there exists a homomorphism from G to H, such that every vertex from G is mapped to a vertex from its list.

First we study the complexity of $\text{LHom}(H)$ in string graphs and exhibit the full complexity dichotomy, i.e., we fully characterize graphs H for which the $\text{LHom}(H)$ problem can be solved in subexponential time. It turns out that the positive cases are precisely the graphs H that not *predacious*. The class of predacious graphs was defined by Okrasa and Rzążewski [29] who studied the complexity of $\text{LHom}(H)$ in P_t-free graphs (i.e., graphs excluding a t-vertex path as an induced subgraph). It is quite surprising that the complexity dichotomies for $\text{LHom}(H)$ in P_t-free graphs and in string graphs coincide; note that the classes are incomparable.

Our approach closely follows the one by Okrasa and Rzążewski [29]. First we show that if H does not belong to the class of predacious graphs, then a combination of branching on a high-degree vertex and divide-&-conquer approach using the string separator theorem yields a subexponential-time algorithm. For the hardness counterpart, we observe that the graphs constructed in [29] are actually string graphs. Summing up, we obtain the following result.[1]

Theorem 1 (♠). *Let H be a fixed graph.*

(a) *If H is not predacious, then* LHOM(H) *can be solved in time* $2^{\mathcal{O}(n^{2/3}\log n)}$ *in n-vertex string graphs, even if a geometric representation is not given.*

(b) *Otherwise, assuming the ETH,* LHOM(H) *cannot be solved in time* $2^{o(n)}$ *in n-vertex string graphs, even if they are given with a geometric representation.*

Then in Sect. 3 we turn our attention to subclasses of string graphs defined by intersections of fat objects. We observe that in this case the parameter of the graph H that seems to have an influence on the (non)existence of subexponential-time algorithms is the size of a maximum reflexive clique, denoted by mrc(H). Here, by a reflexive clique we mean a set of pairwise adjacent vertices, each of which has a loop. We focus on the following question.

Question. *For a class C of geometric objects (subsets of the plane), what is the maximum k (if any), such that for every graph H with* mrc(H) $\leqslant$ *k, the* LHOM(H) *problem admits a subexponential-time algorithm in intersection graphs of objects from C?*

Note that k from the question might not exist, as for example 4-COLORING (and thus LHOM(K_4)) does not admit a subexponential-time algorithm in segment graphs [6], while mrc(K_4) = 0.

First, we show that the existence of clique-based separators of sublinear weight is sufficient to provide subexponential-time algorithms for the case mrc(H) $\leqslant$ 1. In particular, this gives the following result.

Theorem 2 (♠). *Let H be a graph with* mrc(H) $\leqslant$ 1. *Then* LHOM(H) *can be solved in time:*

(a) $2^{\mathcal{O}(\sqrt{n})}$ *in n-vertex intersection graphs of fat convex objects,*

(b) $2^{\mathcal{O}(n^{2/3}\log n)}$ *in n-vertex pseudodisk intersection graphs.*

provided that the instance graph is given along with a geometric representation.

Next, we study intersection graphs of fat, similarly-sized objects. The exact definition of these families is given in Sect. 2, but, intuitively, each object should contain a disk of constant diameter, and be contained in a disk of constant diameter. We show that for such graphs subexponential-time algorithms exist even for the case mrc(H) $\leqslant$ 2. Our proof uses a new separator theorem, which measures the size of the separator in terms of the number of vertices, and the size of the components of the remaining graph in terms of the area.

Theorem 3. *Let H be a graph with* mrc(H) $\leqslant$ 2. *Then* LHOM(H) *can be solved in time* $2^{\mathcal{O}(n^{2/3}\log n)}$ *in n-vertex intersection graphs of fat, similarly-sized objects, provided that the instance graph is given along with a geometric representation.*

---

[1] Proofs of statements marked with (♠) can be found in the full version of the paper [19].

It turns out that both Theorem 2 (a) and Theorem 3 are optimal in terms of the value of mrc(H). More precisely, we prove ($\spadesuit$) that, assuming the ETH, there are graphs H_1 and H_2 with mrc(H_1) = 2 and mrc(H_2) = 3, such that LHom(H_1) does not admit a subexponential-time algorithm in intersection graphs of equilateral triangles, and LHom(H_2) does not admit a subexponential-time algorithm in intersection graphs of fat similarly-sized triangles.

A very natural question is to find the maximum value of mrc(H) that guarantees the existence of subexponential-time algorithms for LHom(H) in *disk graphs*. By Theorem 2 (a) we know that it is at least 1. However, disk graphs admit many nice structural properties that proved very useful in the construction of algorithms. Unfortunately, we were not able to obtain any stronger algorithmic results for disk graphs. For the lower bounds, we note that the constructions in our hardness reductions essentially used that triangles can "pierce each other", which cannot be done with disks. The best lower bound we could provide for disk graphs is as follows.

Theorem 4. *Assume the ETH. There is a graph H with* mrc(H) = 4, *such that* LHom(H) *cannot be solved in time* $2^{o(n/\log n)}$ *in n-vertex disk intersection graphs.*

The paper is concluded with several open questions in Sect. 5.

Full Version. In the full version of this paper [19] in addition to full proofs of statements marked with ($\spadesuit$), we consider the complexity of two weighted generalizations of the list homomorphism problem.

2 Preliminaries

Graph Theory. For a graph G and a vertex $v \in V(G)$, by $N_G(v)$ we denote the set of neighbors of v. If the graph is clear from the context, we simply write $N(v)$. For a graph H with possible loops, by $R(H)$ we denote the set of *reflexive* vertices, i.e., the vertices with a loop, and by $I(H)$ we denote the set of *irreflexive vertices*, i.e., vertices without loops. Clearly $R(H)$ and $I(H)$ form a partition of $V(H)$. By mrc(H) we denote the size of a maximum reflexive clique in H.

String Graphs and their Subclasses. For a set V of subsets of the plane $\mathbb{R}^2$, by IG(V) we denote their *intersection graph*, i.e., the graph with vertex set V where two elements are adjacent if and only if they have nonempty intersection. To avoid confusion, the elements of V will be called *objects*.

A collection V of objects in $\mathbb{R}^2$ is *fat* if there exists a constant $\alpha > 0$, such that each $v \in V$ satisfies $r_{\text{in},v}/r_{\text{out},v} \geq \alpha$, where $r_{\text{in},v}$ and $r_{\text{out},v}$ denote, respectively, the radius of the largest inscribed and smallest circumscribed disk of v. A collection V of objects in $\mathbb{R}^2$ is *similarly-sized* if there is some constant $\beta > 0$, such that $\max_{v \in V} \text{diam}(v)/\min_{v \in V} \text{diam}(v) \leq \beta$, where diam($v$) denotes the diameter of v. If a collection of objects is fat and similarly-sized, then we can set the unit to be the smallest diameter among the maximum inscribed disks of the objects, and as a consequence of the properties each object can be covered by some disk of radius $R = \mathcal{O}(1)$.

List Homomorphisms. A *homomorphism* from a graph G to a graph H is a mapping $f : V(G) \rightarrow V(H)$ such that for every $uv \in E(G)$ it holds that $f(u)f(v) \in E(H)$. If f is a homomorphism from G to H, we denote it shortly by $f : G \rightarrow H$. Note that homomorphisms to the complete graph on k vertices are precisely proper k-colorings. Thus we will often refer to vertices of H as *colors*.

In this paper we consider the LHOM(H) problem, which asks for the existence of *list* homomorphisms. Formally, for a fixed graph H (with possible loops), an instance of LHOM(H) is a pair (G, L), where G is a graph and $L : V(G) \rightarrow 2^{V(H)}$ is a list function. We ask whether there exists a homomorphism $f : G \rightarrow H$, which respects the lists L, i.e., for every $v \in V(G)$ it holds that $f(v) \in L(v)$.

The following straightforward observation will be used several times.

Observation 1. *Let G be an irreflexive graph and let H be a graph with possible loops and let $f : G \rightarrow H$. For every clique C of G we have the following:*

- *at most $|I(H)|$ vertices from C are mapped to vertices of $I(H)$ (each to a distinct vertex of $I(H)$),*
- *the remaining vertices of C are mapped to some reflexive clique of H.*

3   Algorithm for Intersection Graphs of Fat Objects

In this section we consider intersection graphs of fat, similarly-sized objects. The algorithm presented here uses the area occupied by the geometric representation as the measure of the instance. Let us start with introducing some notions.

Let V be a set of n fat, similarly-sized objects in $\mathbb{R}^2$. Recall that there is a constant R, such that each object in V contains a unit diameter disk and is contained in a disk of radius R. In what follows we hide the factors depending on R in the $\mathcal{O}(\cdot)$ notation.

Let us imagine a fine grid partitioning of $\mathbb{R}^2$ into square cells of unit diameter, i.e., of side length $1/\sqrt{2}$. This allows us to use discretized notion of a bounding box and of the area. For an object $v \subseteq \mathbb{R}^2$, by bb(v) we denote the minimum grid rectangle (i.e., rectangle whose sides are contained in grid lines) containing v, and by area(v) we denote the area of bb(v). For a set V of objects, we define bb(V) = bb($\bigcup_{v \in V} v$) and area(V) = area(bb(V)). Let us point out that in general area(V) can be arbitrarily large (unbounded in terms of n). However, it is straightforward to observe that if IG(V) is connected, then area(V) = $\mathcal{O}(n^2)$.

Recall that each object $v \in V$ contains a unit-diameter disk with the center c_v. We assign v to the grid cell containing c_v (if c_v is on the boundary of cells, we choose one arbitrarily). Now note that all objects assigned to a single cell form a clique in IG(V). Consequently, the vertex set of IG(V) can be partitioned into $\mathcal{O}(\text{area}(V))$ subsets, each inducing a clique; we call these subsets *cell-cliques*.

First, we show that intersection graphs of fat, similarly-sized objects admit balanced separators, where the size of instances is measured in terms of the *area* occupied by the geometric representation.

Lemma 1 (♠). *Let $G = \mathsf{IG}(V)$, where V is a set of n fat, similarly-sized objects in $\mathbb{R}^2$ and G is connected. Then either $\mathrm{area}(V) = \mathcal{O}(n^{2/3})$, or there exists a horizontal or vertical separating line ℓ such that:*

- *the number of objects whose convex hull intersects ℓ is $\mathcal{O}(n^{2/3})$, and*
- *the sets V_1, V_2 of objects on each side of ℓ (whose convex hulls are disjoint from ℓ) satisfy $\mathrm{area}(V_1) \leqslant \frac{3}{4}\mathrm{area}(V)$ and $\mathrm{area}(V_2) \leqslant \frac{3}{4}\mathrm{area}(V)$.*

Furthermore ℓ can be found in time polynomial in $\mathrm{area}(V)$ and n.

Let us introduce an auxiliary problem. The $\mathrm{LHoM}_{\mathrm{rc}}(H)$ problem is a restriction of $\mathrm{LHoM}(H)$, where for every instance (G, L), and for every $v \in V(G)$ the set $L(v)$ induces a reflexive clique in H. Note that in this problem we can always focus on the subgraph induced by reflexive vertices of H, as irreflexive vertices do not appear in any lists. Thus, $\mathrm{LHoM}_{\mathrm{rc}}(H)$ is equivalent to $\mathrm{LHoM}_{\mathrm{rc}}(H[R(H)])$.

Lemma 2. *Let V be a set of n similarly-sized fat objects in $\mathbb{R}^2$. Let (G, L) be an instance of $\mathrm{LHoM}(H)$, where $G = \mathsf{IG}(V)$. Then in time $n^{\mathcal{O}(\mathrm{area}(V))}$ we can build a family $\mathcal{Y}$ of instances of $\mathrm{LHoM}_{\mathrm{rc}}(H)$, such that:*

- *$|\mathcal{Y}| = n^{\mathcal{O}(\mathrm{area}(V))}$,*
- *each instance in $\mathcal{Y}$ is an induced subgraph of G,*
- *(G, L) is a yes-instance if and only if $\mathcal{Y}$ contains a yes-instance.*

Proof. Recall that V can be partitioned into $\mathcal{O}(\mathrm{area}(V))$ cell-cliques, and consider one such cell-clique C. By Observation 1 at most $|I(H)|$ vertices from C receive colors from $I(H)$ and the remaining vertices of C must be mapped to some reflexive clique of H. We guess the vertices mapped to $I(H)$ along with their colors and the reflexive clique to which the remaining vertices are mapped. As H is a constant, the total number of branches created for C is $|C|^{\mathcal{O}(|V(H)|)} = n^{\mathcal{O}(1)}$. Repeating this for every clique, we result in $n^{\mathcal{O}(\mathrm{area}(V))}$ branches.

Consider one such a branch. For each vertex v whose color was guessed (i.e., this color is in $I(H)$), we update the lists of neighbors of v. More precisely, if the color guessed for v is a, then we remove every nonneighbor of a from the lists of all neighbors of v. After that we remove v from the graph. Similarly, we update the lists of vertices v that are supposed to be mapped to vertices of $R(H)$: we remove from $L(v)$ every vertex that is not in the guessed reflexive clique.

Note that this way we obtained an instance of $\mathrm{LHoM}_{\mathrm{rc}}(H)$, where the instance graph is an induced subgraph of G. We include such an instance into $\mathcal{Y}$.

As the number of branches is $n^{\mathcal{O}(\mathrm{area}(V))}$, we obtain that $|\mathcal{Y}| = n^{\mathcal{O}(\mathrm{area}(V))}$. It is clear that (G, L) is a yes-instance if and only if $\mathcal{Y}$ contains a yes-instance.

Combining Lemmas 1 and 2 gives the following.

Lemma 3. *Let H be a fixed graph. Suppose that $\mathrm{LHoM}_{\mathrm{rc}}(H)$ can be solved in time $2^{\mathcal{O}(n^{2/3}\log n)}$ in n-vertex intersection graphs of fat, similarly-sized objects, given along with a geometric representation.*

Then LHOM(H) can be solved in time $2^{\mathcal{O}(n^{2/3}\log n)}$ in n-vertex intersection graphs of fat, similarly-sized objects, given along with a geometric representation.

Proof. Let V be a set of n fat, similarly-sized objects in $\mathbb{R}^2$ and let $G = \mathsf{IG}(V)$. Let (G, L) be an instance of LHOM(H). Notice that if G is disconnected, then we can solve the problem for each connected component separately. Thus let us assume that G is connected. We do induction on area(V).

If area(V) = $\mathcal{O}(n^{2/3})$ (the actual constant in $\mathcal{O}(\cdot)$ is the constant from Lemma 1), we call Lemma 2 to obtain a family $\mathcal{Y}$ of instances of $\mathrm{LHOM}_{rc}(H)$, such that $|\mathcal{Y}| = n^{\mathcal{O}(n^{2/3})}$. Each instance in $\mathcal{Y}$ is an induced subgraph of G and thus can be solved in time $2^{\mathcal{O}(n^{2/3}\log n)}$. As solving every instance in $\mathcal{Y}$ is enough to solve (G, L), we can solve the problem in total time $2^{\mathcal{O}(n^{2/3}\log n)}$, as claimed.

In the other case, we apply Lemma 1, let ℓ be the obtained separating line. Let $S \subseteq V$ be the set of objects whose convex hull intersects ℓ; by Lemma 1 the size of S is $\mathcal{O}(n^{2/3})$. Let V_1, V_2 be the partition of $V - S$ into instances of each side of ℓ, as in Lemma 1. Recall that area(V_1) $\leqslant \frac{3}{4}$area(V) and area(V_2) $\leqslant \frac{3}{4}$area(V).

We exhaustively guess the coloring of S, this results in $|V(H)|^{|S|} = 2^{\mathcal{O}(n^{2/3})}$ branches. For each such branch we update the lists of neighbors of vertices whose color was guessed. Now observe that the subinstances induced by V_1 and V_2 can be solved independently. Our initial instance is a yes-instance if and only if for some guess both subinstances are yes-instances.

Denoting by $\mu := \text{area}(V)$, we obtain the recursion for the running time: $F(\mu) \leqslant 2^{\mathcal{O}(n^{2/3})} \cdot F\left(\frac{3}{4}\mu\right)$, which solves to $F(\mu) \leqslant 2^{\mathcal{O}(n^{2/3}\log\mu)}$. As $\mu = \mathcal{O}(n^2)$ (since G is connected), we conclude that the total running time is $2^{\mathcal{O}(n^{2/3}\log n)}$.

Observe that if mrc(H) $\leqslant 2$, then in every instance of $\mathrm{LHOM}_{rc}(H)$, each list is of size at most 2. It follows from the result of Edwards [8] that such instances can be solved in polynomial time. Thus, Lemma 3 immediately implies Theorem 3.

4   Lower Bound for Intersection Graphs of Disks

In this section we show that the assumption that mrc(H) $\leqslant 1$ in Theorem 2 cannot be significantly improved. Our goal is to prove the following theorem.

Theorem 4. *Assume the ETH. There is a graph H with mrc(H) = 4, such that LHOM(H) cannot be solved in time $2^{o(n/\log n)}$ in n-vertex disk intersection graphs.*

We reduce from 3-SAT. Let $\ell_1, \ldots, \ell_t$ be the literals of the formula Φ on N variables and M clauses, each of which contains exactly three variables, i.e., the i-th clause consists of literals ℓ_{3i-2}, ℓ_{3i-1}, and ℓ_{3i}, and $t = 3M$. Let $k = 1 + \lceil \log t \rceil$ be the number of binary digits required to represent numbers up to t.

Construction Overview. We construct an instance of LHOM(H) for the graph H depicted on Fig. 3 (iii). The construction has variable gadgets placed at the

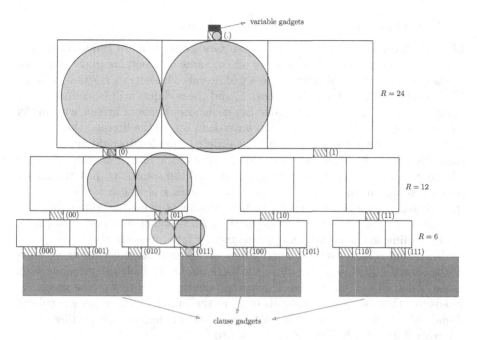

Fig. 1. Overview of the construction, with the path of the literal with binary index '011' highlighted. The lined rectangles are literal cliques, with the common prefix of the literal indices. The triplets of squares represent subset turning and divider gadgets.

top, consisting of two disks with lists $\{T, F\}$, where the value of the first disk corresponds to setting the variable true or false.

The bulk of the construction consists of large cliques of disks of various sizes, and in each clique the disks correspond to some specific subsets of literals. All of these cliques have lists of size 2, where the assigned colors correspond to the literal being true or false. At the top, the initial clique will have all the literals arranged by the index of the corresponding variable, i.e., starting with the positive literals of x_1, then the negative literals of x_1, then the positive literals of x_2, etc.

Represent each literal index i with a binary number of t digits (with leading zeros as necessary). We use a so-called *divider gadget* to partition the set of literals to two subsets: the first subset will contain disks for those literals ℓ_i where the first binary digit of i is 0, and the other subset will contain those where the first binary digit of i is 1. Using two smaller copies of the divider gadget, we further partition both sets according to the second, third, etc. binary digits, creating a structure resembling a binary tree of depth $\log N + \mathcal{O}(1)$. At the leaves, the cliques contain a single disk, and the leaves are ordered in increasing order of the index i, that is, the literals of each clause c_j appear at three consecutive leaves. We attach a gadget on consecutive triplets to check the clauses.

We will now explain the construction in detail.

Literal Cliques, Variable and Clause Gadgets. A literal clique consists of at most t disks of unit radius that are on the same horizontal segment of length 1. Each literal clique will contain the set of literals whose index starts with some fixed binary prefix s of length at most k, and these cliques will be connected by other gadgets, creating a binary tree. Let us denote the set of literals with prefix s by L_s. The initial literal clique will have disks for all the literals (the set $L_\emptyset$). In the initial clique, these literals will be ordered from left to right according to the corresponding variable's index and the sign of the literal. Each later clique will contain the subset L_s of literals, positioned the same way, just translated somewhere else in the plane. Note that for prefixes s of length k, the set L_s is a singleton, it contains the literal of binary index s. These literal cliques will correspond to the leaves of our construction. All literal cliques have lists $\{1, 2\}$, corresponding to the literal being set to true or false, respectively.

In the full version (♠), we describe how a careful arrangement of these disks in the cliques allows us to attach to the top a simple variable gadget, consisting of two small disks, that intersect only the relevant literal disks. The gadget of x_i ensures that the positive literals of x_i in the initial literal clique get color 1 if and only if the gadget of x_i is set to true, and the negative literal disks of x_i get color 2 if and only if the gadget is set to true.

At the bottom of the binary tree, our construction will ensure that the centers of the disks in consecutive singleton literal cliques (at the leaves of the binary tree) have a distance between 11 and 13. We attach a constant-size clause gadget (♠) on the three literal disks that correspond to the clause. The clause gadgets have a valid coloring if and only if not all three literal disks have color 2.

Subset Turning and the Divider Gadget. Our task now is to connect a literal clique to its children by dividing its literals into two subsets, keeping the information carried for each individual literal. First, we show how we can create a turn gadget using disks of any size.

Consider a horizontal segment of length R with its left endpoint at the point o, and let p be a point where a disk in some literal clique touches the segment from above. Suppose moreover that p is somewhere in the length 1 interval at the middle of the segment, see Fig. 2 (i). Then the *turning disk at* p is the unique disk D_p that touches the segment from below and has radius $|op|$. Note that if we draw a vertical segment of length R with top endpoint o, then it will touch the disk on the left at some point p' where $|op| = |op'|$. The *turning gadget* is simply a collection of turning disks for some custom set of points in the middle length-1 interval. We can represent the gadget with a square of side length R whose top side is the initial segment. Note that some turning disks may not be completely covered by the square, but since p is required to be in the middle length-1 interval, the disks can protrude at most distance 1 beyond the boundary of the square. Also note that we can create an analogous gadget with disks that touch any pair of consecutive sides of the side-length R square.

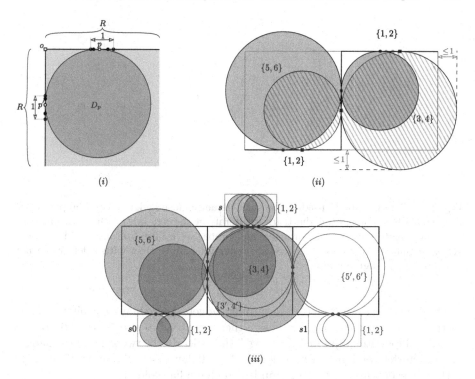

Fig. 2. (i) Unique disk of radius between $\left[\frac{R-1}{2}, \frac{R+1}{2}\right]$ touching two perpendicular lines. (ii) Making two turns with some subset of the literals (iii) Dividing the set of literals into two arbitrary subsets (red versus green disks) using overlaid turns. (Color figure online)

We can glue two turning gadgets together as depicted in Fig. 2 (ii). The disks of the first (right) gadget have lists $\{3,4\}$, and the disks of the second (left) have lists $\{5,6\}$. In the graph H, we have $1,2,3,4$ as well as $3,4,5,6$ and $5,6,1,2$ form induced 4-cycles. The connecting literal cliques have disks with lists $\{1,2\}$, and all of the colors $\{1,2,3,4,5,6\}$ are reflexive vertices of H, see Fig. 3 (ii). It is routine to check that the turning disks receive odd colors if and only if the corresponding disks in the literal cliques have color 1.

Finally, we can overlay such a glued turning gadget with its mirror image, as depicted in Fig. 2 (iii). In the mirror image, the disks of the first turning gadget get the list $\{3',4'\}$, and the disks of the second turning gadget get the list $\{5',6'\}$. The vertices $1,2,3',4',5',6'$ induce the same graph as vertices $1,2,3,4,5,6$. These four turns together define a *divider gadget* of size R. If the literal clique at the top contained the disks of index prefix s, then we use the first two turns (going to the left child, red disks in Fig. 2 (iii)) only on the touching points for literals with prefix $s0$, and the other two turns (going to the right child, green disks in Fig. 2 (iii)) only for the touching points for literals with prefix $s1$.

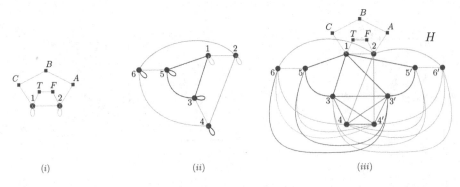

Fig. 3. (i) The part of H used in the variable and clause gadgets. (ii) The part of H responsible for propagating the truth of literals in the left side of the division. Blue edges propagate true literals, orange edges propagate a false literals. (iii) The graph H. Note that all numbered vertices are reflexive, and vertices with a letter are not reflexive. (Color figure online)

Notice however that inside the turning gadgets, there may be arbitrary intersections between red and green disks, therefore $3, 4, 5, 6$ and $3', 4', 5', 6'$ form a complete bipartite graph in H, see Fig. 3 (iii). Clearly the two sides of the gadget do not interfere and disks with colors $3', 4', 5', 6'$ have an odd number if and only if the corresponding disk at the top literal clique has color 1.

Proof of Theorem 4. Recall that our formula has t literals, and each literal index can be represented by a binary string of length $k = \lceil \log_2 t \rceil + 1$.

We place the initial literal clique together with the variable gadgets as described in the construction. At the bottom of this literal clique the disks touch a length 1 interval. We attach a divider gadget of size $R = 6 \cdot 2^{k-1}$ to this interval (see Fig. 1). We then use the divider gadget to propagate the values stored in the literals to the children with prefix $s = 0$ on the left and $s = 1$ on the right. For literal cliques of prefix length $\text{len}(s)$, we attach dividers of size $R = 6 \cdot 2^{k-1-\text{len}(s)}$. At the bottom, we end up with singleton literal cliques hanging off of literal gadgets of size $R = 6$. One can verify that the gaps between literal cliques of consecutive leaves have length 9 (that is, the right side of the 2×3 rectangle covering the first leaf and the left side of the rectangle of the next leaf has distance 9). It is also easy to verify that the turning disks of distinct divider gadgets are disjoint: recall that the disks protrude beyond the boundary of the base square by at most 1, and the literal cliques have height 2.

Based on the formula, the described set can clearly be constructed in polynomial time. Each literal has corresponding disks in $\mathcal{O}(k) = \mathcal{O}(\log t)$ gadgets, and each literal clique and divider has $\mathcal{O}(1)$ disks per represented literal. Additionally, the variable and clause gadgets have constant size. Thus for a 3-CNF formula of N variables and M clauses with $t = 3M$ literals, there are $\mathcal{O}(t \log t + M + N) = \mathcal{O}((M + N) \log(M + N))$ disks in the construction, which implies the desired lower bound under the ETH.

5 Conclusion and Open Problems

One of the best studied classes of intersection graphs are *unit disk intersection graphs*. They are known to admit many nice structural properties that can be exploited in the construction of algorithms [5, 7]. However, we were not able to obtain any better results that the general ones for (pseudo)disk intersection graphs given by Theorem 2. On the other hand we were not able to show that subexponential algorithms for this class cannot exist. We believe that obtaining improved bounds for unit disk graphs is an interesting and natural problem.

Let us point out that there are three natural places where one could try to improve our hardness reduction in Theorem 4: (a) to avoid using disks of unbounded size, (b) to show hardness for some H with $\mathrm{mrc}(H) \in \{2, 3\}$, and (c) to improve the lower bound to $2^{\Omega(n)}$ (instead of $2^{\Omega(n/\log n)}$).

References

1. Alber, J., Fiala, J.: Geometric separation and exact solutions for the parameterized independent set problem on disk graphs. J. Algorithms **52**(2), 134–151 (2004). https://doi.org/10.1016/j.jalgor.2003.10.001
2. de Berg, M., Bodlaender, H.L., Kisfaludi-Bak, S., Marx, D., van der Zanden, T.C.: A framework for exponential-time-hypothesis-tight algorithms and lower bounds in geometric intersection graphs. SIAM J. Comput. **49**(6), 1291–1331 (2020). https://doi.org/10.1137/20M1320870
3. de Berg, M., Kisfaludi-Bak, S., Monemizadeh, M., Theocharous, L.: Clique-based separators for geometric intersection graphs. In: 32nd International Symposium on Algorithms and Computation, ISAAC 2021. LIPIcs, vol. 212, pp. 22:1–22:15. Schloss Dagstuhl - Leibniz-Zentrum für Informatik (2021). https://doi.org/10.4230/LIPIcs.ISAAC.2021.22
4. Biró, C., Bonnet, É., Marx, D., Miltzow, T., Rzążewski, P.: Fine-grained complexity of coloring unit disks and balls. J. Comput. Geom. **9**(2), 47–80 (2018). https://doi.org/10.20382/jocg.v9i2a4
5. Bonamy, M., et al.: EPTAS and subexponential algorithm for maximum clique on disk and unit ball graphs. J. ACM **68**(2), 9:1-9:38 (2021). https://doi.org/10.1145/3433160
6. Bonnet, É., Rzążewski, P.: Optimality program in segment and string graphs. Algorithmica **81**(7), 3047–3073 (2019). https://doi.org/10.1007/s00453-019-00568-7
7. Clark, B.N., Colbourn, C.J., Johnson, D.S.: Unit disk graphs. Discret. Math. **86**(1–3), 165–177 (1990). https://doi.org/10.1016/0012-365X(90)90358-O
8. Edwards, K.: The complexity of colouring problems on dense graphs. Theor. Comput. Sci. **43**, 337–343 (1986). https://doi.org/10.1016/0304-3975(86)90184-2
9. Fishkin, A.V.: Disk graphs: a short survey. In: Solis-Oba, R., Jansen, K. (eds.) WAOA 2003. LNCS, vol. 2909, pp. 260–264. Springer, Heidelberg (2004). https://doi.org/10.1007/978-3-540-24592-6_23
10. Fomin, F.V., Lokshtanov, D., Panolan, F., Saurabh, S., Zehavi, M.: Finding, hitting and packing cycles in subexponential time on unit disk graphs. Discrete Computat. Geom. **62**(4), 879–911 (2019). https://doi.org/10.1007/s00454-018-00054-x

11. Fomin, F.V., Lokshtanov, D., Panolan, F., Saurabh, S., Zehavi, M.: ETH-tight algorithms for long path and cycle on unit disk graphs. In: Cabello, S., Chen, D.Z. (eds.) 36th International Symposium on Computational Geometry, SoCG 2020, 23–26 June 2020, Zürich, Switzerland. LIPIcs, vol. 164, pp. 44:1–44:18. Schloss Dagstuhl - Leibniz-Zentrum für Informatik (2020). https://doi.org/10.4230/LIPIcs.SoCG. 2020.44

12. Fomin, F.V., Lokshtanov, D., Saurabh, S.: Bidimensionality and geometric graphs. In: Rabani, Y. (ed.) Proceedings of the Twenty-Third Annual ACM-SIAM Symposium on Discrete Algorithms, SODA 2012, Kyoto, Japan, 17–19 January 2012, pp. 1563–1575. SIAM (2012). https://doi.org/10.1137/1.9781611973099

13. Golumbic, M.C.: Chapter 8 - interval graphs. In: Golumbic, M.C. (ed.) Algorithmic Graph Theory and Perfect Graphs, Annals of Discrete Mathematics, vol. 57, pp. 171–202. Elsevier (2004). https://doi.org/10.1016/S0167-5060(04)80056-6

14. Huson, M., Sen, A.: Broadcast scheduling algorithms for radio networks. In: Proceedings of MILCOM 1995, vol. 2, pp. 647–651 (1995). https://doi.org/10.1109/ MILCOM.1995.483546

15. Impagliazzo, R., Paturi, R.: On the complexity of k-SAT. J. Comput. Syst. Sci. **62**(2), 367–375 (2001). https://doi.org/10.1006/jcss.2000.1727

16. Impagliazzo, R., Paturi, R., Zane, F.: Which problems have strongly exponential complexity? J. Comput. Syst. Sci. **63**(4), 512–530 (2001). https://doi.org/10.1006/ jcss.2001.1774

17. Jungck, J.R., Viswanathan, R.: Chapter 1 - graph theory for systems biology: interval graphs, motifs, and pattern recognition. In: Robeva, R.S. (ed.) Algebraic and Discrete Mathematical Methods for Modern Biology, pp. 1–27. Academic Press, Boston (2015). https://doi.org/10.1016/B978-0-12-801213-0.00001-0

18. Kaufmann, M., Kratochvíl, J., Lehmann, K.A., Subramanian, A.R.: Max-tolerance graphs as intersection graphs: cliques, cycles, and recognition. In: Proceedings of the Seventeenth Annual ACM-SIAM Symposium on Discrete Algorithms, SODA 2006, Miami, Florida, USA, 22–26 January 2006, pp. 832–841. ACM Press (2006). http://dl.acm.org/citation.cfm?id=1109557.1109649

19. Kisfaludi-Bak, S., Okrasa, K., Rzążewski, P.: Computing list homomorphisms in geometric intersection graphs. CoRR abs/2202.08896 (2022). https://arxiv.org/ abs/2202.08896

20. Kisfaludi-Bak, S., van der Zanden, T.C.: On the exact complexity of Hamiltonian cycle and q-colouring in disk graphs. In: Fotakis, D., Pagourtzis, A., Paschos, V.T. (eds.) CIAC 2017. LNCS, vol. 10236, pp. 369–380. Springer, Cham (2017). https:// doi.org/10.1007/978-3-319-57586-5_31

21. Kratochvíl, J., Matoušek, J.: Intersection graphs of segments. J. Comb. Theory Ser. B **62**(2), 289–315 (1994). https://doi.org/10.1006/jctb.1994.1071

22. Kratochvíl, J.: String graphs. I. The number of critical nonstring graphs is infinite. J. Comb. Theory Ser. B **52**(1), 53–66 (1991). https://doi.org/10.1016/0095- 8956(91)90090-7

23. Kratochvíl, J.: String graphs. II. Recognizing string graphs is NP-hard. J. Comb. Theory Ser. B **52**(1), 67–78 (1991). https://doi.org/10.1016/0095- 8956(91)90091-W

24. Lekkeikerker, C., Boland, J.: Representation of a finite graph by a set of intervals on the real line. Fundam. Math. **51**(1), 45–64 (1962). http://eudml.org/doc/213681

25. Marx, D.: Efficient approximation schemes for geometric problems? In: Brodal, G.S., Leonardi, S. (eds.) ESA 2005. LNCS, vol. 3669, pp. 448–459. Springer, Heidelberg (2005). https://doi.org/10.1007/11561071_41

26. Marx, D.: On the optimality of planar and geometric approximation schemes. In: FOCS 2007 Proceedings, pp. 338–348 (2007). https://doi.org/10.1109/FOCS.2007.50

27. Marx, D., Pilipczuk, M.: Optimal parameterized algorithms for planar facility location problems using voronoi diagrams. In: Bansal, N., Finocchi, I. (eds.) ESA 2015. LNCS, vol. 9294, pp. 865–877. Springer, Heidelberg (2015). https://doi.org/10.1007/978-3-662-48350-3_72

28. Miller, G.L., Teng, S., Thurston, W.P., Vavasis, S.A.: Separators for sphere-packings and nearest neighbor graphs. J. ACM 44(1), 1–29 (1997). https://doi.org/10.1145/256292.256294

29. Okrasa, K., Rzążewski, P.: Subexponential algorithms for variants of the homomorphism problem in string graphs. J. Comput. Syst. Sci. 109, 126–144 (2020). https://doi.org/10.1016/j.jcss.2019.12.004

30. Smith, W.D., Wormald, N.C.: Geometric separator theorems and applications. In: FOCS 1998 Proceedings, pp. 232–243. IEEE Computer Society, Washington, DC, USA (1998)

On Fully Diverse Sets of Geometric Objects and Graphs

Fabian Klute$^{(\boxtimes)}$ and Marc van Kreveld

Department of Information and Computing Sciences, Utrecht University,
Utrecht, The Netherlands
{f.m.klute,m.j.vankreveld}@uu.nl

Abstract. Diversity is a property of sets that shows how varied or different its elements are. We define full diversity in a metric space and study the maximum size of fully diverse sets. A set is *fully diverse* if each pair of elements is as distant as the maximum possible distance between any pair, up to a constant factor. We study metric spaces based on geometry, embeddings of graphs, and graphs themselves. In the geometric cases, we study measures like Hausdorff distance, Frechét distance, and area of symmetric difference between objects in a bounded region. In the embedding cases, we study planar embeddings of trees and planar graphs, and use the number of swaps in the rotation system as the metric. In the graph cases, we use the number of insertions and deletions of leaves or edges as the metric. In most cases, we show (almost) tight lower and upper bounds on the maximum size of fully diverse sets. Our results lead to a very simple randomized algorithm to generate large fully diverse sets in several cases.

Keywords: Diversity · Distance Measures · Diverse Geometric Objects · Diverse Graphs · Diverse Embeddings

1 Introduction

When generating data, for example for benchmarks, it may be important that the generated set is sufficiently diverse. The same is true for systems that assist in choosing a desired layout or configuration by showing various options. For example, in graph drawing this observation has led to systems that present several drawings of a graph. A user can now choose a drawing, or indicate preference, after which more drawings like the preferred one can be generated [2].

But what does diversity mean in this context? We address this question in a formal way. We introduce a framework that allows us to study diversity of "objects", and analyze the maximum number of objects that are pairwise far apart. This framework is applicable in many contexts.

Supported by the Netherlands Organisation for Scientific Research (NWO) under project no. 612.001.651 and the Austrian Science Foundation (FWF) grant J4510.

M. A. Bekos and M. Kaufmann (Eds.): WG 2022, LNCS 13453, pp. 328–341, 2022.
https://doi.org/10.1007/978-3-031-15914-5_24

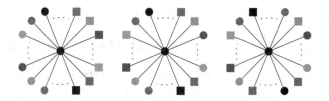

Fig. 1. Fully diverse set of labeled (color+symbol) n-vertex stars. Opposite leaves have the same color but a different symbol. The second embedding has one opposite-pair exchange per triple when compared to the first embedding, and the third embedding has two opposite-pair exchanges per triple. Any two of the three embeddings have distance $\Omega(n^2)$ if distance is measured by the number of swaps of adjacent edges.

Diversity as a Counting Problem. Let $(\mathcal{S}, \mu)$ be a metric space where $\mathcal{S}$ is a base set or class of objects and μ is a distance measure that takes a pair from $\mathcal{S}$ and assigns a distance. We consider the cases where $\mu(a, b)$ is bounded for all $a, b \in \mathcal{S}$; let $M = \sup_{a,b\in\mathcal{S}} \mu(a, b)$ be the highest value that is attained (possibly in the limit) by μ on $\mathcal{S}$.

Definition 1. *For a given $c \geq 1$, a subset $\hat{\mathcal{S}} \subseteq \mathcal{S}$ is called $\frac{1}{c}$-diverse if for all $x, y \in \hat{\mathcal{S}}$, we have $\mu(x, y) \geq \frac{1}{c} \cdot M$. If c can be chosen constant, independent of $|\hat{\mathcal{S}}|$, then $\hat{\mathcal{S}}$ is called* fully diverse.

Intuitively, we relate the distance of all pairs of the subset to the maximum distance within the base set. We are interested in the question how large fully diverse subsets can be. As a simple example, consider all points in a unit square region in the plane and Euclidean distance as the metric. Then the maximum distance is $\sqrt{2}$, so a $\frac{1}{c}$-diverse (sub)set of points must have pairwise distances of at least $\sqrt{2}/c$. It is easy to see that any $\frac{1}{c}$-diverse set has size $O(c^2)$ by a packing argument, and any maximal $\frac{1}{c}$-diverse set has size $\Omega(c^2)$. The maximum size of a fully diverse set of points is $\Theta(c^2) = \Theta(1)$ if c is a constant.

When considering more complex objects, like polygonal lines, triangulations, drawings of graphs, and graphs themselves, we need a distance between any two objects. We consider geometric distance measures for geometric objects and discrete measures for graphs. In several geometric cases, we need to assume that the objects reside in a bounded space to make the metric space bounded. Let U be a unit diameter disk in the plane. Some geometric distance measures are:

- For any two simple polygons inside U, their Hausdorff distance.
- For any two polygonal lines inside U, the Fréchet distance between them.
- For two simple polygons inside U, their symmetric difference.
- For two drawings of a given labeled graph inside U, the total vertex displacement (summed distance between vertices with the same label).
- For any two drawings of a given labeled graph with the same embedding, the L_1-distance of the vector of angles of adjacent edges (sum of differences of corresponding angles).

The maximum distance between any two objects in these cases is bounded by 1, 1, $\pi/4$, $|V|$, and $4\pi|V|$, respectively.

Some discrete measures for two embeddings of a given labeled graph are:

- For two embeddings of a labeled graph, the L_1-distance of the vectors on the edges, where an edge gives 1 if it intersects any other edge and 0 otherwise.
- For two embeddings of a labeled tree, the number of swaps of adjacent edges at a vertex to convert the embedding of one into the other.
- For two planar embeddings of a labeled graph, the number of swaps around cut-vertices and split pairs to convert one into the other.

See Fig. 1 for a fully diverse set of embeddings of a labeled star graph.

For general graphs (independent of embedding), we study measures based on the number of additions and removals of edges or leaves; in other words, the edit distance for a set of possible edits [10].

Relation to Diversity and Similar Notions in Science. Diversity has been studied in a variety of scientific contexts. One well-known example is in ecosystems, specifically, the diversity of species that are represented in a sample of animals or plants, see for instance [12,21]. The Shannon index is commonly used, also known as Shannon entropy in information theory.

In computer science, diversity has been studied in a variety of areas. For example, the diversity of the output in selection tasks in big data [9] or recommender systems [17], the diversity of input data sets for machine learning [16], or the diversity of colored point sets in computational geometry [15].

Diversity without a priori assigned categories is of interest in the study of the diversity of a population in genetic algorithms, e.g. [23]. Following similar ideas, researchers later studied the diversity of sets of solutions in satisfiability problems [14], multicriteria optimization problems [22], and, recently, parameterized algorithms, e.g. [3]. Similar to our work, the diversity measures found in this line of work are commonly based on the Hamming distance. Hebrard et al. [11] introduced the maximization problem to find the set of solutions that maximizes this sum or minimum distance over all sets of solutions.

With respect to drawings of graphs, Biedl et al. [5] studied how to heuristically generate a set of different not necessarily planar straight-line drawings. Their measure of distance between any two drawings is obtained by greedily matching vertices using a composite measure of Euclidean distance and position in the drawing. Bridgeman and Tamassia [7] investigated geometric measures for the distance between two orthogonal drawings.

Counting the overall number of structures such as planar triangulations or crossing-free geometric graphs on a point set (without requiring that they pairwise be far) has been widely studied, e.g. in [13]. See also the blog entry by Sheffer with a list of references on this topic.[1] Finally, for a given planar graph the number of embeddings it admits has been studied in the context of algorithms to count them [8,20].

---

[1] https://adamsheffer.wordpress.com/numbers-of-plane-graphs/.

Table 1. Lower and upper bounds on the maximum size of fully diverse sets in various metric spaces where the measure is geometric. U denotes a unit diameter disk.

Object	Metric	Space	Diameter	Lower bound	Upper bound
Polygons	Hausdorff distance	U	$\Theta(1)$	$\Omega(1)$	$O(1)$
Polylines	Fréchet distance	U	$\Theta(1)$	$2^{\Omega(n)}$	$2^{O(n)}$
Polygons	Area symm. diff.	U	$\Theta(1)$	$2^{\Omega(n)}$	$2^{O(n \log n)}$

The Approach and the Results. We initiate the study of the size of fully diverse sets in bounded metric spaces, as described in this introduction. We believe that such a study is important in all algorithmic problems where different objects are generated, for example, in graph drawing and benchmark construction.

We use a unified approach to obtain our results, which we can explain best on bit strings of length n with the metric the number of bit flips (or Hamming distance). We choose a sufficiently large constant c. Then we show that for any bit string, the number of bit strings that can be obtained by at most n/c bit flips is bounded from above, while the total number of bit strings is 2^n. Dividing the latter quantity (2^n) by the former gives a lower bound on the maximum size of a fully diverse set. When we apply this simple scheme to the various metric spaces listed before, we encounter different kinds of challenges.

In Sect. 2 we investigate the maximum size of a fully diverse set of fair bit strings (fair = equally many 0s as 1s). We show that for fair bit strings of length n, the maximal size of a fully diverse set is exponential. The same is true for fair *cyclic* bit strings. These results are used as a core ingredient in later proofs.

In Sect. 3 we present results for three metric spaces where distance is geometric. Bounds on the maximum size of fully diverse sets are given in Table 1, where diameter specifies the maximum distance in the metric space. The main challenge is the suitable discretization of the space of all possible polylines or polygons, so that the desired distance between pairs can be analyzed.

Next we consider embeddings of trees and planar graphs, that is, the cyclic order of neighboring nodes, in Sect. 4. The metric is the minimum number of swaps of adjacent neighbors to get from the one embedding to the other. Our results are given in Table 2.

In Sect. 5 we consider graphs as combinatorial objects and base the metric on edit distance. We distinguish labeled and unlabeled graphs, and consider trees, planar graphs, and general graphs. Table 3 gives the results.

There exists a very simple randomized algorithm to generate fully diverse sets of size k (provided k is small enough). It works as follows, starting with an empty set S and a constant $c \geq 1$, and a known maximum diameter M of the base set: (i) Generate a random element e from the base set. (ii) Test if e has distance at least M/c to all elements in S. If so, add e to S, and if not, discard it and continue at (i). Stop when set S contains k elements. This algorithm leads to fully diverse sets of large size with high probability for several examples in this paper, if distances can be computed easily.

Table 2. Lower and upper bounds on the maximum size of fully diverse sets in various metric spaces concerning embedded labeled graphs with n nodes. The lower bound for trees holds for any tree, whereas the upper bound holds for some trees (Theorem 4).

Object	Metric	Diameter	Lower bound	Upper bound
Ternary trees	# adjacent swap	$\Theta(n)$	$2^{\Omega(n)}$	$2^{O(n)}$
Star graphs	# adjacent swap	$\Theta(n^2)$	$2^{\Omega(n)}$	$2^{O(n \log n)}$
Trees	# adjacent swap	$\Theta(\sum_{v \in V} \deg^2(v))$	$2^{\Omega(\sqrt{n})}$	$[\,2^{O^*(\sqrt{n})}\,]$
Star graphs	# any swap	$\Theta(n)$	$2^{\Omega(n \log n)}$	$2^{O(n \log n)}$
Planar graphs	# adjacent swap	Theorem 5	Theorem 5	Theorem 5

Table 3. Lower and upper bounds on the maximum size of fully diverse sets in various metric spaces where the measure is edit distance and the objects are graphs with n nodes. Intermediate graphs must be in the same class.

Object	Metric	Diameter	Lower bound	Upper bound
Trees	# reattach leaf	$\Theta(n)$	$2^{\Omega(n \log n)}$	$2^{O(n \log n)}$
Planar graphs	# insert/delete edge	$\Theta(n)$	$2^{\Omega(n \log n)}$	$2^{O(n \log n)}$
Graphs	# insert/delete edge	$\Theta(n^2)$	$2^{\Omega(n^2)}$	$2^{O(n^2)}$
Trees (unlabeled)	# reattach leaf	$\Theta(n)$	$2^{\Omega(n)}$	$2^{O(n)}$
Planar graphs (unlab.)	# insert/delete edge	$\Theta(n)$	$n^{\Omega(1)}$	$2^{O(n)}$
Graphs (unlabeled)	# insert/delete edge	$\Theta(n^2)$	$2^{\Omega(n^2)}$	$2^{O(n^2)}$

2   Fair Bit Strings

Let B be a bit string of length $n \geq 8$. We say that B is a *fair* bit string if it contains at least $\lfloor \frac{n}{2} \rfloor$ ones and at least $\lfloor \frac{n}{2} \rfloor$ zeros. Moreover, we say two fair bit strings B_1 and B_2 of length n are *far* if they differ in at least $\lfloor \frac{n}{8} \rfloor$ positions. Conversely, if B_1 and B_2 are not far we say they are *close*. Since rounding does not influence our results, we omit rounding to integers from now on. We obtain the following lemma using a bound by Robbins [19] and Stirling's approximation.

Lemma 1 ($\star$). *Let B be a fair bit string with n bits, the number of fair bit strings close to B is at most*

$$\frac{2}{3\pi} \cdot \left(\frac{256}{27}\right)^{n/4} = O(1.754...^n).$$

Lemma 1 allows us to show (in Lemma 2) that there are exponentially many fair bit strings of length n that are all pairwise far from each other when we consider the number of bit flips as the distance measure. Since we need at most n bit flips to transform any bit string of length n into any other, upper-bounding M in Definition 1, a set of pairwise far fair bit strings is fully diverse.

Lemma 2. *For fair bit strings of length n, any maximal fully diverse set of fair bit strings, using Hamming distance, has size at least*

$$\Omega\left(n^{-1} \cdot \left(\frac{27}{16}\right)^{n/4}\right) = \Omega\left(1.139...^n\right).$$

Proof. Since $\sum_{i=0}^{n} \binom{n}{i} = 2^n$ and $\binom{n}{n/2}$ is the largest term of $n+1$ terms, it is at least $2^n/n$ (taking the first and last term as one term), which is a lower bound on the number of fair bit strings of length n.

A maximal set of fully diverse fair bit strings can be obtained by starting with the set of all fair bit strings, selecting any member, removing all that are close, and repeating. By Lemma 1, we know how many we maximally delete in one step, so the number of iterations (and size of a maximal set of fair fully diverse bit strings) is at least

$$\frac{2^n/n}{\frac{2}{3\pi} \cdot \left(\frac{256}{27}\right)^{n/4}} = \frac{3\pi \cdot 2^n \cdot 27^{n/4}}{2n \cdot 256^{n/4}} \geq \Omega\left(n^{-1} \cdot \left(\frac{27}{16}\right)^{n/4}\right) = \Omega(1.139...^n). \qquad \Box$$

When considering *cyclic* bit strings (a bit string is equivalent to any of its $n-1$ cyclically shifted versions), the above analysis does not apply directly. For two fair cyclic bit strings B_1 and B_2 of length n, we say that B_1 and B_2 are *far* if they differ in at least $\frac{n}{8}$ positions for all of their cyclically shifted versions. Conversely, if B_1 and B_2 are not far we say they are *close*.

Lemma 3 ($\star$). *For fair cyclic bit strings of length n, any maximal fully diverse set of fair cyclic bit strings, using Hamming distance, has size*

$$\Omega\left(n^{-2} \cdot \left(\frac{27}{16}\right)^{n/4}\right) = \Omega\left(1.139...^n\right).$$

3   Geometric Diversity

Given two closed subsets A and B of a metric space, the *Hausdorff distance* between A and B is defined as the maximum distance of any point in A to its closest point in B or vice versa. For the *Fréchet distance* let A and B be two curves in the plane. Informally, the Fréchet distance between A and B is the minimum length of a leash that allows a person to walk along A and a dog along B with neither of them ever walking backwards. See Alt and Godau [1] for the formal definitions. The area of symmetric difference between two polygons is the total area inside exactly one of the polygons.

Let S be any set of simple polygons inside a unit diameter disk U. Any two polygons inside U have Hausdorff distance ≤ 1. Assume S is fully diverse, so a constant $c \geq 1$ exists such that for any two $P_i, P_j \in S$ $(i \neq j)$, their Hausdorff distance is at least $1/c$. We partition U by horizontal and vertical lines spaced $1/(2c)$, resulting in $O(c^2)$ cells. If P_i and P_j occupy exactly the same cells of this grid, then their Hausdorff distance is at most $1/(\sqrt{2}c) < 1/c$, a contradiction, so there must be a cell occupied by exactly one of P_i and P_j. This property holds for every pair of polygons in S, so S cannot contain more than $2^{O(c^2)} = O(1)$ polygons and be fully diverse. The size of S does not depend on the descriptive complexity of the polygons. The upper bound also applies to polygons with holes or that are disconnected, and to drawings of graphs.

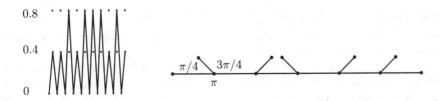

Fig. 2. Left, lower bound construction for Fréchet distance. Right, lower bound construction for sum of angle differences by encoding the bit string 01011.

Theorem 1. *A fully diverse set of polygons inside a bounded region has size $O(1)$ when we measure the distance by the Hausdorff distance.*

Next we show by construction that a fully diverse set of polygonal lines with n vertices inside a unit diameter disk U can have exponential size when using the Fréchet distance. We choose points on three horizontal lines $y = 0$, $y = 0.4$, and $y = 0.8$; on the first line we take points with x-coordinates i/n for $1 \leq i \leq n/2$, and on the second and third line we take points with x-coordinates $i/n + 1/(2n)$ for $1 \leq i \leq n/2$. We make x-monotone polygonal lines by using all points on the line $y = 0$, and between two such points, we choose either the point on $y = 0.4$ or on $y = 0.8$. See Fig. 2(left). Any two of the $2^{n/2}$ different options has Fréchet distance at least 0.4, hence these options together give a set of size $2^{\Omega(n)}$ that is fully diverse. The construction is easily adapted to simple polygon boundaries.

For area of symmetric difference, we can use the construction in Fig. 2(left) if we add one vertex at the bottom right to close the polyline with one straight-line segment to a polygon. Having $\Omega(n)$ spikes different implies an area of symmetric difference of $\Omega(1)$. Hence, the spikes encode the bits of a bit string, and Lemma 2 gives the lower bound.

We obtained $2^{\Omega(n)}$ lower bounds on the size of fully diverse sets in two cases. Is it the right lower bound, or can we also achieve a bound like $2^{\Omega(n \log n)}$?

Concerning the Fréchet distance, assume a unit diameter disk U and let a constant $c \geq 1$ be given. We partition U by a square grid of line spacing $1/(2c)$, so that any two points in the same grid cell have distance $< 1/c$. There are $O(c^2)$ cells, which is constant. We can encode any polyline of n vertices by the sequence of cells in which the vertices lie. It is straightforward to see that two polylines that have the same sequence of cells, have Fréchet distance $< 1/c$, so they cannot be in the same fully diverse set. Consequently, the size of a fully diverse set is bounded by the number of sequences of cells: $(O(c^2))^n = 2^{O(n)}$.

Theorem 2. *A fully diverse set of polygonal lines or simple polygon boundaries with n vertices in a bounded region, may have size $2^{\Omega(n)}$ and has size at most $2^{O(n)}$, if distance is measured by the Fréchet distance.*

For area of symmetric difference we need a much finer grid in order to ensure that visiting the same cells implies a distance of at most $\pi/(4c)$. Consider a grid with cells of diameter $< 1/(2cn)$. Then two simple polygons that have the same vertices in the same cells in the same order have an area of symmetric difference of at most $1/(\sqrt{2}c) < \pi/(4c)$ because each pair of corresponding edges causes a symmetric difference of at most $\sqrt{2}/(2cn)$. This leads to an upper bound of $((2cn)^2)^n = 2^{O(n \log n)}$.

Theorem 3. *A fully diverse set of simple polygons with n vertices, may have size $2^{\Omega(n)}$ and has size at most $2^{O(n \log n)}$, if distance is measured by the area of symmetric difference.*

Remark 1. The techniques presented in Sects. 2 and 3 are quite versatile. Without any new ideas, we can also show that for drawings of labeled star graphs in a bounded region, the maximum size of a fully diverse set is $2^{\Theta(n)}$ when distance is measured as sum of vertex displacements. The lower bound uses an encoding of a fair bit string to generate drawings that are far apart. The construction is in fact the one shown in Fig. 1, used for a different metric space. The upper bound uses the partition of the bounded region into a grid of size $O(c^2)$. Similarly, we can show that for drawings of ternary trees with the same embedding whose distance is measured by the sum of absolute differences of corresponding angles, we also get $2^{\Theta(n)}$ as the maximum size of a fully diverse set. Figure 2(right) shows how a bit string can be converted to a drawing so that far bit strings give far drawings.

4   Embedding Diversity

In this section we investigate the existence of large sets of embedded graphs that are diverse according to a topological measure. We show that there are superpolynomially many fully diverse sets of embedded trees and planar graphs when we use the number of changes in the rotation system as the distance measure. An *adjacent-edge swap* exchanges the position of two edges that are incident to the same vertex and adjacent in its rotation. Notice that degree-2 vertices can be omitted or ignored, since their rotation system is not changed by a swap. In this section, all graphs are assumed to be labeled.

Trees. To start, we consider *ternary trees*, i.e., trees that contain only degree 3 vertices as non-leaf vertices. Let $T = (V, E)$ be such a ternary tree with n leaves and $n - 2$ non-leaf vertices. Observe that at every non-leaf vertex there are exactly two possible cyclic orders of the incident edges. We derive a bit encoding of the possible embeddings of T as a bit string B that contains a bit for every non-leaf vertex of T. For each such vertex we associate its bit set to 0 with one of the cyclic orders, and its bit set to 1 with the other cyclic order.

Lemma 4 ($\star$). *Let T be a labeled ternary tree with n leaves and B_1 and B_2 two bit encodings of embeddings of T, such that B_1 and B_2 are fair bit strings and far from each other, then they correspond to embeddings of T that are $\Omega(n)$ adjacent-edge swaps apart.*

Applying the analysis from Sect. 2 we obtain the following.

Lemma 5 (⋆). *For a labeled ternary tree with n leaves, a fully diverse set of embeddings may have size $2^{\Omega(n)}$ if distance is the number of adjacent-edge swaps.*

Next, we consider labeled star graphs. Let $S = (V, E)$ be a labeled star with central vertex $u \in V$ and leaves $v_1, \ldots, v_n \in V$ incident to edges $e_1, \ldots, e_n \in E$ for some even $n \in \mathbb{N}$. We define a cyclic bit string B describing orders of the edges incident to u as follows. Consider the edges $e_1, \ldots, e_n$ around u, ordered by their indices. For each antipodal pair of edges e_i, e_j in S (where $j = i + \frac{n}{2}$), we add one bit b_i to B and let it be 1 if e_i, e_j have exchanged their positions in the cyclic order and 0 if not; see Fig. 1. Clearly, B has length $\frac{n}{2}$; recall that two cyclic bit strings of length $\frac{n}{2}$ are fair if they contain at least $\frac{n}{4}$ zeros and at least $\frac{n}{4}$ ones, and they are far if they differ in at least $\frac{n}{16}$ positions.

Lemma 6 (⋆). *Let S be a labeled star graph with n leaves and B_1 and B_2 two bit encodings of embeddings of S, such that B_1 and B_2 are fair cyclic bit strings and far from each other, then they correspond to embeddings of S that are $\Omega(n^2)$ adjacent-edge swaps apart.*

Lemma 7 (⋆). *For a labeled star graph with n leaves, a fully diverse set of embeddings may have size $2^{\Omega(n)}$ if distance is the number of adjacent-edge swaps.*

It remains to combine the two previous cases to handle any tree $T = (V, E)$ that does not contain degree 2 vertices. This is non-trivial, and in fact, for some trees, we no longer have a fully diverse set of embeddings of exponential size. First, observe that the maximum distance between two embeddings of a tree whose internal nodes that have degrees $d_1, \ldots, d_k$ is proportional to $\sum_{i=1}^{k} d_i^2$.

Lemma 8 (⋆). *For any labeled tree with n leaves, there exists a fully diverse set of embeddings of size $2^{\Omega(\sqrt{n})}$.*

Proof Sketch. Assume that a labeled tree T is given whose internal vertices $v_1, \ldots, v_k$ are sorted by degrees $d_1 \geq d_2 \geq \cdots \geq d_k$. Let j be the smallest value such that $\sum_{i=1}^{j} d_i^2 \geq \frac{1}{2} \sum_{i=1}^{k} d_i^2$. We distinguish two cases, $d_j \geq \sqrt{n}$ and $d_j < \sqrt{n}$. In the former case, we only use the vertices $v_1, \ldots, v_j$ to make fully diverse sets. Each such vertex already admits a fully diverse set of size $2^{\Omega(d_i)}$ by Lemma 7. This allows us to just combine the embeddings and choose embeddings for the remaining vertices at random.

If $d_j < \sqrt{n}$ we use the vertices $v_j, \ldots, v_k$. We group them into sets $V_1, \ldots, V_z$ with $z = \Theta(\sqrt{n})$ such that for each V_h, $h = 1, \ldots z$, the sum of its squared degrees is in $\Theta\left(\sum_{i=1}^{k} d_i^2/\sqrt{n}\right)$. We then fix two far embeddings for each group V_h. Using these two embeddings to encode a bit string we then get a fully diverse set of sufficient size using Lemma 3 essentially in the same manner as for Lemma 5. □

To prove that no better bound exists that applies to all trees, consider a tree with n leaves, one vertex v with degree $\sqrt{n} \log n$, and all other internal vertices with degree 3. The maximum distance between two embeddings is determined by vertex v only: it is $\Theta(n \log^2 n)$. The linearly many vertices of degree 3 require only

$O(n)$ adjacent-edge swaps, so they play no role in obtaining a fully diverse set. Considering v and its neighbors as a star graph then implies that the maximum size of a fully diverse set is $2^{O(\sqrt{n}\log n)}$. Intuitively, we have just $O(\sqrt{n}\log n)$ bits in an encoding that are effective to realize a fully diverse set. We can give v degree $\sqrt{n}\log\log n$ or even smaller for a slightly better bound.

Theorem 4. *For any labeled tree with n leaves, there is a fully diverse set of embeddings of size $2^{\Omega(\sqrt{n})}$, and there exists a tree whose size of a fully diverse set of embeddings is $2^{O^*(\sqrt{n})}$, where $O^*(\sqrt{n})$ denotes $O(f(n))$ for any function $f(n)$ that is asymptotically larger than $\sqrt{n}$.*

Suppose we consider a different metric, namely the number of edge *relocations* for embedded trees. A relocation on the cyclic order around a vertex places one of its edges anywhere else in the order in a single step. For ternary trees this is equivalent to an adjacent-edge swap, but for a star graph, the maximum distance between any two embeddings of stars is $\Theta(n)$ instead of $\Theta(n^2)$.

Lemma 9 ($\star$). *For a labeled star graph with n leaves, a fully diverse set of embeddings may have size $2^{\Omega(n\log n)}$ if distance is measured by edge relocations.*

Planar Graphs. Here we give a sketch of how the results just given can be extended to planar embeddings of planar graphs. Let $G = (V, E)$ be an embedded labeled planar connected simple graph. Since swapping two adjacent edges in G does not necessarily preserve planarity we instead consider swaps of components separated by *cut-vertices* and *split pairs* [4]. A cut-vertex $u \in V$ is a vertex such that G is not connected after u is removed. Similarly, a split pair $\{u, v\} \subset V$ of G is a pair of vertices such that G is not connected after u and v are both removed from G. The *incident components* of a cut-vertex or a split pair are the connected components obtained after removing this cut-vertex or split pair.

We consider the rotation of the incident components around a cut-vertex or split pair, and so-called *adjacent-component swaps* between them. To ensure that every possible embedding can be reached, we allow the operation of mirroring a triconnected component at no cost. Each cut-vertex or split pair can be treated as the central vertex of a star and its incident components as the leaves. Swapping the order of two leaves corresponds one-to-one to swapping the order of two of its incident components. To ensure this we first resolve nesting components around cut-vertices. Then, it suffices to only consider rotations around cut-vertices and split pairs in which the respective incident components appear one after another. This allows us to derive analogous versions of Lemmas 4 and 6 which in turn enables us to argue in the same fashion as for Theorem 4 to obtain the lower bound in the following theorem.

Moreover, the upper bound of Theorem 4 translates immediately since trees are planar graphs and adjacent-edge swaps in trees are equivalent to swapping the incident components around a cut-vertex.

Theorem 5. *For any labeled planar graph $G = (V, E)$ with n_c cut-vertices and n_p split pairs each with at least 3 incident connected components, a fully diverse* *set of planar embeddings may have size* $2^{\Omega(\sqrt{n_c+n_p})}$ *and there exists a planar graph whose size of a fully diverse set of embeddings is* $2^{O^*(\sqrt{n_c+n_p})}$ *if distance is measured by adjacent-component swaps.*

5   Abstract Graphs

In this section we consider the diversity of abstract graphs of some given graph class and a distance based on edits. Throughout, we require that if a graph is in graph class $\mathcal{G}$, then after applying an operation to it the resulting graph is still in $\mathcal{G}$. We consider trees, planar graphs, and general graphs, and discuss diversity for the labeled and unlabeled cases. As most of the ideas are the same as the ones used earlier, we keep the description short. The results are given in Table 3.

Trees. For trees, we use the following edit operation: Take a leaf, unattach it from its neighbor, and attach it to a different vertex. We consider the edit distance measure: the distance between two trees of n vertices is the number of leaf reattachments needed to convert one tree into the other. Note that any two trees with n vertices have a finite distance, since every tree can easily be turned into the star graph (in the labeled case, a specific node must become the central vertex to use a star as a canonical tree). The maximum distance is $\Theta(n)$, since we need at most $n - 2$ edits to convert any tree into a star.

We start with the labeled case. To construct a large size fully diverse set of labeled trees, we can restrict ourselves to paths. A path essentially encodes a permutation of its labels and reverse permutations are identified. We can use essentially the same proof ideas as for the case of labeled stars and their embedding under swaps (where cyclic shifts were identified). The upper bound is trivial, and we obtain $2^{\Theta(n \log n)}$ for the maximum size of a fully diverse set.

Next we switch to unlabeled trees. The situation is quite different, because there is only one path now, and in fact, it is known that there are only $2^{O(n)}$ different unlabeled trees [18]. To show an exponential lower bound for unlabeled trees, we start out with a path of $2 + n/4$ vertices. We attach either one or two leaves to the middle $n/4$ vertices, encoding a 0 or 1 in a bit string. We attach all remaining vertices equally as paths to the ends of the initial path. These two tails have length at least $n/16$, ensuring that we need $n/16$ operations to operate on the bit string in unwanted ways. Using our knowledge on the full diversity of bit strings, we obtain $2^{\Theta(n)}$ as the bound.

General Graphs. We consider general graphs of n vertices with edge insertion or deletion as the elementary operation. The edit distance is the distance between two graphs. We again distinguish in the labeled and unlabeled cases. The two graphs furthest apart are the empty graph and the complete graph in both cases.

We start with the labeled case. Every labeled edge can be seen as a bit in a bit string, where absence encodes 0 and presence 1. We immediately get a bound of $2^{\Theta(n^2)}$ by Sect. 2. For the unlabeled case, we observe that there are at least $2^{n(n-1)/2}/n!$ graphs, since we can assign labels in at most $n!$ ways. This is still $2^{\Omega(n^2)}$. The upper bound follows from the labeled case.

Planar Graphs. For planar graphs we use the same edit operation as for general graphs. Since we can always first remove edges and then insert them, any edit sequence can be turned into an edit sequence that stays within the class of planar graphs. An upper bound on the number of operations needed is clearly $6n - 12$.

For labeled planar graphs, we obtain a lower bound by analyzing how many labeled planar graphs are within $(6n - 12)/c$ edits from a given graph. This number is certainly bounded by $(n^2)^{(6n-12)/c} = n^{O(n/c)}$. At the same time, the number of labeled paths is already $n!/2$. By choosing c sufficiently large, we obtain $2^{\Theta(n \log n)}$ as the maximum size of a fully diverse set.

The most intriguing case turns out to be unlabeled planar graphs. It is known that the number of unlabeled planar graphs is bounded by $2^{O(n)}$ [6], which is obviously also an upper bound on the size of a fully diverse set.

For a lower bound, the idea is to consider graphs that are unions of stars. We can connect them into one connected graph if needed, but the argument is cleanest for these unconnected graphs. We consider only stars with 2^i vertices, $0 \le i \le \log n - \log \log n$. Suppose we have $n/(2^i \log n)$ stars of size 2^i, then it takes $n/(2 \log n)$ edge insertions (and a number of edge deletions) to convert this into $n/(2^{i+1} \log n)$ stars of size 2^{i+1}. Converting in the other direction also takes at least $n/(2 \log n)$ operations.

In the fully diverse set we construct, we choose stars with either 2^i or 2^{i+1} vertices, for $i = 0, 2, 4, \ldots, (\log n) - (\log \log n) - 1$, the latter value rounded down to the nearest even number, henceforth denoted by m. We then have roughly $m/2$ different sizes in any single set, out of the twice as many sizes used in the whole construction. Notice that a set indeed has size n. We can see the choice between stars of size 2^i and 2^{i+1} as an encoding of a bit, and hence we have a bit string of length roughly $m/2$. We choose a fully diverse set of bit strings, which implies the choice of stars in a graph in the set. By Lemma 2, a fully diverse set of fair bit strings of this length has maximum size $2^{\Omega(m/2)}$, which is $n^{\Omega(1)}$.

6   Conclusions and Open Problems

We introduced the concept of a fully diverse set of objects, like polygons and graphs, in a metric space, by relating the inter-distance between any two objects in that set to the maximum distance possible. We then studied a number of distance measures, both geometric and combinatorial, and proved bounds on the maximum size of fully diverse sets. There are two cases where the lower and upper bounds do not match, giving rise to the two main open problems of this paper. We also sketched a simple randomized algorithm to generate fully diverse sets of a certain type of objects.

As our full diversity definition can be applied to any class of objects in a metric space provided the maximum distance is bounded, there are many other cases to be explored. For example, 2-dimensional distributions with the Wasserstein distance, or graphs with different edit distances than the ones used in this paper. Furthermore, a definition of full diversity that does not require the metric space to be bounded is worth examination.

References

1. Alt, H., Godau, M.: Computing the Fréchet distance between two polygonal curves. Int. J. Comput. Geom. Appl. **5**, 75–91 (1995). https://doi.org/10.1142/S0218195995000064

2. Bach, B., Spritzer, A., Lutton, E., Fekete, J.-D.: Interactive random graph generation with evolutionary algorithms. In: Didimo, W., Patrignani, M. (eds.) GD 2012. LNCS, vol. 7704, pp. 541–552. Springer, Heidelberg (2013). https://doi.org/10.1007/978-3-642-36763-2_48

3. Baste, J., et al.: Diversity of solutions: an exploration through the lens of fixed-parameter tractability theory. In: Proceedings of the 29th International Joint Conference on Artificial Intelligence (IJCAI 2020), pp. 1119–1125 (2020). https://doi.org/10.24963/ijcai.2020/156

4. Battista, G.D., Tamassia, R.: On-line planarity testing. SIAM J. Comput. **25**(5), 956–997 (1996). https://doi.org/10.1137/S0097539794280736

5. Biedl, T., Marks, J., Ryall, K., Whitesides, S.: Graph multidrawing: finding nice drawings without defining nice. In: Whitesides, S.H. (ed.) GD 1998. LNCS, vol. 1547, pp. 347–355. Springer, Heidelberg (1998). https://doi.org/10.1007/3-540-37623-2_26

6. Bonichon, N., Gavoille, C., Hanusse, N., Poulalhon, D., Schaeffer, G.: Planar graphs, via well-orderly maps and trees. Graphs Combin. **22**(2), 185–202 (2006). https://doi.org/10.1007/s00373-006-0647-2

7. Bridgeman, S., Tamassia, R.: Difference metrics for interactive orthogonal graph drawing algorithms. In: Whitesides, S.H. (ed.) GD 1998. LNCS, vol. 1547, pp. 57–71. Springer, Heidelberg (1998). https://doi.org/10.1007/3-540-37623-2_5

8. Cai, J.: Counting embeddings of planar graphs using DFS trees. SIAM J. Discret. Math. **6**(3), 335–352 (1993). https://doi.org/10.1137/0406027

9. Drosou, M., Jagadish, H.V., Pitoura, E., Stoyanovich, J.: Diversity in big data: a review. Big Data **5**(2), 73–84 (2017). https://doi.org/10.1089/big.2016.0054

10. Gao, X., Xiao, B., Tao, D., Li, X.: A survey of graph edit distance. Pattern Anal. Appl. **13**(1), 113–129 (2010)

11. Hebrard, E., Hnich, B., O'Sullivan, B., Walsh, T.: Finding diverse and similar solutions in constraint programming. In: Proceedings of 20th National Conference on Artificial Intelligence (AAAI 2005), pp. 372–377 (2005)

12. Hill, M.O.: Diversity and evenness: a unifying notation and its consequences. Ecology **54**(2), 427–432 (1973). https://doi.org/10.2307/1934352

13. Huemer, C., Pilz, A., Silveira, R.I.: A new lower bound on the maximum number of plane graphs using production matrices. Comput. Geom. **84**, 36–49 (2019). https://doi.org/10.1016/j.comgeo.2019.07.005

14. Ingmar, L., de la Banda, M.G., Stuckey, P.J., Tack, G.: Modelling diversity of solutions. In: Proceedings of 34th AAAI Conference on Artificial Intelligence (AAAI 2020), pp. 1528–1535 (2020)

15. van Kreveld, M., Speckmann, B., Urhausen, J.: Diverse partitions of colored points. In: Lubiw, A., Salavatipour, M. (eds.) WADS 2021. LNCS, vol. 12808, pp. 641–654. Springer, Cham (2021). https://doi.org/10.1007/978-3-030-83508-8_46

16. Kulesza, A., Taskar, B.: Determinantal point processes for machine learning. Found. Trends Mach. Learn. **5**(2–3), 123–286 (2012). https://doi.org/10.1561/2200000044

17. Kunaver, M., Pozrl, T.: Diversity in recommender systems - a survey. Knowl. Based Syst. **123**, 154–162 (2017). https://doi.org/10.1016/j.knosys.2017.02.009

18. Otter, R.: The number of trees. Ann. Math. **49**(3), 583–599 (1948)
19. Robbins, H.: A remark on Stirling's formula. Am. Math. Mon. **62**(1), 26–29 (1955)
20. Stallmann, M.F.M.: On counting planar embeddings. Discret. Math. **122**(1–3), 385–392 (1993). https://doi.org/10.1016/0012-365X(93)90316-L
21. Tuomisto, H.: A consistent terminology for quantifying species diversity? Yes, it does exist. Oecologia **164**(4), 853–860 (2010). https://doi.org/10.1007/s00442-010-1812-0
22. Ulrich, T., Bader, J., Thiele, L.: Defining and optimizing indicator-based diversity measures in multiobjective search. In: Schaefer, R., Cotta, C., Kołodziej, J., Rudolph, G. (eds.) PPSN 2010. LNCS, vol. 6238, pp. 707–717. Springer, Heidelberg (2010). https://doi.org/10.1007/978-3-642-15844-5_71
23. Wineberg, M., Oppacher, F.: The underlying similarity of diversity measures used in evolutionary computation. In: Cantú-Paz, E., et al. (eds.) GECCO 2003. LNCS, vol. 2724, pp. 1493–1504. Springer, Heidelberg (2003). https://doi.org/10.1007/3-540-45110-2_21

Polynomial-Delay and Polynomial-Space Enumeration of Large Maximal Matchings

Yasuaki Kobayashi[1], Kazuhiro Kurita[2(✉)], and Kunihiro Wasa[3]

[1] Graduate School of Information Science and Technology, Hokkaido University,
Sapporo, Japan
koba@ist.hokudai.ac.jp
[2] Graduate School of Informatics, Nagoya University, Nagoya, Japan
kurita@i.nagoya-u.ac.jp
[3] Department of Computer Science and Engineering, Toyohashi University
of Technology, Aichi, Japan
wasa@cs.tut.ac.jp, wasa@hosei.ac.jp

Abstract. Enumerating matchings is a classical problem in the field of enumeration algorithms. There are polynomial-delay enumeration algorithms for several settings, such as enumerating perfect matchings, maximal matchings, and (weighted) matchings in specific orders. In this paper, we present polynomial-delay enumeration algorithms for maximal matchings with cardinality at least given threshold t. Our algorithm enumerates all such matchings in $O(nm)$ delay with exponential space, where n and m are the number of vertices and edges of an input graph, respectively. We also present a polynomial-delay and polynomial-space enumeration algorithm for this problem. As a variant of this algorithm, we give an algorithm that enumerates k-best maximal matchings that runs in polynomial-delay.

Keywords: Maximal matching · Cardinality constraint enumeration · K-best enumeration

1 Introduction

Computing a maximum cardinality matching in graphs is a fundamental problem in combinatorial optimization and has numerous applications in many theoretical and practical contexts. This problem is well known to be solvable in polynomial time by the famous blossom algorithm due to Edmonds [9]. This algorithm runs in time $O(n^2 m)$, where n and m are the numbers of vertices and edges of an input graph, and the running time is improved to $O(n^{1/2} m)$ [21] and $O(n^\omega)$ [22], where $\omega < 2.37$ is the matrix multiplication exponent.

Enumerating matchings in graphs is also a well-studied problem in the literature [4,11,27–29]. In this problem, we are given a graph $G = (V, E)$ and the goal is to compute all matchings of G satisfying some prescribed conditions. This is motivated by a typical situation that a single optimal matching can be

M. A. Bekos and M. Kaufmann (Eds.): WG 2022, LNCS 13453, pp. 342–355, 2022.
https://doi.org/10.1007/978-3-031-15914-5_25

inadequate for real-world problems since intricate constraints and preferences emerging in real-world problems are overly simplified or even ignored to solve the problem efficiently. In this situation, multiple near optimal solutions are preferable rather than a single optimal solution.

There are two lines of research for enumerating matchings. The problem of enumerating inclusion-wise maximal matchings is a special case of enumerating maximal independent sets or cliques in graphs, which is one of the most prominent problems in the field of enumeration algorithms [6,7,12,20,26]. Tsukiyama et al. [26] showed that the problem of enumerating all maximal independent sets in graphs is solvable in $O(nm)$ delay and polynomial space. Johnson et al. [12] also discussed a similar algorithm for this problem. Makino and Uno [20] and Comin and Rizzi [6] improved the running time for dense graphs via fast matrix multiplication algorithms. For maximal matching enumeration, Uno [28] gave an $O(n + m + \Delta N)$-time algorithm for enumerating all maximal matchings of graphs, which substantially improves the known algorithm for enumerating maximal independent sets in general graphs [26] when input graphs are restricted to line graphs. Here, Δ is the maximum degree and N is the number of maximal matchings in an input graph.

The other line of work is to enumerate matchings with cardinality or weight constraints. One of the best known results along this line is based on k-best enumeration [10]. Here, we say that an enumeration algorithm for (weighted) matchings is a k-best enumeration algorithm if given an integer k, the algorithm enumerates k distinct matchings $\mathcal{M} = \{M_1, \ldots, M_k\}$ of G such that every matching in $\mathcal{M}$ has cardinality (or weight) not smaller than that not in $\mathcal{M}$. In the 1960s, Murty developed a k-best enumeration algorithm based on a simple binary partition technique [23]. Lawler [19] generalized Murty's algorithm to many combinatorial problems, and then k-best enumeration algorithms for other problems have been discussed in various fields (see [10] for a survey). Chegireddy and Hamacher [4] developed an $O(kn^3)$-time k-best enumeration algorithm for weighted perfect matchings in general graphs.

In this paper, we focus on enumerating matchings satisfying both maximality and cardinality conditions. More specifically, we address the following problems.

Definition 1. *Given a graph $G = (V, E)$ and a non-negative integer t, LARGE MAXIMAL MATCHING ENUMERATION asks to enumerate all maximal matchings of G with cardinality at least t.*

Definition 2. *Given a graph $G = (V, E)$ and a non-negative integer k, k-BEST MAXIMAL MATCHING ENUMERATION asks to compute a set $\mathcal{M}$ of k maximal matchings of G such that the cardinality of any maximal matching in $\mathcal{M}$ is not smaller than that not in $\mathcal{M}$.*

These kind of problems are recently focused in several work [15,17], where they considered the problems of enumerating minimal solutions with weight or cardinality constraints. We would like to mention that satisfying both weight or cardinality and maximal/minimal constraints makes enumeration problems even more difficult: Korhonen [17] showed that the problem of enumerating minimal

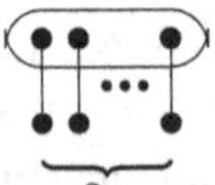

$2n$

Fig. 1. The graph is obtained from the complete graph K_{2n} (depicted as the rounded rectangle) with $2n$ vertices by adding a pendant vertex to each vertex. Since K_{2n} has $\frac{(2n)!}{2^n \cdot n!}$ perfect matchings, the graph contains exactly one maximal matchings of cardinality $2n$ and at least $\frac{(2n)!}{2^n \cdot n!}$ maximal matchings.

separators of cardinality at most k is solvable in incremental polynomial time or FPT delay, while the problem of enumerating minimal separators without cardinality constraint k is solvable in polynomial delay [25].

The results of our paper are as follows. We observe that a straightforward application of the binary partition technique [2,11] would not yield polynomial-delay algorithms for LARGE MAXIMAL MATCHING ENUMERATION: This technique is typically based on the *extension problem*, which will be defined in Sect. 3, and we prove that this problem for LARGE MAXIMAL MATCHING ENUMERATION is NP-hard even if $t = 0$. This result is independently shown by Casel et al. [3]. See Theorem 10 in [3]. As algorithmic results, we present $O(nm)$-delay enumeration algorithms for LARGE MAXIMAL MATCHING ENUMERATION and k-BEST MAXIMAL MATCHING ENUMERATION. These algorithms run in exponential space. Note that for k-BEST MAXIMAL MATCHING ENUMERATION, our algorithm requires $\Omega(k)$ space, while this is indeed exponential when k is exponential in n. We also present an $O(n^2\Delta^2)$-delay and polynomial-space enumeration algorithm for LARGE MAXIMAL MATCHING ENUMERATION.

Whereas our algorithms are slower than the known algorithm of Uno [28], our algorithms only enumerate matchings that are maximal and have cardinality at least t, which would be more efficient with respect to the overall performance when the number of "large" maximal matchings are sufficiently smaller than that of all maximal matchings. See Fig. 1 for such an example.

Our algorithms are based on the supergraph technique, which is frequently used in designing enumeration algorithms [5,8,13,14,18,24]. In this technique, we define a directed graph on the set of all solutions and the enumeration algorithm simply traverses this directed graph from an arbitrary solution. To enumerate all solutions, we need to carefully design this directed graph so that all the nodes can be traversed from an arbitrary solution. We basically follow the technique due to Cohen et al. [5], which allows to define a suitable directed graph for enumerating maximal matchings. We carefully analyze this directed graph and prove that this directed graph has a "monotone" path from an arbitrary maximum matching to any maximal matching of G, where we mean by a monotone path a sequence of maximal matchings $(M_1, M_2, \ldots, M_k)$ with $|M_1| \geq |M_2| \geq \cdots \geq |M_k|$. This also enables us to enumerate all maximal matchings in a non-decreasing order of its cardinality. Let us note that our approach is different from those in the maximal matching enumeration [28] and the k-best enumeration for matchings [4, 23]. Our polynomial-space enumeration algorithm also exploits this monotone path. Due to the space limitation, we omit the proofs marked with star $\star$, which can be found in the full version [16].

2 Preliminaries

Let $G = (V, E)$ be a graph. Let $n = |V|$ and $m = |E|$. Throughout this paper, we assume that G has no self-loops and parallel edges. We also assume that G has no isolated vertices and hence we have $n = O(m)$. The vertex set and edge set of G are denoted by $V(G)$ and $E(G)$, respectively. For a vertex $v \in V$, the set of edges incident to v is denoted by $\Gamma(v)$. To simplify the notation, we also use $\Gamma(e)$ to denote $(\Gamma(u) \cup \Gamma(v)) \setminus \{e\}$ for each edge $e = \{u, v\} \in E$. A sequence of vertices $P = (u_1, u_2, \ldots, u_k)$ is called a *path* if u_i is adjacent to u_{i+1} for any $1 \le i < k$ and all the vertices are distinct. A sequence of vertices $C = (u_1, u_2, \ldots, u_k)$ is called a *cycle* if u_i is adjacent to u_{i+1} for any $1 \le i \le k$, where u_{k+1} is considered as u_1, and all the vertices except for pair $\{u_1, u_{k+1}\}$ are distinct. For $F \subseteq E$, we denote by $G[F]$ the subgraph consisting of all end vertices of F and edges in F. For two sets X and Y, we denote by $X \triangle Y$ the symmetric difference between X and Y (i.e., $X \triangle Y = (X \setminus Y) \cup (Y \setminus X)$).

Let M be a set of edges in G. We say that M is a *matching* of G if for any pair of distinct $e, f \in M$, they does not share their end vertices (i.e., $e \cap f = \emptyset$ holds). Moreover, M is a *maximal matching* of G if M is a matching and $M \cup \{e\}$ is not a matching of G for every $e \in E \setminus M$. The maximum cardinality of a matching of G is denoted by $\nu(G)$. Every matching with cardinality $\nu(G)$ is called a *maximum matching* of G. For a matching M, we say that a vertex v is *matched* in M if M has an edge incident to v. Otherwise, v is *unmatched* in M.

In this paper, we measure the running time of enumeration algorithms in an *output-sensitive manner* [12]. In particular, we focus on the delay of enumeration algorithms: The *delay* of a enumeration algorithm is the maximum time interval between two consecutive outputs (including both preprocessing time and postprocessing time).

3 Hardness of the Extension Problem

We show that a direct application of the binary partition technique or the k-best enumeration framework to LARGE MAXIMAL MATCHING ENUMERATION seems to be impossible.

In enumeration algorithms based on the binary partition technique [2, 11], we solve a certain decision or optimization problem, called an *extension problem*, to enumerate solutions. Basically, the extension problem asks to decide whether, given disjoint sets I and O, there is a solution that includes all elements in I and excludes every element in O. For enumerating *maximum* matchings, the extension problem is tractable: For $I, O \subseteq E$ with $I \cap O = \emptyset$, the extension problem simply asks for a matching M in a graph obtained by removing all endpoints in I and edges in O from G with $|M| = \nu(G) - |I|$. However, for LARGE

MAXIMAL MATCHING ENUMERATION, the extension problem is intractable. The formal definition of the extension problem is as follows: Given a graph $G = (V, E)$ and $I, O \subseteq E$ with $I \cap O = \emptyset$, MAXIMAL MATCHING EXTENSION asks to determine whether G has a maximal matching M with $I \subseteq M$ and $M \cap O = \emptyset$.

Theorem 1 ($\star$). MAXIMAL MATCHING EXTENSION *is* NP-*complete even on planar bipartite graphs with maximum degree three.*

As for k-best enumeration algorithms based on Lawler's framework [19], we need to solve an optimization version of MAXIMAL MATCHING EXTENSION. The above theorem also rules out the applicability of Lawler's framework to obtain a polynomial-delay algorithm for LARGE MAXIMAL MATCHING ENUMERATION, assuming that P $\neq$ NP.

4   Enumeration of Maximal Matchings

4.1   LARGE MAXIMAL MATCHING ENUMERATION

Our enumeration algorithm is based on the *supergraph technique*, which is frequently used in many enumeration algorithms [5,8,13,14,18,24]. In particular, our algorithm is highly related to the enumeration algorithm for maximal independent sets with the *input-restricted problem* due to [5]. The basic idea of the supergraph technique is quite simple. We define a directed graph $\mathcal{G}$ whose node set corresponds to all the solutions we wish to enumerate. The enumeration algorithm solely traverses this directed graph and outputs a solution at each node. To this end, we need to carefully design the arc set of $\mathcal{G}$ so that all the nodes in $\mathcal{G}$ are reachable from a specific node.

For maximal matchings (without cardinality constraints), we can enumerate those in polynomial delay with this technique. Let $G = (V, E)$ be a graph. For a (not necessarily maximal) matching M of G, we denote by $\mu(M)$ an arbitrary maximal matching of G that contains M. This maximal matching can be computed from M by greedily adding edges in $E \setminus M$. Let M be a maximal matching of G. For $e \in E \setminus M$, $(M \setminus \Gamma(e)) \cup \{e\}$ is a matching of G, and then $M_e = \mu((M \setminus \Gamma(e)) \cup \{e\})$ is a maximal matching of G. We define the (out-)neighbors of a maximal matching M in $\mathcal{G}$, denoted $\mathcal{N}_{\mathcal{G}}(M)$, as the set of maximal matchings $\{M_e : e \in E \setminus M\}$. To avoid a confusion, each maximal matching in $\mathcal{N}_{\mathcal{G}}(M)$ is called a $\mathcal{G}$-*neighbor* of M. The arc set of $\mathcal{G}$ is defined by this neighborhood relation. With this definition, we can show that every maximal matching M_2 is reachable from any other maximal matching M_1, that is, $\mathcal{G}$ is strongly connected. To see this, we consider the value $m(M, M') = |M \cap M'|$ defined between two (maximal) matchings M and M' of G. Since M_1 and M_2 are maximal matchings of G, there is an edge $e \in M_2 \setminus M_1$. Then, we have

$$m(M_1, M_2) = |M_1 \cap M_2| < |((M_1 \setminus \Gamma(e)) \cup \{e\}) \cap M_2| \leq |\mu((M_1 \setminus \Gamma(e)) \cup \{e\}) \cap M_2|,$$

where the first inequality follows from $e \in M_2$ and $\Gamma(e) \cap M_2 = \emptyset$ and the second inequality follows from $M \subseteq \mu(M)$ for any matching M of G. This indicates that M_1 has a $\mathcal{G}$-neighbor $M' = \mu((M_1 \setminus \Gamma(e)) \cup \{e\})$ such that $m(M_1, M_2) < m(M', M_2)$. Moreover, the following proposition holds.

Algorithm 1: Given a graph G and an integer t, the algorithm enumerates all maximal matchings of G with cardinality at least t.

---

1 **Procedure** Traverse(G, t)
2   Let M^* be a maximum matching of G
3   Add M^* to a queue $\mathcal{Q}$ and to set $\mathcal{S}$
4   **while** $\mathcal{Q}$ *is not empty* **do**
5   Let M be a maximal matching in $\mathcal{Q}$
6   Output M and delete M from $\mathcal{Q}$
7   **foreach** $M' \in \mathcal{N}_{\mathcal{G}}(M)$ **do**
8   **if** $M' \notin \mathcal{S}$ *and* $|M'| \geq t$ **then** Add M' to $\mathcal{Q}$ and to $\mathcal{S}$

Proposition 1. *Let M_1 and M_2 be maximal matchings of G. Then, $m(M_1, M_2) \leq |M_2|$. Moreover, $m(M_1, M_2) = |M_2|$ if and only if $M_1 = M_2$.*

By induction on $k = |M_2| - m(M_1, M_2)$, there is a directed path from M_1 to M_2 in $\mathcal{G}$ for every pair of maximal matchings of G, which proves the strong connectivity of $\mathcal{G}$.

Our algorithm for LARGE MAXIMAL MATCHING ENUMERATION also traverses $\mathcal{G}$ from an arbitrary *maximum* matching of G in a breadth-first manner but truncates all maximal matchings of cardinality less than given threshold t. The pseudocode is shown in Algorithm 1. In the following, we show that Algorithm 1 enumerates all the maximal matchings of G with cardinality at least t, provided $t < \nu(G)$. We discuss later for the other case $t = \nu(G)$.

For a non-negative integer k, we say that a directed path in $\mathcal{G}$ is k-*thick* if every maximal matching on the path has cardinality at least k. To show the correctness of Algorithm 1, it is sufficient to prove that (1) for any pair of maximum matching M^* of G and a maximal matching M of G with $|M| < \nu(G)$, there is a directed $|M|$-thick path from M^* to M in $\mathcal{G}$ and (2) for a pair of maximum matchings M and M' of G, there is a directed $(\nu(G) - 1)$-thick path from M to M' in $\mathcal{G}$.

In the rest of this subsection, fix distinct maximal matchings M_1 and M_2 of G. Since M_1 and M_2 are matchings of G, each component of the graph $G[M_1 \triangle M_2]$ is either a path or a cycle. We say that a path component P in $G[M_1 \triangle M_2]$ is *even-alternating* if exactly one end vertices of P is unmatched in M_1. Let us note that P is even-alternating if and only if it has an even number of edges. We say P is M_1-*augmenting* (resp. M_2-augmenting) if the both end vertices of P are unmatched in M_1 (resp. M_2). Since both M_1 and M_2 are maximal matchings of G, the following proposition holds.

Proposition 2. *Every component in $G[M_1 \triangle M_2]$ is either a path with at least two edges or a cycle with at least four edges.*

For M_2-augmenting path component P in $G[M_1 \triangle M_2]$, it holds that $|M_1 \cap E(P)| > |M_2 \cap E(P)|$. Thus, the following proposition holds.

Proposition 3. *Suppose that $G[M_1 \triangle M_2]$ has no M_1-augmenting path components but has at least one M_2-augmenting path component. Then, $|M_1| > |M_2|$.*

Lemma 2. *If $G[M_1 \triangle M_2]$ has an M_1-augmenting or even-alternating path component P, then there is a directed $|M_1|$-thick path from M_1 to a maximal matching M' in $\mathcal{G}$ such that $m(M_1, M_2) < m(M', M_2)$ and $|M'| \geq |M_1|$. Moreover, if P is M_1-augmenting, then $|M'| > |M_1|$.*

Proof. Let $P = (v_1, v_2, \ldots, v_\ell)$ and let $e_i = \{v_i, v_{i+1}\}$ for $1 \leq i < \ell$. By Proposition 2, P contains at least two edges. Assume, without loss of generality, we have $e_1 \in M_2$ and $e_2 \in M_1$. Define $\hat{M} = (M_1 \setminus \{e_2\}) \cup \{e_1\}$. As v_1 is unmatched in M_1, $\hat{M}$ is a matching of G, and hence $M' = \mu(\hat{M})$ is a $\mathcal{G}$-neighbor of M_1. Then,

$$m(M', M_2) \geq m(\hat{M}, M_2) = |((M_1 \setminus \{e_2\}) \cup \{e_1\}) \cap M_2| > |M_1 \cap M_2| = m(M_1, M_2),$$

as $e_1 \in M_2$ and $e_2 \in M_1 \setminus M_2$. Moreover, we have $|\mu(\hat{M})| \geq |\hat{M}| = |M_1|$. Thus, the arc $(M_1, \mu(\hat{M}))$ is the desired directed path in $\mathcal{G}$.

Suppose moreover that P is M_1-augmenting. In this case, we have $\ell \geq 4$. We prove the claim by induction on ℓ. Let $\hat{M}$ be as above. If $\hat{M} \subset \mu(\hat{M})$, we are done. Suppose otherwise. If $\ell = 4$, as v_4 is unmatched in M_1, $\hat{M} \cup \{e_3\}$ is a matching of G, implying that $\hat{M} \subset \mu(\hat{M})$. Otherwise, that is, $\hat{M} = \mu(\hat{M})$, the subpath $(v_3, v_4, \ldots, v_\ell)$ of P is a path component in $G[\hat{M} \triangle M_2]$ and is $\hat{M}$-augmenting. Applying the induction hypothesis to this subpath, there is a directed $|M'|$-thick path from M' to a maximal matching M'' in $\mathcal{G}$ such that $m(M', M_2) < m(M'', M_2)$ and $|M''| > |M'| = |M_1|$. As $m(M_1, M_2) < m(M', M_2) < m(M'', M_2)$, the lemma follows. $\square$

If $G[M_1 \triangle M_2]$ has neither M_1-augmenting path components nor cycle components, the maximal matching M' of $\mathcal{G}$ in Lemma 2 satisfies the following additional property.

Corollary 3 ($\star$). *Let M' be the maximal matching of $\mathcal{G}$ obtained in Lemma 2. If $G[M_1 \triangle M_2]$ has neither M_1-augmenting path components nor cycle components, then $G[M' \triangle M_2]$ has no cycle components.*

Lemma 4. *If $G[M_1 \triangle M_2]$ has a cycle component C, then there is a directed $(|M_1| - 1)$-thick path from M_1 to a maximal matching M' of $\mathcal{G}$ in $\mathcal{G}$ such that $m(M_1, M_2) < m(M', M_2)$ and $|M'| \geq |M_1|$.*

Proof. Let $C = (v_1, v_2, \ldots, v_\ell)$ and for each $1 \leq i \leq \ell$, let $e_i = \{v_i, v_{i+1}\}$, where $v_{\ell+1} = v_1$. Assume without loss of generality that $e_i \in M_1$ for odd i. Define $\hat{M} = (M_1 \setminus \{e_1, e_3\}) \cup \{e_2\}$. Clearly, $\hat{M}$ is a matching of G and then $M' = \mu(\hat{M})$ is a $\mathcal{G}$-neighbor of M_1. Similarly to Lemma 2, we have $m(M_1, M_2) < m(M', M_2)$. If $|M_1| \leq |M'|$, we are done. Moreover, if $\ell = 4$, $\hat{M} \cup \{e_4\}$ is a matching of G and hence we have $|M_1| \leq |M'|$ as well. Thus, suppose that $\ell \geq 6$ and $|M'| = |\hat{M}| = |M_1| - 1$. The subpath $P = (v_4, v_5, \ldots, v_{\ell+1})$ is an M'-augmenting path component in $G[M' \triangle M_2]$. By Lemma 2, there is a directed path from M' to a maximal matching M'' of $\mathcal{G}$ in $\mathcal{G}$ such that $m(M', M_2) < m(M'', M_2)$ and $|M'| < |M''|$. Moreover, each maximal matching on the directed path has cardinality at least $|M'|$. This completes the proof of lemma. $\square$

Lemma 5. *If $G[M_1 \triangle M_2]$ has an M_2-augmenting path component P, then there is a directed $(|M_1| - 1)$-thick path from M_1 to a maximal matching M' of G in $\mathcal{G}$ such that $m(M_1, M_2) < m(M', M_2)$ and $|M'| \geq |M_1| - 1$.*

Proof. The proof is almost analogous to that in Lemma 2. Let $P = (v_1, v_2, \ldots, v_\ell)$ and let $e_i = \{v_i, v_{i+1}\}$ for $1 \leq i < \ell$. By Proposition 2, P contains at least two edges. Moreover, as P is M_2-augmenting, P contains at least three edges and $e_1, e_3 \in M_1$ and $e_2 \in M_2$. Define $\hat{M} = (M_1 \setminus \{e_1, e_3\}) \cup \{e_2\}$. Then, $\hat{M}$ is a matching of G, and hence $M' = \mu(\hat{M})$ is a $\mathcal{G}$-neighbor of M_1. As $e_1, e_3 \in M_1 \setminus M_2$ and $e_2 \in M_2 \setminus M_1$, we have $m(M_1, M_2) < m(M', M_2)$. Moreover, $|\hat{M}| = |M_1| - 1$. the lemma follows. □

Corollary 6 ($\star$). *Let M' be the maximal matching of G obtained in Lemma 5. If $G[M_1 \triangle M_2]$ has neither M_1-augmenting path components nor cycle components, then $G[M' \triangle M_2]$ has no cycle components.*

Now, we are ready to prove the main claims.

Lemma 7. *Let M_1 and M_2 be maximal matchings of G with $|M_1| > |M_2|$. Then, there is a directed $|M_2|$-thick path from M_1 to M_2 in $\mathcal{G}$.*

Proof. We prove the lemma by induction on $k = |M_2| - m(M_1, M_2)$. Note that, by Proposition 1, it holds that $k \geq 0$. Let $G' = G[M_1 \triangle M_2]$. We prove a slightly stronger claim: If either (1) $|M_1| > |M_2|$ or (2) $|M_1| = |M_2|$ and G' has no cycle components, then there is a directed $|M_2|$-thick path from M_1 to M_2 in $\mathcal{G}$. The base case $k = 0$ follows from Proposition 1.

We assume that $k > 0$ and the lemma holds for all $k' < k$. As $k > 0$, G' has at least one connected component. Suppose that G' has a cycle component. In this case, it holds that $|M_1| > |M_2|$ from the assumption. By Lemma 4, $\mathcal{G}$ has a directed $(|M_1| - 1)$-thick path from M_1 to a maximal matching M' of G such that $m(M_1, M_2) < m(M', M_2)$ and $|M'| \geq |M_1|$. Applying the induction hypothesis to M' and M_2, $\mathcal{G}$ has a directed $|M_2|$-thick path from M' to M_2. As $|M_1| > |M_2|$, the directed path obtained by concatenating these two paths (from M_1 to M' and from M' to M_2) is a directed $|M_2|$-thick path from M_1 to M_2 and hence the claim follows in this case.

Suppose that G' has an M_1-augmenting path component. By Lemma 2, $\mathcal{G}$ has a directed $|M_1|$-thick path from M_1 to M' such that $m(M_1, M_2) < m(M', M_2)$ and $|M_1| < |M'|$. Since pair M' and M_2 satisfies (1), by the induction hypothesis, $\mathcal{G}$ has a directed $|M_2|$-thick path from M' to M_2, and hence the claim holds for this case.

In the following, we assume that G' has neither cycle components nor M_1-augmenting path components. Suppose that G' has an even-alternating path component. By Lemma 2, $\mathcal{G}$ has a directed $|M_1|$-thick path from M_1 to M' such that $m(M_1, M_2) < m(M', M_2)$ and $|M_1| \leq |M'|$. Moreover, by Corollary 3, $G[M' \triangle M_2]$ has no cycle components. Thus, applying the induction hypothesis to M' and M_2 proves the claim.

Finally, suppose that G' has only M_2-augmenting path components. By Proposition 3, $|M_1| > |M_2|$. By Lemma 5, $\mathcal{G}$ has a directed $(|M_1| - 1)$-thick path from M_1 to M' such that $m(M_1, M_2) < m(M', M_2)$ and $|M'| \geq |M_1| - 1$. More-over, by Corollary 6, $G[M' \triangle M_2]$ has no cycle components. Thus, as $|M'| \geq |M_2|$, applying the induction hypothesis to M' and M_2 proves the claim as well.   □

Lemma 8. *Let M_1 and M_2 be maximum matchings of G. Then, there is a directed $(\nu(G) - 1)$-thick path from M_1 to M_2 in $\mathcal{G}$.*

Proof. We prove the lemma by induction on $k = |M_2| - m(M_1, M_2)$. The base case $k = 0$ follows from Proposition 1. We assume that $k > 0$ and the lemma holds for all $k' < k$. Let $G' = G[M_1 \triangle M_2]$. As $k > 0$, G' has at least one connected component. Observe that every component of G' is either an even-alternating component or a cycle component. This follows from the fact that if G' has an M_1- or M_2-augmenting path component, then this path is an augmenting path for M_1 or M_2, respectively, which contradicts to the assumption that M_1 and M_2 are maximum matchings of G. By Lemmas 2 and 4, $\mathcal{G}$ has a directed $(|M_1| - 1)$-thick path from M_1 to a maximal matching M' of G such that $m(M_1, M_2) < m(M', M_2)$ and $|M_1| \leq |M'|$. As M_1 is a maximum matching of G, M' is also a maximum matching of G. Applying the induction hypothesis to pair M' and M_2 proves the lemma.   □

Thus, we can enumerate all large maximal matchings in polynomial delay. By simply traversing $\mathcal{G}$, we can enumerate all neighbor of $\mathcal{G}$ in $O(nm)$ time. However, to determine whether each neighbor has already been output, we need a data structure.

Lemma 9 ($\star$). *Let $\mathcal{M}$ be a collection of maximal matchings. There is a data structure for representing $\mathcal{M}$ that supports the following operations: (1) Decide if $\mathcal{M}$ contains a given matching M in $O(n)$ time; (2) Insert a matching M into $\mathcal{M}$ in $O(n)$ time.*

Theorem 10. *Algorithm 1 enumerates all maximal matchings of G with cardi-nality at least t in $O(nm)$ delay and exponential space, provided that $t < \nu(G)$.*

Proof. The correctness of the algorithm directly follows from Lemmas 7 and 8. We analyze the delay of the algorithm. We first compute a maximum matching M^* of G. This can be done in time $O(n^{1/2}m)$ using the algorithm of [21]. Each maximal matching has $\mathcal{G}$-neighbors at most m. For each such $\mathcal{G}$-neighbor M' of M, we can check whether $M' \in \mathcal{S}$ at line 8 in $O(n)$ time with the data structure given in Lemma 9. Thus, it suffices to show that $\mu((M \setminus \Gamma(e)) \cup \{e\})$ can be computed in $O(n)$ time from given a maximal matching M and $e \in E \setminus M$. Observe that $M \cap \Gamma(e)$ consists of at most two edges $f_1, f_2 \in E$, each of which is incident to one of the end vertices of e. Since at least one of end vertices of each edge in $E \setminus (\Gamma(f_1) \cup \Gamma(f_2))$ is matched in $(M \setminus \{f_1, f_2\}) \cup \{e\}$, we can compute $\mu((M \setminus \Gamma(e)) \cup \{e\})$ from $(M \setminus \Gamma(e)) \cup \{e\}$ by greedily adding edges in $\Gamma(f_1) \cup \Gamma(f_2)$. Since $|\Gamma(f_1) \cup \Gamma(f_2)| = O(n)$, this can be done in $O(n)$ time. Therefore, the theorem follows.   □

To reduce the space complexity, we follow another well-known strategy, called the *reverse search technique*, due to Avis and Fukuda [1]. The basic idea of this technique is to define a rooted tree $\mathcal{T}$ over the set of solutions instead of a directed graph. The reverse search technique solely traverses this rooted tree and outputs a solution on each node in the tree. A crucial difference from the supergraph technique is that we do not need exponential-space data structures used in the supergraph technique to avoid duplicate outputs.

Let $\mathcal{G}_{\geq t}$ be the subgraph of $\mathcal{G}$ induced by the maximal matchings of G with cardinality at least t. Our rooted tree $\mathcal{T}$ is in fact defined as a spanning tree of the underlying undirected graph of $\mathcal{G}_{\geq t}$. To define the rooted tree $\mathcal{T}$, we select an arbitrary maximum matching R^* of G as a root. In the following, we fix R^* and define a parent function **par** with respect to R^*. We assume that the edges in G are totally ordered with respect to some edge ordering. Let M be a maximal matching of G with $|M| \geq t$ and $M \neq R^*$. If $G[M \triangle R^*]$ contains a path component, then there is an edge $e \in R^* \setminus M$ that is incident to an end vertex of a path component in $G[M \triangle R^*]$. Then, we choose the minimum edge (with respect to the edge ordering) satisfying this condition as e, and we define $\mathbf{par}(M) = \mu((M \setminus \Gamma(e)) \cup \{e\})$. Otherwise, we choose the minimum edge in $R^* \setminus M$ as e, and we define $\mathbf{par}(M) = \mu((M \setminus \Gamma(e)) \cup \{e\})$. Note that as R^* is a maximum matching of G, there is at least one path component in $G[M \triangle R^*]$ whose end vertex is matched in R^* under the assumption that $G[M \triangle R^*]$ has a path component. Similarly to the proofs of Lemmas 2 and 4, we have $m(\mathbf{par}(M), R^*) > m(M, R^*)$ and $|\mathbf{par}(M)| \geq \min\{|M|, \nu(G)-1\}$. This implies that the parent function defines a rooted tree $\mathcal{T}$ in $\mathcal{G}_{\geq t}$ with root R^* by considering $\mathbf{par}(M)$ is the parent of M.

Next, we consider the time complexity for computing $\mathbf{par}(M)$. As we have seen in the proof of Theorem 10, $\mu((M \setminus \Gamma(e)) \cup \{e\})$ can be computed in $O(n)$ time. Given M and R^*, we can compute the minimum edge e in $G[M \triangle R^*]$ in $O(n)$ time as $|M| + |R^*| \leq n$. Thus, $\mathbf{par}(M)$ can be computed in $O(n)$ time and the following lemma holds.

Lemma 11. *Given a maximal matching M of G with $M \neq R^*$, we can compute* $\mathbf{par}(M)$ *in $O(n)$ time.*

Now, we are ready to describe our polynomial-space enumeration algorithm for LARGE MAXIMAL MATCHING ENUMERATION. For maximal matchings M and M' of G with cardinality at least t, M' is a *child* of M if $\mathbf{par}(M') = M$. Algorithm 2 recursively generates the set of children of a given maximal matching, which enable us to traverse all nodes in $\mathcal{T}$.

The following lemma is vital to bound the delay of the algorithm.

Lemma 12. *Let M be a maximal matching of G with $|M| \geq t$. Then, there are $O(n\Delta^2)$ children of M. Moreover, the children of M can be enumerated in total time $O(n^2\Delta^2)$.*

Proof. Let M' be a child of M. From the definition of the parent-child relation, we have

$$|M \triangle M'| = |\mu((M' \setminus \Gamma(e)) \cup \{e\}) \triangle M'| \leq 5.$$

Algorithm 2: Given a graph G and an integer $t < \nu(G)$, `Reverse-Search` enumerates all maximal matchings of G with cardinality at least t.

---

1 **Procedure** `Reverse-Search`(G, t)
2 | Let R^* be a maximum matching of G;
3 | `Traverse-tree`(G, t, R^*);
4 **Procedure** `Traverse-tree`(G, t, M)
5 | Output M;
6 | Let $\mathcal{C}$ be the children of M;
7 | **foreach** $M' \in \mathcal{C}$ **do**
8 | | `Traverse-tree`(G, t, M')

This inequality follows from the facts that $|M' \cap \Gamma(e)| \leq 2$ and $|M \setminus \mu((M' \setminus \Gamma(e)) \cup \{e\})| \leq 2$ (as observed in the proof of Theorem 10). Thus, we can enumerate all the children $\mathcal{C} = \{M' \mid \mathbf{par}(M') = M\}$ of M in polynomial time. To improve the running time of enumerating children in $\mathcal{C}$, we take a closer look at both M and M'.

Let e be the edge such that $M = \mathbf{par}(M') = \mu((M' \setminus \Gamma(e)) \cup \{e\})$ and let $F = M' \cap \Gamma(e)$. As observed above, $|F| \leq 2$. Let A be the edge set such that $M = (M' \setminus F) \cup \{e\} \cup A$. Each $f \in A$ is incident to an edge f' in F as M' is a maximal matching of G. This implies that $A \subseteq M$ is uniquely determined from $e \in M$ and $F \subseteq \Gamma(e)$ with $|F| \leq 2$ (i.e., $A = \{f \in (M \setminus \{e\}) \cap \Gamma(f') \mid f' \in F\}$). Thus, we have $|\mathcal{C}| = O(n\Delta^2)$. For each $e \in M$ and $F \subseteq \Gamma(e)$ with $|F| \leq 2$, we can compute M' as $M' = M \setminus (A \cup \{e\}) \cup F$ in $O(n)$ time and, by Lemma 11, check if $\mathbf{par}(M') = M$ in $O(n)$ time as well. Hence we can enumerate all children in $\mathcal{C}$ in $O(n^2\Delta^2)$ time. □

We output a solution for each recursive call. This implies that the delay of the algorithm is upper bounded by the running time of Line 6 in Algorithm 2. Since $m \leq n\Delta$, the delay of the algorithm is bounded by $O(n^{1/2}m + n^2\Delta^2) = O(n^2\Delta^2)$ as well. Moreover, as the depth of $\mathcal{T}$ is at most $|R^*|$, the algorithm runs in polynomial space.

Finally, when $t = \nu(G)$, we can enumerate all maximum matchings using the binary partition technique in $O(nm)$ delay and polynomial space. By combining these results, we obtain the following theorem.

Theorem 13. *One can enumerate all maximal matchings of G with cardinality at least t in $O(n^2\Delta^2)$ delay with polynomial space.*

4.2 k-Best Maximal Matching Enumeration

In the previous subsection, we define the graph $\mathcal{G}$, allowing us to enumerate all the maximal matchings of G with cardinality at least given threshold t. The key to this result is Lemma 7, which states that, for any $t < \nu(G)$, there is a directed t-thick path from a maximum matching M_1 to a maximal matching M_2 with cardinality t. This implies that every maximal matching of cardinality at least

---

Algorithm 3: Given a graph G and a non-negative integer k, the algorithm solves k-BEST MAXIMAL MATCHING ENUMERATION

---

1 **Procedure** k-best(G, k)
2   Let $\mathcal{A}$ be a polynomial delay enumeration algorithm for maximum matchings.;
3   **foreach** M *generated by* $\mathcal{A}(G)$ **do**
4   Output M and add M to queue $\mathcal{Q}$;
5   **if** k *solutions are output* **then halt**;
6   **foreach** $M' \in \mathcal{N}_\mathcal{G}(M)$ *with* $M' \notin \mathcal{S}$ **do**
7 | Add M' to $\mathcal{Q}$ and to $\mathcal{S}$
8   **while** $\mathcal{Q}$ *is not empty* **do**
9   Let M be a largest maximal matching in $\mathcal{Q}$;
10   Output M and delete M from $\mathcal{Q}$;
11   **if** k *solutions are output* **then halt**;
12   **foreach** $M' \in \mathcal{N}_\mathcal{G}(M)$ *with* $M' \notin \mathcal{S}$ **do**
13 | Add M' to $\mathcal{Q}$ and to $\mathcal{S}$

---

t is "reachable" from a maximum matching in the subgraph of $\mathcal{G}$ induced by the node set $\{M : M$ is a maximal matching of $G, |M| \geq t\}$ for every $t < \nu(G)$. From this fact, we can extend Algorithm 1 to an algorithm for k-BEST MAXIMAL MATCHING ENUMERATION, which is shown in Algorithm 3. The algorithm first enumerates all maximum matchings of G with the algorithm $\mathcal{A}$ and outputs those maximum matchings as long as at most k solutions are output. The remaining part of the algorithm is almost analogous to Algorithm 1 and the essential difference from it is that the algorithm chooses a largest maximal matching in the priority queue $\mathcal{Q}$ at line 9. Intuitively, we traverse a forest based on best-first manner.

Theorem 14. *We can solve k-BEST MAXIMAL MATCHING ENUMERATION in $O(nm)$ delay.*

Proof. The delay of the algorithm follows from a similar analysis in Theorem 10. Note that, at line 9, we can choose in time $O(1)$ a largest maximal matching in $\mathcal{Q}$ by using $\nu(G)$ linked lists $L_1, L_2, \ldots, L_{\nu(G)}$ for $\mathcal{Q}$, where L_i is used for maximal matchings of cardinality i.

Let $\mathcal{M}$ be the set of maximal matchings of G that are output by the algorithm. To show the correctness of the algorithm, suppose for contradiction that there are maximal matchings $M \notin \mathcal{M}$ and $M' \in \mathcal{M}$ of G such that $|M| > |M'|$. Since $\mathcal{G}$ has a directed $|M|$-thick path from a maximum matching of G to M, we can choose such M so that every maximal matching on the path except for M belongs to $\mathcal{M}$. Let M'' be the immediate predecessor of M on the path. As $M'' \in \mathcal{M}$, M must be in the queue $\mathcal{Q}$ at some point. Hence as $|M| > |M'|$, the algorithm outputs M before M'. □

References

1. Avis, D., Fukuda, K.: Reverse search for enumeration. Discret. Appl. Math. **65**(1), 21–46 (1996). https://doi.org/10.1016/0166-218X(95)00026-N
2. Birmelé, E., et al.: Optimal listing of cycles and st-paths in undirected graphs. In: Proceedings of the SODA 2013, pp. 1884–1896 (2013). https://doi.org/10.1137/1.9781611973105.134
3. Casel, K., Fernau, H., Khosravian Ghadikolaei, M., Monnot, J., Sikora, F.: Extension of some edge graph problems: standard and parameterized complexity. In: Gąsieniec, L.A., Jansson, J., Levcopoulos, C. (eds.) FCT 2019. LNCS, vol. 11651, pp. 185–200. Springer, Cham (2019). https://doi.org/10.1007/978-3-030-25027-0_13
4. Chegireddy, C.R., Hamacher, H.W.: Algorithms for finding k-best perfect matchings. Discret. Appl. Math. **18**(2), 155–165 (1987). https://doi.org/10.1016/0166-218X(87)90017-5
5. Cohen, S., Kimelfeld, B., Sagiv, Y.: Generating all maximal induced subgraphs for hereditary and connected-hereditary graph properties. J. Comput. Syst. Sci. **74**(7), 1147–1159 (2008). https://doi.org/10.1016/j.jcss.2008.04.003
6. Comin, C., Rizzi, R.: An improved upper bound on maximal clique listing via rectangular fast matrix multiplication. Algorithmica **80**(12), 3525–3562 (2017). https://doi.org/10.1007/s00453-017-0402-5
7. Conte, A., Grossi, R., Marino, A., Versari, L.: Sublinear-space bounded-delay enumeration for massive network analytics: maximal cliques. In: Proceedings of the ICALP 2016. LIPIcs, vol. 55, pp. 148:1–148:15. Schloss Dagstuhl-Leibniz-Zentrum fuer Informatik (2016). https://doi.org/10.4230/LIPIcs.ICALP.2016.148
8. Conte, A., Uno, T.: New polynomial delay bounds for maximal subgraph enumeration by proximity search. In: Proceedings of the STOC 2019, pp. 1179–1190 (2019)
9. Edmonds, J.: Paths, trees, and flowers. Canadian J. Math. **17**, 449–467 (1965). https://doi.org/10.4153/CJM-1965-045-4
10. Eppstein, D.: k-best enumeration. In: Kao, M.Y. (ed.) Encyclopedia of Algorithms, pp. 1003–1006. Springer, New York (2016). https://doi.org/10.1007/978-1-4939-2864-4_733
11. Fukuda, K., Matsui, T.: Finding all minimum-cost perfect matchings in bipartite graphs. Networks **22**(5), 461–468 (1992). https://doi.org/10.1002/net.3230220504
12. Johnson, D.S., Yannakakis, M., Papadimitriou, C.H.: On generating all maximal independent sets. Inf. Process. Lett. **27**(3), 119–123 (1988). https://doi.org/10.1016/0020-0190(88)90065-8
13. Khachiyan, L., Boros, E., Borys, K., Elbassioni, K., Gurvich, V., Makino, K.: Enumerating spanning and connected subsets in graphs and matroids. In: Azar, Y., Erlebach, T. (eds.) ESA 2006. LNCS, vol. 4168, pp. 444–455. Springer, Heidelberg (2006). https://doi.org/10.1007/11841036_41
14. Khachiyan, L., Boros, E., Borys, K., Elbassioni, K., Gurvich, V., Makino, K.: Generating cut conjunctions in graphs and related problems. Algorithmica **51**(3), 239–263 (2008)
15. Kobayashi, Y., Kurita, K., Wasa, K.: Efficient constant-factor approximate enumeration of minimal subsets for monotone properties with weight constraints. CoRR, abs/2009.08830 (2020). https://arxiv.org/abs/2009.08830
16. Kobayashi, Y., Kurita, K., Wasa, K.: Polynomial-delay enumeration of large maximal matchings. CoRR, abs/2105.04146 (2021). https://arxiv.org/abs/2105.04146

17. Korhonen, T.: Listing small minimal separators of a graph. CoRR, abs/2012.09153 (2020). https://arxiv.org/abs/2012.09153

18. Kurita, K., Kobayashi, Y.: Efficient enumerations for minimal multicuts and multiway cuts. In: Proceedings of the MFCS 2020, pp. 60:1–60:14 (2020). https://doi.org/10.4230/LIPIcs.MFCS.2020.60

19. Lawler, E.L.: A procedure for computing the k best solutions to discrete optimization problems and its application to the shortest path problem. Manage. Sci. **18**(7), 401–405 (1972). https://doi.org/10.1287/mnsc.18.7.401

20. Makino, K., Uno, T.: New algorithms for enumerating all maximal cliques. In: Hagerup, T., Katajainen, J. (eds.) SWAT 2004. LNCS, vol. 3111, pp. 260–272. Springer, Heidelberg (2004). https://doi.org/10.1007/978-3-540-27810-8_23

21. Micali, S., Vazirani, V.V.: An $O(\sqrt{|V|}|E|)$ algorithm for finding maximum matching in general graphs. In: Proceedings of the FOCS 1980, pp. 17–27 (1980)

22. Mucha, M., Sankowski, P.: Maximum matchings via Gaussian elimination. In: Proceedings of the FOCS 2004, pp. 248–255. IEEE Computer Society (2004). https://doi.org/10.1109/FOCS.2004.40

23. Murty, K.G.: Letter to the editor - an algorithm for ranking all the assignments in order of increasing cost. Oper. Res. **16**(3), 682–687 (1968). https://doi.org/10.1287/opre.16.3.682

24. Schwikowski, B., Speckenmeyer, E.: On enumerating all minimal solutions of feedback problems. Discret. Appl. Math. **117**(1–3), 253–265 (2002)

25. Takata, K.: Space-optimal, backtracking algorithms to list the minimal vertex separators of a graph. Discret. Appl. Math. **158**(15), 1660–1667 (2010)

26. Tsukiyama, S., Ide, M., Ariyoshi, H., Shirakawa, I.: A new algorithm for generating all the maximal independent sets. SIAM J. Comput. **6**(3), 505–517 (1977). https://doi.org/10.1137/0206036

27. Uno, T.: Algorithms for enumerating all perfect, maximum and maximal matchings in bipartite graphs. In: Proceedings of the ISAAC 1997, pp. 92–101 (1997)

28. Uno, T.: A fast algorithm for enumerating non-bipartite maximal matchings. NII J. **3**, 89–97 (2001)

29. Uno, T.: Constant time enumeration by amortization. In: Dehne, F., Sack, J.-R., Stege, U. (eds.) WADS 2015. LNCS, vol. 9214, pp. 593–605. Springer, Cham (2015). https://doi.org/10.1007/978-3-319-21840-3_49

The Complexity of Contracting Bipartite Graphs into Small Cycles

R. Krithika[1], Roohani Sharma[2], and Prafullkumar Tale[3]($\boxtimes$)

[1] Indian Institute of Technology Palakkad, Palakkad, India
krithika@iitpkd.ac.in
[2] Max Planck Institute for Informatics, Saarland Informatics Campus,
Saarbrücken, Germany
rsharma@mpi-inf.mpg.de
[3] CISPA Helmholtz Center for Information Security, Saarbrücken, Germany
prafullkumar.tale@cispa.de

Abstract. For a positive integer $\ell \geq 3$, the C_ℓ-CONTRACTIBILITY problem takes as input an undirected simple graph G and determines whether G can be transformed into a graph isomorphic to C_ℓ (the induced cycle on ℓ vertices) using only edge contractions. Brouwer and Veldman [JGT 1987] showed that C_4-CONTRACTIBILITY is NP-complete in general graphs. It is easy to verify that that C_3-CONTRACTIBILITY is polynomial-time solvable. Dabrowski and Paulusma [IPL 2017] showed that C_ℓ-CONTRACTIBILITY is NP-complete on bipartite graphs for $\ell = 6$ and posed as open problems the status of C_ℓ-CONTRACTIBILITY when ℓ is 4 or 5. In this paper, we show that both C_5-CONTRACTIBILITY and C_4-CONTRACTIBILITY are NP-complete on bipartite graphs.

Keywords: C_5-CONTRACTIBILITY · C_4-CONTRACTIBILITY · bipartite graphs

1 Introduction

Operations on graphs produce new graphs from existing ones. Elementary editing operations include deleting vertices, deleting and/or adding edges, subdividing edges and contracting edges. Due to the ubiquitous presence of graphs in modeling real-world networks, many problems of practical importance may be posed as editing problems on graphs. In this work, we focus on modifying a graph by only performing edge contractions. Contracting an edge in a graph results in the addition of a new vertex adjacent to the neighbors of its endpoints followed by the deletion of the endpoints. As graphs typically represent binary relationships among a collection of objects, edge contractions naturally correspond to merging

The full version of this paper is at https://arxiv.org/abs/2206.07358

P. Tale—The author has received funding from the European Research Council (ERC) under the European Union's Horizon 2020 research and innovation programme under grant agreement SYSTEMATICGRAPH (No. 725978).

M. A. Bekos and M. Kaufmann (Eds.): WG 2022, LNCS 13453, pp. 356–369, 2022.
https://doi.org/10.1007/978-3-031-15914-5_26

two objects into a single entity or to treating two objects as indistinguishable. Contractions can therefore be seen as a way of 'simplifying' the graph and they have applications in clustering, compression, sparsification and computer graphics [1,3,6,7,13,19]. Edge contractions also play an imporant role in Hamiltonian graph theory, planar graph theory and graph minor theory [5,15,24].

Given graphs G and H, the GRAPH CONTRACTIBILITY problem decides whether G can be transformed into a graph isomorphic to H using only edge contractions. GRAPH CONTRACTIBILITY is known to be NP-complete [10, GT51]. This led to the study of the problem on special graph classes and for restricted choices of H. When H is a fixed graph, the GRAPH CONTRACTIBILITY problem is called H-CONTRACTIBILITY. Intuitively, this problem of determining whether G is contractible to H may be seen as the task of determining if the 'underlying structure' of G is H. One of the related graph parameters in this context is cyclicity. The cyclicity of a graph is the largest integer ℓ for which the graph is contractible to the induced cycle on ℓ vertices (denoted as C_ℓ). This parameter was introduced in the study of another important graph invariant called circularity [4]. Ever since, there have been efforts towards understanding the complexity of computing cyclicity and expressing it in terms of some structural property of the graph. Brouwer and Veldman [5] showed that C_4-CONTRACTIBILITY is NP-complete, hence proving that determining cyclicity is NP-hard in general. This result led to the study of the problem on special graph classes including bipartite graphs, claw-free graphs and planar graphs [8,9,11].

Hammack [11] showed that the cyclicity of planar graphs can be computed in polynomial time and in another work [12], he showed that C_ℓ-CONTRACTIBILITY is NP-complete for every $\ell \geq 5$ in general. Later, Kaminski et al. [18] showed that H-CONTRACTIBILITY is polynomial-time solvable on planar graphs for any H. Levin et al. [21] showed that H-CONTRACTIBILITY is polynomial-time solvable on general graphs if H is a graph on at most 5 vertices containing a universal vertex. However, the presence of a universal vertex in H on more than 5 vertices does not guarantee that H-CONTRACTIBILITY can be solved in polynomial time [16]. Fiala et al. [9] showed that C_ℓ-CONTRACTIBILITY is NP-complete for claw-free graphs for every $\ell \geq 6$. Heggernes et al. [14] proved that P_ℓ-CONTRACTIBILITY is polynomial-time solvable on chordal graphs for every $\ell \geq 1$, where P_ℓ denotes the induced path on ℓ vertices. Later, Belmonte et al. [2] proved that H-CONTRACTIBILITY is polynomial-time solvable on chordal graphs for every H. Dabrowski and Paulusma [8] showed that C_6-CONTRACTIBILITY is NP-complete for bipartite graphs. It is easy to verify that that C_3-CONTRACTIBILITY is polynomial-time solvable in general graphs. In this paper, we show that both C_5-CONTRACTIBILITY and C_4-CONTRACTIBILITY are NP-complete on bipartite graphs.

Theorem 1. C_5-CONTRACTIBILITY *is* NP-*complete on bipartite graphs.*

Theorem 2. C_4-CONTRACTIBILITY *is* NP-*complete on bipartite graphs.*

Theorems 1 and 2 involve reductions from the POSITIVE NOT ALL EQUAL SAT (POSITIVE NAE-SAT) problem where given a formula ψ in conjunctive normal form with no negative literals, the objective is to determine if there is an assignment of True or False to each of the variables such that for each clause at least one but not all variables in it are set to True. Such an assignment is called a *not-all-equal satisfying assignment*. POSITIVE NAE-SAT (also referred to as MONOTONE NAE-SAT) is known to be NP-complete [25]. Also, a straightforward reduction from SET SPLITTING or HYPERGRAPH 2-COLORABILITY [10, SP4] to POSITIVE NAE-SAT ascertains this fact.

Preliminaries. For a positive integer q, $[q]$ denotes the set $\{1, 2, \ldots, q\}$. $\mathbb{N}$ denotes the collection of all non-negative integers. A partition of a set S is a set of disjoint subsets of S whose union is S. Standard graph-theoretic terminology is omitted here due to space constraints and is given in the full version. We now formally define the notion of graph contractibility.

Definition 1. *G is said to be contractible to H if there is a surjective function $\psi : V(G) \to V(H)$ such that the following properties hold.*

1. *For each $h \in V(H)$, $\psi^{-1}(h)$, called the* witness set *corresponding to h, is connected.*
2. *For each $h, h' \in V(H)$, $hh' \in E(H)$ if and only if $E(\psi^{-1}(h), \psi^{-1}(h')) \neq \emptyset$.*

Then, we say that G is contractible to H *via the function ψ and that G has a H-witness structure $\mathcal{W} = \{\psi^{-1}(h) \mid h \in V(H)\}$ which is the collection of all witness sets.*

In Definition 1, a witness set that contains more than one vertex is called a *big witness set* and the one that is a singleton set is called a *small witness set* or *singleton witness set*. Note that a witness structure $\mathcal{W}$ is a partition of $V(G)$. Also, if a vertex v is in some big witness set W, then at least one neighbor of v is also in W. Recall that the H-CONTRACTIBILITY problem takes as input a graph G and decides whether G is contractible to H or not. Observe that this task is equivalent to determining if G has a H-witness structure or not.

Now, we proceed to proving Theorems 1 and 2 in Sects. 2 and 3, respectively. Proofs of results labelled with a [⋆] have been deferred to the full version of the paper due to space constraints.

2 C_5-Contractibility on Bipartite Graphs

In this section, we prove Theorem 1. It is easy to verify that C_5-CONTRACTIBILITY is in NP. Given an instance ψ of POSITIVE NAE-SAT with N variables and M clauses, we give a polynomial-time algorithm that outputs a bipartite graph G equivalent to ψ. For the sake of simplicity, we describe the algorithm in two steps. In the first step, the algorithm constructs a non-bipartite graph H equivalent to ψ (Lemmas 3 and 4) and then in the second step, the algorithm constructs a bipartite graph G that is equivalent to H (Lemma 1). We remark that G is obtained from H by dividing some (and not all) of the edges of H.

2.1   Construction of H and G

Let $\{X_1, X_2, \ldots, X_N\}$ and $\{C_1, C_2, \ldots, C_M\}$ be the sets of variables and clauses, respectively, in ψ. The non-bipartite graph H is constructed as follows. Refer to Fig. 1 for an illustration.

1. Add a set $V_\alpha = \{\alpha^0, \alpha^1, \alpha^2, \alpha^3, \alpha^4\}$ of five vertices that induce the 5-cycle $(\alpha^0, \alpha^1, \alpha^2, \alpha^3, \alpha^4)$. This set forms the "base cycle" in the witness structure.
2. For every $i \in [N]$, add a set of five vertices that induce a 5-cycle $C^i = (x_i^0, x_i^1, x_i^2, x_i^3, x_i^4)$ and two sets of edges $\{x_i^0 \alpha^0, \ x_i^1 \alpha^1, \ x_i^2 \alpha^2, \ x_i^3 \alpha^3, \ x_i^4 \alpha^4\}$ and $\{x_i^0 \alpha^1, \ x_i^1 \alpha^2, \ x_i^2 \alpha^3, \ x_i^3 \alpha^4, \ x_i^4 \alpha^0\}$. The variable gadget is designed so that there are two choices for C^i to co-exist (in a C_5-witness structure) with the C_5 induced by V_α. We will associate these two choices with a True or False assignment to the corresponding variable.
3. For every $j \in [M]$, add vertices c_j and b_j and a set $\{c_j \alpha^0, c_j \alpha^2, b_j \alpha^2, b_j \alpha^4\}$ of edges. The neighbours of c_j and b_j are defined so that c_j will be in the same witness set as α^1 (a non-neighbor of c_j) and b_j will be in the same witness set as α^3 (a non-neighbor of b_j).
4. Finally, for every $i \in [N]$ and $j \in [M]$ such that X_i appears in C_j, add edges $x_i^1 c_j$ and $x_i^2 b_j$. This step is the one that encodes the clause-variable relationship. Relevant variables are expected to help c_j (and b_j) to be connected to witness sets containing α^1 (and α^3).

This completes the construction of H. For $p \in \{0, 1, 2, 3, 4\}$, define $X^p := \{x_i^p \mid i \in [N]\}$. Also, define $Y^c := \{c_j \mid j \in [M]\}$ and $Y^b := \{b_j \mid j \in [M]\}$. For an edge $uv \in E(H)$, let $\lambda(u, v)$ denote the new vertex added while subdividing uv in the construction of G. Let $L = \{\alpha^0, \alpha^2, \alpha^4\} \cup X^1 \cup X^3$ and $R = \{\alpha^1, \alpha^3\} \cup X^0 \cup X^2 \cup X^4 \cup Y^c \cup Y^b$. Then, $\{L, R\}$ is a partition of H into two parts where there are certain edges with both endpoints in the same part. We subdivide exactly these edges to obtain G.

5. Subdivide the edge $\alpha^0 \alpha^4$.
6. For every $i \in [N]$, subdivide the edges $x_i^0 x_i^4, \ x_i^0 \alpha^1, \ x_i^1 \alpha^2, \ x_i^2 \alpha^3$, and $x_i^3 \alpha^4$.
7. For every $i \in [N]$ and $j \in [M]$, subdivide the edge $x_i^2 b_j$ if it exists.

This completes the construction of G.

We now argue that G is a bipartite graph. Observe that L and R are independent sets in G. We will extend this partition $\{L, R\}$ of H into a bipartition of G as follows: $\lambda(\alpha^0, \alpha^4) \in R$ and for every $i \in [N]$, $\lambda(x_i^0, x_i^4) \in L$, $\lambda(x_i^0, \alpha^1) \in L$, $\lambda(x_i^1, \alpha^2) \in R$, $\lambda(x_i^2, \alpha^3) \in L$ and $\lambda(x_i^3, \alpha^4) \in R$. For every $i \in [N], j \in [M]$, if $x_i^2 b_j \in E(H)$, then $\lambda(x_i^2, b_j) \in L$. See Fig. 1 for an illustration. It is easy to verify that $\{L, R\}$ is a bipartition of G and hence G is a bipartite graph. We remark that the natural bipartite graph obtained from H by subdividing all the edges may not be equivalent to H in the context of C_5-CONTRACTIBLITY. In Lemma 1, we show that the set of edges of H that are subdivided to obtain G are safe (in preserving contractiblity to C_5) to subdivide.

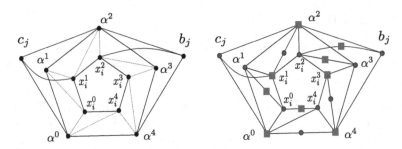

Fig. 1. (Left) The graph H with certain edges highlighted as purple (dotted) edges denote setting variable X_i to `True` and as green (dashed) edges denote setting X_i to `False`, respectively. (Right) The bipartite graph G where blue (round) and red (squares) vertices denote a bipartition. (Color figure online)

2.2 Equivalence of H and G

We show that G and H are equivalent in the context of C_5-CONTRACTIBLITY. As G is obtained from H by subdividing some edges, one can obtain H from G by contracting some edges. Hence, if one can obtain a C_5 by contracting edges in H, then one can also obtain a C_5 by contracting edges in G by first contracting G to H and then contracting H to C_5. To prove the converse, we first argue that no vertex in $V(G) \setminus V(H)$ is a singleton witness set in any C_5-witness structure $\mathcal{W}$ of G. Then, we show that deleting vertices of $V(G) \setminus V(H)$ from $\mathcal{W}$ results in a C_5-witness structure $\mathcal{W}'$ of H.

Lemma 1. [⋆] H is a YES-instance of C_5-CONTRACTIBILITY if and only if G is a YES-instance of C_5-CONTRACTIBILITY.

2.3 Properties of a C_5-Witness Structure of H

Before we state properties of H, we mention the following observation.

Observation 1. In any partition $\{X, Y\}$ of the vertices of an induced 5-cycle into 2 non-empty parts, $E(X, Y) \neq \emptyset$.

Now, we state certain properties of vertex subsets in H that we later use to show properties of a C_5-witness structure of H.

Observation 2. $X^0, X^1, X^2, X^3, X^4, Y^c$ and Y^b are independent sets and V_α is a dominating set in H. Further, $X^0 \cup X^4 \cup Y^c \subseteq N(\alpha^0)$, $X^1 \cup X^0 \subseteq N(\alpha^1)$, $X^2 \cup X^1 \cup Y^c \cup Y^b \subseteq N(\alpha^2)$, $X^3 \cup X^2 \subseteq N(\alpha^3)$ and $X^4 \cup X^3 \cup Y^b \subseteq N(\alpha^4)$.

Next, we show a property of a C_5-witness structure of H that will be crucial to proving the correctness of the reduction. As we have indicated in the construction of H, we need a handle on the base cycle of the C_5-witness structure (for YES-instances) which Lemma 2 provides.

Lemma 2. *In any C_5-witness structure of G, every pair of vertices in V_α are in different witness sets.*

Proof. Suppose $\mathcal{W} = \{W^i \mid i \in [4] \cup \{0\}\}$ is a C_5-witness structure of H where $E(W^i, W^j) \neq \emptyset$ if and only if $j = (i \pm 1) \mod 5$. We argue that V_α has a non-empty intersection with each W^i. Suppose $V_\alpha \subseteq W^i$ for some $0 \leq i \leq 4$. Then, $W^{(i+2) \mod 5} = \emptyset$ and $W^{(i+3) \mod 5} = \emptyset$ leading to a contradiction. Suppose V_α intersects exactly two witness sets. We will consider the cases when these sets are W^0, W^1 and W^0, W^2. The other cases are similar to these cases. If V_α intersects only with W^0 and W^1, then since V_α is a dominating set in H it follows that $W^3 = \emptyset$ and this leads to a contradiction. Suppose V_α intersects only with W^0 and W^2. From Observation 1, this implies that $E(W^0, W^2) \neq \emptyset$ leading to a contradiction. Suppose V_α intersects exactly four witness sets, say W^0, W^1, W^2, and W^3. Without loss of generality, assume $\alpha^0 \in W^0$. As α^1 and α^4 are adjacent to α^0, we have $\{\alpha^1, \alpha^4\} \subseteq W^0 \cup W^1$. Then, one of α^2 or α^3 is in W^2 and the other is in W^3. However, as $\alpha^1\alpha^2, \alpha^3\alpha^4 \in E(H)$, neither α^2 nor α^3 can be in W^3 implying that $W^3 = \emptyset$ and leading to a contradiction.

Suppose V_α intersects exactly three witness sets. Without loss of generality, let $\alpha^0 \in W^0$. We consider the following cases.

- Case (i) V_α intersects W^0, W^1 and W^2.
- Case (ii) V_α intersects W^0, W^1 and W^4.
- Case (iii) V_α intersects W^0, W^2 and W^3. This leads to contradiction as Observation 1 implies $E(W^0, W^2 \cup W^3) \neq \emptyset$.
- Case (iv) V_α intersects W^0, W^2 and W^4. This leads to contradiction as Observation 1 implies $E(W^2, W^0 \cup W^4) \neq \emptyset$.
- Case (v) V_α intersects W^0, W^4 and W^3. This is similar to Case (i).
- Case (vi) V_α intersects W^0, W^1 and W^3. This is similar to Case (iv).

Consider Case (i). As α^1 and α^4 are adjacent to α^0, we have $\{\alpha^1, \alpha^4\} \subseteq W^0 \cup W^1$. Then, at least one of α^2 or α^3 is in W^2 and since $\alpha^2\alpha^3 \in E(H)$, neither α^2 nor α^3 can be in W^0. Thus, we have $\{\alpha^2, \alpha^3\} \subseteq W^1 \cup W^2$. Since $E(W^0, W^3) = \emptyset$ and $E(W^1, W^3) = \emptyset$, we have $W^3 \cap N(\alpha^0) = \emptyset$, $W^3 \cap N(\alpha^1) = \emptyset$ and $W^3 \cap N(\alpha^4) = \emptyset$. From Observation 2, this implies that $W^3 \subseteq X^2$. Similarly, since $E(W^1, W^4) = \emptyset$ and $E(W^2, W^4) = \emptyset$, we have $W^4 \cap N(\alpha^2) = \emptyset$ and $W^4 \cap N(\alpha^3) = \emptyset$. From Observation 2, this implies $W^4 \subseteq (X^0 \cup X^4)$. However, by the construction, $E(X^2, X^0 \cup X^4) = \emptyset$ implying that $E(W^3, W^4) = \emptyset$ which leads to a contradiction.

Let us now consider Case (ii). Recall that $\alpha^0 \in W^0$. Then, either $\alpha^1 \in W^0 \cup W^1$ or $\alpha^1 \in W^0 \cup W^4$. As both these cases are similar, we consider the case when $\alpha^1 \in W^0 \cup W^1$. Suppose $\alpha^1 \in W^1$. Then, we have $\{\alpha^1, \alpha^2\} \subseteq W^0 \cup W^1$ since $\alpha^1\alpha^2 \in E(H)$. We will show that this leads to a contradiction. At least one of α^3 or α^4 is in W^4 and since $\alpha^3\alpha^4 \in E(H)$, neither α^3 nor α^4 can be in W^1. Thus, we have $\{\alpha^3, \alpha^4\} \subseteq W^0 \cup W^4$. Since $E(W^0, W^3) = \emptyset$ and $E(W^1, W^3) = \emptyset$, we have $W^3 \cap N(\alpha^0) = \emptyset$, $W^3 \cap N(\alpha^1) = \emptyset$ and $W^3 \cap N(\alpha^2) = \emptyset$. From Observation 2, this implies $W^3 \subseteq X^3$. Similarly, since $E(W^0, W^2) = \emptyset$ and $E(W^4, W^2) = \emptyset$, we have $W^2 \cap N(\alpha^0) = \emptyset$, $W^2 \cap N(\alpha^3) = \emptyset$ and $W^2 \cap N(\alpha^4) = \emptyset$. From

Observation 2, this implies $W^2 \subseteq X^1$. However, by construction, $E(X^1, X^3) = \emptyset$ implying that $E(W^2, W^3) = \emptyset$ which leads to a contradiction.

Suppose $\alpha^1 \in W^0$. If $\alpha^2 \in W^0$, then one of α^3 or α^4 is in W^1 and the other is in W^4 resulting in an edge between W^1 and W^4. Thus, $\alpha^2 \in W^1$ or $\alpha^2 \in W^4$. As these cases are similar, we only consider $\alpha^2 \in W^1$. Then we once again have $\{\alpha^1, \alpha^2\} \subseteq W^0 \cup W^1$ which leads to a contradiction.   □

2.4   Equivalence of H and ψ

Now, we are ready to establish the equivalence of ψ and H.

Lemma 3. *If ψ is a* YES-*instance of* POSITIVE NAE-SAT *then H is a* YES-*instance of C_5-*CONTRACTIBILITY.

Proof. Suppose $\pi : \{X_1, X_2, \ldots, X_N\} \mapsto \{\texttt{True}, \texttt{False}\}$ is a not-all-equal satisfying assignment of ψ. Define the following partition of $V(H)$.

$$W^0 := \{\alpha^0\} \cup \{x_i^0 \mid i \in [N], \pi(X_i) = \texttt{True}\} \cup \{x_i^4 \mid i \in [N], \pi(X_i) = \texttt{False}\},$$

$$W^1 := \{\alpha^1\} \cup \{x_i^1 \mid i \in [N], \pi(X_i) = \texttt{True}\} \cup \{x_i^0 \mid i \in [N], \pi(X_i) = \texttt{False}\}$$
$$\cup \{c_j \mid j \in [M]\},$$

$$W^2 := \{\alpha^2\} \cup \{x_i^2 \mid i \in [N], \pi(X_i) = \texttt{True}\} \cup \{x_i^1 \mid i \in [N], \pi(X_i) = \texttt{False}\},$$

$$W^3 := \{\alpha^3\} \cup \{x_i^3 \mid i \in [N], \pi(X_i) = \texttt{True}\} \cup \{x_i^2 \mid i \in [N], \pi(X_i) = \texttt{False}\}$$
$$\cup \{b_j \mid j \in [M]\},$$

$$W^4 := \{\alpha^4\} \cup \{x_i^4 \mid i \in [N], \pi(X_i) = \texttt{True}\} \cup \{x_i^3 \mid i \in [N], \pi(X_i) = \texttt{False}\},$$

Clearly W^0, W^2, and W^4 are connected sets. For any $j \in [M]$, there exists $i \in [N]$ such that $x_i^1 \in W^1$ (since π sets at least one of the variables in C_j to $\texttt{True}$) and $i' \in [N]$ such that $x_{i'}^2 \in W^3$ (since π sets at least one of the variables in C_j to $\texttt{False}$). Also, $c_j x_i^1, b_j x_{i'}^2 \in E(H)$. As for every $i \in [N]$, α^1 is adjacent to x_i^1 and α^3 is adjacent to x_i^2, it follows that W^1 and W^3 are connected sets. Now, it is easy to verify that $\{W^0, W^1, W^2, W^3, W^4\}$ is a C_5-witness structure.   □

In the proof of the converse of Lemma 3, we will crucially use Lemma 2. That is, if H is contractible to a 5-cycle, then in any C_5-witness structure $\{W^0, W^1, W^2, W^3, W^4\}$ with $E(W^i, W^j) \neq \emptyset$ if and only if $j = (i \pm 1) \mod 5$, each of the five witness sets has a non-empty intersection with V_α. This structure along with a couple of other properties translates to a not-all-equal satisfying assignment of ψ.

Lemma 4. *If H is a* YES-*instance of C_5-*CONTRACTIBILITY *then ψ is a* YES-*instance of* POSITIVE NAE-SAT.

Proof. Suppose $\mathcal{W} = \{W^0, W^1, W^2, W^3, W^4\}$ is a C_5-witness structure of H where $E(W^i, W^j) \neq \emptyset$ if and only if $j = (i \pm 1) \mod 5$. Then, by Lemma 2, V_α has a non-empty intersection with each W^i. Without loss of generality, let

$\alpha^p \in W^p$ for every $p \in \{0, 1, 2, 3, 4\}$. We first argue that for any $i \in [N]$, the set $S_i = \{x_i^0, x_i^1, x_i^2, x_i^3, x_i^4\}$ also has a non-empty intersection with each W^j. Suppose $S_i \cap W^0 = \emptyset$. Then, as α^0 is adjacent to x_i^0, x_i^4 and $x_i^0 x_i^4 \in E(H)$, either $\{x_i^0, x_i^4\} \subseteq W^1$ or $\{x_i^0, x_i^4\} \subseteq W^4$. As $\alpha^4 x_i^4, \alpha^1 x_i^0 \in E(H)$, both these cases contradict the fact that $E(W^1, W^4) = \emptyset$. Using the similar arguments, it follows that S_i has a non-empty intersection with each W^j.

Next, we claim that for each $i \in [N]$ and $0 \le p \le 4$, $x_i^p \in W^p \cup W^{(p+1) \bmod 5}$. This is due to the fact that x_i^p is adjacent with α^p and $\alpha^{p+1 \pmod 5}$. Now, we show that for each $i \in [N]$ and $0 \le p \le 4$, $x_i^p \in W^p$ if and only if $x_i^{(p+1) \bmod 5} \in W^{(p+1) \bmod 5}$ and $x_i^p \in W^{(p+1) \bmod 5}$ if and only if $x_i^{(p+1) \bmod 5} \in W^{(p+2) \bmod 5}$. If $x_i^0 \in W^0$ and $x_i^1 \notin W^1$, then $E(W^0, W^2) \cup E(W^0, W^3) \cup E(W^2, W^4) \ne \emptyset$ leading to a contradiction. If $x_i^0 \in W^1$ and $x_i^1 \notin W^2$, then $E(W^1, W^3) \cup E(W^0, W^2) \cup E(W^1, W^4) \ne \emptyset$ leading to a contradiction. Similar arguments hold for x_i^1, x_i^2, x_i^3 and x_i^4. This is indicated by the collections of purple (dotted) edges and green (dashed) edges in Fig. 1. We will associate these two choices with setting X_i to True and to False, respectively.

We now construct an assignment $\pi : \{X_1, X_2, \ldots, X_N\} \mapsto \{\text{True}, \text{False}\}$. Consider the witness set W^1. For each $i \in [N]$, if $x_i^1 \in W^1$ then set $\pi(X_i) = \text{True}$, otherwise $(x_i^1 \in W^2)$ set $\pi(X_i) = \text{False}$. We argue that π is a not-all-equal satisfying assignment for ψ. We show that for each $j \in [M]$, $c_j \in W^1$ and $b_j \in W^3$, further, the clause C_j has variables X_i and $X_{i'}$ such that $x_i^1 \in W^1$ and $x_{i'}^2 \in W^3$. Observe that c_j (being adjacent with α^0 and α^2) is in the same witness set that has α^1 and b_j (being adjacent with α^2 and α^4) is in the same witness set that has α^3. Thus, for each $j \in [M]$, $c_j \in W^1$ and $b_j \in W^3$. By the property of witness structures, W^1 and W^3 are connected sets. As the only vertices outside V_α that are adjacent to c_j are vertices x_i^1 corresponding to variables X_i appearing in C_j, it follows that C_j has a variable X_i such that $x_i^1 \in W^1$. Similarly, as the only vertices outside V_α that are adjacent to b_j are vertices x_i^2 corresponding to variables X_i appearing in C_j, it follows that C_j has a variable $X_{i'}$ such that $x_{i'}^2 \in W^3$. □

3 C_4-Contractiblity on Biparitite Graphs

In this section, we prove Theorem 2. It is easy to verify that C_4-CONTRACTIBILITY is in NP. Given an instance ψ of POSITIVE NAE-SAT with N variables and M clauses, we give a polynomial-time algorithm that outputs a bipartite graph G equivalent to ψ (Lemmas 7 and 8).

3.1   Construction of G

Let $\{X_1, X_2, \ldots, X_N\}$ and $\{C_1, C_2, \ldots, C_M\}$ be the sets of variables and clauses, respectively, in ψ. The graph G with a partition $\{V, V'\}$ of its vertex set is constructed as follows. See Fig. 2 for an illustration.

1. Add vertices t, f to V, vertices t', f' to V' and edges tt', ff' to $E(G)$. This set would eventually form the "base cycle" in the witness structure.

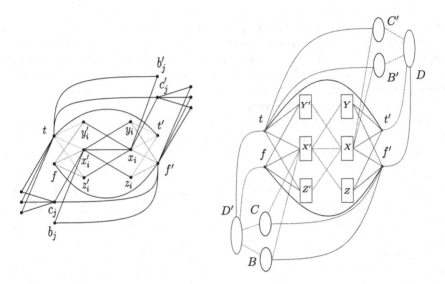

Fig. 2. (Left) The graph G where only three vertices each in D and D' shown with purple (dotted) edges denote setting variable X_i to True and green (dashed) edges denote setting X_i to False. (Right) Adjacency relation between different subsets of vertices. (Color figure online)

2. For every $i \in [N]$, add vertices x_i, y_i, z_i to V and x'_i, y'_i, z'_i to V' corresponding to the variable X_i. Further, make every vertex in $\{x'_i, y'_i, z'_i\}$ adjacent to every vertex in $\{x_i, t, f\}$ and every vertex in $\{x_i, y_i, z_i\}$ adjacent to every vertex in $\{x'_i, t', f'\}$. Let $X = \{x_i \mid i \in [N]\}$, $X' = \{x'_i \mid i \in [N]\}$, $Y = \{y_i \mid i \in [N]\}$, $Y' = \{y'_i \mid i \in [N]\}$, $Z = \{z_i \mid i \in [N]\}$, $Z' = \{z'_i \mid i \in [N]\}$. The neighborhood of X' is set so that every element of X' is in the witness set containing t or f. This forces every element of X to be respectively in the witness set containing t' or f'. These binary choices would be associated with setting the corresponding variable to True or False. The sets Y, Y', Z, Z' are added for technical reasons.

3. For every $j \in [M]$, add vertices c_j, b_j to V, c'_j, b'_j to V' and edges $c_j f', b_j f'$, $c'_j t, b'_j t$ to $E(G)$ corresponding to clause C_j. Let $C = \{c_j \mid j \in [M]\}$, $C' = \{c'_j \mid j \in [M]\}$, $B = \{b_j \mid j \in [M]\}$, $B' = \{b'_j \mid j \in [M]\}$. Subsequently, we will add more vertices (sets D and D' defined subsequently) adjacent to vertices in $C \cup B \cup C' \cup B'$ so that no vertex in $B \cup C$ is in a witness set that is non-adjacent to the one containing t and no vertex in $B' \cup C'$ is in a witness set that is non-adjacent to the one containing f'.

4. For every $i \in [N]$ and $j \in [M]$, if X_i appears in C_j then add edges $c_j x'_i, b_j x'_i$, $x_i c'_j$, and $x_i b'_j$ to $E(G)$. This step is the one that encodes the clause-variable relationship. Relevant variables are expected to help clause vertices to be connected to witness sets containing them.

5. Let $\mathcal{D}$ denote the following collection of pairs of vertices: $\{\{t, f\}, \{t', f'\}\}$ $\bigcup \{\{t, c_j\}, \{t, b_j\}, \{f', c'_j\}, \{f', b'_j\} \mid j \in [M]\}$. Note that for any pair of vertices

in $\mathcal{D}$, either both elements of the pair are in V or both are in V'. For every pair $\{u, v\}$ of vertices in $\mathcal{D}$ that are in V, add three vertices $d'_{u,v,1}$, $d'_{u,v,2}$, $d'_{u,v,3}$ to V' and make them adjacent to both u, v. For every pair $\{u, v\}$ of vertices in $\mathcal{D}$ that are in V', add three vertices $d_{u,v,1}$, $d_{u,v,2}$, $d_{u,v,3}$ to V and make them adjacent to both u, v. The pairs in $\mathcal{D}$ are the ones that should not be in non-adjacent witness sets and the common neighbors are added to achieve this property.

This completes the construction of G. As the reduction always adds edges with one of its endpoints in V and the other endpoint in V', G is a bipartite graph with bipartition $\{V, V'\}$. Let $D = \{d_{u,v,p} \mid \{u,v\} \in \mathcal{D}, u, v \in V \text{ and } p \in [3]\}$ and $D' = \{d'_{u,v,p} \mid \{u,v\} \in \mathcal{D}, u, v \in V' \text{ and } p \in [3]\}$.

3.2   Properties of a Nice C_4-Witness Structure of G

Now, we show that if G is contractible to a 4-cycle, then there is a C_4-witness structure of G satisfying certain nice properties. For this purpose, we introduce the following notion of a *nice C_4-witness structure*.

Definition 2. *A C_4-witness structure of G is a nice C_4-witness structure if the following properties hold.*
(P1) For every pair $\{u, v\}$ in $\mathcal{D}$, u and v are in the same or adjacent witness sets.
(P2) Every vertex in $D \cup D'$ is in a big witness set. Further, every vertex in D' is in the same witness set as t and every vertex in D is in the same witness set as f'.

Next, we show the existence of a nice C_4-witness structure for YES-instances.

Lemma 5. [⋆] *If G is contractible to a 4-cycle, then there is a nice C_4-witness structure of G.*

Now, we show a property of a nice C_4-witness structure of G that will be crucial to proving the correctness of the reduction.

Lemma 6. [⋆] *In any nice C_4-witness structure of G, every pair of vertices in $\{t, t', f, f'\}$ are in different witness sets.*

3.3   Equivalence of G and ψ

Now, we are ready to establish the equivalence of ψ and G.

Lemma 7. *If ψ is a YES-instance of POSITIVE NAE-SAT then G is a YES-instance of C_4-CONTRACTIBILITY.*

Proof. Suppose $\pi : \{X_1, X_2, \ldots, X_N\} \mapsto \{\texttt{True}, \texttt{False}\}$ is a not-all-equal satisfying assignment of ψ. Define the following partition of $V(G)$.

$$W^0 := \{t\} \cup \{x'_i, y'_i, z'_i \mid i \in [N] \text{ and } \pi(X_i) = \texttt{True}\} \cup D',$$
$$W^1 := \{t'\} \cup \{x_i, y_i, z_i \mid i \in [N] \text{ and } \pi(X_i) = \texttt{True}\} \cup B' \cup C',$$
$$W^2 := \{f'\} \cup \{x_i, y_i, z_i \mid i \in [N] \text{ and } \pi(X_i) = \texttt{False}\} \cup D, \text{ and}$$
$$W^3 := \{f\} \cup \{x'_i, y'_i, z'_i \mid i \in [N] \text{ and } \pi(X_i) = \texttt{False}\} \cup B \cup C.$$

As t is adjacent to every vertex in $X' \cup Y' \cup Z' \cup D'$, and f' is adjacent to every vertex in $X \cup Y \cup Z \cup D$, W^0 and W^2 are connected sets in G. Further, by construction, $E(W^0, W^2) = \emptyset$ and $E(W^1, W^3) = \emptyset$. W^1 is a connected set since $X \cup Y \cup Z \subseteq N(t')$ and for each $j \in [M]$, there exists $i \in [N]$ such that $x_i \in W^1$ (corresponding to a variable in C_j set to $\texttt{True}$) and $c'_j x_i, b'_j x_i \in E(G)$. Similarly, W^3 is also a connected set. The edges tt' and ff', respectively, ensure that W^0 is adjacent to W^1 and W^3 is adjacent to W^2. As for any $i \in [N]$, x'_i is adjacent with t and f and $x'_i \in W^0 \cup W^3$, it follows that W^0 and W^3 are adjacent. Similarly, W^1 and W^2 are adjacent. Hence, $\{W^0, W^1, W^2, W^3\}$ is a C_4-witness structure. $\qquad\square$

Now, we proceed to show the converse of Lemma 7. We crucially use the properties of a nice C_4-witness structure. This structure along with certain other properties help to obtain a not-all-equal satisfying assignment of ψ.

Lemma 8. *If G is a* YES*-instance of C_4*-CONTRACTIBILITY *then ψ is a* YES*-instance of* POSITIVE NAE-SAT.

Proof. Suppose $\mathcal{W} = \{W^0, W^1, W^2, W^3\}$ is a C_4-witness structure of G where $E(W^i, W^j) \neq \emptyset$ if and only if $j = (i \pm 1) \mod 4$. From Lemmas 5 and 6, we may assume that $\mathcal{W}$ is a nice C_4-witness structure in which every pair of vertices in $\{t, t', f, f'\}$ are in different witness sets. As $\{t, f\}$ and $\{t', f'\}$ are in $\mathcal{D}$, by Property (P1) of a nice C_4-witness structure of G, t and f are in adjacent witness sets and t' and f' are in adjacent witness sets. Hence, without loss of generality, we may assume that $t \in W^0$, $t' \in W^1$, $f' \in W^2$, and $f \in W^3$. Also, by Property (P2) of a nice C_4-witness structure of G, we have $D' \subseteq W^0$ and $D \subseteq W^2$.

For each $i \in [N]$, x'_i is adjacent to t, f and x_i is adjacent to t', f'. Therefore, $x_i \notin W^0 \cup W^3$, $x'_i \notin W^1 \cup W^2$ and we have $X' \subseteq W^0 \cup W^3$ and $X \subseteq W^1 \cup W^2$. Further, since $x_i x'_i \in E(G)$, it follows that $x_i \in W^1$ if and only if $x'_i \in W^0$ and $x_i \in W^2$ if and only if $x'_i \in W^3$. Refer to Fig. 2 for an illustration where these two choices are indicated by the purple (dotted) edges and green (dashed) edges. We will associate these two choices with setting the variable X_i to $\texttt{True}$ or $\texttt{False}$, respectively. Consider a vertex $c_j \in C$ for some $j \in [M]$. As $f' \in W^2$ and $f'c_j \in E(G)$, it follows that $c_j \notin W^0$. Also, since $t \in W^0$ and $\{t, c_j\}$ is in $\mathcal{D}$, by Property (P1) of a nice C_4-witness structure of G, it follows that c_j is not in W^2. As $N(c_j) \subseteq W^0 \cup W^2 \cup W^3$ and $t' \in W^1$, if $c_j \in W^1$, then W^1 cannot be a connected set. Hence, $c_j \in W^3$. As c_j is an arbitrary vertex of C in this reasoning, we have $C \subseteq W^3$. Similarly, $B \subseteq W^3$. This implies $C \cup B \subseteq W^3$. By a symmetric argument, we have $C' \cup B' \subseteq W^1$.

We now construct an assignment $\pi : \{X_1, X_2, \ldots, X_N\} \mapsto \{\text{True}, \text{False}\}$ using $\mathcal{W}$. For every $i \in [N]$, set $\pi(X_i) = \text{True}$ if $x_i \in W^1$ (or equivalently $x_i' \in W^0$) and set $\pi(X_i) = \text{False}$ if $x_i' \in W^3$ (or equivalently $x_i \in W^2$). As mentioned before, $x_i \in W^1$ if and only if $x_i' \in W^0$ and $x_i' \in W^3$ if and only if $x_i \in W^2$. As W^3 is connected and $f, c_j \in W^3$, for every $j \in [M]$, there exists $i \in [N]$, such that $x_i' \in W^3$ and $c_j x_i \in E(G)$. Similarly, as W^1 is connected, for every $j \in [M]$, there exists $i \in [N]$ such that $x_i \in W^1$ and $c_j' x_i \in E(G)$.   □

4   Conclusion and Future Directions

In this work, we showed that C_ℓ-CONTRACTIBILITY is NP-complete on bipartite graphs for $\ell \in \{4, 5\}$ by giving polynomial-time reductions from POSITIVE NAE-SAT.

POSITIVE NAE-SAT (or equivalently, HYPERGRAPH 2-COLORABILITY) has been one of the canonical NP-complete problems in many intractability results on C_ℓ-CONTRACTIBILITY [5,8,9]. In general, in most contraction problems, it is a non-trivial task to forbid certain edges from being contracted in a solution. The simultaneous property of requiring a variable to be True and a variable to be False in every clause of a YES-instance of POSITIVE NAE-SAT helps to encode that certain edges in the output graph of the reduction cannot be contracted, hence, giving a handle on the required structure of the witness sets. This is one of the reasons that makes POSITIVE NAE-SAT an amenable choice in many reductions for graph contractibility problems. However, the sophistication level of the gadgets involved in the reduction increases with the restriction required on the input graph (e.g. bipartite graphs, claw-free graphs). In contrast, the sophistication decreases with increase in the size of the target graph, for instance, the gadgets required for the NP-hardness of C_4-CONTRACTIBILITY are more complex than those needed for C_5-CONTRACTIBILITY, which are more complex that what are required for C_6-CONTRACTIBILITY.

Continuing along the direction of solving cycle contractibility in restricted graph classes, we can also show the following result.

Theorem 3. [⋆] *C_4-CONTRACTIBILITY is NP-complete on K_4-free graphs of diameter 2.*

Theorem 3 can be generalized to show that $K_{p,q}$-CONTRACTIBILITY (the problem of determining if a graph is contractible to the complete bipartite graph with p vertices in one part and q vertices in the other part) is also NP-complete for each $p, q \geq 2$ on K_4-free graphs of diameter 2. Our interest in this restricted case stems from its relationship with DISCONNECTED CUT, the problem of determining if a connected graph G contains a subset $U \subseteq V(G)$ such that both $G[U]$ and $G - U$ are disconnected [17,22,23]. If the diameter of G is 2, then G has a disconnected cut if and only if G is contractible to $K_{p,q}$ for some $p, q \geq 2$ [17,

Proposition 1]. Martin et al. proved that DISCONNECTED CUT is polynomial-time solvable for H-free graphs when $H \neq K_4$ is a graph on at most 4 vertices [23, Theorem 7]. Theorem 3 (and its generalization to $p, q \geq 2$) implies that (p, q)-DISCONNECTED CUT (see [17]) is NP-complete for all $p, q \geq 2$ on K_4-free graphs. Although this falls short of completing the dichotomy of [23, Theorem 7], we believe that it strongly suggests that there is no polynomial-time algorithm for DISCONNECTED CUT on K_4-free graphs.

Finally, determining the longest cycle to which an H-free graph (for a fixed H) is contractible is another interesting future direction. A similar study on H-free graphs in the context of longest paths is known [20]. Note that assuming P$\neq$NP, the complexities of contracting to a longest path and longest cycle do not coincide on H-free graphs.

References

1. Andersson, M., Gudmundsson, J., Levcopoulos, C.: Restricted mesh simplification using edge contractions. Int. J. Comput. Geom. Appl. **19**(3), 247–265 (2009). https://doi.org/10.1142/S0218195909002940
2. Belmonte, R., Golovach, P.A., Heggernes, P., van 't Hof, P., Kamiński, M., Paulusma, D.: Detecting Fixed Patterns in Chordal Graphs in Polynomial Time. Algorithmica **69**(3), 501–521 (2013). https://doi.org/10.1007/s00453-013-9748-5
3. Bernstein, A., Däubel, K., Disser, Y., Klimm, M., Mütze, T., Smolny, F.: Distance-preserving graph contractions. SIAM J. Discrete Math. **33**(3), 1607–1636 (2019). https://doi.org/10.1137/18M1169382
4. Blum, D.J.: Circularity of graphs. Ph.D. thesis, Virginia Polytechnic Institute and State University (1982)
5. Brouwer, A.E., Veldman, H.J.: Contractibility and NP-completeness. J. Graph Theory **11**(1), 71–79 (1987). https://doi.org/10.1002/jgt.3190110111
6. Cheng, S., Dey, T.K., Poon, S.: Hierarchy of surface models and irreducible triangulations. Comput. Geom. **27**(2), 135–150 (2004). https://doi.org/10.1016/j.comgeo.2003.07.001
7. Cong, J., Lim, S.K.: Edge separability-based circuit clustering with application to multilevel circuit partitioning. IEEE Trans. Comput. Aided Des. Integr. Circuits Syst. **23**(3), 346–357 (2004). https://doi.org/10.1109/TCAD.2004.823353
8. Dabrowski, K.K., Paulusma, D.: Contracting bipartite graphs to paths and cycles. Inf. Process. Lett. **127**, 37–42 (2017). https://doi.org/10.1016/j.ipl.2017.06.013
9. Fiala, J., Kaminski, M., Paulusma, D.: A note on contracting claw-free graphs. Discrete Math. Theor. Comput. Sci. **15**(2), 223–232 (2013)
10. Garey, M.R., Johnson, D.S.: Computers and Intractability: A Guide to the Theory of NP-Completeness. W.H Freeman, New York (1979)
11. Hammack, R.H.: Cyclicity of graphs. J. Graph Theory **32**(2), 160–170 (1999). https://doi.org/10.1002/(SICI)1097-0118(199910)32:2⟨160::AID-JGT6⟩3.0.CO;2-U
12. Hammack, R.H.: A note on the complexity of computing cyclicity. Ars Comb. **63**, 89–95 (2002)
13. Harel, D., Koren, Y.: On clustering using random walks. In: Hariharan, R., Vinay, V., Mukund, M. (eds.) FSTTCS 2001. LNCS, vol. 2245, pp. 18–41. Springer, Heidelberg (2001). https://doi.org/10.1007/3-540-45294-X_3

14. Heggernes, P., van 't Hof, P., Lévêque, B., Paul, C.: Contracting chordal graphs and bipartite graphs to paths and trees. Discrete Appl, Math. **164**, 444–449 (2014). https://doi.org/10.1016/j.dam.2013.02.025

15. Hoede, C., Veldman, H.J.: Contraction theorems in Hamiltonian graph theory. Discrete Math. **34**(1), 61–67 (1981). https://doi.org/10.1016/0012-365X(81)90022-4

16. van 't Hof, P., Kaminski, M., Paulusma, D., Szeider, S., Thilikos, D.M.: On graph contractions and induced minors. Discrete Appl. Math. **160**(6), 799–809 (2012). https://doi.org/10.1016/j.dam.2010.05.005

17. Ito, T., Kaminski, M., Paulusma, D., Thilikos, D.M.: Parameterizing cut sets in a graph by the number of their components. Theor. Comput. Sci. **412**(45), 6340–6350 (2011). https://doi.org/10.1016/j.tcs.2011.07.005

18. Kamiński, M., Paulusma, D., Thilikos, D.M.: Contractions of planar graphs in polynomial time. In: de Berg, M., Meyer, U. (eds.) ESA 2010. LNCS, vol. 6346, pp. 122–133. Springer, Heidelberg (2010). https://doi.org/10.1007/978-3-642-15775-2_11

19. Karypis, G., Kumar, V.: A fast and high quality multilevel scheme for partitioning irregular graphs. SIAM J. Sci. Comput. **20**(1), 359–392 (1998). https://doi.org/10.1137/S1064827595287997

20. Kern, W., Paulusma, D.: Contracting to a longest path in H-free graphs. In: Cao, Y., Cheng, S., Li, M. (eds.) 31st International Symposium on Algorithms and Computation, ISAAC 2020, 14–18 December 2020, Hong Kong, China (Virtual Conference). LIPIcs, vol. 181, pp. 22:1–22:18. Schloss Dagstuhl - Leibniz-Zentrum für Informatik (2020). https://doi.org/10.4230/LIPIcs.ISAAC.2020.22

21. Levin, A., Paulusma, D., Woeginger, G.J.: The computational complexity of graph contractions I: polynomially solvable and NP-complete cases. Networks **51**(3), 178–189 (2008). https://doi.org/10.1002/net.20214

22. Martin, B., Paulusma, D.: The computational complexity of disconnected cut and $2K_2$-partition. J. Comb. Theory Ser. B **111**, 17–37 (2015). https://doi.org/10.1016/j.jctb.2014.09.002

23. Martin, B., Paulusma, D., van Leeuwen, E.J.: Disconnected cuts in claw-free graphs. J. Comput. Syst. Sci. **113**, 60–75 (2020). https://doi.org/10.1016/j.jcss.2020.04.005

24. Robertson, N., Seymour, P.D.: Graph Minors. XIII. The Disjoint Paths Problem. J. Comb. Theory Ser. B **63**(1), 65–110 (1995). https://doi.org/10.1006/jctb.1995.1006

25. Schaefer, T.J.: The complexity of satisfiability problems. In: Lipton, R.J., Burkhard, W.A., Savitch, W.J., Friedman, E.P., Aho, A.V. (eds.) Proceedings of the 10th Annual ACM Symposium on Theory of Computing, 1–3 May 1978, San Diego, California, USA. pp. 216–226. ACM (1978). https://doi.org/10.1145/800133.804350

Algorithmic Aspects of Small Quasi-Kernels

Hélène Langlois[1]([✉]) [ID], Frédéric Meunier[1] [ID], Romeo Rizzi[2] [ID],
and Stéphane Vialette[3] [ID]

[1] CERMICS, École des Ponts ParisTech, 77455 Marne-la-Vallée, France
{helene.langlois,frederic.meunier}@enpc.fr
[2] Department of Computer Science, Università di Verona, 37129 Verona, Italy
romeo.rizzi@univr.it
[3] LIGM, Univ Gustave Eiffel, CNRS, 77454 Marne-la-Vallée, France
stephane.vialette@univ-eiffel.fr

Abstract. In a digraph, a quasi-kernel is a subset of vertices that is
independent and such that every vertex can reach some vertex in that
subset via a directed path of length at most two. Whereas Chvátal and
Lovász proved in 1974 that every digraph has a quasi-kernel, very little
is known so far about the complexity of computing small quasi-kernels.
In 1976, Erdős and Székely conjectured that every sink-free digraph has
a quasi-kernel containing at most half of the vertices. Obviously, if a
digraph has two disjoint quasi-kernels then it has such a quasi-kernel
and in 2001, Gutin, Koh, Tay and Yeo conjectured that every sink-free
digraph has two disjoint quasi-kernels. Yet, they constructed in 2004 a
counterexample, thereby disproving this stronger conjecture.

We shall show that not only do sink-free digraphs occasionally fail
to contain two disjoint quasi-kernels, but it is computationally hard to
distinguish those that do from those that do not. We also prove that the
problem of computing a smallest quasi-kernel is computationally hard,
even for restricted classes of acyclic digraphs and for orientations of split
graphs. Finally, we observe that this latter problem is polynomial-time
solvable for graphs with bounded treewidth and identify a class of graphs
with unbounded treewidth for which the problem is also polynomial-time
solvable, namely orientations of complete split graphs.

Keywords: Quasi-kernel · Digraph · Computational complexity

1 Introduction

Let $D = (V, A)$ be a digraph. A *kernel* K is a subset of vertices that is independent (*i.e.*, all pairs of distinct vertices of K are non-adjacent) and such that, for every vertex $v \notin K$, there exists $w \in K$ with $(v, w) \in A$. Kernels were introduced by von Neumann and Morgenstern [20]. It is now a central notion in graph theory and has important applications in relation with colorings [11], perfect graphs [4], game theory and economics [15], logic [21], etc. Clearly, not every digraph has

© Springer Nature Switzerland AG 2022
M. A. Bekos and M. Kaufmann (Eds.): WG 2022, LNCS 13453, pp. 370–382, 2022.
https://doi.org/10.1007/978-3-031-15914-5_27

a kernel (for instance, a directed cycle of odd length does not contain a kernel) and a digraph may have several kernels. Chvátal proved that deciding whether a digraph has a kernel is NP-complete [6] and the problem is equally hard for planar digraphs with bounded degree [10].

Chvátal and Lovász [5] later introduced the notion of quasi-kernels. A *quasi-kernel* in a digraph is a subset of vertices that is independent and such that every vertex can reach some vertex in that set via a directed path of length at most two. Defining the (directed) distance $d(v, w)$ from a vertex v to a vertex w as the minimum length of a directed path from v to w, a quasi-kernel Q is a subset of vertices that is independent and such that for every vertex $v \notin Q$ there exists $w \in Q$ such that $d(v, w) \leqslant 2$. In particular, any kernel is a quasi-kernel. Yet, unlike kernels, every digraph has a quasi-kernel. Chvátal and Lovász provided a proof of this fact, which can be turned into a simple polynomial-time algorithm (alternative simple proofs exist [3]).

In 1976, Erdős and Székely [9] conjectured that a sink-free digraph $D = (V, A)$ (*i.e.*, every vertex of D has positive outdegree) has a quasi-kernel of size at most $|V|/2$. Very recently, Kostochka et al. [17] renewed the interest in the small quasi-kernel conjecture and proved that the conjecture holds for orientations of 4-colorable graphs (in particular, for all planar graphs). The conjecture is however still wide open.

In 2001, Gutin et al. [13] conjectured that every sink-free digraph has two disjoint quasi-kernels (this stronger conjecture implies the original small quasi-kernel conjecture). In 2004, in an update of their paper, the authors constructed a counterexample with 14 vertices [12]. The conjecture about disjoint quasi-kernels holds however for special classes of digraphs; see Heard and Huang [14].

As we shall prove, not only do sink-free digraphs occasionally fail to contain two disjoint quasi-kernels, but it is actually computationally hard to distinguish those that do from those that do not. Whereas the small quasi-kernel conjecture has been established for planar sink-free digraphs, the systematic existence of two disjoint quasi-kernels in such graphs is still unsettled (the counterexample constructed by Gutin et al. does contain a directed K_7). We shall show however that deciding whether a planar digraph has three disjoint quasi-kernels is NP-complete.

In addition to these results on the complexity of deciding the existence of disjoint quasi-kernels, we initiate the study of computing quasi-kernels of minimum size. Surprisingly enough, whereas every digraph has a quasi-kernel, very little was known so far about this problem, which we call MIN-QUASI-KERNEL (and we let QUASI-KERNEL stand for the related decision problem). The main message is that these problems are computationally hard even for very simple digraph classes, e.g., for acyclic orientations of bipartite graphs or for orientations of split graphs. Courcelle's theorem [8] ensures that the problem is polynomial-time solvable for orientations of graphs with bounded treewidth (and a similar result holds for deciding the existence of disjoint quasi-kernels). We show that orientations of complete split graphs form a nontrivial class of graphs with unbounded treewidth for which the problem is polynomial, but identifying other classes of this sort seems to be quite challenging.

We assume that the readers are familiar with standard terms of directed graphs [2], parametrized complexity [8] and approximability theory [19].

2 Disjoint Quasi-Kernels

Our main complexity result about the existence of disjoint quasi-kernels is the following one. Our proof uses the counterexample constructed by Gutin et al. [12].

Theorem 1. *Deciding if a digraph has two disjoint quasi-kernels is* NP-*complete, even for digraphs with maximum outdegree six.*

Proof. Given a Boolean expression F in conjunctive normal form (CNF) where each clause is the disjunction of at most three distinct literals, 3-SAT asks to decide whether F is satisfiable. We reduce from 3-SAT, which is known to be NP-complete [16].

Consider an instance of 3-SAT. Let $X = \{x_1, x_2, \ldots, x_n\}$ be its variables, and let $F = C_1 \vee C_2 \vee \cdots \vee C_m$ be its CNF-formula. We construct a digraph $D = (V, A)$ as follows.

- We start with the gadget D_0 shown on the top part of Fig. 1 which contains the specified vertex b'.
- For every Boolean variable $x_i \in X$ we introduce the gadget D_i shown in the middle part of Fig. 1 which contains two specified vertices $\mathtt{f}_i$ and $\mathtt{t}_i$. Furthermore, we connect D_i to D_0 with two arcs $(\mathtt{f}_i, b')$ and $(\mathtt{t}_i, b')$.
- For every clause $C = \ell_i \vee \ell_j \vee \ell_k$ of F we introduce the gadget D_C shown in the bottom part of Fig. 1 which contains one specified vertex $k_{C,1}$. Furthermore, we connect D_C to the gadgets D_i, D_j, D_k with three arcs $(k_{C,1}, \lambda_i)$, $(k_{C,1}, \lambda_j)$ and $(k_{C,1}, \lambda_k)$, where $\lambda_i = \mathtt{t}_i$ (resp. $\lambda_j = \mathtt{t}_j$ & $\lambda_k = \mathtt{t}_k$) if ℓ_i (resp. ℓ_j & ℓ_k) is a positive literal, and $\lambda_i = \mathtt{f}_i$ (resp. $\lambda_j = \mathtt{f}_j$ & $\lambda_k = \mathtt{f}_k$) if ℓ_i (resp. ℓ_j & ℓ_k) is a negative literal.

Note that for every clause C of F, the digraph D_C is the counterexample constructed by Gutin et al. [12]. It has the important property that any two distinct vertices of $\{k_{C,i} : 1 \leqslant i \leqslant 7\}$ have a common outneighbor in $\{k_{C,i} : 1 \leqslant i \leqslant 7\}$.

It is clear that $|V| = 14m + 6n + 6$ and $|A| = 31m + 11n + 9$. Moreover, D has maximum outdegree six (but it has unbounded indegree; see vertex b'). We claim that the Boolean formula F is satisfiable if and only if the digraph D has two disjoint quasi-kernels.

Suppose that the Boolean formula F is satisfiable and consider any satisfying assignment φ. Construct two subsets $Q_1, Q_2 \subseteq V$ as follows.

- The elements of Q_1 are the following vertices: the vertices b and b' from D_0, the vertex A'_i from D_i for every variable $x_i \in X$, and the vertices $k_{C,6}, s_{C,1}, s_{C,3}$ and $s_{C,7}$ from D_C for every clause C of F.
- The elements of Q_2 are the following vertices: the vertices c and c' from D_0, the vertices A''_i and $\mathtt{t}_i$ from D_i for every variable $x_i \in X$ with $\varphi(x_i) = \mathtt{true}$, or the vertices A''_i and $\mathtt{f}_i$ from D_i with $\varphi(x_i) = \mathtt{false}$ and the vertices $k_{C,7}, s_{C,2}$ and $s_{C,4}$ from D_C for every clause C of F.

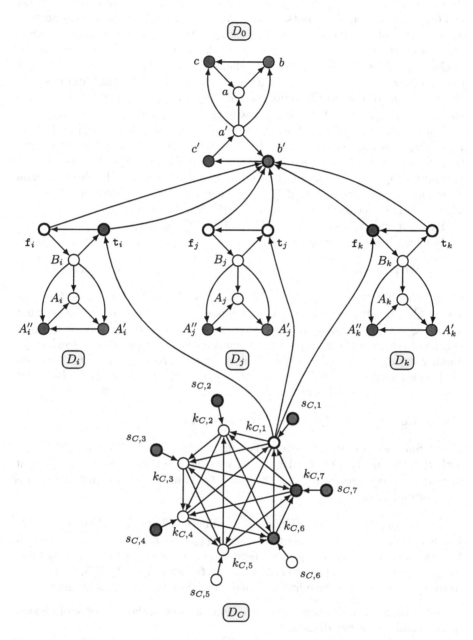

Fig. 1. Proof of Theorem 1: Connecting the gadgets for clause $c = x_i \vee x_j \vee \neg x_k$. Red (resp. Blue) vertices denote vertices in Q_1 (resp. Q_2). Shown here is the case $\varphi(x_i) = \texttt{true}$, $\varphi(x_j) = \texttt{false}$ and $\varphi(x_k) = \texttt{false}$ (*i.e.*, $\mathbf{t}_i \in Q_2$, $\mathbf{f}_j \in Q_2$ and $\mathbf{f}_k \in Q_2$). Note that $\mathbf{f}_j \notin Q_2$ and $\mathbf{t}_j \notin Q_2$ implies $\varphi(x_j) = \texttt{false}$. (Color figure online)

It is a simple matter to check that Q_1 and Q_2 are disjoint and that both Q_1 and Q_2 are independent subsets. Furthermore, we claim that Q_1 and Q_2 are two quasi-kernels of D. The claim is clear for Q_1. As for Q_2, it is enough to show that, for every clause C, the vertex $s_{C,1}$ is at distance at most two of some vertex in Q_2. Indeed, let $C = \ell_i \vee \ell_j \vee \ell_k$ be a clause where ℓ_i, ℓ_j and ℓ_k are positive or negative literals. Since φ is a satisfying assignment, there exists one literal, say ℓ_i, that evaluates to true in the clause C. Therefore, if $\varphi(x_i) = \texttt{true}$ then $\mathbf{t}_i \in Q_2$ and $(k_{C,1}, \mathbf{t}_i) \in A$, and if $\varphi(x_i) = \texttt{false}$ then $\mathbf{f}_i \in Q_2$ and $(k_{C,1}, \mathbf{f}_i) \in A$.

Conversely, suppose that there exist two disjoint quasi-kernels Q_1 and Q_2 in D. We first observe that $Q_1 \cap \{a, b, c\} \neq \varnothing$ and $Q_2 \cap \{a, b, c\} \neq \varnothing$. Then it follows that $a' \notin Q_1 \cup Q_2$ (by independence), and hence $b' \in Q_1 \cup Q_2$. Without loss of generality, suppose $b' \in Q_1$. Define an assignment φ for the Boolean formula F as follows: for $1 \leqslant i \leqslant n$, if $\mathbf{t}_i \in Q_2$ then set $\varphi(x_i) = \texttt{true}$; otherwise set $\varphi(x_i) = \texttt{false}$. Let us show that φ is a satisfying assignment.

By independence, we have $\mathbf{t}_i \notin Q_1$ and $\mathbf{f}_i \notin Q_1$ for $1 \leqslant i \leqslant n$.

We need the following claim.

Claim. We have $\{k_{C,1}, k_{C,2}, k_{C,3}, k_{C,5}\} \cap (Q_1 \cup Q_2) = \varnothing$ for every clause C of F.

Proof. We only prove $k_{C,1} \notin Q_1 \cup Q_2$ (the proof is similar for $k_{C,2} \notin Q_1 \cup Q_2$, $k_{C,3} \notin Q_1 \cup Q_2$ and $k_{C,5} \notin Q_1 \cup Q_2$.) Suppose, aiming at a contradiction, that $k_{C,1} \in Q_1 \cup Q_2$. Without loss of generality we may assume $k_{C,1} \in Q_1$ (the argument is symmetric if $k_{C,1} \in Q_2$). Then it follows that $\{s_{C,2}, s_{C,3}, s_{C,5}\} \subseteq Q_1$, and hence $\{s_{C,2}, s_{C,3}, s_{C,5}\} \cap Q_2 = \varnothing$. But, for any vertex $k_{C,i}$, $2 \leqslant i \leqslant 7$, we can easily check that either $d(s_{C,2}, k_{C,i}) > 2$, or $d(s_{C,3}, k_{C,i}) > 2$, or $d(s_{C,5}, k_{C,i}) > 2$. Hence Q_2 is not a quasi-kernel of D. This is the sought contradiction.   □

Claim. We have $s_{C,1} \in Q_1$ for every clause C of F.

Proof. Suppose, aiming at a contradiction, that $s_{C,1} \notin Q_1$. Combining Claim 2 with $\mathbf{t}_i \notin Q_1$ and $\mathbf{f}_i \notin Q_1$ for $1 \leqslant i \leqslant n$, we conclude that no vertex in Q_1 is at distance at most two from $s_{C,1}$. Therefore, Q_1 is not a quasi-kernel of D. This is a contradiction.   □

Let $C = \ell_i \vee \ell_j \vee \ell_k$ be a clause. According to Claim 2, we have $s_{C,1} \in Q_1$. Furthermore, according to Claim 2, $\{k_{C,1}, k_{C,2}, k_{C,3}, k_{C,5}\} \cap Q_2 = \varnothing$. Then it follows that $\{\lambda_i, \lambda_j, \lambda_k\} \cap Q_2 \neq \varnothing$ where $\lambda_i = \mathbf{t}_i$ (resp. $\lambda_j = \mathbf{t}_j$ & $\lambda_k = \mathbf{t}_k$) if ℓ_i (resp. ℓ_j & ℓ_k) is a positive literal, and $\lambda_i = \mathbf{f}_i$ (resp. $\lambda_j = \mathbf{f}_j$ & $\lambda_k = \mathbf{f}_k$) if ℓ_i (resp. ℓ_j & ℓ_k) is a negative literal. Therefore φ is a satisfying assignment.   □

As noted in the introduction, very little is known about the existence of disjoint quasi-kernels in planar digraphs.

Theorem 2. *Deciding if a digraph has three disjoint quasi-kernels is* NP-*complete, even for bounded degree planar digraphs.*

3   Acyclic Digraphs

In this section, we address the complexity status of QUASI-KERNEL and MIN-QUASI-KERNEL for acyclic orientations of various classes of graphs. The next two theorems show that there is not so much room for extending the polynomiality result about orientations of graphs with bounded treewidth.

We recall that a *cubic* graph is a graph in which every vertex has degree three.

Theorem 3. QUASI-KERNEL *is* NP-*complete, even for acyclic orientations of cubic graphs.*

Assuming FPT $\neq$ W[2], our next result shows that one cannot confine the seemingly inevitable combinatorial explosion of computational difficulty to an additive function of the size of the quasi-kernel, even for restricted digraph classes.

Theorem 4. QUASI-KERNEL *is* W[2]-*complete when the parameter is the size of the sought quasi-kernel, even for acyclic orientations of bipartite graphs.*

We finish the section with a series of propositions providing complementary evidence for the versatile hardness of computing small quasi-kernels.

Recall that a kernel is a quasi-kernel. Actually we have more: a kernel is an inclusion-wise maximal quasi-kernel. Inclusion-wise minimal quasi-kernels are easy to find with a greedy algorithm. Though, finding a minimum-size quasi-kernel included in a kernel is hard as shown by the following result, whose proof is identical to the one of Theorem 4 ($\mathcal{F} \cup \{t\}$ is actually a kernel of the digraph D).

Proposition 1. *Let* $D = (V, A)$ *be an acyclic orientation of a bipartite graph,* $K \subseteq V$ *be a kernel of* D *and* k *be a positive integer. Deciding whether there exists a quasi-kernel included in* K *of size* k *is* W[2]-*complete for parameter* k.

Dinur and Steuer [7] have shown that SET COVER cannot be approximated in polynomial time within a factor of $(1 - \epsilon) \ln(|U|)$ for some constant $\epsilon > 0$ unless P = NP. Moreover, they built an instance of SET COVER where the number of subsets is a polynomial of the universe size. Therefore, the construction used in the proof of Theorem 4 allows us to state the following inapproximability result.

Proposition 2. MIN-QUASI-KERNEL *cannot be approximated in polynomial time within a factor of* $(1 - \epsilon) \ln(|V|)$ *for some constant* $\epsilon > 0$ *unless* P = NP, *even for acyclic orientations of bipartite graphs.*

Our last result focuses on another restricted classes of digraphs, namely acyclic digraphs with bounded degrees. We need a preliminary lemma which we state for general digraphs.

Lemma 1. MIN-QUASI-KERNEL *belongs to* APX *for digraphs with fixed maximum indegrees.*

Proof. Let $D = (V, A)$ be a digraph and $Q \subseteq V$ be a quasi-kernel.

It is clear that $(d^2 + d + 1)|Q| \geqslant |V|$, where d is the maximum indegree of D. Then it follows that any polynomial-time algorithm that computes a quasi-kernel (such as the algorithm proposed by Chvátal and Lovász [5]) is a $(d^2 + d + 1)$-approximation algorithm. □

Proposition 3. MIN-QUASI-KERNEL *is* APX-*complete for acyclic digraphs with maximum indegree three and maximum outdegree two.*

Proof. Membership in APX for acyclic digraphs with fixed indegrees follows from Lemma 1. Specifically, MIN-QUASI-KERNEL for acyclic digraphs with maximum indegree three can be approximated in polynomial time within a factor of 13.

To prove hardness, we L-reduce from VERTEX COVER in cubic graphs which is known to be APX-complete [1]. As defined in [18], letting P and P' be two optimization problems, we say that P L-reduces to P' if there are two polynomial-time alogirhtms f, g, and constants $\alpha, \beta > 0$ such that for each instance I of P : algorithm f produces an instance $I' = f(I)$ of P, such that the optima of I and I', $OPT(I)$ and $OPT(I')$, respectively, satisfy $OPT(I') \leqslant \alpha OPT(I)$ and given any solution of I' with cost c', algorithm g produces a solution of I with cost c such that $|c - OPT(I)| \leqslant \beta |c' - OPT(I')|$. Let f be the following L-reduction from VERTEX COVER in cubic graphs to MIN-QUASI-KERNEL with maximum indegree three. Given a cubic graph $G = (V, E)$ with $V = [n]$ and m edges, we construct a digraph $D = (V', A)$ as follows:

$$V' = \{w_i, w_i', w_i'' : 1 \leqslant i \leqslant n\} \cup \{z_e, z_e' : e \in E\},$$
$$A = \{(w_i, w_i'), (w_i', w_i'') : 1 \leqslant i \leqslant n\} \cup \{(z_e', z_e), (z_e, w_i), (z_e, w_j) : e = ij \in E\}.$$

Note that the vertices w_i'' are sinks in D. It is clear that $|V'| = 3n + 2m$, $|A| = 2n + 3m$ and, since G is a cubic graph, that every vertex has maximum indegree three in D. We also observe that the maximum outdegree is two in D. See Fig. 2 for an example.

Consider a quasi-kernel $Q \subseteq V'$ of $D = f(G)$. We claim that it can be transformed in polynomial time into a vertex cover $C \subseteq V$ of G such that $|C| \leqslant |Q|$. To see this, observe first that Q can be transformed in polynomial time into a quasi-kernel $Q' \subseteq V'$ such that (i) $|Q'| \leqslant |Q|$ and (ii) $z_e' \notin Q'$ and $z_e \notin Q'$ for every $e \in E$. Indeed, repeated applications of the following two procedures enable us to achieve the claimed quasi-kernel.

- Suppose that there exists $z_e' \in Q$ for some $e = ij \in E$. Then it follows that $z_e \notin Q$ (by independence). Furthermore, we have $w_i'' \in Q$ and $w_j'' \in Q$, and hence $w_i' \notin Q$ and $w_j' \notin Q$. Therefore, $w_i \in Q$ or $w_j \in Q$ (possibly both). On account of the above remarks, $Q' = Q \setminus \{z_e'\}$ is a quasi-kernel of D and $|Q'| < |Q|$.
- Let $Z_i \subseteq Q$ stand for the set of vertices $z_e \in Q$, where e is an edge incident to the vertex i in G. Suppose that there exists some set $Z_i \neq \varnothing$. Then it follows that $w_i \notin Q$ (by independence). Furthermore, we have $w_i'' \in Q$, and hence $w_i' \notin Q$. On account of the above remarks, $Q' = (Q \setminus Z_i) \cup \{w_i\}$ is a quasi-kernel of D and $|Q'| \leqslant |Q|$.

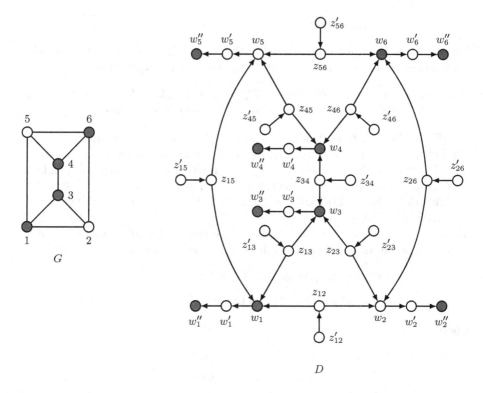

D

Fig. 2. Example of the construction presented in the proof of Proposition 3.

From such a Q', construct then a vertex cover $C \subseteq V$ of G as follows: for $1 \leqslant i \leqslant n$, add the vertex i to C if $w_i \in Q'$. By construction, C is a vertex cover of G of size $|C| = |Q'| - |V|$

Finally, it is easy to see that from a vertex cover $C \in V$ of G we can construct a quasi-kernel $Q \subseteq V'$ of $D = f(G)$ of size exactly $|C| + |V|$: for every $1 \leqslant i \leqslant n$, add w_i'' to Q and add w_i to Q if $i \in C$. Since G is a cubic graph, we have $|C| \geqslant |V|/4$, and hence $|Q| = |C| + |V| \leqslant |C| + 4|C| = 5|C|$.

Thus $\mathrm{opt}(f(G)) \leqslant 5\,\mathrm{opt}(G)$ and we have shown that f is an L-reduction with parameters $\alpha = 5$ and $\beta = 1$.   □

4   Orientations of Split Graphs

In this section, we focus on orientations of split graphs. A graph is a split graph if it can be partitioned in an independent set and a clique. This class seems to play an important role in the study of small quasi-kernels since the only examples of oriented graphs having no two disjoint quasi-kernels contain the orientation of a split graph constructed by Gutin et al. [12].

4.1   Computational Hardness

We first show that one cannot confine the seemingly inevitable combinatorial explosion of computational difficulty to the size of the sought quasi-kernel.

Proposition 4. QUASI-KERNEL *is* W[2]*-complete when the parameter is the size of the sought quasi-kernel even for orientations of split graphs.*

Proof. Membership in W[2] is clear. Given a digraph $D = (V, A)$ and an integer q, DIRECTED DOMINATING SET is the problem of deciding if there exists a subset $S \subseteq V$ of size q such that every vertex $v \in V$ is either in S or has an outneighbor in S. DIRECTED DOMINATING SET is W[2]-complete for parameter q [8].

We reduce DIRECTED DOMINATING SET to QUASI-KERNEL. Let $D = (V, A)$ be a digraph and q be a positive integer. Write $n = |V|$, $V = \{v_i : 1 \leqslant i \leqslant n\}$ and $m = |A|$, and set $b = 2q + 3$. Let $\prec$ be some arbitrary total order on A. Define an orientation of a split graph $D' = (V', A')$ as follows:

$$V' = \{s\} \cup S^1 \cup S^2 \cup K^1 \cup K^2$$
$$A' = A(s) \cup A\left(S^1\right) \cup A\left(S^2\right) \cup A\left(K^1\right) \cup A\left(K^2\right)$$

where

$$S^1 = \left\{s_i^1 : 1 \leqslant i \leqslant n\right\}$$
$$S^2 = \left\{s_i^2 : 1 \leqslant i \leqslant b\right\}$$
$$K^1 = \left\{k_{i,j}^1 : (v_i, v_j) \in A\right\}$$
$$K^2 = \left\{k_i^2 : 1 \leqslant i \leqslant b\right\}$$

and

$$A(s) = \left\{\left(s, k_{i,j}^1\right) : (v_i, v_j) \in A\right\} \cup \left\{\left(k_i^2, s\right) : 1 \leqslant i \leqslant b\right\}$$
$$A\left(S^1\right) = \left\{\left(s_i^1, k_{i,j}^1\right) : (v_i, v_j) \in A\right\} \cup \left\{\left(k_{i,j}^1, s_j^1\right) : (v_i, v_j) \in A\right\}$$
$$A\left(S^2\right) = \left\{\left(s_i^2, k_i^2\right) : 1 \leqslant i \leqslant b\right\}$$
$$A\left(K^1\right) = \left\{\left(k_{i,j}^1, k_{i',j'}^1\right) : (v_i, v_j) \in A, (v_{i'}, v_{j'}) \in A, (v_i, v_j) \prec (v_{i'}, v_{j'})\right\} \cup$$
$$\qquad \left\{(k_{i,j}^1, k_l^2) : (v_i, v_j) \in A, 1 \leqslant l \leqslant b\right\}$$
$$A\left(K^2\right) = \left\{(k_i^2, k_j^2) : 1 \leqslant i < j \leqslant b, i \equiv j \pmod 2\right\} \cup$$
$$\qquad \left\{(k_j^2, k_i^2) : 1 \leqslant i < j \leqslant b, i \not\equiv j \pmod 2\right\}.$$

Clearly, D' is an orientation of a split graph (*i.e.*, $\{s\} \cup S^1 \cup S^2$ is an independent set and $K^1 \cup K^2$ induces a tournament), $|V'| = n + m + 2b + 1$ and $|A'| = \binom{m+b}{2} + 2m + 2b$.

We claim that there exists a dominating set of size q in D if and only if D' has a quasi-kernel of size $q + 1$.

Suppose first that there exists a directed dominating set $M \subseteq V$ of size q in D. Define $Q = \{s\} \cup \left\{s_i^1 : v_i \in M\right\}$. We note that $Q \subseteq \{s\} \cup S^1$, and hence Q is an independent set. Furthermore, by construction, the vertex s is at distance at most two from every vertex in $S^2 \cup K^1 \cup K^2$. Since M is a directed dominating set, it is now clear that Q is a quasi-kernel of D' of size $q+1$.

Conversely, suppose that there exists a quasi-kernel $Q \subseteq V'$ of size $q+1$ in D'. By independence of Q, we have $|Q \cap (K^1 \cup K^2)| \leqslant 1$. We first claim that $s \in Q$. Indeed, suppose, aiming at a contradiction, that $s \notin Q$. Let $X = S^2 \setminus Q$. By construction, $N^+(X) = \{k_i^2 \in K^2 : s_i^2 \in X\}$. Furthermore, $|X| \geqslant |S^2| - |Q| = b - (q+1) = q+2$, and hence $|N^+(X)| \geqslant q+2$. Since $|X|$ is strictly positive, there exists $k_j^2 \in K^2 \cap Q$ such that $\{k_j^2 \in K^2 : s_i^2 \in X\} \subseteq N^-[k_j^2]$. But, according to the definition of $A(K^2)$, $|N^-[k_j^2] \cap K^2| \leqslant \lfloor b/2 \rfloor < q+2$ for every $k_j^2 \in K^2$. This is a contradiction and hence $s \in Q$. We now observe that $k_i^1 \in N^+(s)$ for every $k_i^1 \in K^1$ and $s \in N^+(k_j^2)$ for every $k_j^2 \in K^2$. Combining this observation with $s \in Q$ and the independence of Q, we obtain $Q \cap (K^1 \cup K^2) = \varnothing$. Furthermore, since the vertex s is at distance at most two from every vertex in S^2, we may safely assume that $(S^2 \cup K^1 \cup K^2) \cap Q = \varnothing$ and hence $|S^1 \cap Q| = q$. We now turn to S^1. It is clear that s is at distance three from every vertex $s_i^1 \in S_1$. Therefore, by definition of quasi-kernels, for every vertex $s_i^1 \in S^1 \setminus Q$, there exists one vertex $s_j^1 \in S^1 \cap Q$ such that $(s_i^1, k_{i,j}^1) \in A'$ and $(k_{i,j}^1, s_j^1) \in A'$. Note that, by construction, $(s_i^1, k_{i,j}^1)$ and $(k_{i,j}^1, s_j^1)$ are two arcs of D' if and only if (v_i, v_j) is an arc of D. Then it follows that $M = \{v_i : s_i^1 \in Q\}$ is a directed dominating set in D. $\qquad\square$

Proposition 5. QUASI-KERNEL *for orientations of split graphs is* FPT *for parameter $|K|$ or parameter $k + |I|$, where K is the set of vertices in the clique-part, I is the set of vertices in the independent-part and k is the size of the sought quasi-kernel.*

Proof. Let $D = (K \cup I, A)$ be an orientation of a split graph, and write $n = |K \cup I|$. Let M be the adjacency matrix of D. It is clear that, after having computed M^2, one can decide in linear time if any given subset $Q \subseteq K \cup I$ is a quasi-kernel of D. This preprocessing step is $O(n^3)$ time (a better running time can be achieved by fast matrix multiplication but is not relevant here). Furthermore, by independence of quasi-kernels, we have $|Q \cap K| \leqslant 1$ for every quasi-kernel Q of D. This straightforward observation is the first step of the two algorithms.

Algorithm for parameter $|K|$. Select (including none) a vertex of K. Define the equivalence relation $\sim$ on I as follows: $s \sim s'$ if and only if $N^-(s) = N^-(s')$ and $N^+(s) = N^+(s')$. The key point is to observe that in any minimum cardinality quasi-kernel Q of D, for every equivalence class $I' \in I/\sim$, either $I' \cap Q = \varnothing$, $I' \subseteq Q$ or any vertex of I' is in Q. For any combination, check if the selected vertices of I together with the selected vertex of K (if any) is a quasi-kernel of D of size k. The size of $I/\sim$ is bounded by $4^{|K|}$ since each equivalence class is determined by its out and inneighborhood. The algorithm is $O(n^3 + k|K| 3^{|I/\sim|}) = O(n^3 + k|K| 3^{(4^{|K|})})$ time.

Algorithm for parameter $k + |I|$. Select (including none) a vertex of K. For every subset $I' \subseteq I$ of size $k-1$ (or k, if no vertex of K is selected), check if I' together with the selected vertex of K is a quasi-kernel of D. The algorithm is $O(n^3 + k|K| \binom{|I|}{k})$ time. $\qquad\square$

4.2   Complete Split Graphs

A split graph is *complete* if every vertex in the independent-part is adjacent to every vertex in the clique-part. Before stating our main result about this class we start with a lemma that will play a role in its proof.

Lemma 2. *Let $D = (V, A)$ be an orientation of a complete split graph with no quasi-kernel of size one. Let x be a vertex with the maximum number of inneighbors in the clique-part. Denote by S the set of vertices with the same inneighborhood as x (including x). Then,*

- *the set S is included in the independent-part.*
- *every vertex v in S forms a quasi-kernel of $D[(V \setminus S) \cup \{v\}]$.*

Proof. Denote by K (resp. I) the set of vertices in the clique-part (resp. independent-part) of D. Suppose, aiming for a contradiction, that there is a vertex v in K with the maximum number of inneighbors in K. Then $\{v\}$ is a quasi-kernel because every vertex in $N^+(v)$ has an outneighbor in $N^-(v)$, by the maximality of v; a contradiction with D having no quasi-kernel of size one. This proves the first item.

Consider a vertex v in S and a vertex u not in S. Suppose first that u is in K. We have just seen that u has fewer inneighbors in K than v. Thus, u has an outneighbor in $N^-(v) \cup \{v\}$. Suppose now that u is in I. Since u is not in S, it has an outneighbor in $N^-(v)$. In any case, there is a path of length at most two from u to v. This proves the second item.   □

A consequence of the next theorem is that QUASI-KERNEL is polynomial-time solvable for complete split digraphs.

Theorem 5. *Let D be an orientation of a complete split graph. If D has a sink, then there is a unique minimum-size quasi-kernel, which is formed by all sinks. If D has no sink, then the minimum size of a quasi-kernel is at most two.*

Proof. Observe that in an orientation of a complete split graph, if a vertex is a sink, then there is a path of length two from every other non-sink vertex to this sink. Thus, if D has at least one sink, then there is a unique inclusionwise minimal quasi-kernel, which is formed by all sinks. Assume from now on that D has no sink and no quasi-kernel of size one. We are going to show that D has a quasi-kernel of size two.

Denote by K the vertices in the clique-part of D and by I the vertices in the independent-part of D. Let x be a vertex maximizing $|N^-(x) \cap K|$. We know from Lemma 2 that x is in I.

Suppose now, aiming for a contradiction, that every vertex v in I is such that $N^+(x) \subseteq N^+(v)$. Choose any vertex y in $N^+(x)$. The singleton $\{y\}$ is no quasi-kernel of $D[K]$, since otherwise it would be a quasi-kernel of D of size one. A well-known consequence of the proof of Chvátal and Lovász is that in a digraph every vertex is in a quasi-kernel or has an outneighbor in a quasi-kernel. Thus, there exists a vertex z in $N^+(y) \cap K$ that forms a quasi-kernel of $D[K]$.

The singleton $\{z\}$ is then a quasi-kernel of D as well since every vertex of I has y as outneighbor; a contradiction.

Hence, there is a vertex t in I with $N^+(x) \cap N^-(t) \neq \varnothing$. We claim that $\{x, t\}$ is a quasi-kernel of D. It is an independent set. Let S be the set of vertices having the same inneighborhood as x. Consider a vertex v in $V \setminus \{x, t\}$. If v is in S, then by definition of t there is a directed path of length two from v to t. If v is in $V \setminus S$, Lemma 2 ensures that there is a directed path of length at most two from v to x. □

Orientations of complete split graphs always have two disjoint quasi-kernels when there is no sink. This is a consequence of a result by Heard and Huang [14]. The existence of two disjoint quasi-kernels for this class of digraphs can thus trivially be decided in polynomial time. Their proof provides actually a polynomial-time algorithm for finding such quasi-kernels.

5 Concluding Remarks

We mentioned in the introduction that Gutin et al. conjectured in 2001 that every sink-free digraph has two disjoint quasi-kernel and that they disproved this conjecture with a counterexample a few years later. Yet, the key element of the counterexample is the presence of a K_7. On the other hand, the small quasi-kernel conjecture is true for sink-free orientations of 4-colorable graphs, which have no K_5. (This is the result of Kostochka et al., also mentioned in the introduction.) This raises the question on whether every sink-free K_5-free digraph has two disjoint quasi-kernels, and, more generally, on how disjoint quasi-kernels and the clique number relate.

References

1. Alimonti, P., Kann, V.: Hardness of approximating problems on cubic graphs. In: Bongiovanni, G., Bovet, D.P., Di Battista, G. (eds.) CIAC 1997. LNCS, vol. 1203, pp. 288–298. Springer, Heidelberg (1997). https://doi.org/10.1007/3-540-62592-5_80
2. Bang-Jensen, J., Gutin, G.Z.: Digraphs: Theory Algorithms and Applications. Springer, Berlin (2008)
3. Bondy, J.A.: Short proofs of classical theorems. J. Graph Theor. 44(3), 159–165 (2003)
4. Boros, E., Gurvich, V.: Perfect graphs, kernels, and cores of cooperative games. Discrete Math. 306(19–20), 2336–2354 (2006)
5. Chvátal, V., Lovász, L.: Every directed graph has a semi-kernel. In: Berge, C., Ray-Chaudhuri, D. (eds.) Hypergraph Seminar. LNM, vol. 411, pp. 175–175. Springer, Heidelberg (1974). https://doi.org/10.1007/BFb0066192
6. Chvátal, V.: On the computational complexity of finding a kernel. Technical Report CRM300, Centre de Recherches Mathématiques, Université de Montréal (1973)
7. Dinur, I., Steurer, D.: Analytical approach to parallel repetition. In: Proceedings of the Forty-sixth Annual ACM Symposium on Theory of Computing, pp. 624–633 (2014)

8. Downey, R.G., Fellows, M.R.: Fundamentals of Parameterized Complexity. TCS, Springer, London (2013). https://doi.org/10.1007/978-1-4471-5559-1
9. Erdős, P.L., Székely, L.A.: Two conjectures on quasi-kernels, open problems no. 4. in fete of combinatorics and computer science. Bolyai Society Mathematical Studies (2010)
10. Fraenkel, A.S.: Planar kernel and Grundy with $d \leq 3$, $d_{out} \leq 2$, $d_{in} \leq 2$ are NP-complete. Discrete Appl. Math. **3**(4), 257–262 (1981)
11. Galvin, F.: The list chromatic index of a bipartite multigraph. J. Combin. Theory Ser. B **63**, 153–158 (1995)
12. Gutin, G., Koh, K.M., Tay, E.G., Yeo, A.: On the number of quasi-kernels in digraphs, Rep. Ser. 01-7 (2001)
13. Gutin, G., Koh, K.M., Tay, E.G., Yeo, A.: On the number of quasi-kernels in digraphs. J. Graph Theor. **46**(1), 8–56 (2004)
14. Heard, S., Huang, J.: Disjoint quasi-kernels in digraphs. J. Graph Theor. **58**(3), 251–260 (2008)
15. Igarashi, A.: Coalition formation in structured environments. In: Proceedings of the 16th Conference on Autonomous Agents and Multiagent Systems, pp. 1836–1837 (2017)
16. Karp, R.: Reducibility among combinatorial problems. In: Miller, R., Thatcher, J. (eds.) Complexity of Computer Computations, pp. 85–103. Plenum Press (1972). https://doi.org/10.1007/978-1-4684-2001-2_9
17. Kostochka, A., Luo, R., Shan, S.: Towards the Small Quasi-Kernel Conjecture. arXiv:2001.04003 [math] (2020)
18. Papadimitriou, C.H., Yannakakis, M.: Optimization, approximation, and complexity classes. J. Comput. Syst. Sci. **43**(3), 425–440 (1991)
19. Vazirani, V.V.: Approximation Algorithms. Springer, Heidelberg (2001). https://doi.org/10.1007/978-3-662-04565-7
20. Morgenstern, O., Von Neumann, J.: Theory of Games and Economic Behavior. Princeton University Press, Princeton (1947)
21. Walicki, M., Dyrkolbotn, S.: Finding kernels or solving SAT. J. Discret. Algorithms **10**, 146–164 (2012)

Parameterized Complexity of Graph Planarity with Restricted Cyclic Orders

Giuseppe Liotta[1] , Ignaz Rutter[2] , and Alessandra Tappini[1(✉)]

[1] Dipartimento di Ingegneria, Università degli Studi di Perugia, Perugia, Italy
{giuseppe.liotta,alessandra.tappini}@unipg.it
[2] Faculty of Computer Science and Mathematics, University of Passau,
Passau, Germany
rutter@fim.uni-passau.de

Abstract. We study the complexity of testing whether a biconnected graph $G = (V, E)$ is planar with the additional constraint that some cyclic orders of the edges incident to its vertices are allowed while some others are forbidden. The allowed cyclic orders are conveniently described by associating every vertex v of G with a set $D(v)$ of FPQ-trees. Let tw be the treewidth of G and let $D_{\max} = \max_{v \in V} |D(v)|$, *i.e.*, the maximum number of FPQ-trees per vertex. We show that the problem is FPT when parameterized by $tw + D_{\max}$; for a contrast, we prove that the problem is paraNP-hard when parameterized by $D_{\max}$ only and it is W[1]-hard when parameterized by tw only. We also apply our techniques to the problem of testing whether a clustered graph is NodeTrix planar with fixed sides. We extend a result by Di Giacomo et al. [Algorithmica, 2019] and prove that NodeTrix planarity with fixed sides is FPT when parameterized by the size of the clusters plus the treewidth of the graph obtained by collapsing these clusters to single vertices, provided that this graph is biconnected.

Keywords: Planarity Testing · Embedding Constraints · NodeTrix

1 Introduction

The study of graph planarity testing and of its variants is at the heart of graph algorithms. Mostly motivated by graph drawing applications, constrained versions of graph planarity testing have been extensively studied in the literature. They include, for example, rectilinear planarity testing (see, *e.g.*, [23,24,30,32]), upward planarity testing (see, *e.g.*, [7,22,29,30]), and clustered planarity testing (see, *e.g.*, [8,9,13,28]). See also [37,38] for more results and references.

This paper studies the complexity of a fundamental, but not yet completely explored, constrained planarity testing problem: Given a graph G such that

This work was partially supported by: (*i*) MIUR, grant 20174LF3T8; (*ii*) Dipartimento di Ingegneria - Università degli Studi di Perugia, grants RICBA20EDG and RICBA21LG; (*iii*) German Science Foundation (DFG), grant Ru 1903/3-1.

M. A. Bekos and M. Kaufmann (Eds.): WG 2022, LNCS 13453, pp. 383–397, 2022.
https://doi.org/10.1007/978-3-031-15914-5_28

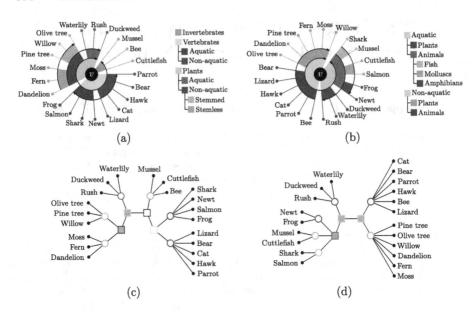

Fig. 1. (a)–(b) Two allowed cyclic orders for the neighbors of a vertex representing different living things and tree-like structures describing different hierarchical groupings of the attributes; (c)–(d) two FPQ-trees representing the allowed cyclic orders, where F-nodes are gray-filled boxes, Q-nodes are white-filled boxes, P-nodes are white-filled disks, and leaves are small black disks. (Color figure online)

each vertex v of G is equipped with a set of allowed cyclic orders for its incident edges, we want to test whether G admits a planar embedding that uses the allowed orders. Besides its theoretical interest, the question is motivated by those applications of information visualization where different cyclic orders around the vertices of a network help to convey the semantic properties of its vertices and edges. For instance, in a knowledge graph vertices and edges are typically equipped with several attributes and different cyclic orders around the vertices correspond to different hierarchical groupings based on the meaning of such attributes.

Consider, for example, vertex v in Figs. 1(a) and 1(b) whose neighbors represent different living things and are equipped with attributes that specify whether they are animals or plants, whether they are aquatic or not, and whether they are stemless plants or vertebrate animals. In the visualization of Fig. 1(a) the cyclic order around v groups together the neighbors corresponding to invertebrate animals which precede the vertebrates and follow the plants in clockwise order; among the plants, the aquatic plants follow the non-aquatic plants, and among the non-aquatic plants those that are stemless precede the stemmed ones; within each subgroup, any permutation is allowed as long as it makes it possible to construct a planar drawing of a larger graph that includes v and its neighbors. Conversely, the cyclic order of Fig. 1(b) is such that aquatic living things follow the non-aquatic ones and, for each of these two groups, animals precede plants; again, within each subgroup any permutation that guarantees planarity is possible.

The allowed cyclic orders for each vertex v of G can be conveniently represented by associating v with a set of FPQ-trees [36], a data structure that generalizes the classical PQ-trees [11] and effectively encodes edge permutations. The leaves of each FPQ-tree associated with v are the edges incident to v, while each non-leaf node is either a *P-node*, or a *Q-node*, or an *F-node*. The children of a P-node can be arbitrarily permuted, the order of the children of a Q-node is fixed up to reversal, while an F-node corresponds to a permutation of its children that cannot be changed. For example, the two FPQ-trees of Figs. 1(c) and 1(d) encode two allowed sets of cyclic orders for the vertex v of Figs. 1(a) and 1(b). In Figs. 1(c) and 1(d), F-nodes are gray-filled boxes, Q-nodes are white-filled boxes, P-nodes are white-filled disks, and leaves are small black disks. Observe that the cyclic order of Fig. 1(a) is one of those described by the FPQ-tree of Fig. 1(c), while the cyclic order of Fig. 1(b) is one of those described by the FPQ-tree of Fig. 1(d). More formally, we study the following problem.

FPQ-CHOOSABLE PLANARITY
Input: A pair (G, D) where $G = (V, E)$ is a (multi-)graph and D is a mapping that associates each vertex $v \in V$ with a set $D(v)$ of FPQ-trees whose leaves represent the edges incident to v.
Question: Does there exist a planar embedding of G such that, for each $v \in V$, the cyclic order of the edges incident to v is encoded by an FPQ-tree in $D(v)$?

We study the parameterized complexity of FPQ-CHOOSABLE PLANARITY with respect to two natural parameters: The treewidth tw of G and the maximum number $D_{\max}$ of FPQ-trees per vertex. We remark that the special case where $D_{\max} = 1$ is linear-time solvable even for graphs of unbounded treewidth [31]. The following theorem summarizes our main contribution.

Theorem 1. FPQ-CHOOSABLE PLANARITY *for biconnected graphs is paraNP-hard when parameterized by the maximum number* $D_{\max}$ *of FPQ-trees per vertex, it is* W[1]-*hard when parameterized by the treewidth* tw *of the graph, and it is fixed-parameter tractable when parameterized by* tw + $D_{\max}$.

As an application of Theorem 1, we shed new light on the complexity of another constrained planarity testing problem, namely NODETRIX PLANARITY WITH FIXED SIDES, which we briefly describe below.

Let G be a clustered graph, *i.e.*, a graph whose vertex set is partitioned into subsets called clusters. A NODETRIX representation of G represents clusters as adjacency matrices, while the edges connecting different clusters are Jordan arcs. NODETRIX PLANARITY asks whether G admits a NODETRIX representation without edge crossings. The question can be asked in the "fixed sides" scenario and in the "free sides" scenario. The former specifies, for each edge e between two matrices M and M', the sides ("top", "bottom", "left", "right") of M and M' to which e must be incident; in the free sides scenario the algorithm can choose the sides to which e is incident. We focus on the scenario with fixed sides and, by combining results of [15,21] with our techniques, we derive the following.

Theorem 2. NODETRIX PLANARITY WITH FIXED SIDES *is paraNP-hard when parameterized by the maximum size* k *of the clusters; it is also paraNP-hard* *when parameterized by the treewidth* tw *of the graph* G_C *obtained by collapsing every cluster into a single vertex; the problem is fixed-parameter tractable when parameterized by* tw $+ k$, *provided that* G_C *is biconnected.*

Theorem 2 extends a result of [21], where it is proved that NODETRIX PLANARITY WITH FIXED SIDES can be solved in polynomial time when the size of the clusters is bounded by a constant and the treewidth of the graph obtained by collapsing each cluster into a single vertex is two.

We remark that Theorems 1 and 2 contribute to the flourishing literature that studies the parameterized complexity of graph drawing problems (see, *e.g.*, [6,12, 20,26,27,34]). Our paper can also be related to [8,10,36], which study planarity problems where vertices are equipped with one FPQ-tree but several constraints in addition to those imposed by the planar embedding are taken into account.

The rest of the paper is organized as follows. Section 2 reports preliminary definitions; Sect. 3 introduces FPQ-CHOOSABLE PLANARITY and studies its computational complexity; Sect. 4 describes an FPT approach for FPQ-CHOOSABLE PLANARITY; Sect. 5 analyzes the interplay between FPQ-CHOOSABLE PLANARITY and NODETRIX PLANARITY; open problems are in Sect. 6. Some proofs are omitted or sketched; their statements are marked with [*].

2   Preliminaries

We assume familiarity with graph theory and algorithms (see, *e.g.*, [4,16]).

FPQ-Tree. A *PQ-tree* is a tree-based data structure that represents a family of permutations on a set of elements [11]. In a PQ-tree, each element is represented by one of the leaf nodes and each non-leaf node is a *P-node* or a *Q-node*. The children of a P-node can be permuted arbitrarily, while the order of the children of a Q-node is fixed up to reversal. An *FPQ-tree* is a PQ-tree where, for some of the Q-nodes, the reversal of the permutation described by their children is not allowed. To distinguish these special Q-nodes, we call them *F-nodes*.

SPQR-Tree. Let G be a biconnected planar multi-graph. The *SPQR-tree* $\mathcal{T}$ of G describes the structure of G in terms of its triconnected components (see, *e.g.*, [16,17]). The tree $\mathcal{T}$ can be computed in linear time and it has three types of internal nodes that correspond to different arrangements of the components of G. If the components are arranged in a cycle, they correspond to an *S-node* of $\mathcal{T}$; if they share two vertices and are arranged in parallel, they correspond to a *P-node* of $\mathcal{T}$; if they are arranged in a triconnected graph, they correspond to an *R-node* of $\mathcal{T}$. The leaves of $\mathcal{T}$ are *Q-nodes*, and each of them corresponds to an edge of G. For each node μ of $\mathcal{T}$, the *skeleton* of μ is an auxiliary graph that represents the arrangement of the triconnected components of G corresponding to μ, and it is denoted by skel(μ). Each edge of skel(μ) corresponds to one such triconnected component and is called a *virtual edge*; the end-points of a virtual edge are called *poles*. The tree $\mathcal{T}$ encodes all possible planar combinatorial embeddings of G. These embeddings are determined by P- and R-nodes, since the skeletons of S- and Q-nodes have a unique embedding. Indeed, the skeleton of a P-node

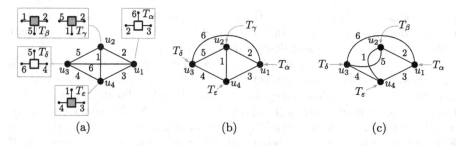

Fig. 2. (a) An FPQ-choosable planar graph (G, D). (b) A planar embedding of G that is consistent with assignment $\{A(u_1) = T_\alpha, A(u_2) = T_\gamma, A(u_3) = T_\delta, A(u_4) = T_\varepsilon\}$; the assignment is compatible with G. (c) A non-planar embedding of G that is consistent with assignment $\{A(u_1) = T_\alpha, A(u_2) = T_\beta, A(u_3) = T_\delta, A(u_4) = T_\varepsilon\}$; there is no planar embedding that is consistent with A.

consists of parallel edges that can be arbitrarily permuted, while the skeleton of an R-node is triconnected, and hence it has a unique embedding up to a flip.

Embedding Tree. The planar combinatorial embeddings that are represented by the SPQR-tree of a biconnected graph G define constraints on the cyclic order of edges around each vertex of G. Such constraints can be encoded by associating an FPQ-tree with each vertex v of G, called *embedding tree* of v and denoted by T_v^ϵ.

3 FPQ-Choosable Planarity and Its Complexity

Let $G = (V, E)$ be a multi-graph, let $v \in V$, and let T_v be an FPQ-tree whose leaf set is $E(v)$, *i.e.*, the set of the edges incident to v. Let σ_v be a cyclic order of the edges incident to v. If σ_v is in a bijection with a permutation of the leaves of T_v, we say that T_v *represents* σ_v or, equivalently, that σ_v *is represented by* T_v. We define *consistent*(T_v) as the set of all cyclic orders of the edges incident to v that are represented by T_v. Given a planar embedding $\mathcal{E}$ of G, we denote by $\mathcal{E}(v)$ the cyclic order of edges incident to v in $\mathcal{E}$.

An *FPQ-choosable graph* is a pair (G, D) where $G = (V, E)$ is a multi-graph, and D is a mapping that associates each vertex $v \in V$ with a set $D(v)$ of FPQ-trees whose leaf set is $E(v)$. An *assignment* A is a function that assigns to each vertex $v \in V$ an FPQ-tree in $D(v)$. We say that A is *compatible with G* if there exists a planar embedding $\mathcal{E}$ of G such that $\mathcal{E}(v) \in consistent(A(v))$ for all $v \in V$. In this case, we also say that $\mathcal{E}$ is *consistent with A*. An FPQ-choosable graph (G, D) is *FPQ-choosable planar* if there exists an assignment of FPQ-trees that is compatible with G. Figure 2(a) shows an FPQ-choosable planar graph G. It has two possible assignments that differ by the FPQ-tree chosen from $D(u_2)$. As Figs. 2(b) and 2(c) show, one of them is compatible with G, while there is no planar embedding that is consistent with the other assignment.

The FPQ-CHOOSABLE PLANARITY problem asks whether an FPQ-choosable graph (G, D) is FPQ-choosable planar, *i.e.*, whether there exists an assignment that is compatible with G. Clearly, G must be planar or else the problem becomes trivial. Also, any assignment that is compatible with G must define a planar embedding of G among those described by an SPQR-tree of G.

Therefore, a preliminary step for an algorithm that tests whether (G, D) is FPQ-choosable planar is to intersect each FPQ-tree $T_v \in D(v)$ with the embedding tree T_v^ϵ of v, so that the cyclic order of the edges incident to v satisfies both the constraints given by T_v and the ones given by T_v^ϵ (see, e.g., [10] for details on the intersection operation). Thus, from now on we assume that each FPQ-tree of D has been intersected with the corresponding embedding tree. If no permutation is possible, the intersection returns a *null-tree*, which formally represents the empty set of permutations. For ease of notation, we denote by $D(v)$ the set of FPQ-trees associated with v and resulting from the intersections after the removal of null-trees, if any. Clearly, a necessary condition for the FPQ-choosable planarity of (G, D) is that $D(v) \neq \emptyset$ for each vertex v.

As we will show, FPQ-CHOOSABLE PLANARITY is fixed-parameter tractable when parameterized by the treewidth of the input graph plus the number of FPQ-trees per vertex. One may wonder whether the problem remains FPT when parameterized by treewidth or by the number of FPQ-trees per vertex only.

Theorem 3. [*] FPQ-CHOOSABLE PLANARITY *for biconnected graphs is paraNP-hard with respect to the maximum number of FPQ-trees per vertex, even when the FPQ-trees have only P-nodes.*

Theorem 3 uses a reduction from 3-EDGECOLORING, which is NP-complete for cubic graphs [35]. For the reduction, we replace each edge of the input graph by a bundle of parallel edges, so that the ordering of the edges inside the bundle encodes the color of the corresponding edge. The possible choices of the FPQ-trees ensure that edges incident to the same vertex must receive different colors. One complication is that all planar cubic graphs are 3-edge colorable. We therefore start from a non-planar graph and planarize it.

Theorem 4. [*] FPQ-CHOOSABLE PLANARITY *for biconnected graphs is W[1]-hard with respect to treewidth, even when the FPQ-trees have only P-nodes.*

Theorem 4 uses a reduction from LISTCOLORING, which is W[1]-hard parameterized by treewidth [14], even for biconnected planar graphs. The colors of the vertices correspond to the choices of the FPQ-trees, which allows us to naturally encode the list restriction. The key idea is to replace each edge (u, v) of the input graph by a bundle that consists of a sub-bundle of three edges for each color. The FPQ-tree that encodes color c for vertex u is constructed so to impose a fixed ordering on the sub-bundles of color c for each edge (u, v) such that this choice is incompatible with a choice of the FPQ-tree that encodes color c for vertex v.

4   FPT Algorithm for FPQ-Choosable Planarity

In this section we show that FPQ-CHOOSABLE PLANARITY is FPT with respect to $tw + D_{\max}$. Our algorithm is based on studying the interplay between different data structures, namely SPQR-trees, FPQ-trees, and sphere-cut decompositions.

Boundaries and Extensible Orders. Let T be an FPQ-tree, let $leaves(T)$ denote the set of its leaves, and let L be a proper subset of leaves(T). We say that L is a *consecutive set* if the leaves in L are consecutive in every cyclic order represented by T. Let e be an edge of T, and let T' and T'' be the two subtrees obtained by removing e from T. If either leaves(T') or leaves(T'') are a subset of a consecutive set L, then we say that e is a *split edge for L*. The subtree that contains the leaves in L is the *split subtree* of e for L. A split edge e is *critical* for L if there exists no split edge e' such that the split subtree of e' contains e.

Lemma 1. [*] *Let T be an FPQ-tree, let L be a consecutive proper subset of* leaves(T), *and let S be the set of critical split edges for L. Then either $|S| = 1$, or $|S| > 1$ and there exists a Q-node or an F-node χ of T such that χ has degree at least $|S| + 2$ and the elements of S appear consecutively around χ.*

If $|S| = 1$, the split edge in S is called the *boundary of L*. If $|S| > 1$, the Q-node or the F-node χ defined in the statement of Lemma 1 is the *boundary of L*. See Fig. 3(a) for an example, where the three red edges b, c, and d of G define a consecutive set L_u in T_u, and the red edges e and f define a consecutive set L_v in T_v. The boundary of L_u in T_u is a Q-node, while the boundary of L_v in T_v is a split edge. We denote by $\mathcal{B}(L)$ the boundary of a set of leaves L. If $\mathcal{B}(L)$ is a Q-node, we associate $\mathcal{B}(L)$ with a default orientation (*i.e.*, a flip) that arbitrarily defines one of the two possible permutations of its children. We call this default orientation the *clockwise orientation* of $\mathcal{B}(L)$. The other possible permutation of the children of $\mathcal{B}(L)$ corresponds to the *counter-clockwise orientation*. If $\mathcal{B}(L)$ is an F-node, its fixed orientation is clockwise. Since F-nodes are a more constrained version of Q-nodes, when we refer to boundary Q-nodes we also take into account the case in which they are F-nodes.

Let $L' = L \cup \{\ell\}$, where ℓ is a new element. Let $\sigma \in consistent(T)$, and let $\sigma|_{L'}$ be a cyclic order obtained from σ by replacing the elements of the consecutive set leaves(T)$\setminus L$ by the single element ℓ. We say that a cyclic order σ' of L' is *extensible* if there exists a cyclic order $\sigma \in consistent(T)$ with $\sigma|_{L'} = \sigma'$. In this case, we say that σ is an *extension* of σ'. Note that if the boundary of L is a Q-node χ, any two extensions of σ' induce the same clockwise or counter-clockwise orientation of the edges incident to χ. An extensible order σ is *clockwise* if the orientation of χ is clockwise; σ is *counter-clockwise* otherwise. If the boundary of L is an edge, we consider any extensible order as both clockwise and counter-clockwise.

Let L and $\hat{L}$ be disjoint consecutive sets of leaves that have the same boundary Q-node χ in T and let ℓ and $\hat{\ell}$ be new elements. Let σ and $\hat{\sigma}$ be the extensions of two extensible orders of $L \cup \ell$ and $\hat{L} \cup \hat{\ell}$, respectively. We say that σ and $\hat{\sigma}$ are *incompatible* if one of them is clockwise and the other one is counter-clockwise.

Lemma 2. [*] *Let T be an FPQ-tree, let $L_1, L_2, \ldots, L_k$ be a partition of* leaves(T) *into consecutive sets, and let σ_i be an extensible order of L_i, for $1 \leq i \leq k$. There exists an order Σ of* leaves(T) *represented by T such that $\Sigma|_{L_i} = \sigma_i$ if and only if no pair σ_i, σ_j is incompatible, for $1 \leq i, j \leq k$.*

Pertinent FPQ-Trees, Skeletal FPQ-Trees, and Admissible Tuples. Let (G, D) be an FPQ-choosable graph, let T be an SPQR-tree of G rooted at an

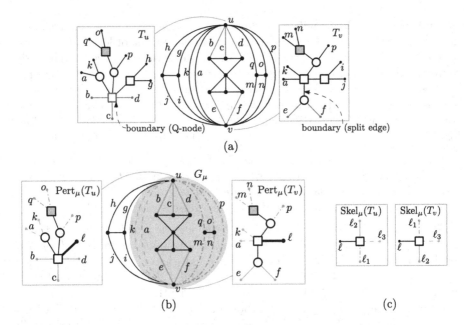

Fig. 3. (a) Two different types of boundaries: A boundary Q-node in T_u and a boundary edge in T_v. (b) The pertinent FPQ-trees $\mathrm{Pert}_\mu(T_u)$ of T_u and $\mathrm{Pert}_\mu(T_v)$ of T_v. (c) The skeletal FPQ-trees $\mathrm{Skel}_\mu(T_u)$ of $\mathrm{Pert}_\mu(T_u)$ and $\mathrm{Skel}_\mu(T_v)$ of $\mathrm{Pert}_\mu(T_v)$.

arbitrary Q-node, let μ be a node of T, and let T_μ be the subtree of T rooted at μ. The *pertinent graph of* μ, denoted as G_μ, is the subgraph of G induced by the edges represented by the leaves of T_μ. Let v be a pole of a node μ of T, let $T_v \in D(v)$ be an FPQ-tree associated with v, let E_{ext} be the set of edges that are incident to v and not contained in the pertinent graph G_μ, and let $E_\mu^\star(v) = E(v) \setminus E_{\mathrm{ext}}$. Note that there is a bijection between the edges $E(v)$ of G and the leaves of T_v, hence we shall refer to the set of leaves of T_v as $E(v)$. Also note that $E_\mu^\star(v)$ is represented by a consecutive set of leaves in T_v, because in every planar embedding of G the edges in $E_\mu^\star(v)$ must appear consecutively in the cyclic order of the edges incident to v.

The *pertinent FPQ-tree* of T_v, denoted as $\mathrm{Pert}_\mu(T_v)$, is the FPQ-tree obtained from T_v by replacing the consecutive set E_{ext} with a single leaf ℓ. Informally, the pertinent FPQ-tree of v describes the embedding constraints for the pole v within G_μ. For example, in Fig. 3(b) a pertinent graph G_μ with poles u and v is highlighted by a shaded region; the pertinent FPQ-trees $\mathrm{Pert}_\mu(T_u)$ of T_u and $\mathrm{Pert}_\mu(T_v)$ of T_v are obtained by the FPQ-trees T_u and T_v of Fig. 3(a).

Let $\nu_1, \ldots, \nu_k$ be the children of μ in T. Observe that the edges $E_\mu^\star(v)$ of each G_{ν_i} ($1 \le i \le k$) form a consecutive set of leaves of $\mathrm{Pert}_\mu(T_v)$. The *skeletal FPQ-tree* $\mathrm{Skel}_\mu(T_v)$ of $\mathrm{Pert}_\mu(T_v)$ is the tree obtained from $\mathrm{Pert}_\mu(T_v)$ by replacing each of the consecutive sets $E_{\nu_i}^\star(v)$ ($1 \le i \le k$) by a single leaf ℓ_i. See for example, Fig. 3(c). Observe that each Q-node of $\mathrm{Skel}_\mu(T_v)$ corresponds to a

Q-node of $\mathrm{Pert}_\mu(T_v)$, and thus to a Q-node of T_v; also, distinct Q-nodes of $\mathrm{Skel}_\mu(T_v)$ correspond to distinct Q-nodes of $\mathrm{Pert}_\mu(T_v)$, and thus to distinct Q-nodes of T_v. Each Q-node χ of T_v that is a boundary of μ or of one of its children ν_i inherits its default orientation from the corresponding Q-node iner, there is a corresponding Q-node in $\mathrm{Skel}_\mu(T_v)$ that inherits its default orientation from T_v.

Let (G, D) be an FPQ-choosable graph, let T be an SPQR-tree of G, let μ be a node of T, and let u and v be the poles of μ. We denote with (G_μ, D_μ) the FPQ-choosable graph consisting of the pertinent graph G_μ and the set D_μ that is defined as follows: $D_\mu(z) = D(z)$ for each vertex z of G_μ that is not a pole, and $D_\mu(v) = \{\mathrm{Pert}_\mu(T_v) \mid T_v \in D(v)\}$ if v is a pole of μ. A tuple $\langle T_u, T_v, o_u, o_v \rangle \in D(u) \times D(v) \times \{0,1\} \times \{0,1\}$ is *admissible for* G_μ if there exist an assignment A_μ of (G_μ, D_μ) and a planar embedding $\mathcal{E}_\mu$ of G_μ consistent with A_μ such that $A_\mu(u) = \mathrm{Pert}_\mu(T_u)$, $A_\mu(v) = \mathrm{Pert}_\mu(T_v)$, $\mathcal{B}(E_\mu^\star(u))$ is clockwise (counter-clockwise) in T_u if $o_u = 0$ ($o_u = 1$), and $\mathcal{B}(E_\mu^\star(v))$ is clockwise (counter-clockwise) in T_v if $o_v = 0$ ($o_v = 1$). We say that a tuple is *admissible for* μ if it is admissible for G_μ. We denote by $\Psi(\mu)$ the set of admissible tuples for G_μ.

FPT Algorithm. In order to test if (G, D) is FPQ-choosable planar, we root the SPQR-tree T at an arbitrary Q-node and we visit T from the leaves to the root. To simplify the description and without loss of generality, we shall assume that every S-node of T has exactly two children. Indeed, we iteratively replace every S-node μ having children $\nu_1, \ldots, \nu_k$ in T ($k > 2$) by an S-node μ' whose children are ν_1 and an S-node μ'' with children $\nu_2, \ldots, \nu_k$. By repeating this operation until every S-node has exactly two children, we obtain a rooted SPQR-tree T' which implicitly represents the same planar embeddings as those represented by T. At each step of the visit, we equip the currently visited node μ with the set $\Psi(\mu)$. If we encounter a node μ such that $\Psi(\mu) = \emptyset$, we return that (G, D) is not FPQ-choosable planar; otherwise the planarity test returns an affirmative answer. If the currently visited node μ is a leaf of T, we set $\Psi(\mu) = D(u) \times D(v) \times \{0,1\} \times \{0,1\}$, because its pertinent graph is a single edge. If μ is an internal node, $\Psi(\mu)$ is computed from the sets of admissible tuples of the children of μ. Let $D_{\max}$ be the maximum number of FPQ-trees per vertex, *i.e.*, $D_{\max} = \max_{v \in V} |D(v)|$. The next lemmas describe how to compute $\Psi(\mu)$ depending on whether μ is an S-, P-, or R-node. Due to space constraints, we report here only the proof for the R-nodes.

Lemma 3. [*] *Let μ be a node of the SPQR-tree with children $\nu_1, \ldots, \nu_k$. Given $\Psi(\nu_1), \ldots, \Psi(\nu_k)$, the set $\Psi(\mu)$ can be computed in $O(D_{\max}^2 \log(D_{\max}))$ time if μ is an S-node (with two children), and in $O(D_{\max}^2 \cdot n)$ time if μ is a P-node.*

The next lemma uses branchwidth as a parameter. Recall that, for a graph G with treewidth tw and branchwidth $bw > 1$, $bw - 1 \leq tw \leq \lfloor \frac{3}{2} bw \rfloor - 1$ holds [39].

Lemma 4. [*] *Let μ be an R-node with children $\nu_1, \nu_2, \ldots, \nu_k$. Given $\Psi(\nu_1)$, $\Psi(\nu_2), \ldots, \Psi(\nu_k)$, the set $\Psi(\mu)$ can be computed in $O(D_{\max}^{\frac{3}{2} bw} \cdot n_\mu^2 + n_\mu^3)$ time, where bw is the branchwidth of G_μ, and n_μ is the number of vertices of G_μ.*

Proof (Sketch). Since μ is an R-node, skel(μ) has only two possible planar embeddings, which we denote by $\mathcal{E}_\mu$ and $\mathcal{E}'_\mu$. Let u and v be the poles of μ. Let ν_i $(1 \leq i \leq k)$ be a child of μ that corresponds to a virtual edge (x, y) of $\mathcal{T}$ and let $T_x \in D_\mu(x)$. Recall that $E^\star_{\nu_i}(x)$ is a consecutive set of leaves in T_x. If $\mathcal{B}(E^\star_{\nu_i}(x))$ in T_x is a Q-node χ, by Lemma 1 there are at least two edges incident to χ that do not belong to $E^\star_{\nu_i}(x)$. Hence, an orientation o_x of χ determines an embedding of skel(μ). We call the pair (T_x, o_x) *compliant* with a planar embedding $\mathcal{E}_\mu$ of skel(μ) if either the boundary is an edge, or if the orientation of the boundary Q-node χ determines the embedding $\mathcal{E}_\mu$ of skel(μ). We denote by $\Psi_{\mathcal{E}_\mu}(\nu_i)$ the subset of tuples $\langle T_x, T_y, o_x, o_y \rangle \in \Psi(\nu_i)$ such that T_x with orientation o_x and T_y with orientation o_y are both compliant with $\mathcal{E}_\mu$; $\Psi_{\mathcal{E}_\mu}(\mu)$ is the subset of tuples $\langle T_u, T_v, o_u, o_v \rangle \in \Psi(\mu)$ whose pairs (T_u, o_u) and (T_v, o_v) are compliant with $\mathcal{E}_\mu$.

We show how to compute $\Psi_{\mathcal{E}_\mu}(\mu)$ from the sets $\Psi_{\mathcal{E}_\mu}(\nu_i)$ of the children ν_i of μ $(1 \leq i \leq k)$. Set $\Psi_{\mathcal{E}'_\mu}(\mu)$ is computed analogously. Note that the set $\Psi_{\mathcal{E}_\mu}(\nu_i)$ can be extracted by scanning $\Psi(\nu_i)$ and selecting only those admissible tuples whose pairs (T_x, o_x) and (T_y, o_y) are both compliant with $\mathcal{E}_\mu$. Since G_μ has branchwidth bw, skel(μ) is planar, it has branchwidth at most bw, and we can compute a sphere-cut decomposition of width at most bw [25] of the planar embedding $\mathcal{E}_\mu$ of skel(μ). Such a decomposition recursively divides skel(μ) into two subgraphs, each of which is embedded inside a topological disk having at most bw vertices on its boundary. The decomposition is described by a rooted binary tree, called the *sphere-cut decomposition tree* and denoted as T_{sc}. The root of T_{sc} is associated with skel(μ); the leaves of T_{sc} are the edges of skel(μ); any internal node β of T_{sc} is associated with the subgraph of skel(μ) induced by the leaves of the subtree rooted at β. Tree T_{sc} is such that when removing any of its internal edges, the two subgraphs induced by the leaves in the resulting subtrees share at most bw vertices. We denote as skel(β) the subgraph associated with a node β of T_{sc} and with $\mathcal{D}_\beta$ the topological disk that separates skel(β) from the rest of skel(μ). Note that skel(β) has at most bw vertices on the boundary of $\mathcal{D}_\beta$. In particular, if β is the root of T_{sc}, skel(β) coincides with skel(μ) and the vertices of skel(β) on the boundary of $\mathcal{D}_\beta$ are exactly the poles u and v of μ.

We compute $\Psi_{\mathcal{E}_\mu}(\mu)$ by visiting T_{sc} bottom-up. We equip each node β of T_{sc} with a set of tuples $\Psi_{\mathcal{E}_\mu}(\beta)$, each one consisting of at most bw pairs of elements (T_x, o_x) such that (T_x, o_x) is compliant with $\mathcal{E}_\mu$, and (T_x, o_x) belongs to some $\Psi_{\mathcal{E}_\mu}(\nu_i)$. The set of tuples associated with the root of T_{sc} is therefore the set $\Psi_{\mathcal{E}_\mu}(\mu)$. Let β be the currently visited node of T_{sc}. If β is a leaf, it is associated with an edge representing a child ν_i of μ in $\mathcal{T}$ and $\Psi_{\mathcal{E}_\mu}(\beta) = \Psi_{\mathcal{E}_\mu}(\nu_i)$.

If β is an internal node of T_{sc}, we compute $\Psi_{\mathcal{E}_\mu}(\beta)$ from the sets of tuples $\Psi_{\mathcal{E}_\mu}(\beta_1)$ and $\Psi_{\mathcal{E}_\mu}(\beta_2)$ associated with the two children β_1 and β_2 of β. Let $B_1 = \{w_1^1, \ldots, w_1^i, w_c^1, \ldots, w_c^r\}$ be the set of vertices of skel(β_1) that lie on the boundary of $\mathcal{D}_{\beta_1}$, and let $B_2 = \{w_2^1, \ldots, w_2^j, w_c^1, \ldots, w_c^r\}$ be the set of vertices of skel(β_2) that lie on the boundary of $\mathcal{D}_{\beta_2}$; see Fig. 4. Let $\{w_1^1, \ldots, w_1^i, w_c^1, \ldots, w_c^r, w_2^1, \ldots, w_2^j\}$ be the set of vertices of $B_1 \cup B_2$. Also, let $B = \{w_c^1, \ldots, w_c^r\}$ be the set of vertices that lie on the boundary of $\mathcal{D}_{\beta_1} \cap \mathcal{D}_{\beta_2}$;

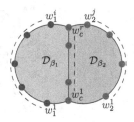

Fig. 4. An example illustrating two topological disks $\mathcal{D}_{\beta_1}$ and $\mathcal{D}_{\beta_2}$ containing two subgraphs $\mathrm{skel}(\beta_1)$ and $\mathrm{skel}(\beta_2)$. $B_1 = \{w_1^1, \ldots, w_1^i, w_c^1, \ldots, w_c^r\}$, $B_2 = \{w_2^1, \ldots, w_2^j, w_c^1, \ldots, w_c^r\}$, $B = \{w_c^1, \ldots, w_c^r\}$.

note that B consists of at most bw vertices, i.e., $r \leq bw$, and $B \subseteq B_1 \cup B_2$. A tuple $\langle T_{w_1^1}, \ldots, T_{w_1^i}, T_{w_c^1}, \ldots, T_{w_c^r}, o_{w_1^1}, \ldots, o_{w_1^i}, o_{w_c^1}, \ldots, o_{w_c^r} \rangle \in \Psi_{\mathcal{E}_\mu}(\beta_1)$ consists of pairs $(T_{w_1^l}, o_{w_1^l})$ and pairs $(T_{w_c^h}, o_{w_c^h})$ $(1 \leq l \leq i, 1 \leq h \leq r)$ that are compliant with $\mathcal{E}_\mu$. Similarly, a tuple $\langle T_{w_2^1}, \ldots, T_{w_2^j}, T_{w_c^1}, \ldots, T_{w_c^r}, o_{w_2^1}, \ldots, o_{w_2^j}, o_{w_c^1}, \ldots, o_{w_c^r} \rangle \in \Psi_{\mathcal{E}_\mu}(\beta_2)$ consists of pairs $(T_{w_2^q}, o_{w_2^q})$ and pairs $(T_{w_c^h}, o_{w_c^h})$ $(1 \leq q \leq j, 1 \leq h \leq r)$ that are compliant with $\mathcal{E}_\mu$. Let B' be the vertices on the boundary of $\mathcal{D}_\beta$. We can assume $B' = \{w_1^1, \ldots, w_1^i, w_2^1, \ldots, w_2^j, w_c^1, w_c^r\}$. An admissible tuple of $\Psi_{\mathcal{E}_\mu}(\beta)$ is constructed by combining two admissible tuples that agree on the pairs for $w_c^1, \ldots, w_c^r$. Therefore, $\Psi_{\mathcal{E}_\mu}(\beta)$ can be computed from $\Psi_{\mathcal{E}_\mu}(\beta_1)$ and $\Psi_{\mathcal{E}_\mu}(\beta_2)$ by a join operation between two sorted tables τ_1 and τ_2. Observe that τ_1 has $O(D_{\max}^{(i+r)})$ tuples and τ_2 has $O(D_{\max}^{(j+r)})$ tuples. The join operation between τ_1 and τ_2 gives rise to a table τ that has $O(D_{\max}^{(i+j+r)})$ tuples; since $i + r \leq bw$, $j + r \leq bw$, and $i + j \leq bw$, we have that $2i + 2j + 2r \leq 3bw$ and thus $i + j + r \leq \frac{3}{2}bw$. □

By means of Lemmas 3 and 4, we can prove the following result which, together with Theorems 3 and 4, implies Theorem 1.

Theorem 5. [*] FPQ-CHOOSABLE PLANARITY *parameterized by* $D_{\max} + tw$ *is FPT for biconnected graphs. Precisely, it can be solved in* $O(D_{\max}^{\frac{3}{2}bw} \cdot n^2 + n^3)$ *time, where* $bw \leq tw + 1$ *is the branchwidth of the graph.*

5   FPQ-Choosable Planarity and NodeTrix Planarity

A *clustered graph* is a graph G for which subsets of its vertices are grouped into sets $C_1, \ldots, C_{n_C}$, called *clusters*, and no vertex belongs to two clusters. An edge (u, v) with $u \in C_i$ and $v \in C_j$ is an *inter-cluster edge* if $i \neq j$, $(1 \leq i, j \leq n_C)$. In a NODETRIX *representation* Γ of G each cluster C_i is represented as an adjacency matrix M_i of the graph induced by the vertices of C_i [5,15,21,33]. Let C_i and C_j be two clusters represented by matrices M_i and M_j, respectively. Each inter-cluster edge (u, v) with $u \in C_i$ and $v \in C_j$ is represented in Γ as a Jordan arc γ connecting a point on the boundary of M_i belonging to the row or

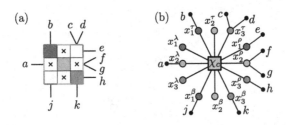

Fig. 5. (a) A matrix M_i; (b) the matrix FPQ-tree T_{M_i}.

the column corresponding to u to a point on the boundary of M_j belonging to the row or the column corresponding to v; also, γ is such that it does not cross any matrix of Γ. A NODETRIX representation is *planar* if no inter-cluster edges cross. A NODETRIX *graph with fixed sides* is a clustered graph G that admits a NODETRIX representation where, for each inter-cluster edge e, the sides of the matrices to which e is incident are specified [21]. Let G be a NODETRIX graph with fixed sides and clusters $C_1, \ldots, C_{n_C}$. Each permutation of the vertices of C_i ($1 \leq i \leq n_C$) corresponds to a matrix M_i in some NODETRIX representation of G. Note that even if the side of M_i to which each inter-cluster edge is incident is fixed, it is still possible to permute the edges incident to a same side and to a same vertex. For example, the edges f and g incident to the right side of the matrix in Fig. 5(a) can be permuted. Each of the possible cyclic orders of the edges incident to M_i can be described by means of an FPQ-tree, that we call the *matrix FPQ-tree* of M_i, denoted as T_{M_i}. Tree T_{M_i} consists of an F-node χ_c connected to $4|M_i|$ P-nodes representing the vertices of C_i; see Fig. 5(b). The P-nodes around χ_c appear in the clockwise order defined by M_i, namely $x_1^\tau, \ldots, x_{|M_i|}^\tau, x_1^\rho, \ldots, x_{|M_i|}^\rho,$ $x_{|M_i|}^\beta, \ldots, x_1^\beta, x_{|M_i|}^\lambda, \ldots, x_1^\lambda$ (τ, ρ, β, and λ represent the top, right, bottom, and left side of M_i, respectively). Any inter-cluster edge incident to a vertex v of M_i corresponds to a leaf of T_{M_i} adjacent to x_v^s ($1 \leq v \leq |M_i|$, $s \in \{\tau, \rho, \beta, \lambda\}$).

The *constraint graph* of a NODETRIX graph with fixed sides G, denoted as G_C, is the FPQ-choosable multi-graph defined as follows. Graph G_C has n_C vertices, each corresponding to one of the clusters of G, and in G_C there is an edge (u, v) for each inter-cluster edge that connects the two clusters corresponding to u and to v in G. Each vertex v of G_C is associated with a set $D(v)$ of $|C_v|!$ FPQ-trees. More precisely, for each permutation π of the vertices of C_v, let M_v^π be the matrix associated with C_v. For each such a permutation, we equip v with $T_{M_v^\pi}$.

Da Lozzo et al. [15] proved that NODETRIX PLANARITY WITH FIXED SIDES is NP-complete even when G_C has only two vertices, and thus bounded treewidth; Di Giacomo et al. [21] proved that NODETRIX PLANARITY WITH FIXED SIDES is NP-complete when the size k of the clusters is larger than 2. These results imply that the problem is paraNP-hard parameterized by either k or the treewidth of G_C. Also, by the above described relationship between FPQ-CHOOSABLE PLANARITY and NODETRIX PLANARITY WITH FIXED SIDES, we can prove that the last problem is FPT parameterized by both parameters and obtain Theorem 2.

6   Open Problems

It would be interesting to: (i) Improve the time complexity stated by Theorem 5. (ii) Study the complexity of FPQ-CHOOSABLE PLANARITY for simply connected instances. (iii) Apply our approach to other problems of planarity testing related with hybrid representations including, for example, intersection-link, ChordLink, and (k, p) representations (see, *e.g.*, [1–3,18,19]).

References

1. Angelini, P., Da Lozzo, G., Di Battista, G., Frati, F., Patrignani, M., Rutter, I.: Intersection-link representations of graphs. J. Graph Algorithms Appl. **21**(4), 731–755 (2017). https://doi.org/10.7155/jgaa.00437
2. Angelini, P., et al.: Graph planarity by replacing cliques with paths. Algorithms **13**(8), 194 (2020). https://doi.org/10.3390/a13080194
3. Angori, L., Didimo, W., Montecchiani, F., Pagliuca, D., Tappini, A.: Hybrid graph visualizations with ChordLink: algorithms, experiments, and applications. IEEE Trans. Vis. Comput. Graph. **28**(2), 1288–1300 (2022). https://doi.org/10.1109/TVCG.2020.3016055
4. Arumugam, S., Brandstädt, A., Nishizeki, T., Thulasiraman, K.: Handbook of Graph Theory, Combinatorial Optimization, and Algorithms. Chapman and Hall/CRC (2016)
5. Batagelj, V., Brandenburg, F., Didimo, W., Liotta, G., Palladino, P., Patrignani, M.: Visual analysis of large graphs using (X, Y)-clustering and hybrid visualizations. IEEE Trans. Vis. Comput. Graph. **17**(11), 1587–1598 (2011). https://doi.org/10.1109/TVCG.2010.265
6. Bhore, S., Ganian, R., Montecchiani, F., Nöllenburg, M.: Parameterized algorithms for book embedding problems. J. Graph Algorithms Appl. **24**(4), 603–620 (2020). https://doi.org/10.7155/jgaa.00526
7. Binucci, C., Di Giacomo, E., Liotta, G., Tappini, A.: Quasi-upward planar drawings with minimum curve complexity. In: Purchase, H.C., Rutter, I. (eds.) GD 2021. LNCS, vol. 12868, pp. 195–209. Springer, Cham (2021). https://doi.org/10.1007/978-3-030-92931-2_14
8. Bläsius, T., Fink, S.D., Rutter, I.: Synchronized planarity with applications to constrained planarity problems. In: 29th Annual European Symposium on Algorithms, ESA 2021, Lisbon, Portugal, 6–8 September 2021 (Virtual Conference), pp. 19:1–19:14 (2021). https://doi.org/10.4230/LIPIcs.ESA.2021.19
9. Bläsius, T., Rutter, I.: A new perspective on clustered planarity as a combinatorial embedding problem. Theor. Comput. Sci. **609**, 306–315 (2016). https://doi.org/10.1016/j.tcs.2015.10.011
10. Bläsius, T., Rutter, I.: Simultaneous PQ-ordering with applications to constrained embedding problems. ACM Trans. Algorithms **12**(2), 16:1–16:46 (2016). https://doi.org/10.1145/2738054
11. Booth, K.S., Lueker, G.S.: Testing for the consecutive ones property, interval graphs, and graph planarity using PQ-tree algorithms. J. Comput. Syst. Sci. **13**(3), 335–379 (1976). https://doi.org/10.1016/S0022-0000(76)80045-1
12. Chaplick, S., Di Giacomo, E., Frati, F., Ganian, R., Raftopoulou, C.N., Simonov, K.: Parameterized algorithms for upward planarity. In: 38th International Symposium on Computational Geometry, SoCG 2022, Berlin, Germany, 7–10 June 2022, pp. 26:1–26:16 (2022). https://doi.org/10.4230/LIPIcs.SoCG.2022.26

13. Cortese, P.F., Di Battista, G.: Clustered planarity. In: Proceedings of the 21st ACM Symposium on Computational Geometry, Pisa, Italy, 6–8 June 2005, pp. 32–34 (2005). https://doi.org/10.1145/1064092.1064093

14. Cygan, M., et al.: Parameterized Algorithms. Springer, Cham (2015). https://doi.org/10.1007/978-3-319-21275-3

15. Da Lozzo, G., Di Battista, G., Frati, F., Patrignani, M.: Computing NodeTrix representations of clustered graphs. J. Graph Algorithms Appl. **22**(2), 139–176 (2018). https://doi.org/10.7155/jgaa.00461

16. Di Battista, G., Eades, P., Tamassia, R., Tollis, I.G.: Graph Drawing. Prentice Hall, Upper Saddle River (1999)

17. Di Battista, G., Tamassia, R.: On-line planarity testing. SIAM J. Comput. **25**(5), 956–997 (1996). https://doi.org/10.1137/S0097539794280736

18. Di Giacomo, E., Didimo, W., Montecchiani, F., Tappini, A.: A user study on hybrid graph visualizations. In: Purchase, H.C., Rutter, I. (eds.) GD 2021. LNCS, vol. 12868, pp. 21–38. Springer, Cham (2021). https://doi.org/10.1007/978-3-030-92931-2_2

19. Di Giacomo, E., Lenhart, W.J., Liotta, G., Randolph, T.W., Tappini, A.: (k, p)-planarity: a relaxation of hybrid planarity. Theor. Comput. Sci. **896**, 19–30 (2021). https://doi.org/10.1016/j.tcs.2021.09.044

20. Di Giacomo, E., Liotta, G., Montecchiani, F.: Orthogonal planarity testing of bounded treewidth graphs. J. Comput. Syst. Sci. **125**, 129–148 (2022). https://doi.org/10.1016/j.jcss.2021.11.004

21. Di Giacomo, E., Liotta, G., Patrignani, M., Rutter, I., Tappini, A.: NodeTrix planarity testing with small clusters. Algorithmica **81**(9), 3464–3493 (2019). https://doi.org/10.1007/s00453-019-00585-6

22. Didimo, W., Giordano, F., Liotta, G.: Upward spirality and upward planarity testing. SIAM J. Discret. Math. **23**(4), 1842–1899 (2009). https://doi.org/10.1137/070696854

23. Didimo, W., Liotta, G., Ortali, G., Patrignani, M.: Optimal orthogonal drawings of planar 3-graphs in linear time. In: Proceedings of the 2020 ACM-SIAM Symposium on Discrete Algorithms, SODA 2020, Salt Lake City, UT, USA, 5–8 January 2020, pp. 806–825 (2020). https://doi.org/10.1137/1.9781611975994.49

24. Didimo, W., Liotta, G., Patrignani, M.: HV-planarity: algorithms and complexity. J. Comput. Syst. Sci. **99**, 72–90 (2019). https://doi.org/10.1016/j.jcss.2018.08.003

25. Dorn, F., Penninkx, E., Bodlaender, H.L., Fomin, F.V.: Efficient exact algorithms on planar graphs: exploiting sphere cut decompositions. Algorithmica **58**(3), 790–810 (2010). https://doi.org/10.1007/s00453-009-9296-1

26. Dujmovic, V., et al.: A fixed-parameter approach to 2-layer planarization. Algorithmica **45**(2), 159–182 (2006). https://doi.org/10.1007/s00453-005-1181-y

27. Dujmovic, V., et al.: On the parameterized complexity of layered graph drawing. Algorithmica **52**(2), 267–292 (2008). https://doi.org/10.1007/s00453-007-9151-1

28. Feng, Q.-W., Cohen, R.F., Eades, P.: Planarity for clustered graphs. In: Spirakis, P. (ed.) ESA 1995. LNCS, vol. 979, pp. 213–226. Springer, Heidelberg (1995). https://doi.org/10.1007/3-540-60313-1_145

29. Garg, A., Tamassia, R.: Upward planarity testing. Order **12**(2), 109–133 (1995)

30. Garg, A., Tamassia, R.: On the computational complexity of upward and rectilinear planarity testing. SIAM J. Comput. **31**(2), 601–625 (2001). https://doi.org/10.1137/S0097539794277123

31. Gutwenger, C., Klein, K., Mutzel, P.: Planarity testing and optimal edge insertion with embedding constraints. J. Graph Algorithms Appl. **12**(1), 73–95 (2008). https://doi.org/10.7155/jgaa.00160

32. Hasan, M.M., Rahman, M.S.: No-bend orthogonal drawings and no-bend orthogonally convex drawings of planar graphs (extended abstract). In: Du, D.-Z., Duan, Z., Tian, C. (eds.) COCOON 2019. LNCS, vol. 11653, pp. 254–265. Springer, Cham (2019). https://doi.org/10.1007/978-3-030-26176-4_21

33. Henry, N., Fekete, J., McGuffin, M.J.: NodeTrix: a hybrid visualization of social networks. IEEE Trans. Vis. Comput. Graph. **13**(6), 1302–1309 (2007). https://doi.org/10.1109/TVCG.2007.70582

34. Hliněný, P., Sankaran, A.: Exact crossing number parameterized by vertex cover. In: Archambault, D., Tóth, C.D. (eds.) GD 2019. LNCS, vol. 11904, pp. 307–319. Springer, Cham (2019). https://doi.org/10.1007/978-3-030-35802-0_24

35. Holyer, I.: The NP-completeness of edge-coloring. SIAM J. Comput. **10**(4), 718–720 (1981). https://doi.org/10.1137/0210055

36. Liotta, G., Rutter, I., Tappini, A.: Simultaneous FPQ-ordering and hybrid planarity testing. Theor. Comput. Sci. **874**, 59–79 (2021). https://doi.org/10.1016/j.tcs.2021.05.012

37. Nishizeki, T., Rahman, M.S.: Planar Graph Drawing. Lecture Notes Series on Computing, vol. 12. World Scientific (2004). https://doi.org/10.1142/5648

38. Patrignani, M.: Planarity testing and embedding. In: Handbook on Graph Drawing and Visualization, pp. 1–42 (2013)

39. Robertson, N., Seymour, P.D.: Graph minors. X. obstructions to tree-decomposition. J. Comb. Theory Ser. B **52**(2), 153–190 (1991). https://doi.org/10.1016/0095-8956(91)90061-N

Induced Disjoint Paths and Connected Subgraphs for H-Free Graphs

Barnaby Martin[1], Daniël Paulusma[1], Siani Smith[1]([⊠]),
and Erik Jan van Leeuwen[2]

[1] Department of Computer Science, Durham University, Durham, UK
{barnaby.d.martin,daniel.paulusma,siani.smith}@durham.ac.uk
[2] Department of Information and Computing Sciences, Utrecht University,
Utrecht, The Netherlands
e.j.vanleeuwen@uu.nl

Abstract. Paths $P^1,\ldots,P^k$ in a graph $G = (V,E)$ are mutually induced if any two distinct P^i and P^j have neither common vertices nor adjacent vertices. The INDUCED DISJOINT PATHS problem is to decide if a graph G with k pairs of specified vertices (s_i, t_i) contains k mutually induced paths P^i such that each P^i starts from s_i and ends at t_i. This is a classical graph problem that is NP-complete even for $k = 2$. We introduce a natural generalization, INDUCED DISJOINT CONNECTED SUBGRAPHS: instead of connecting pairs of terminals, we must connect sets of terminals. We give almost-complete dichotomies of the computational complexity of both problems for H-free graphs, that is, graphs that do not contain some fixed graph H as an induced subgraph. Finally, we give a complete classification of the complexity of the second problem if the number k of terminal sets is fixed, that is, not part of the input.

Keywords: induced subgraphs · connectivity · H-free graph · complexity dichotomy

1 Introduction

The well-known DISJOINT PATHS problem is one of the problems in Karp's list of NP-complete problems. It is to decide if a graph has pairwise vertex-disjoint paths $P^1,\ldots,P^k$ where each P^i connects two pre-specified vertices s_i and t_i. Its generalization, DISJOINT CONNECTED SUBGRAPHS, plays a crucial role in the graph minor theory of Robertson and Seymour. This problem asks for connected subgraphs $D^1,\ldots,D^k$, where each D^i connects a pre-specified set of vertices Z_i. In a recent paper [18] we classified, subject to a small number of open cases, the complexity of both these problems for H-*free* graphs, that is, for graphs that do not contain some fixed graph H as an *induced* subgraph.

Our Focus. We consider the *induced* variants of DISJOINT PATHS and DISJOINT CONNECTED SUBGRAPHS. These problems behave differently. Namely, DISJOINT PATHS for fixed k, or more generally, DISJOINT CONNECTED SUBGRAPHS, after

© Springer Nature Switzerland AG 2022
M. A. Bekos and M. Kaufmann (Eds.): WG 2022, LNCS 13453, pp. 398–411, 2022.
https://doi.org/10.1007/978-3-031-15914-5_29

fixing both k and $\ell = \max\{|Z_1|, \ldots, |Z_k|\}$, is polynomial-time solvable [29]. In contrast, INDUCED DISJOINT PATHS is NP-complete even when $k = 2$, as shown both by Bienstock [2] and Fellows [5]. Just as for the classical problems [18], we perform a systematic study and focus on H-free graphs. As it turns out, for the restriction to H-free graphs, the induced variants actually become computationally easier for an infinite family of graphs H. We first give some definitions.

Terminology. For a subset $S \subseteq V$ in a graph $G = (V, E)$, let $G[S]$ denote the *induced* subgraph of G by S, that is, $G[S]$ is the graph obtained from G after removing every vertex not in S. Let $G_1 + G_2$ be the disjoint union of two vertex-disjoint graphs G_1 and G_2. We say that paths $P^1, \ldots, P^k$, for some $k \geq 1$, are *mutually induced paths* of G if there exists a set $S \subseteq V$ such that $G[S] = P^1 + \ldots + P^k$; note that every P^i is an induced path and that there is no edge between two vertices from different paths P^i and P^j. A path P is an *s-t-path* (or *t-s-path*) if the end-vertices of P are s and t.

A *terminal pair* (s, t) is an unordered pair of two distinct vertices s and t in a graph G, which we call *terminals*. A set $T = \{(s_1, t_1), \ldots, (s_k, t_k)\}$ of terminal pairs of G is a *terminal pair collection* if the terminals pairs are pairwise disjoint, so, apart from $s_i \neq t_i$ for $i \in \{1, \ldots, k\}$, we also have $\{s_i, t_i\} \cap \{s_j, t_j\} = \emptyset$ for every $1 \leq i < j \leq k$. We now define the following decision problem:

INDUCED DISJOINT PATHS
   Instance: a graph G and terminal pair collection $T = \{(s_1, t_1) \ldots, (s_k, t_k)\}$.
   Question: does G have a set of mutually induced paths $P^1, \ldots, P^k$ such that
         P^i is an s_i-t_i path for $i \in \{1, \ldots, k\}$?

Note that as every path between two vertices s and t contains an induced path between s and t, the condition that every P^i must be induced is not strictly needed in the above problem definition. We say that the paths $P^1, \ldots, P^k$, if they exist, form a *solution* of INDUCED DISJOINT PATHS.

We now generalize the above notions from pairs and paths to sets and connected subgraphs. Subgraphs $D^1, \ldots, D^k$ of a graph $G = (V, E)$ are *mutually induced subgraphs* of G if there exists a set $S \subseteq V$ such that $G[S] = D^1 + \ldots + D^k$. A connected subgraph D of G is a *Z-subgraph* if $Z \subseteq V(D)$. A *terminal set* Z is an unordered set of distinct vertices, which we again call *terminals*. A set $\mathcal{Z} = \{Z_1, \ldots, Z_k\}$ is a *terminal set collection* if $Z_1, \ldots, Z_k$ are pairwise disjoint terminal sets. We now introduce the generalization:

INDUCED DISJOINT CONNECTED SUBGRAPHS
   Instance: a graph G and terminal set collection $\mathcal{Z} = \{Z_1, \ldots, Z_k\}$.
   Question: does G have a set of mutually induced connected subgraphs
         $D^1, \ldots, D^k$ such that D^i is a Z_i-subgraph for $i \in \{1, \ldots, k\}$?

The subgraphs $D^1, \ldots, D^k$, if they exist, form a *solution*. We write INDUCED DISJOINT CONNECTED ℓ-SUBGRAPHS if $\ell = \max\{|Z_1|, \ldots, |Z_k|\}$ is fixed. Note that INDUCED DISJOINT CONNECTED 2-SUBGRAPHS is exactly INDUCED DISJOINT PATHS.

1.1 Known Results

Only results for INDUCED DISJOINT PATHS are known and these hold for a slightly more general problem definition (see Sect. 6). Namely, INDUCED DISJOINT PATHS is linear-time solvable for circular-arc graphs [10]; polynomial-time solvable for chordal graphs [1], AT-free graphs [11], graph classes of bounded mim-width [15]; and NP-complete for claw-free graphs [6], line graphs of triangle-free chordless graphs [28] and thus for (theta,wheel)-free graphs, and for planar graphs; the last result follows from a result of Lynch [23] (see [11]). Moreover, INDUCED DISJOINT PATHS is XP with parameter k for (theta,wheel)-free graphs [28] and even FPT with parameter k for claw-free graphs [9] and planar graphs [17]; the latter can be extended to graph classes of bounded genus [20].

1.2 Our Results

Let P_r be the path on r vertices. A *linear forest* is the disjoint union of one or more paths. We write $F \subseteq_i G$ if F is an induced subgraph of G and sG for the disjoint union of s copies of G. We can now present our first two results: the first one includes our dichotomy for INDUCED DISJOINT PATHS (take $\ell = 2$).

Theorem 1. *Let $\ell \geq 2$. For a graph H, INDUCED DISJOINT CONNECTED ℓ-SUBGRAPHS on H-free graphs is polynomial-time solvable if $H \subseteq_i sP_3 + P_6$ for some $s \geq 0$; NP-complete if H is not a linear forest; and quasipolynomial-time solvable otherwise.*

Theorem 2. *For a graph H such that $H \neq sP_1 + P_6$ for some $s \geq 0$, INDUCED DISJOINT CONNECTED SUBGRAPHS on H-free graphs is polynomial-time solvable for H-free graphs if $H \subseteq_i sP_1 + P_3 + P_4$ or $H \subseteq_i sP_1 + P_5$ for some $s \geq 0$, and it is NP-complete otherwise.*

Note the complexity jumps if we no longer fix ℓ. We will show that all open cases in Theorem 2 are equivalent to exactly **one** open case, namely $H = P_6$.

Comparison. The DISJOINT CONNECTED SUBGRAPHS problem restricted to H-free graphs is polynomial-time solvable if $H \subseteq_i P_4$ and else it is NP-complete, even if the maximum size of the terminal sets is $\ell = 2$, except for the three unknown cases $H \in \{3P_1, 2P_1 + P_2, P_1 + P_3\}$ [18]. Perhaps somewhat surprisingly, Theorems 1 and 2 show the induced variant is computationally easier for an infinite number of linear forests H (if P $\neq$ NP).

Fixing k. If the number k of terminal sets is fixed, we write k-INDUCED DISJOINT CONNECTED SUBGRAPHS and prove the following complete dichotomy.

Theorem 3. *Let $k \geq 2$. For a graph H, k-INDUCED DISJOINT CONNECTED SUBGRAPHS on H-free graphs is polynomial-time solvable for H-free graphs if $H \subseteq_i sP_1 + 2P_4$ or $H \subseteq_i sP_1 + P_6$ for some $s \geq 0$, and it is NP-complete otherwise.*

Comparison. We note a complexity jump between Theorems 2 and 3 when $H = sP_1 + 2P_4$ for some $s \geq 0$.

Paper Outline. Sect. 2 contains terminology, known results and auxiliary results that we will use as lemmas. Hardness results for Theorem 1 transfer to Theorem 2, whereas the reverse holds for polynomial results. As such, we show all our polynomial-time algorithms in Sect. 3 and all our hardness reductions in Sect. 4. The cases $H = sP_3 + P_6$ in Theorem 1 and $H = sP_1 + P_5$ in Theorem 2 are proven by a reduction to INDEPENDENT SET via so-called *blob graphs*, just as the quasipolynomial-time result if H is a linear forest. Hence, we also include the proof of the latter result in Sect. 3. In Sect. 5 we combine the results from the previous two sections to prove Theorems 1–3.

In our theorems we have infinite families of polynomial cases related to nearly H-free graphs. For a graph H, a graph G is *nearly H-free* if G is $(P_1 + H)$-free. It is easy to see (cf [3]) that INDEPENDENT SET is polynomial-time solvable on nearly H-free graphs if it is so on H-free graphs. However, for many other graph problems, this might either not be true or less easy to prove (see, for example, [16]). In Sect. 3 we show that it holds for the relevant cases in Theorem 2, in particular for the case $H = P_6$ (see Lemma 7). The latter result yields no algorithm but shows that essentially $H = P_6$ is the only one open case left in Theorem 2.

In Sect. 6 we consider a number of directions for future work. In particular we consider the restriction k-DISJOINT CONNECTED ℓ-SUBGRAPHS where *both* k and ℓ are fixed and discuss some open problems.

2   Preliminaries

Let $G = (V, E)$ be a graph. A subset $S \subseteq V$ is *connected* if $G[S]$ is connected. A subset $D \subseteq V(G)$ is *dominating* if every vertex of $V(G) \setminus D$ is adjacent to least one vertex of D; if $D = \{v\}$ then v is a *dominating* vertex. The *open* and *closed neighbourhood* of a vertex $u \in V$ are $N(u) = \{v \mid uv \in E\}$ and $N[u] = N(u) \cup \{u\}$. For a set $U \subseteq V$ we define $N(U) = \bigcup_{u \in U} N(u) \setminus U$ and $N[U] = N(U) \cup U$.

For a graph $G = (V, E)$ and a subset $S \subseteq U$, we write $G - S = G[V \setminus S]$. If $S = \{u\}$ for some $u \in V$, we write $G - u$ instead of $G - \{u\}$. A vertex u is a *cut-vertex* of a connected graph G if $G - u$ is disconnected.

The *contraction* of an edge $e = uv$ in a graph G replaces the vertices u and v by a new vertex w that is adjacent to every vertex previously adjacent to u or v; note that the resulting graph G/e is still *simple*, that is, G/e contains no multi-edges or self-loops. The following lemma is easy to see (see, for example, [19]).

Lemma 1. *For a linear forest H, let G be an H-free graph. Then G/e is H-free for every $e \in E(G)$.*

In a solution $(D^1, \ldots, D^k)$ for an instance $(G, \mathcal{Z})$ of INDUCED DISJOINT CONNECTED SUBGRAPHS, if D^i is minimal and X_i is a minimum connected dominating set of D^i, then $X_i \cup Z_i = D^i$ or, equivalently, $D^i \setminus X_i \subseteq Z_i$. This will be relevant in our proofs, where we use the following result of Camby and Schaudt, in particular for the case $r = 6$ (alternatively, we could use the slightly weaker characterization of P_6-free graphs in [13] but the below characterization gives a faster algorithm).

Theorem 4 ([4]). *Let $r \geq 4$ and G be a connected P_r-free graph. Let X be any minimum connected dominating set of G. Then $G[X]$ is either P_{r-2}-free or isomorphic to P_{r-2}.*

Let $G = (V, E)$ be a graph. Two sets $X_1, X_2 \subseteq V$ are *adjacent* if $X_1 \cap X_2 \neq \emptyset$ or there exists an edge with one end-vertex in X_1 and the other in X_2. The *blob graph* G° of G has vertex set $\{X \subseteq V(G) \mid X \text{ is connected}\}$ and edge set $\{X_1 X_2 \mid X_1 \text{ and } X_2 \text{ are adjacent}\}$. Note that blob graphs may have exponential size, but in our proofs we will only construct parts of blob graphs that have polynomial size. We need the following known lemma that generalizes a result of Gartland et al. [8] for paths.

Lemma 2 ([26]). *For every linear forest H, a graph G is H-free if and only if G° is H-free.*

The INDEPENDENT SET problem is to decide if a graph G has an *independent set* (set of pairwise non-adjacent vertices) of size at least k for some given integer k. We need the following two known results for INDEPENDENT SET. The first one is due to Grzesik, Klimošová, Pilipczuk and Pilipczuk [12]. The second one is due to Pilipczuk, Pilipczuk and Rzążewski [27], who improved the previous quasipolynomial-time algorithm for INDEPENDENT SET on P_t-free graphs, due to Gartland and Lokshtanov [7] (whose algorithm runs in $n^{O(\log^3 n)}$ time).

Theorem 5 ([12]). *The INDEPENDENT SET problem is polynomial-time solvable for P_6-free graphs.*

Theorem 6 ([7]). *For every $r \geq 1$, the INDEPENDENT SET problem can be solved in $n^{O(\log^2 n)}$ time for P_r-free graphs.*

Two instances of a decision problem are *equivalent* if one is a yes-instance if and only if the other one is. We frequently use the following lemmas (proofs omitted).

Lemma 3. *From an instance $(G, \mathcal{Z})$ of INDUCED DISJOINT CONNECTED SUB-GRAPHS we can in linear time, either find a solution for $(G, \mathcal{Z})$ or obtain an equivalent instance $(G', \mathcal{Z}')$ with $|V(G')| \leq |V(G)|$, such that the following holds:*

1. *$|\mathcal{Z}'| \geq 2$;*
2. *every $Z_i' \in \mathcal{Z}'$ has size at least 2; and*
3. *the union of the sets in $\mathcal{Z}'$ is an independent set.*

Moreover, if G is H-free for some linear forest H, then G' is also H-free.

Lemma 4. *Let H be a linear forest. If $(G, \mathcal{Z})$ is a yes-instance of INDUCED DISJOINT CONNECTED SUBGRAPHS and G is H-free, then $(G, \mathcal{Z})$ has a solution $(D^1, \ldots, D^k)$, where each D^i has size at most $(2|V(H)| - 1)|Z_i|$.*

3 Algorithms

In this section we show all the polynomial-time and quasipolynomial-time results needed to prove our main theorems. We start with the following result.

Lemma 5. *Let $\ell \geq 2$. For every $s \geq 0$, INDUCED DISJOINT CONNECTED ℓ-SUBGRAPHS is polynomial-time solvable for $(sP_3 + P_6)$-free graphs.*

Proof. Let $(G, \mathcal{Z})$ be an instance of the INDUCED DISJOINT CONNECTED ℓ-SUBGRAPHS problem, where G is $(sP_3 + P_6)$-free for some $s \geq 0$. By Lemma 3, we may assume the union of the sets in $\mathcal{Z} = \{Z_1, \ldots, Z_k\}$ is independent.

First suppose that $k \leq s$. By Lemma 4 we may assume that each D^i in a solution $(D^1, \ldots, D^k)$ has size at most $t = (6s+11)\ell$. So $|D^1| + \ldots + |D^k|$ has size at most $kt \leq st$. Hence, we can consider all $O(n^{st})$ options of choosing a solution. As s and t are constants, this takes polynomial time in total. Now suppose that $k \geq s+1$. We consider all $O(n^{(s-1)t})$ options of choosing the first s subgraphs D^i, discarding those with an edge between distinct D^i or between some D^i and Z_j for some $j \geq s+1$. For each remaining option, let $G' = G - N[V(D^1) \cup \cdots \cup V(D^s)]$ and $\mathcal{Z}' = \{Z_{s+1}, \ldots, Z_k\}$. Note that G' is P_6-free.

Let F be the subgraph of the blob graph G'° induced by all connected subsets X in G' that have size at most 11ℓ, such that X contains all vertices of one set from $\mathcal{Z}'$ and no vertices from any other set of $\mathcal{Z}'$. Then F has polynomial size, as it has $O(n^{11\ell})$ vertices, so we can construct F in polynomial time. By Lemma 2, F is P_6-free.

We claim that $(G', \mathcal{Z}')$ has a solution if and only if F has an independent set of size $k - s$. First suppose that $(G', \mathcal{Z}')$ has a solution. Then, by Lemma 4, it has a solution $(D^{s+1}, \ldots, D^k)$, where each D^i has size at most 11ℓ. Such a solution corresponds to an independent set of size $k - s$ in F. For the reverse implication, two vertices in F that each contain vertices of the same set Z_i are adjacent. Hence, an independent set of size $k - s$ in F is a solution for $(G', \mathcal{Z}')$.

Due to the above, it remains to apply Theorem 5 to find in polynomial time whether G'° has an independent set of size $k - s$.   □

By replacing Theorem 5 by Theorem 6 in the above proof and repeating the arguments of the second part we obtain the following result.

Lemma 6. *Let $\ell \geq 2$. For every $r \geq 1$, INDUCED DISJOINT CONNECTED ℓ-SUBGRAPHS is quasipolynomial-time solvable for P_r-free graphs.*

We no prove a crucial lemma on nearly H-free graphs.

Lemma 7. *For $k \geq 2$, $r \leq 6$ and $s \geq 1$, if $(k-)$INDUCED DISJOINT CONNECTED SUBGRAPHS is polynomial-time solvable for P_r-free, graphs, then it is so for $(sP_1 + P_r)$-free graphs.*

Proof. First let $r = 6$ and k be part of the input. Let $(G, \mathcal{Z})$ be an instance of INDUCED DISJOINT CONNECTED SUBGRAPHS, where G is an $(sP_1 + P_6)$-free graph for some integer $s \geq 1$ and $\mathcal{Z} = \{Z_1, \ldots, Z_k\}$. We may assume without loss of generality that $|Z_1| \geq |Z_2| \geq \cdots \geq |Z_k|$. By Lemma 3, we may assume that $k \geq 2$; every $Z_i \in \mathcal{Z}$ has size at least 2; and the union of the sets in $\mathcal{Z}$ is an independent set. We assume that INDUCED DISJOINT CONNECTED SUBGRAPHS is polynomial-time solvable for P_6-free graphs.

Case 1. For every $i \geq 2$, $|Z_i| \leq s - 1$.

Let $D^1, \ldots, D^k$ be a solution for $(G, \mathcal{Z})$ (assuming it exists). By Lemma 4, we may assume without loss of generality that for $i \geq 2$, the number of vertices of D^i is at most $(2s + 11)|Z_i| \leq (2s + 11)(s - 1)$.

First assume $k \leq s$. Then $V(D^2) \cup \cdots \cup V(D^k)$ has size at most t, where $t = (s-1)(2s+11)(s-1)$ is a constant. Hence, we can do as follows. We consider all $O(n^t)$ options for choosing the subgraphs $D^2, \ldots, D^k$. For each choice we check in polynomial time if $D^2, \ldots, D^k$ are mutually induced and connected, and if each D^i contains Z^i. We then check in polynomial time if the graph $G - N[(V(D^2) \cup \cdots V(D^k)]$ has a connected component containing Z_1. As the number of choices is polynomial, the total running time is polynomial.

Now assume $k \geq s + 1$. We consider all $O(n^{s(2s+11)(s-1)})$ options of choosing the s subgraphs $D^2, \ldots, D^{s+1}$. We discard an option if for some $i \in \{1, \ldots, s\}$, the graph D^i is disconnected. We also discard an option if there is an edge between two vertices from two different subgraphs D^h and D^i for some $2 \leq h < i \leq s + 1$, or if there is an edge between a vertex from some subgraph D^h ($2 \leq h \leq s$) and a vertex from some set Z_i ($i = 1$ or $i \geq s + 2$). If we did not discard the option, then we solve INDUCED DISJOINT CONNECTED SUBGRAPHS on instance $(G - \bigcup_{i=2}^{s+2} N[V(D^i)], \mathcal{Z} \setminus \{Z_2, \ldots, Z_{s+1}\})$. The latter takes polynomial time as $G - \bigcup_{i=2}^{s+1} N[D^i]$ is P_6-free. As the number of branches is polynomial as well, the total running time is polynomial.

Case 2. $|Z_2| \geq s$ (and thus also $|Z_1| \geq s$).

Let $D^1, \ldots, D^k$ be a solution for $(G, \mathcal{Z})$ (assuming it exists). As $|Z_1| \geq s$, we find that for every $i \geq 2$, D^i is P_6-free. As $|Z_2| \geq s$, we also find that D^1 is P_6-free. Then, by setting $r = 6$ in Theorem 4, every D^i ($i \in \{1, \ldots, k\}$) has a connected dominating set X_i such that $G[X_i]$ is either P_4-free or isomorphic to P_4. We may assume that every X_i is inclusion-wise minimal (as else we could just replace X_i by a smaller connected dominating set of D^i).

Case 2a. There exist some X_i with size at least $7s + 2$.

As $s \geq 1$, we have that $G[X_i]$ is P_4-free. We now set $r = 4$ in Theorem 4 and find that $G[X_i]$ has a connected dominating set Y_i of size at most 2. Hence, $G[X_i]$ contains a set R of $7s$ vertices that are not cut-vertices of $G[X_i]$. As X_i is minimal, this means that in D^i, each $r \in R$ has at least one neighbour $z \in Z_i$ that is not adjacent to any vertex of $X_i \setminus \{r\}$. We say that z is a *private* neighbour of r. We now partition R into sets $R_1, \ldots, R_7$, each of exactly s vertices. For $h = 1, \ldots, 7$, let $R_h = \{r_h^1, \ldots, r_h^s\}$ and pick a private neighbour z_h^j of r_h^j. For $h = 1, \ldots, 7$, let $Q_h = \{z_h^1, \ldots, z_h^s\}$. Each Q_h is independent, as Z_i is independent and $Q_h \subseteq Z_i$.

We claim that there exists an index $h \in \{1, \ldots, 7\}$ such that $G - (N[Q_h] \setminus R_h)$ is P_6-free. For a contradiction, assume that for every $h \in \{1, \ldots, 7\}$, we have that $G - (N[Q_h] \setminus R_h)$ is not P_6-free. As G is $(sP_1 + P_6)$-free and every Q_h is an independent set of size s, we have that $G - N[Q_h]$ is P_6-free. We conclude that every induced P_6 of G contains a vertex of R_h for every $h \in \{1, \ldots, 7\}$. This is contradiction, as every induced P_6 only has six vertices. Hence, there exists an index $h \in \{1, \ldots, 7\}$ such that $G - (N[Q_h] \setminus R_h)$ is P_6-free.

We exploit the above structural claim algorithmically as follows. We consider all $k = O(n)$ options that one of the sets X_i has size at least $7s + 2$. For each choice of index i we do as follows. We consider all $O(n^{2s})$ options of choosing a set Q_h of s vertices from the independent set Z_i together with a set R_h of s vertices from $N(Q_h)$. We discard the option if a vertex of Q_h has more than one neighbour in R_h, or if $G' = G - (N[Q_h] \setminus R_h)$ is not P_6-free. Otherwise, we solve INDUCED DISJOINT CONNECTED SUBGRAPHS on instance $(G', \mathcal{Z}')$, where $\mathcal{Z}' = (\mathcal{Z} \setminus \{Z_i\}) \cup \{(Z_i \setminus Q_h) \cup R_h\}$. As G' is P_6-free, the latter takes polynomial time by our initial assumption. Hence, as the total number of branches is $O(n^{2s+1})$ the total running time of this check takes polynomial time.

Case 2b. Every X_i has size at most $7s + 1$.
First assume $k \le s$. We consider all $O(n^{s(7s+1)})$ options of choosing the sets $X_1, \ldots, X_k$. For each option we check if $(X_1 \cup Z_1, \ldots, X_k \cup Z_k)$ is a solution for $(G, \mathcal{Z})$. As the latter takes polynomial time and the total number of branches is polynomial, this takes polynomial time.

Now assume $k \ge s + 1$. We consider all $O(n^{s(7s+1)})$ options of choosing the first s sets $X_1, \ldots, X_s$. We discard an option if for some $i \in \{1, \ldots, s\}$, the set $X_i \cup Z_i$ is disconnected. We also discard an option if there is an edge between two vertices from two different sets $X_h \cup Z_h$ and $X_i \cup Z_i$ for some $1 \le h < i \le s$, or if there is an edge between a vertex from some set $X_h \cup Z_h$ ($h \le s$) and a vertex from some set Z_i ($i \ge s + 1$). If we did not discard the option, then we solve INDUCED DISJOINT CONNECTED SUBGRAPHS on instance $(G - \bigcup_{i=1}^s N[X_i \cup Z_i], \{Z_{s+1}, \ldots, Z_k\})$. The latter takes polynomial time as $G - \bigcup_{i=1}^s N[X_i \cup Z_i]$ is P_6-free. As the number of branches is polynomial as well, the total running time is polynomial.

From the above case analysis we conclude that the running time of our algorithm is polynomial. If $r \le 5$ and/or k is fixed we use exactly the same arguments. $\square$

Remark 1. Due to Lemma 7, the missing cases $H = sP_1 + P_6$ in Theorem 2 are all equivalent to the case $H = P_6$.

We will use Lemma 7 for the case where $r = 5$. We also make use of the blob approach again.

Lemma 8. *For every $s \ge 0$,* INDUCED DISJOINT CONNECTED SUBGRAPHS *is polynomial-time solvable for $(sP_1 + P_5)$-free graphs.*

Proof. Due to Lemma 7 it suffices to prove the statement for P_5-free graphs only. Let $(G, \mathcal{Z})$ be an instance of INDUCED DISJOINT CONNECTED SUBGRAPHS, where G is a P_5-free graph and $\mathcal{Z} = \{Z_1, \ldots, Z_k\}$. By Lemma 3, we may assume that $k \geq 2$; every $Z_i \in \mathcal{Z}$ has size at least 2; and the union of the sets in $\mathcal{Z}$ is an independent set. We may also delete every vertex from G that is not in a terminal set from $\mathcal{Z}$ but that is adjacent to two terminals in different sets Z_h and Z_i (such a vertex cannot be used in any subgraph of a solution). We now make a structural observation that gives us a procedure for safely contracting edges; recall that edge contraction preserves P_5-freeness by Lemma 1.

Consider a solution $(D^1 \ldots D^k)$ that is *maximal* in the sense that any vertex v outside $V(D^1) \cup \cdots \cup V(D^k)$ must have a neighbour in at least two distinct subgraphs D^i and D^j. As G is P_5-free, v must be adjacent to all vertices of at least one of D^i and D^j. As v has no neighbours in both $Z_i \subseteq V(D^i)$ and $Z_j \subseteq V(D^j)$, v must be adjacent to all vertices of exactly one of D^i and D^j.

The above gives rise to the following algorithm. Let v be a vertex that is adjacent to at least one vertex $z \in Z_i$ but not to all vertices of Z_i. As v is adjacent to z and z is in Z_i, it hold that v does not belong to any D^h with $h \neq i$ for every (not necessarily maximal) solution $(D^1, \ldots, D^k)$. The observation from the previous paragraph tells us that if v is not in any D^h and $(D^1, \ldots, D^k)$ is a maximal solution, then v must be adjacent to all vertices of some D^j. As v is adjacent to $z \in Z_i$, it holds by construction that v is not adjacent to any vertex of any $Z_h \subseteq V(D^h)$ with $h \neq i$. Hence, $i = j$ must hold. However, this is not possible, as we assumed that v is not adjacent to all vertices of $Z_i \subseteq V(D^i)$. Hence, we may assume without loss of generality that v belongs to D^i (should a solution exist). This means that we can safely contract the edge vz and put the resulting vertex in Z_i. Then we apply Lemma 3 again and also remove all common neighbours of vertices from Z_i and vertices from other sets Z_j. This takes polynomial time and the resulting graph has one vertex less. Hence, by applying this procedure exhaustively we have, in polynomial time, either solved the problem or obtained an equivalent but smaller instance.

Suppose the latter case holds. For simplicity we denote the obtained instance by $(G, \mathcal{Z})$ again, where G is a P_5-free graph and $\mathcal{Z} = \{Z_1, \ldots, Z_k\}$ with $k \geq 2$. Due to our procedure, every $Z_i \in \mathcal{Z}$ has size at least 2; the union of the sets in $\mathcal{Z}$ is an independent set. Moreover, every non-terminal vertex is adjacent either to no terminal vertex or is adjacent to all terminals of exactly one terminal set. We let S be the set of vertices of the latter type. Observe that it follows from the preceding that only vertices of S need to be used for a solution.

We now construct the subgraph F of the blob graph G° that is induced by all connected subsets X of the form $X = Z_i \cup \{s\}$ for some $1 \leq i \leq k$ and $s \in S$. Note that F has $O(kn)$ vertices. Hence, constructing F takes polynomial time. Moreover, F is P_5-free due to Lemma 2. As in the proof of Lemma 5, we observe that $(G, \mathcal{Z})$ has a solution if and only if F has an independent set of size k. It now remains to apply (in polynomial time) Theorem 5. □

We now show a stronger result when k is fixed (proof omitted).

Lemma 9. *For every $s \geq 0$, k-INDUCED DISJOINT CONNECTED SUBGRAPHS is polynomial-time solvable for $(sP_1 + P_6)$-free graphs.*

We now present our final two polynomial-time algorithms (proofs omitted).

Lemma 10. *For every $k \geq 2$ and $s \geq 0$, k-INDUCED DISJOINT CONNECTED SUBGRAPHS is polynomial-time solvable for $(sP_1 + 2P_4)$-free graphs.*

Lemma 11. *For every $s \geq 0$, INDUCED DISJOINT CONNECTED SUBGRAPHS is polynomial-time solvable for $(sP_1 + P_3 + P_4)$-free graphs.*

4   NP-Completeness Results

In this section we present a number of NP-completeness results; we omitted all proofs except one. If $\ell = 2$, we write INDUCED DISJOINT PATHS instead of INDUCED DISJOINT CONNECTED ℓ-SUBGRAPHS. The *girth* of a graph G that is not a forest is the length of a shortest cycle of G.

Lemma 12. *For every $g \geq 3$, INDUCED DISJOINT PATHS is NP-complete for graphs of girth at least g.*

Lemma 13. *For every $g \geq 3$, 2-INDUCED DISJOINT CONNECTED SUBGRAPHS is NP-complete for graphs of girth at least g.*

The *line graph $L(G)$* of a graph G has vertex set $\{v_e \mid e \in E(G)\}$ and an edge between v_e and v_f if and only if e and f are incident on the same vertex in G. The following two lemmas show NP-completeness for line graphs. Lemma 14 is due to Fiala et al. [6]. They consider a more general variant of INDUCED DISJOINT PATHS, but their reduction holds in our setting as well. Lemma 15 can be derived from the NP-completeness of 2-DISJOINT CONNECTED SUBGRAPHS [14].

Lemma 14 ([6]). INDUCED DISJOINT PATHS *is NP-complete for line graphs.*

Lemma 15. 2-INDUCED DISJOINT CONNECTED SUBGRAPHS *is NP-complete for line graphs.*

Finally, we show two lemmas for graphs without certain induced linear forests.

Lemma 16. 2-INDUCED DISJOINT CONNECTED SUBGRAPHS *is NP-complete for $(3P_2, P_7)$-free graphs.*

Proof. We reduce from NOT-ALL-EQUAL-3-SAT, known to be NP-complete [30]. Let $(\mathcal{X}, \mathcal{C})$ be an instance of NOT-ALL-EQUAL-3-SAT containing n variables $x_1, \ldots, x_n$ and m clauses $C_1, \ldots, C_m$. We construct a graph G as follows. Let X be a clique of size n on vertices $v_1, \ldots, v_n$. Introduce a copy v_i' of each v_i in X. Call the new set X' and make it a clique. Add the edges $v_i v_i'$ for each v_i in X. Let C be an independent set of size m on vertices $c_1, \ldots, c_m$. Introduce a copy c_j' of each vertex c_j in C. Call the new set C' (and keep it an independent set). Now for all $1 \leq i \leq n$ and $1 \leq j \leq m$, add an edge $v_i c_j$ and an edge $v_i' c_j'$ if clause C_j contains variable x_i. Set $Z_1 = C$ and $Z_2 = C'$. Then, (G, Z_1, Z_2) is an instance of 2-INDUCED DISJOINT CONNECTED SUBGRAPHS.

Observe that G is P_7-free. Indeed, let P be any longest induced path in G. Then P can contain at most two vertices from X and at most two vertices from X'. If P contains at most one vertex from C and at most one vertex from C', then P has length at most $2 + 2 + 1 + 1 = 6$. On the other hand, if P contains two vertices from C or two vertices from C', then P has length at most 3.

We also observe that G is $3P_2$-free, as any P_2 must contain at least one vertex from X or from X', and X and X' are cliques. So we are done after proving the following claim: $(\mathcal{X}, \mathcal{C})$ is a yes-instance of NOT-ALL-EQUAL-3-SAT if and only if (G, Z_1, Z_2) is a yes-instance of 2-INDUCED DISJOINT CONNECTED SUBGRAPHS.

In the forward direction, let τ be a satisfying truth assignment. We put in A every vertex of X for which the corresponding variable is set to true. We put in A' every vertex of X' for which the corresponding variable is set to false. As each clause C_j contains at least one true variable, c_j is adjacent to a vertex in A. Similarly, each clause C_j contains at least one false variable, so each c'_j is adjacent to a vertex in A'. As X and X' are cliques, A and A' are cliques. Hence, $G[C \cup A]$ and $G[C' \cup A']$ are connected.

Now suppose there is an edge between a vertex of $C \cup A$ and a vertex of $C' \cup A'$. Then, by construction, this edge must be equal to some $v_i v'_i$, which means that v_i is in A and v'_i is in A', so x_i must be true and false at the same time, a contradiction. Hence, there exists no edge between a vertex from $C \cup A$ and a vertex from $C' \cup A'$. We conclude that $(C \cup A, C' \cup A')$ is a solution.

In the backwards direction, let $(C \cup A, C' \cup A')$ be a solution. Then, by definition, there is no edge between $C \cup A$ and $C' \cup A'$, which means that there is no edge between A and A'. Then $A \subseteq X$ and $A' \subseteq X'$, since X and X' are cliques and A (A') needs to contain at least one vertex of X (X'). Also, there is no variable x_i such that v_i is in A and v'_i is in A'. This means we can define a truth assignment τ by setting all variables corresponding to vertices in A to be true, all variables corresponding to vertices in A' to be false, and all remaining vertices in $\mathcal{X}$ to be true (or false, it does not matter).

As C is an independent set and $C \cup A$ is connected, each c_j has a neighbour in A. So each C_j contains a true literal. As C' is an independent set and $C' \cup A'$ is connected, each c'_j has a neighbour in A'. So each C_j contains a false literal. Hence, τ is a satisfying truth assignment. This completes the proof. $\square$

Lemma 17. INDUCED DISJOINT CONNECTED SUBGRAPHS *is* NP-*complete for* $2P_4$-*free graphs.*

5 The Proofs of Theorems 1–3

We are now ready to prove Theorems 1–3, which we restate below.

Proof of Theorem 1. We prove the theorem for $\ell = 2$; extending the proof to $\ell \geq 3$ is trivial. If H contains a cycle C_s, then we use Lemma 12 by setting the girth to $g = s + 1$. Suppose that H contains no cycle, that is, H is a forest. If H contains a vertex of degree at least 3, then we use Lemma 14, as in that case the class of H-free graphs contains the class of $K_{1,3}$-free graphs, which in turn contains the class of line graphs. In the remaining cases, H is a linear forest. If $H \subseteq_i sP_3 + P_6$ for some $s \geq 0$ we use Lemma 5. Else we use Lemma 6. □

Proof of Theorem 2. If H is not a linear forest, we use Theorem 1. Suppose H is a linear forest. If $H \subseteq_i sP_1 + P_5$ for some $s \geq 0$ we use Lemma 8. If $H \subseteq_i sP_1 + P_3 + P_4$ for some $s \geq 0$ we use Lemma 11. If $3P_2 \subseteq_i H$ or $P_7 \subseteq_i H$ we use Lemma 16. Otherwise $2P_4 \subseteq_i H$ and we use Lemma 17. □

Proof of Theorem 3. If H contains a cycle C_s, then we use Lemma 13 by setting the girth to $g = s + 1$. Suppose that H contains no cycle, that is, H is a forest. If H contains a vertex of degree at least 3, then we use Lemma 15, as in that case the class of H-free graphs contains the class of $K_{1,3}$-free graphs, which in turn contains the class of line graphs. In the remaining cases, H is a linear forest. If $H \subseteq_i sP_1 + P_6$ for some $s \geq 0$ we use Lemma 9. If $H \subseteq_i sP_1 + 2P_4$ for some $s \geq 0$ we use Lemma 10. Otherwise $3P_2 \subseteq_i H$ or $P_7 \subseteq_i H$ and we use Lemma 16. □

6   Future Work

Our results naturally lead to some open problems. First of all, can we find polynomial-time algorithms for the quasipolynomial cases in Theorem 1? This is a challenging task that is also open for INDEPENDENT SET; note that we reduce to the latter problem to solve the case where $H = sP_1 + P_6$ for some $s \geq 0$.

We also recall that the case $H = P_6$ is essentially the only remaining open case left in Theorem 2, which is for the setting where k and ℓ are both part of the input. As shown in Theorems 1 and 3, respectively, we have a positive answer for the settings where ℓ is fixed (and k is part of the input) and where k is fixed (and ℓ is part of the input), respectively. However, it seems challenging to combine the techniques when both k and ℓ are part of the input.

We did not yet discuss the k-INDUCED DISJOINT CONNECTED ℓ-SUBGRAPHS problem, which is the variant where both k and ℓ are fixed; note that if $\ell = 2$, then we obtain the k-INDUCED DISJOINT PATHS problem. The latter problem restricted to $k = 2$ is closely related to the problem of deciding if a graph contains a cycle passing through two specified vertices and has been studied for hereditary graph classes as well; see [21]. Recently, we made some more progress. A *subdivided claw* is obtained from a claw after subdividing each edge zero or more times. In particular, the *chair* is the graph obtained from the claw by subdividing one of its edges exactly once. The set $\mathcal{S}$ consists of all graphs with the property that each of their connected components is a path or a subdivided claw. We proved in [24] that for every integer $k \geq 2$ and graph H, k-INDUCED DISJOINT PATHS is polynomial-time solvable if H is a subgraph of the disjoint union of a linear forest and a chair, and it is NP-complete if H is not in $\mathcal{S}$.

From the above it follows in particular that k-INDUCED DISJOINT PATHS is polynomial-time solvable for claw-free graphs (just like INDEPENDENT SET [25, 31]) in contrast to the other three variants, which are NP-complete for claw-free graphs (see Theorems 1–3). We leave completing the classification of k-INDUCED DISJOINT PATHS as future work and refer to [24] for a more in-depth discussion.

Acknowledgments. We thank Paweł Rzążewski for the argument using blob graphs, which simplified two of our proofs and led to the case $H = P_6$ in Theorem 1.

References

1. Belmonte, R., Golovach, P.A., Heggernes, P., van't Hof, P., Kaminski, M., Paulusma, D.: Detecting fixed patterns in chordal graphs in polynomial time. Algorithmica **69**, 501–521 (2014)
2. Bienstock, D.: On the complexity of testing for odd holes and induced odd paths. Discret. Math. **90**, 85–92 (1991)
3. Brandstädt, A., Hoàng, C.T.: On clique separators, nearly chordal graphs, and the maximum weight stable set problem. Theoret. Comput. Sci. **389**, 295–306 (2007)
4. Camby, E., Schaudt, O.: A new characterization of P_k-free graphs. Algorithmica **75**, 205–217 (2016)
5. Fellows, M.R.: The Robertson-Seymour theorems: a survey of applications. Proc. AMS-IMS-SIAM Joint Summer Res. Conf. Contemp. Math. **89**, 1–18 (1989)
6. Fiala, J., Kamiński, M., Lidický, B., Paulusma, D.: The k-in-a-Path problem for claw-free graphs. Algorithmica **62**, 499–519 (2012)
7. Gartland, P., Lokshtanov, D.: Independent set on P_k-free graphs in quasi-polynomial time. In: Proceedings of the FOCS 2020, pp. 613–624 (2020)
8. Gartland, P., Lokshtanov, D., Pilipczuk, M., Pilipczuk, M., Rzążewski, P.: Finding large induced sparse subgraphs in C_t-free graphs in quasipolynomial time. In: Proceedings of the STOC 2021, pp. 330–341 (2021)
9. Golovach, P.A., Paulusma, D., van Leeuwen, E.J.: Induced disjoint paths in claw-free graphs. SIAM J. Discret. Math. **29**, 348–375 (2015)
10. Golovach, P.A., Paulusma, D., van Leeuwen, E.J.: Induced disjoint paths in circular-arc graphs in linear time. Theoret. Comput. Sci. **640**, 70–83 (2016)
11. Golovach, P.A., Paulusma, D., van Leeuwen, E.J.: Induced disjoint paths in AT-free graphs. J. Comput. Syst. Sci. **124**, 170–191 (2022)
12. Grzesik, A., Klimosová, T., Pilipczuk, M., Pilipczuk, M.: Polynomial-time algorithm for maximum weight independent set on P_6-free graphs. In: Proceedings of the SODA 2019, pp. 1257–1271 (2019)
13. van't Hof, P., Paulusma, D.: A new characterization of P_6-free graphs. Discrete Appl. Math. **158**, 731–740 (2010)
14. van't Hof, P., Paulusma, D., Woeginger, G.J.: Partitioning graphs into connected parts. Theoret. Comput. Sci. **410**, 4834–4843 (2009)
15. Jaffke, L., Kwon, O., Telle, J.A.: Mim-width I. induced path problems. Discrete Appl. Math. **278**, 153–168 (2020)
16. Johnson, M., Paesani, G., Paulusma, D.: Connected Vertex Cover for $(sP_1 + P_5)$-free graphs. Algorithmica **82**, 20–40 (2020)
17. Kawarabayashi, K., Kobayashi, Y.: A linear time algorithm for the induced disjoint paths problem in planar graphs. J. Comput. Syst. Sci. **78**, 670–680 (2012)
18. Kern, W., Martin, B., Paulusma, D., Smith, S., van Leeuwen, E.J.: Disjoint paths and connected subgraphs for H-free graphs. Theoret. Comput. Sci. **898**, 59–68 (2022)
19. Kern, W., Paulusma, D.: Contracting to a longest path in H-free graphs. Proc. ISAAC 2020, LIPIcs **181**, 22:1–22:18 (2020)
20. Kobayashi, Y., Kawarabayashi, K.: Algorithms for finding an induced cycle in planar graphs and bounded genus graphs. In: Proceedings of the SODA 2009, pp. 1146–1155 (2009)

21. Lévêque, B., Lin, D.Y., Maffray, F., Trotignon, N.: Detecting induced subgraphs. Discret. Appl. Math. **157**, 3540–3551 (2009)
22. Li, W.N.: Two-segmented channel routing is strong NP-complete. Discret. Appl. Math. **78**, 291–298 (1997)
23. Lynch, J.: The equivalence of theorem proving and the interconnection problem. SIGDA Newsl. **5**, 31–36 (1975)
24. Martin, B., Paulusma, D., Smith, S., van Leeuwen, E.J.: Few induced disjoint paths for H-free graphs. Proc. ISCO 2022, LNCS (to appear)
25. Minty, G.J.: On maximal independent sets of vertices in claw-free graphs. J. Comb. Theor. Ser. B **28**, 284–304 (1980)
26. Paesani, G., Paulusma, D., Rzążewski, P.: Feedback Vertex Set and Even Cycle Transversal for H-free graphs: finding large block graphs. SIAM J. Discret. Math. (to appear)
27. Pilipczuk, M., Pilipczuk, M., Rzążewski, P.: Quasi-polynomial-time algorithm for independent set in P_t-free graphs via shrinking the space of induced paths. In: Proceedings of the SOSA 2021, pp. 204–209 (2021)
28. Radovanović, M., Trotignon, N., Vušković, K.: The (theta, wheel)-free graphs Part IV: induced paths and cycles. J. Comb. Theor. Ser. B **146**, 495–531 (2021)
29. Robertson, N., Seymour, P.D.: Graph minors. XIII. The disjoint paths problem. J. Comb. Theor. Ser. B **63**, 65–110 (1995)
30. Schaefer, T.J.: The complexity of satisfiability problems. In: STOC, pp. 216–226 (1978)
31. Shibi, N.: Algorithme de recherche d'un stable de cardinalité maximum dans un graphe sans étoile. Discret. Math. **29**, 53–76 (1980)

Classifying Subset Feedback Vertex Set for H-Free Graphs

Giacomo Paesani[1]([✉]) [ID], Daniël Paulusma[2] [ID], and Paweł Rzążewski[3,4] [ID]

[1] School of Computing, University of Leeds, Leeds, UK
g.paesani@leeds.ac.uk
[2] Department of Computer Science, Durham University, Durham, UK
daniel.paulusma@durham.ac.uk
[3] Faculty of Mathematics and Information Science, Warsaw University
of Technology, Warsaw, Poland
pawel.rzazewski@pw.edu.pl
[4] Faculty of Mathematics, Informatics, and Mechanics, University of Warsaw,
Warsaw, Poland

Abstract. In the FEEDBACK VERTEX SET problem, we aim to find a small set S of vertices in a graph intersecting every cycle. The SUBSET FEEDBACK VERTEX SET problem requires S to intersect only those cycles that include a vertex of some specified set T. We also consider the WEIGHTED SUBSET FEEDBACK VERTEX SET problem, where each vertex u has weight $w(u) > 0$ and we ask that S has small weight. By combining known NP-hardness results with new polynomial-time results we prove full complexity dichotomies for SUBSET FEEDBACK VERTEX SET and WEIGHTED SUBSET FEEDBACK VERTEX SET for H-free graphs, that is, graphs that do not contain a graph H as an induced subgraph.

Keywords: Feedback vertex set · H-free graph · Complexity dichotomy

1 Introduction

In a *graph transversal* problem the aim is to find a small set of vertices within a given graph that must intersect every subgraph that belongs to some specified family of graphs. Apart from the VERTEX COVER problem, the FEEDBACK VERTEX SET problem is perhaps the best-known graph transversal problem. A vertex subset S is a *feedback vertex set* of a graph G if S intersects every cycle of G. In other words, the graph $G - S$ obtained by deleting all vertices of S is a forest. We can now define the problem:

FEEDBACK VERTEX SET
    Instance: a graph G and an integer k.
    Question: does G have a feedback vertex set S with $|S| \leq k$?

P. Rzążewski—Supported by Polish National Science Centre grant no. 2018/31/D/ST6/00062.

M. A. Bekos and M. Kaufmann (Eds.): WG 2022, LNCS 13453, pp. 412–424, 2022.
https://doi.org/10.1007/978-3-031-15914-5_30

The FEEDBACK VERTEX SET problem is well-known to be NP-complete even under input restrictions. For example, by Poljak's construction [14], the FEEDBACK VERTEX SET problem is NP-complete even for graphs of finite girth at least g (the girth of a graph is the length of its shortest cycle). To give another relevant example, FEEDBACK VERTEX SET is also NP-complete for line graphs [10].

In order to understand the computational hardness of FEEDBACK VERTEX SET better, other graph classes have been considered as well, in particular those that are closed under vertex deletion. Such graph classes are called *hereditary*. It is readily seen that a graph class $\mathcal{G}$ is hereditary if and only if $\mathcal{G}$ can be characterized by a (possibly infinite) set $\mathcal{F}$ of forbidden induced subgraphs. From a systematic point of view it is natural to first consider the case where $\mathcal{F}$ has size 1, say $\mathcal{F} = \{H\}$ for some graph H. This leads to the notion of H-freeness: a graph G is H-*free* for some graph H if G does not contain H as an *induced* subgraph, that is, G cannot be modified into H by a sequence of vertex deletions.

As FEEDBACK VERTEX SET is NP-complete for graphs of finite girth at least g for every $g \geq 1$, it is NP-complete for H-free graphs whenever H has a cycle. As it is NP-complete for line graphs and line graphs are claw-free, FEEDBACK VERTEX SET is NP-complete for H-free graphs whenever H has an induced claw (the *claw* is the 4-vertex star). In the remaining cases, the graph H is a *linear forest*, that is, the disjoint union of one or more paths. When H is a linear forest, several positive results are known even for the weighted case. Namely, for a graph G, we can define a *(positive) weighting* as a function $w : V \to \mathbb{Q}^+$. For $v \in V$, $w(v)$ is the *weight* of v, and for $S \subseteq V$, we define the weight $w(S) = \sum_{u \in S} w(u)$ of S as the sum of the weights of the vertices in S. This brings us to the following generalization of FEEDBACK VERTEX SET:

WEIGHTED FEEDBACK VERTEX SET
> *Instance:* a graph G, a positive vertex weighting w of G and a rational number k.
> *Question:* does G have a feedback vertex set S with $w(S) \leq k$?

Note that if w is a constant weighting function, then we obtain the FEEDBACK VERTEX SET problem. We denote the r-vertex path by P_r, and the *disjoint union* of two vertex-disjoint graphs G_1 and G_2 by $G_1 + G_2 = ((V(G_1) \cup V(G_2), E(G_1) \cup E(G_2)))$, where we write sG for the disjoint union of s copies of G. It is known that WEIGHTED FEEDBACK VERTEX SET is polynomial-time solvable for sP_3-free graphs [11] and P_5-free graphs [1]. The latter result was recently extended to $(sP_1 + P_5)$-free graphs for every $s \geq 0$ [11]. We write $H \subseteq_i G$ to denote that H is an *induced* subgraph of G. We can now summarize all known results [1,10,11,14] as follows.

Theorem 1. (WEIGHTED) FEEDBACK VERTEX SET *for the class of H-free graphs is polynomial-time solvable if $H \subseteq_i sP_3$ or $H \subseteq_i sP_1 + P_5$ for some $s \geq 1$, and is* NP-*complete if H is not a linear forest.*

Note that the open cases of Theorem 1 are when H is a linear forest with $P_2 + P_4 \subseteq_i H$ or $P_6 \subseteq_i H$.

The (WEIGHTED) FEEDBACK VERTEX SET problem can be further generalized in the following way. Let T be some specified subset of vertices of a graph G. A T-*cycle* of G is a cycle that intersects T. A set $S_T \subseteq V$ is a T-*feedback vertex set* of G if S_T contains at least one vertex of every T-cycle; see also Fig. 1. We now consider the following generalizations of FEEDBACK VERTEX SET:

SUBSET FEEDBACK VERTEX SET
> *Instance:* a graph G, a subset $T \subseteq V(G)$ and an integer k.
> *Question:* does G have a T-feedback vertex set S_T with $|S_T| \leq k$?

WEIGHTED SUBSET FEEDBACK VERTEX SET
> *Instance:* a graph G, a subset $T \subseteq V(G)$, a positive vertex weighting w
> of G and a rational number k.
> *Question:* does G have a T-feedback vertex set S_T with $w(S_T) \leq k$?

The NP-complete cases in Theorem 1 carry over to (WEIGHTED) SUBSET FEEDBACK VERTEX SET; just set $T := V(G)$ in both cases. However, this is no longer true for the polynomial-time cases: Fomin et al. [7] proved NP-completeness of SUBSET FEEDBACK VERTEX SET for split graphs, which form a subclass of $2P_2$-free graphs. Interestingly, Papadopoulos and Tzimas [13] proved that WEIGHTED SUBSET FEEDBACK VERTEX SET is NP-complete for $5P_1$-free graphs, whereas Brettell et al. [4] proved that SUBSET FEEDBACK VERTEX SET can be solved in polynomial time even for $(sP_1 + P_3)$-free graphs for every $s \geq 1$ [4]. Hence, in contrast to many other transversal problems, the complexities on the weighted and unweighted subset versions do not coincide for H-free graphs.

It is also known that WEIGHTED SUBSET FEEDBACK VERTEX SET can be solved in polynomial time for permutation graphs [12] and thus for its subclass of P_4-free graphs. The latter result also follows from a more general result related to the graph parameter mim-width [15]. Namely, Bergougnoux, Papadopoulos and Telle [3] proved that WEIGHTED SUBSET FEEDBACK VERTEX SET is

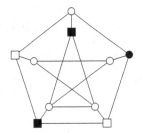 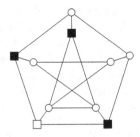

Fig. 1. Two examples of a slightly modified Petersen graph with the set T indicated by square vertices. In both examples, the set S_T of black vertices is a T-feedback vertex set. On the left, $S_T \setminus T \neq \emptyset$. On the right, $S_T \subseteq T$.

polynomial-time solvable for graphs for which we can find a decomposition of constant mim-width in polynomial time [3]; the class of P_4-free graphs is an example of such a class. Brettell et al. [5] extended these results by proving that WEIGHTED SUBSET FEEDBACK VERTEX SET, restricted to H-free graphs, is polynomial-time solvable if $H \subseteq_i 3P_1 + P_2$ or $H \subseteq_i P_1 + P_3$.

The above results leave open a number of unresolved cases for both problems, as identified in [4] and [5], where the following open problems are posed:

Fig. 2. The graph $2P_1 + P_4$.

Open Problem 1. *Determine the complexity of* WEIGHTED SUBSET FEEDBACK VERTEX SET *for H-free graphs if $H \in \{2P_1 + P_3, P_1 + P_4, 2P_1 + P_4\}$.*

Open Problem 2. *Determine the complexity of* SUBSET FEEDBACK VERTEX SET *for H-free graphs if $H = sP_1 + P_4$ for some integer $s \geq 1$.*

1.1 Our Results

We completely solve Open Problems 1 and 2.

In Sect. 3, we prove that WEIGHTED SUBSET FEEDBACK VERTEX SET is polynomial-time solvable for $(2P_1 + P_4)$-free graphs. This result generalizes all known polynomial-time results for WEIGHTED FEEDBACK VERTEX SET. It also immediately implies polynomial-time solvability for the other two cases in Open Problem 1, as $(2P_1 + P_3)$-free graphs and $(P_1 + P_4)$-free graphs form subclasses of $(2P_1 + P_4)$-free graphs. Combining the aforementioned NP-completeness results of [7] and [13] for $2P_2$-free graphs and $5P_1$-free graphs, respectively, with the NP-completeness results in Theorem 1 for the case where H has a cycle or a claw and this new result gives us the following complexity dichotomy (see also Fig. 2).

Theorem 2. *For a graph H, the* WEIGHTED SUBSET FEEDBACK VERTEX SET *problem on H-free graphs is polynomial-time solvable if $H \subseteq_i 2P_1 + P_4$, and is NP-complete otherwise.*

In Sect. 4 we solve Open Problem 2 by proving that SUBSET FEEDBACK VERTEX SET can be solved in polynomial time for $(sP_1 + P_4)$-free graphs, for every $s \geq 1$. This result generalizes all known polynomial-time results for WEIGHTED FEEDBACK VERTEX SET. After combining it with the aforementioned NP-completeness results of [7] and Theorem 1 we obtain the following complexity dichotomy.

Theorem 3. *For a graph H, the* SUBSET FEEDBACK VERTEX SET *problem on H-free graphs is polynomial-time solvable if $H \subseteq_i sP_1 + P_4$ for some $s \geq 0$, and is* NP-*complete otherwise.*

Due to Theorems 2 and 3 we now know where exactly the complexity jump between the weighted and unweighted versions occurs.

Our proof technique for these results is based on the following two ideas. First, if the complement F_T of a T-feedback vertex set contains s vertices of small degree in F_T, then we can "guess" these vertices and their neighbours in F_T. We then show that after removing all the other neighbours of small-degree vertices, we will obtain a graph of small mim-width. If F_T does not contain s small-degree vertices, we will argue that F_T contains a bounded number of vertices of T. We guess these vertices and exploit their presence. This is straightforward for SUBSET FEEDBACK VERTEX SET but more involved for WEIGHTED SUBSET FEEDBACK VERTEX SET. The latter was to be expected from the hardness construction for WEIGHTED SUBSET FEEDBACK VERTEX SET on $5P_1$-free graphs, in which $|T| = 1$ (but as we will show our algorithm is able to deal with that construction due to the $(2P_1 + P_4)$-freeness of the input graph).

We finish our paper with a brief discussion on related graph transversal problems and some open questions in Sect. 5.

2 Preliminaries

Let $G = (V, E)$ be a graph. If $S \subseteq V$, then $G[S]$ denotes the subgraph of G induced by S, and $G - S$ is the graph $G[V \setminus S]$. We say that S is *independent* if $G[S]$ has no edges, and that S is a *clique* and $G[S]$ is *complete* if every pair of vertices in S is joined by an edge. A *(connected) component* of G is a maximal connected subgraph of G. The *neighbourhood* of a vertex $u \in V$ is the set $N(u) = \{v \mid uv \in E\}$. A graph is *bipartite* if its vertex set can be partitioned into at most two independent sets.

Recall that for a graph $G = (V, E)$ and a subset $T \subseteq V$, a T-feedback vertex set is a set $S \subseteq V$ that intersects all T-cycles. Note that $G - S$ is a graph that has no T-cycles; we call such a graph a T-*forest*. Thus the problem of finding a T-feedback vertex set of minimum size is equivalent to finding a T-forest of maximum size. Similarly, the problem of finding a T-feedback vertex set of minimum weight is equivalent to finding a T-forest of maximum weight. These maximisation problems are actually the problems that we will solve. Consequently, any T-forest will be called a *solution* for an instance (G, T) or (G, w, T), respectively, and our aim is to find a solution of maximum size or maximum weight, respectively.

Throughout our proofs we will need to check if some graph F is a solution. The following lemma shows that we can recognize solutions in linear time. The lemma combines results claimed but not proved in [9,13]. It is easy to show but for an explicit proof we refer to [4, Lemma 3].

Lemma 1. *It is possible to decide in $O(n + m)$ time if a graph F is a T-forest for some given set $T \subseteq V(F)$.*

In our proofs we will not refer to Lemma 1 explicitly, but we will use it implicitly every time we must check if some graph F is a solution.

3   The Weighted Variant

In this section, we present our polynomial-time algorithm for WEIGHTED SUB-SET FEEDBACK VERTEX SET on $(2P_1 + P_4)$-free graphs.

Outline. Our algorithm is based on the following steps. We first show in Sect. 3.1 how to compute a solution F that contains at most one vertex from T, which moreover has small degree in F. In Sect. 3.2 we then show that if two vertices of small degree in a solution are non-adjacent, we can exploit the $(2P_1 + P_4)$-freeness of the input graph G to reduce to a graph G' of bounded mim-width. The latter enables us to apply the algorithm of Bergougnoux, Papadopoulos and Telle [3]. In Sect. 3.3 we deal with the remaining case, where all the vertices of small degree in a solution F form a clique and F contains at least two vertices of T. We first show that every vertex of T that belongs to F must have small degree in F. Hence, as the vertices in $T \cap V(F)$ must also induce a forest, F has exactly two adjacent vertices of T, each of small degree in F. This structural result enables us to do a small case analysis. We combine this step together with our previous algorithmic procedures into one algorithm.

Remark. Some of the lemmas in the following three subsections hold for $(sP_1 + P_4)$-free graphs, for every $s \geq 2$, or even for general graphs. In order to re-use these lemmas in Sect. 4, where we consider SUBSET FEEDBACK VERTEX SET for $(sP_1 + P_4)$-free graphs, we formulate these lemmas as general as possible.

3.1   Three Special Types of Solutions

In this section we will show how we can find three special types of solutions in polynomial time for $(2P_1 + P_4)$-free graphs. These solutions have in common that they contain at most one vertex from the set T and moreover, this vertex has small degree in F.

Let $G = (V, E)$ be a graph and let $T \subseteq V$ be a subset of vertices of G. A T-forest F is a ≤ 1-part solution if F contains at most one vertex from T and moreover, if F contains a vertex u from T, then u has degree at most 1 in F. The following lemma holds for general graphs and is easy to see.

Lemma 2. *For a graph $G = (V, E)$ with a positive vertex weighting w and a set $T \subseteq V$, it is possible to find a ≤ 1-part solution of maximum weight in polynomial time.*

Let $G = (V, E)$ be a graph and let $T \subseteq V$ be a subset of vertices of G. A T-forest F is a 2-part solution if F contains exactly one vertex u of T and u has exactly two neighbours v_1 and v_2 in F. We say that u is the *center* of F and that v_1 and v_2 are the *center neighbours*. Let A be the connected component of F that contains u. Then we say that A is the *center component* of F. We will prove how to find 2-part solutions in polynomial time even for general graphs. In order to do this, we will reduce to a classical problem, namely:

WEIGHTED VERTEX CUT

    Instance: a graph $G = (V, E)$, two distinct non-adjacent terminals t_1 and t_2, and a positive vertex weighting w.

    Task: determine a set $S \subseteq V \setminus \{t_1, t_2\}$ of minimum weight such that t_1 and t_2 are in different connected components of $G - S$.

The WEIGHTED VERTEX CUT problem is well known to be polynomial-time solvable by standard network flow techniques.

Lemma 3. WEIGHTED VERTEX CUT *is polynomial-time solvable.*

We use Lemma 3 in several of our proofs, including in the (omitted) proof of the next lemma.

Lemma 4. *For a graph $G = (V, E)$ with a positive vertex weighting w and a set $T \subseteq V$, it is possible to find a 2-part solution of maximum weight in polynomial time.*

Let $G = (V, E)$ be a graph and let $T \subseteq V$ be a subset of vertices of G. A T-forest F is a 3-*part solution* if F contains exactly one vertex u of T and u has exactly three neighbours v_1, v_2, v_3 in F. Again we say that u is the *center* of F; that v_1, v_2, v_3 are the *center neighbours*; and that the connected component of F that contains u is the *center component* of F. We can show the following lemma (proof omitted).

Lemma 5. *For a $(2P_1 + P_4)$-free graph $G = (V, E)$ with a positive vertex weighting w and a set $T \subseteq V$, it is possible to find a 3-part solution of maximum weight in polynomial time.*

3.2   Mim-Width

We also need some known results that involve the mim-width of a graph. This width parameter was introduced by Vatshelle [15]. For the definition of mim-width we refer to [15], as we do not need it here. A graph class $\mathcal{G}$ has *bounded* mim-width if there exists a constant c such that every graph in $\mathcal{G}$ has mim-width at most c. The mim-width of a graph class $\mathcal{G}$ is *quickly computable* if it is possible to compute in polynomial time a so-called branch decomposition for a graph $G \in \mathcal{G}$ whose mim-width is bounded by some function in the mim-width of G. We can now state the aforementioned result of Bergougnoux, Papadopoulos and Telle in a more detailed way.

Theorem 4 ([3]). WEIGHTED SUBSET FEEDBACK VERTEX SET *is polynomial-time solvable for every graph class whose mim-width is bounded and quickly computable.*

Belmonte and Vatshelle [2] proved that the mim-width of the class of permutation graphs is bounded and quickly computable. As P_4-free graphs form a subclass of the class of permutation graphs, we immediately obtain the following lemma[1].

Lemma 6. *The mim-width of the class of P_4-free graphs is bounded and quickly computable.*

For a graph class $\mathcal{G}$ and an integer $p \geq 0$, we let $\mathcal{G} + pv$ be the graph class that consists of all graphs that can be modified into a graph from $\mathcal{G}$ by deleting at most p vertices. The following lemma follows in a straightforward way from a result of Vatshelle [15].

Lemma 7. *If $\mathcal{G}$ is a graph class whose mim-width is bounded and quickly computable, then the same holds for the class $\mathcal{G} + pv$, for every constant $p \geq 0$.*

Let $G = (V, E)$ be an $(sP_1 + P_4)$-free graph for some $s \geq 2$ and let $T \subseteq V$. Let F be a T-forest of G. We define the *core* of F as the set of vertices of F that have at most $2s - 1$ neighbours in F. We say that F is *core-complete* if the core of F has no independent set of size at least s; otherwise F is *core-incomplete*[2]. We use the above results to show the following algorithmic lemma (proof omitted).

Lemma 8. *Let $s \geq 2$. For an $(sP_1 + P_4)$-free graph $G = (V, E)$ with a positive vertex weighting w and a set $T \subseteq V$, it is possible to find a core-incomplete solution of maximum weight in polynomial time.*

3.3 The Algorithm

In this section we present our algorithm for WEIGHTED SUBSET FEEDBACK VERTEX SET restricted to $(2P_1 + P_4)$-free graphs. We first need to prove one more structural lemma for core-complete solutions. We prove this lemma for any value $s \geq 2$, such that we can use this lemma in the next section as well. However, for $s = 2$ we have a more accurate upper bound on the size of the core.

Lemma 9. *For some $s \geq 2$, let $G = (V, E)$ be an $(sP_1 + P_4)$-free graph. Let $T \subseteq V$. Let F be a core-complete T-forest of G such that $T \cap V(F) \neq \emptyset$. Then the core of F contains every vertex of $T \cap V(F)$, and $T \cap V(F)$ has size at most $2s - 2$. If $s = 2$, the core of F is a clique of size at most 2 (in this case $T \cap V(F)$ has size at most 2 as well).*

Proof. Consider a vertex $u \in T \cap V(F)$. For a contradiction, assume that u does not belong to the core of F. Then u has at least $2s$ neighbours in F. Let $V_u = \{v_1, \ldots, v_p\}$ for some $p \geq 2s$ be the set of neighbours of u in F.

Let A be the connected component of F that contains u. As F is a T-forest, $A - u$ consists of p connected components $D_1, \ldots, D_p$ such that $v_i \in V(D_i)$ for

---

[1] It is well-known that P_4-free graphs have clique-width at most 2, and instead of Theorem 4 we could have used a corresponding result for clique-width. We chose to formulate Theorem 4 in terms of mim-width, as mim-width is a more powerful parameter than clique-width [15] and thus bounded for more graph classes.

[2] These notions are not meaningful if $s \in \{0, 1\}$. Hence, we defined them for $s \geq 2$.

$i \in \{1, \ldots, p\}$. In particular, this implies that $V_u = \{v_1, \ldots, v_p\}$ must be an independent set. As the core of F has no independent set of size s, this means that at most $s-1$ vertices of V_u may belong to the core of F. Recall that $p \geq 2s$. Hence, we may assume without loss of generality that $v_1, \ldots, v_{s+1}$ do *not* belong to the core of F. This means that $v_1, \ldots, v_{s+1}$ each have degree at least $2s$ in A. Hence, for $i \in \{1, \ldots, s+1\}$, vertex v_i is adjacent to some vertex w_i in D_i. As $s \geq 2$, we have that $2s > s+1$ and hence, vertex v_{s+2} exists. However, now the vertices $w_1, v_1, u, v_{s+2}, w_2, w_3, \ldots, w_{s+1}$ induce an $sP_1 + P_4$, a contradiction (see also Fig. 3).

From the above, we conclude that every vertex of $T \cap V(F)$ belongs to the core of F. As F is a T-forest, $T \cap V(F)$ induces a forest, and thus a bipartite graph. As F is core-complete, every independent set in the core has size at most $s - 1$. Hence, $T \cap V(F)$ has size at most $2(s - 1) = 2s - 2$.

Now suppose that $s = 2$. As F is core-complete, the core of F must be a clique. As the core of F contains $T \cap V(F)$ and $T \cap V(F)$ induces a forest, this means that the core of F, and thus also $T \cap V(F)$, has size at most 2. This completes the proof of the lemma.   □

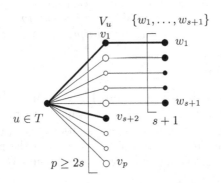

Fig. 3. An example of the contradiction obtained in Lemma 9: the assumption that a vertex $u \in T$ does not belong to the core of a core-complete solution leads to the presence of an induced $sP_1 + P_4$ (highlighted by the black vertices and thick edges). (Color figure online)

By using the above results and the results from Sects. 3.1 and 3.2, we are now able to prove our main result.

Theorem 5. WEIGHTED SUBSET FEEDBACK VERTEX SET *is polynomial-time solvable for* $(2P_1 + P_4)$*-free graphs.*

Proof. Let $G = (V, E)$ be a $(2P_1 + P_4)$-free graph, and let T be some subset of V. Let w be a positive vertex weighting of G. We aim to find a maximum weight T-forest F for (G, T, w) (recall that we call T-forests solutions for our problem). As $s = 2$, the core of F is, by definition, the set of vertices of F that have maximum degree at most 3 in F.

We first compute a core-incomplete solution of maximum weight; this takes polynomial time by Lemma 8 (in which we set $s = 2$). We will now compute in polynomial time a core-complete solution F of maximum weight for (G, T, w). We then compare the weights of the two solutions found to each other and pick one with the largest weight.

By Lemma 9, it holds for every core-complete solution F that $T \cap V(F)$ belongs to the core of F, and moreover that $|T \cap V(F)| \leq 2$. We first compute a core-complete solution F with $|T \cap V(F)| \leq 1$ of maximum weight. As $T \cap V(F)$ belongs to the core of F, we find that if $|T \cap V(F)| = 1$, say $T \cap V(F) = \{u\}$ for some $u \in T$, then u has maximum degree at most 3 in F. Hence, in the case where $|T \cap V(F)| \leq 1$, it suffices to compute a ≤ 1-part solution, 2-part solution and 3-part solution for (G, T, w) of maximum weight and to remember one with the largest weight. By Lemmas 2, 4 and 5, respectively, this takes polynomial time.

It remains to compute a core-complete solution F with $|T \cap V(F)| = 2$ of maximum weight. By Lemma 9, it holds for every such solution F that both vertices of $T \cap V(F)$ are adjacent and are the only vertices that belong to the core of F.

We consider all $O(n^2)$ possibilities of choosing two adjacent vertices of T to be the two core vertices of F. Consider such a choice of adjacent vertices u_1, u_2. So, u_1 and u_2 are the only vertices of degree at most 3 in the solution F that we are looking for and moreover, all other vertices of T do not belong to F.

Suppose one of the vertices u_1, u_2 has degree 1 in F. First let this vertex be u_1. Then we remove u_1 and all its neighbours except for u_2 from G. Let G' be the resulting graph. Let $T' = T \setminus (\{u_1\} \cup (N(u_1) \setminus \{u_2\}))$, and let w' be the restriction of w to G'. We now compute for (G', w', T'), a ≤ 1-part solution and 2-part solution of maximum weight with u_2 as center. By Lemmas 2 and 4, respectively, this takes polynomial time[3]. We then add u_1 back to the solution to get a solution for (G, w, T). We do the same steps with respect to u_2. In the end we take a solution with largest weight.

So from now on, assume that both u_1 and u_2 have degree at least 2 in F. We first argue that in this case both u_1 and u_2 have degree exactly 2 in F. For a contradiction, suppose u_1 has degree 3 in F (recall that u_1 has degree at most 3 in F). Let v_1 and v_1' be two distinct neighbours of u_1 in $V(F) \setminus \{u_2\}$. Let v_2 be a neighbour of u_2 in $V(F) \setminus \{u_1\}$. As F is a T-forest, v_1, v_1', v_2 belong to distinct connected components D_1, D_1' and D_2, respectively, of $F - \{u_1, u_2\}$. As the core of F consists of u_1 and u_2 only, v_1, v_1', v_2 each have a neighbour x_1, x_1', x_2 in D_1, D_1' and D_2, respectively. However, now $x_2, v_2, u_2, u_1, x_1, x_1'$ induce a $2P_1 + P_4$ in F and thus also in G, a contradiction; see also Fig. 4.

---

[3] Strictly speaking, this statement follows from the proofs of these two lemmas, as we have fixed u_2 as the center.

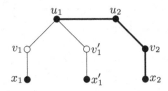

Fig. 4. The situation in Theorem 5 where u_1 has degree at least 3 in F and u_2 has degree 2 in F; this leads to the presence of an induced $2P_1 + P_4$ (highlighted by the black vertices and thick edges). (Color figure online)

From the above we conclude that each of u_1 and u_2 has exactly one other neighbour in F. Call these vertices v_1 and v_2, respectively. We consider all $O(n^2)$ possibilities of choosing v_1 and v_2. As F is a T-forest, $G - \{u_1, u_2\}$ consists of two connected components D_1 and D_2, such that v_1 belongs to D_1 and v_2 belongs to D_2.

Let G' be the graph obtained from G by removing every vertex of T, every neighbour of u_1 except v_1 and every neighbour of u_2 except v_2. Let w' be the restriction of w to G'. Then, it remains to solve WEIGHTED VERTEX CUT for the instance (G', v_1, v_2, w'). By Lemma 3, this can be done in polynomial time. Out of all the solutions found for different pairs u_1, u_2 we take one with the largest weight. Note that we found this solution in polynomial time, as the number of branches is $O(n^4)$.

As mentioned we take a solution of maximum weight from all the solutions found in the above steps. The correctness of our algorithm follows from the fact that we exhaustively considered all possible situations. Moreover, the number of situations is polynomial and processing each situation takes polynomial time. Hence, the running time of our algorithm is polynomial.   □

4   The Unweighted Variant

In this section, we present our polynomial-time algorithm for SUBSET FEEDBACK VERTEX SET on $(sP_1 + P_4)$-free graphs for every $s \geq 0$. As this problem is a special case of WEIGHTED SUBSET FEEDBACK VERTEX SET (namely when $w \equiv 1$), we can use some of the structural results from the previous section.

Theorem 6. SUBSET FEEDBACK VERTEX SET *is polynomial-time solvable on* $(sP_1 + P_4)$-*free graphs for every* $s \geq 0$.

Proof. Let $G = (V, E)$ be an $(sP_1 + P_4)$-free graph for some integer s, and let $T \subseteq V$. Let $|V| = n$. As the class of $(sP_1 + P_4)$-free graphs is a subclass of the class of $((s + 1)P_1 + P_4)$-free graphs, we may impose any lower bound on s; we set $s \geq 2$. We aim to find a T-forest F of G of maximum size (recall that we call T-forests solutions for our problem).

We first compute a maximum-size core-incomplete solution for (G, T). By Lemma 8, this takes polynomial time. It remains to compare the size of this solution with a maximum-size core-complete solution, which we compute below.

By Lemma 9, we find that $T \cap V(F)$ has size at most $2s - 2$ for every core-complete solution F. We consider all $O(n^{2s-2})$ possibilities of choosing the vertices of $T \cap V(F)$. For each choice of $T \cap V(F)$ we do as follows. We note that the set of vertices of $G - T$ that do not belong to F has size at most $|T \cap V(F)|$; otherwise $F' = V \setminus T$ would be a larger solution than F. Hence, we can consider all $O(n^{|T \cap V(F)|}) = O(n^{2s-2})$ possibilities of choosing the set of vertices of $G - T$ that do not belong to F, or equivalently, of choosing the set of vertices of $G - T$ that *do* belong to F. In other words, we guessed F by brute force, and the number of guesses is $O(n^{4s-4})$. In the end we found in polynomial time a maximum-size core-complete solution. We compare it with the maximum-size core-incomplete solution found above and pick one with the largest size. $\qquad\square$

5   Conclusions

By combining known hardness results with new polynomial-time results, we completely classified the complexities of WEIGHTED SUBSET FEEDBACK VERTEX SET and SUBSET FEEDBACK VERTEX SET for H-free graphs. We recall that the classical versions WEIGHTED FEEDBACK VERTEX SET and FEEDBACK VERTEX SET are not yet completely classified (see Theorem 1).

We now briefly discuss the variant where instead of intersecting every T-cycle, a solution only needs to intersect every T-cycle of *odd* length. These two problems are called WEIGHTED SUBSET ODD CYCLE TRANSVERSAL and SUBSET ODD CYCLE TRANSVERSAL, respectively. So far, these problems behave in exactly the same way on H-free graphs as their feedback vertex set counterparts (see [4] and [5]). So, the only open cases for WEIGHTED SUBSET ODD CYCLE TRANSVERSAL on H-free graphs are the ones where $H \in \{2P_1 + P_3, P_1 + P_4, 2P_1 + P_4\}$ and the only open cases for SUBSET ODD CYCLE TRANSVERSAL on H-free graphs are the ones where $H = sP_1 + P_4$ for some $s \geq 1$. As solutions F for these problems may only contain vertices of T of high degree, we can no longer use our proof technique, and new ideas are needed.

We note, however, that complexity dichotomies of WEIGHTED SUBSET ODD CYCLE TRANSVERSAL and SUBSET ODD CYCLE TRANSVERSAL do not have to coincide with those in Theorems 2 and 3 for their feedback vertex set counterparts. After all, the complexities of the corresponding classical versions may not coincide either. Namely, it is known that ODD CYCLE TRANSVERSAL is NP-complete for $(P_2 + P_5, P_6)$-free graphs [6], and thus for $(P_2 + P_5)$-free graphs and P_6-free graphs, whereas for FEEDBACK VERTEX SET such a hardness result is unlikely: for every linear forest H, FEEDBACK VERTEX SET is quasipolynomial-time solvable on H-free graphs [8].

References

1. Abrishami, T., Chudnovsky, M., Pilipczuk, M., Rzążewski, P., Seymour, P.: Induced subgraphs of bounded treewidth and the container method. Proc. SODA **2021**, 1948–1964 (2021)

2. Belmonte, R., Vatshelle, M.: Graph classes with structured neighborhoods and algorithmic applications. Theoret. Comput. Sci. **511**, 54–65 (2013)
3. Bergougnoux, B., Papadopoulos, C., Telle, J.A.: Node multiway cut and subset feedback vertex set on graphs of bounded mim-width. Proc. WG 2020, LNCS 12301, 388–400 (2020)
4. Brettell, N., Johnson, M., Paesani, G., Paulusma, D.: Computing subset transversals in H-free graphs. Theoret. Comput. Sci. **898**, 59–68 (2022)
5. Brettell, N., Johnson, M., Paulusma, D.: Computing weighted subset transversals in H-free graphs. J. Comput. Syst. Sci. **128**, 71–85 (2022). https://doi.org/10.1007/978-3-030-83508-8_17
6. Dabrowski, K.K., Feghali, C., Johnson, M., Paesani, G., Paulusma, D., Rzążewski, P.: On cycle transversals and their connected variants in the absence of a small linear forest. Algorithmica **82**, 2841–2866 (2020). https://doi.org/10.1007/s00453-020-00706-6
7. Fomin, F.V., Heggernes, P., Kratsch, D., Papadopoulos, C., Villanger, Y.: Enumerating minimal subset feedback vertex sets. Algorithmica **69**, 216–231 (2014). https://doi.org/10.1007/s00453-012-9731-6
8. Gartland, P., Lokshtanov, D., Pilipczuk, M., Pilipczuk, M., Rzążewski, P.: Finding large induced sparse subgraphs in C_t-free graphs in quasipolynomial time. Proc. STOC, **2021**, 330–341 (2021)
9. Lokshtanov, D., Misra, P., Ramanujan, M.S., Saurabh, S.: Hitting selected (odd) cycles. SIAM J. Discret. Math. **31**, 1581–1615 (2017)
10. Munaro, A.: On line graphs of subcubic triangle-free graphs. Discret. Math. **340**, 1210–1226 (2017)
11. Paesani, G., Paulusma, D., Rzążewski, P.: Feedback vertex set and even cycle transversal for H-free graphs: finding large block graphs. SIAM Journal on Discrete Mathematics, to appear
12. Papadopoulos, C., Tzimas, S.: Polynomial-time algorithms for the subset feedback vertex set problem on interval graphs and permutation graphs. Discret. Appl. Math. **258**, 204–221 (2019)
13. Papadopoulos, C., Tzimas, S.: Subset feedback vertex set on graphs of bounded independent set size. Theoret. Comput. Sci. **814**, 177–188 (2020)
14. Poljak, S.: A note on stable sets and colorings of graphs. Comment. Math. Univ. Carol. **15**, 307–309 (1974)
15. Vatshelle, M.: New Width Parameters of Graphs. PhD thesis, University of Bergen (2012)

Linearizing Partial Search Orders

Robert Scheffler[(⊠)]

Institute of Mathematics, Brandenburg University of Technology,
Cottbus, Germany
robert.scheffler@b-tu.de

Abstract. In recent years, questions about the construction of special orderings of a given graph search were studied by several authors. On the one hand, the so called end-vertex problem introduced by Corneil et al. in 2010 asks for search orderings ending in a special vertex. On the other hand, the problem of finding orderings that induce a given search tree was introduced already in the 1980s s by Hagerup and received new attention most recently by Beisegel et al. Here, we introduce a generalization of some of these problems by studying the question whether there is a search ordering that is a linear extension of a given partial order on a graph's vertex set. We show that this problem can be solved in polynomial time on chordal bipartite graphs for LBFS, which also implies the first polynomial-time algorithms for the end-vertex problem and two search tree problems for this combination of graph class and search. Furthermore, we present polynomial-time algorithms for LBFS and MCS on split graphs, which generalize known results for the end-vertex and search tree problems.

Keywords: Graph search · Partial order · End-vertex problem · Search tree recognition · LBFS · MCS

1 Introduction

The graph searches *Breadth First Search* (BFS) and *Depth First Search* (DFS) are considered as some of the most basic algorithms in both graph theory and computer science. Taught in many undergraduate courses around the world, they are an elementary component of several graph algorithms. There are also many other more sophisticated graph searches, e.g., the *Lexicographic Breadth First Search* (LBFS) [21] or the *Maximum Cardinality Search* (MCS) [23] which are also used to solve several graph problems among them the recognition problems of graph classes [6,8,12,13], the computation of minimal separators [19] as well as the computation of minimal triangulations [5].

In recent years, different problems of finding special search orderings have gained attention from several researchers. One of these problems is the end-vertex problem introduced in 2010 by Corneil et al. [10]. It asks whether a given vertex in a graph can be visited last by some graph search. The problem was motivated by *multi-sweep algorithms* where a search is applied several times to a

© Springer Nature Switzerland AG 2022
M. A. Bekos and M. Kaufmann (Eds.): WG 2022, LNCS 13453, pp. 425–438, 2022.
https://doi.org/10.1007/978-3-031-15914-5_31

graph such that every application starts in the vertex where the preceding search ordering has ended. Corneil et al. [10] showed that the end-vertex problem is $\mathcal{NP}$-complete for LBFS on weakly chordal graphs. Similar results were obtained for other searches such as BFS, DFS and MCS [2, 7, 20], while for several graph classes, among them split graphs, polynomial-time algorithms were presented in [2, 7, 10, 14, 20].

An important structure closely related to a graph search is the corresponding search tree. Such a tree contains all the vertices of the graph and for every vertex different from the start vertex exactly one edge to a vertex preceding it in the search ordering. Such trees can be of particular interest as for instance the tree obtained by a BFS contains the shortest paths from the root to all other vertices in the graph and the trees generated by DFS can be used for fast planarity testing [17]. The problem of deciding whether a given spanning tree of a graph can be obtained by a particular search was introduced by Hagerup [15] in 1985, who presented a linear-time algorithm for recognizing DFS-trees. In the same year, Hagerup and Nowak [16] gave a similar result for the BFS-tree recognition.

Recently, Beisegel et al. [3, 4] introduced a more general framework for the search tree recognition problem. They introduced the term $\mathcal{F}$-tree for search trees where a vertex is connected to its first visited neighbor, i.e., BFS-like trees, and $\mathcal{L}$-trees for search trees where a vertex is connected to its most recently visited neighbor, i.e., DFS-like trees. They showed, among other things, that the $\mathcal{F}$-tree recognition is $\mathcal{NP}$-complete for LBFS and MCS on weakly chordal graphs, while the problem can be solved in polynomial time for both searches on chordal graphs.

Our Contribution. There seems to be a strong relationship between the complexity of the end-vertex problem and the recognition problem of $\mathcal{F}$-trees. There are many combinations of graph classes and graph searches where both problems are $\mathcal{NP}$-complete or both problems are solvable in polynomial time. To further study this relationship, we present a generalization of these two problems by introducing the *Partial Search Order Problem (PSOP)* of a graph search $\mathcal{A}$. Given a graph G and a partial order π on its vertex set, it asks whether there is a search ordering produced by $\mathcal{A}$ which is a linear extension of π.

We show that a greedy algorithm solves the PSOP of Generic Search, i.e., the search where every vertex can be visited next as long as it has an already visited neighbor. Furthermore, we present a polynomial-time algorithm for the PSOP of LBFS on chordal bipartite graphs, i.e., bipartite graphs without induced cycles of length larger than four. This result also implies a polynomial-time algorithm for the end-vertex problem on chordal bipartite graphs, a generalization of the result by Gorzny and Huang [14] on the end-vertex problem of LBFS on AT-free bipartite graphs. For split graphs, we will give polynomial-time algorithms for the PSOP of LBFS and MCS that generalize the results on the end-vertex problem [2, 7] and the $\mathcal{F}$-tree problem [3] of these searches on this graph class.

Due to lack of space, the proofs of the results are omitted here. They can be found in the full version [22].

2   Preliminaries

General Notation. The graphs considered in this paper are finite, undirected, simple and connected. Given a graph G, we denote by $V(G)$ the set of vertices and by $E(G)$ the set of edges. For a vertex $v \in V(G)$, we denote by $N(v)$ the *neighborhood* of v in G, i.e., the set $N(v) = \{u \in V \mid uv \in E\}$, where an edge between u and v in G is denoted by uv. The *neighborhood of a set* $A \subset V(G)$ is the union of the neighborhoods of the vertices in A. The *distance* of a vertex v to a vertex w is the number of edges of the shortest path from v to w. The set $N^{\ell}(v)$ contains all vertices whose distance to the vertex v is equal to ℓ.

A *clique* in a graph G is a set of pairwise adjacent vertices and an *independent set* in G is a set of pairwise nonadjacent vertices. A *split graph* G is a graph whose vertex set can be partitioned into sets C and I, such that C is a clique in G and I is an independent set in G. We call such a partition a *split partition*. A graph is *bipartite* if its vertex set can be partitioned into two independent sets X and Y. A bipartite graph G is called *chordal bipartite* if every induced cycle contained in G has a length of four. Note that there is a strong relationship between split graphs and bipartite graphs. Every bipartite graph is a spanning subgraph of a split graph and every split graph can be made to a bipartite graph by removing the edges between the clique vertices.

A *tree* is an acyclic connected graph. A *spanning tree* of a graph G is an acyclic connected subgraph of G which contains all vertices of G. A tree together with a distinguished *root vertex* r is said to be *rooted*. In such a rooted tree T, a vertex v is the *parent* of vertex w if v is an element of the unique path from w to the root r and the edge vw is contained in T. A vertex w is called the *child* of v if v is the parent of w.

A *vertex ordering* of G is a bijection $\sigma : \{1, 2, \dots, |V(G)|\} \to V(G)$. We denote by $\sigma^{-1}(v)$ the position of vertex $v \in V(G)$. Given two vertices u and v in G we say that u is *to the left* (resp. *to the right*) of v if $\sigma^{-1}(u) < \sigma^{-1}(v)$ (resp. $\sigma^{-1}(u) > \sigma^{-1}(v)$) and we denote this by $u \prec_{\sigma} v$ (resp. $u \succ_{\sigma} v$).

A *partial order* π on a set X is a reflexive, antisymmetric and transitive binary relation on X. We also denote $(x, y) \in \pi$ by $x \prec_{\pi} y$ if $x \neq y$. A *linear extension* of π is a total order σ of the elements of X that fulfills all conditions of π, i.e., if $x \prec_{\pi} y$, then $x \prec_{\sigma} y$. We will often use the term "σ extends π". For a binary relation π' on X we say that the *reflexive and transitive closure* of π' is the smallest binary relation $\pi \supseteq \pi'$ that is reflexive and transitive.

Graph Searches. A *graph search* is an algorithm that, given a graph G as input, outputs a vertex ordering of G. All graph searches considered in this paper can be formalized adapting a framework introduced by Corneil et al. [9] (a similar framework is given in [18]). This framework uses subsets of $\mathbb{N}^+$ as vertex labels. Whenever a vertex is visited, its index in the search ordering is added to the labels of its unvisited neighbors. The search $\mathcal{A}$ is defined via a strict partial order $\prec_{\mathcal{A}}$ on the elements of $\mathcal{P}(\mathbb{N}^+)$ (see Algorithm 1). For a given graph search $\mathcal{A}$ we say that a vertex ordering σ of a graph G is an $\mathcal{A}$-*ordering* of G if σ can be the output of $\mathcal{A}$ with input G.

Algorithm 1. Label Search($\prec_{\mathcal{A}}$)

Input: A graph G
Output: A search ordering σ of G
1 **begin**
2   **foreach** $v \in V(G)$ **do** $label(v) \leftarrow \emptyset$;
3   **for** $i \leftarrow 1$ **to** $|V(G)|$ **do**
4          $Eligible \leftarrow \{x \in V(G) \mid x$ unnumbered and $\nexists$ unnumbered $y \in V(G)$
5   such that $label(x) \prec_{\mathcal{A}} label(y)\}$;
6   let v be an arbitrary vertex in $Eligible$;
7          $\sigma(i) \leftarrow v$;   /* assigns to v the number i */
8   **foreach** *unnumbered vertex* $w \in N(v)$ **do** $label(w) \leftarrow label(w) \cup \{i\}$;

In the following, we define the searches considered in this paper by presenting suitable partial orders $\prec_{\mathcal{A}}$ (see [9]). The *Generic Search* (GS) is equal to the Label Search($\prec_{GS}$) where $A \prec_{GS} B$ if and only if $A = \emptyset$ and $B \neq \emptyset$. Thus, any vertex with a numbered neighbor can be numbered next.

The partial label order $\prec_{BFS}$ for *Breadth First Search* (BFS) is defined as follows: $A \prec_{BFS} B$ if and only if $A = \emptyset$ and $B \neq \emptyset$ or $\min(A) > \min(B)$. For the *Lexicographic Breadth First Search* (LBFS) [21] we consider the partial order $\prec_{LBFS}$ with $A \prec_{LBFS} B$ if and only if $A \subsetneq B$ or $\min(A \setminus B) > \min(B \setminus A)$.

The *Maximum Cardinality Search* (MCS) [23] uses the partial order $\prec_{MCS}$ with $A \prec_{MCS} B$ if and only if $|A| < |B|$. The *Maximal Neighborhood Search* (MNS) [11] is defined using $\prec_{MNS}$ with $A \prec_{MNS} B$ if and only if $A \subsetneq B$. If $A \prec_{MNS} B$, then it also holds that $A \prec_{LBFS} B$ and $A \prec_{MCS} B$. Thus, any ordering produced by LBFS or MCS is also an MNS ordering.

In the search algorithms following the framework given in Algorithm 1, any of the vertices in the set *Eligible* can be chosen as the next vertex. Some applications use special variants of these searches that involve tie-breaking. For any instantiation $\mathcal{A}$ of Algorithm 1, we define the graph search $\mathcal{A}^+$ as follows: Add a vertex ordering ρ of graph G as additional input and replace line 6 in Algorithm 1 with "let v be the vertex in *Eligible* that is leftmost in ρ". Note that this corresponds to the algorithm TBLS given in [9]. The search ordering $\mathcal{A}^+(\rho)$ is unique since there are no ties to break.

3   The Partial Search Order Problem

We start this section by introducing the problem considered in this paper.

Problem 1. *Partial Search Order Problem (PSOP) of graph search* $\mathcal{A}$

Instance: *A graph* G, *a partial order* π *on* $V(G)$.
Task: *Decide whether there is an* $\mathcal{A}$-*ordering of* G *that extends* π.

We will also consider a special variant, where the start vertex of the search ordering is fixed. We call this problem the *rooted partial search order problem*.

Note that the general problem and the rooted problem are polynomial time equivalent. If we have a polynomial time algorithm to solve the rooted problem we can apply it $|V(G)|$ times to solve the general problem. On the other hand, the rooted problem with fixed start vertex r can be solved by a general algorithm. To this end, we add all the tuples (r, v), $v \in V(G)$, to the partial order π. Note that in the following we always assume that a given start vertex r is a minimal element of the partial order π since otherwise we can reject the input immediately.

The *end-vertex problem* of a graph search $\mathcal{A}$ introduced by Corneil et al. [10] in 2010 asks whether the vertex t can be the last vertex of an $\mathcal{A}$-ordering of a given graph G. This question can be encoded by the partial order $\pi := \{(u, v) \mid u, v \in V(G), u = v \text{ or } v = t\}$, leading to the following observation.

Observation 2. *The end-vertex problem of a graph search $\mathcal{A}$ on a graph G can be solved by solving the PSOP of $\mathcal{A}$ on G for a partial order of size $\mathcal{O}(|V(G)|)$.*

From this observation it follows directly that the partial search order problem is $\mathcal{NP}$-complete for BFS, DFS, LBFS, LDFS, MCS and MNS [2,7,10].

In [3], Beisegel et al. introduced the terms $\mathcal{F}$-*tree* and $\mathcal{L}$-*tree* of a search ordering. For this we only consider search orderings produced by a *connected graph searches*, i.e., a graph search that outputs search orderings of the Generic Search. In the $\mathcal{F}$-tree of such an ordering, every vertex different from the start vertex is connected to its leftmost neighbor in the search ordering. In the $\mathcal{L}$-tree, any vertex v different from the start vertex is connected to its rightmost neighbor that is to the left of v in the search ordering. The problem of deciding whether a given spanning tree of a graph can be the $\mathcal{F}$-tree ($\mathcal{L}$-tree) of a search ordering of a given type is called $\mathcal{F}$-*tree ($\mathcal{L}$-tree) recognition problem*. If the start vertex is fixed, it is called the *rooted $\mathcal{F}$-tree ($\mathcal{L}$-tree) recognition problem*. The rooted $\mathcal{F}$-tree recognition problem is a special case of the (rooted) PSOP, as the following proposition shows.

Proposition 3. *Let $\mathcal{A}$ be a connected graph search. Given a graph G and a spanning tree T of G rooted in r, we define π to be the reflexive, transitive closure of the relation $\mathcal{R} := \{(x, y) \mid x \text{ is parent of } y \text{ in } T \text{ or there is child } z \text{ of } x \text{ in } T \text{ with } yz \in E(G)\}$. The tree T is the $\mathcal{F}$-tree of an $\mathcal{A}$-ordering σ of G if and only if π is a partial order and σ extends π.*

Therefore, the rooted $\mathcal{F}$-tree problem of a graph search $\mathcal{A}$ on a graph G can be solved by solving the (rooted) PSOP of $\mathcal{A}$ on G.

Note that the general $\mathcal{F}$-tree recognition problem without fixed start vertex can be solved by deciding the partial search order problem for any possible root. The $\mathcal{L}$-tree recognition problem, however, is not a special case of the partial search order problem. For a vertex w, its parent v and another neighbor z of w, it must either hold that $v \prec_\sigma w \prec_\sigma z$ or that $z \prec_\sigma v \prec_\sigma w$. These constraints cannot be encoded using a partial order. Nevertheless, we will see in Sect. 5 that on bipartite graphs the PSOP of (L)BFS is a generalization of the $\mathcal{L}$-tree recognition problem of (L)BFS.

Algorithm 2. Rooted PSOP of Generic Search

Input: Connected graph G, a vertex $r \in V(G)$, a partial order π on $V(G)$
Output: GS ordering σ of G extending π or "π cannot be linearized"

```
1  begin
2  │  S ← {r};      i ← 1;
3  │  while S ≠ ∅ do
4  │  │  let v be an arbitrary element of S;
5  │  │  remove v from S and from π;
6  │  │  σ(i) ← v;      i ← i + 1;
7  │  │  foreach w ∈ N(v) do mark w;
8  │  └  foreach marked x ∈ V(G) which is minimal in π do  S ← S ∪ {x};
9  │  if i = |V(G)| + 1 then return σ;
10 └  else return "π cannot be linearized";
```

We conclude this section with a simple algorithm for the rooted PSOP of Generic Search (see Algorithm 2 for the pseudocode). First the algorithm visits the given start vertex r. Afterwards it looks for a vertex with an already visited neighbor among all vertices that are minimal in the remaining partial order. If no such vertex exists, then it rejects. Otherwise, it visits one of these vertices next.

Theorem 4. *Algorithm 2 solves the rooted partial search order problem of Generic Search for a graph G and a partial order π in time $\mathcal{O}(|V(G)|+|E(G)|+|\pi|)$.*

4   One-Before-All Orderings

Before we present algorithms for the PSOP we introduce a new ordering problem that will be used in the following two sections to solve the PSOP of LBFS on both chordal bipartite graphs and split graphs.

Problem 5. *One-Before-All Problem (OBAP)*

Instance: *A set M, a set $\mathcal{Q} \subseteq \mathcal{P}(M)$, a relation $\mathcal{R} \subseteq \mathcal{Q} \times \mathcal{Q}$*
Task: *Decide whether there is a linear ordering σ of M fulfilling the One-Before-All property, i.e., for all $A, B \in \mathcal{Q}$ with $(A, B) \in \mathcal{R}$ and $B \neq \emptyset$ there is an $x \in A$ such that for all $y \in B$ it holds that $x \prec_\sigma y$.*

Note that every partial order π on a set X can be encoded as an OBAP instance by setting $M = X$, $\mathcal{Q} = \{\{x\} \mid x \in X\}$ and $\mathcal{R} = \{(\{x\}, \{y\}) \mid x \prec_\pi y\}$. Thus, the OBAP generalizes the problem of finding a linear extension of an partial order.

In the following we describe how we can solve the one-before-all problem in time linear in the input size $|M| + |\mathcal{R}| + \sum_{A \in \mathcal{Q}} |A|$ (see Algorithm 3 for the pseudocode). For every set $A \in \mathcal{Q}$ we introduce a counter $r(A)$ containing the number of tuples $(X, A) \in \mathcal{R}$. For every element $x \in M$ the variable $t(x)$ counts

---

Algorithm 3. OBAP

Input: A set M, a set $\mathcal{Q} \subseteq \mathcal{P}(M)$, a relation $\mathcal{R} \subseteq \mathcal{Q} \times \mathcal{Q}$.
Output: An OBA-ordering σ of the elements in M or "No ordering".

```
1  begin
2  │  r(A) ← 0  ∀A ∈ Q;     t(x) ← 0  ∀x ∈ M;     S ← ∅;     i ← 1;
3  │  foreach (A, B) ∈ R do r(B) ← r(B) + 1;
4  │  foreach A ∈ Q with r(A) > 0 do
5  │  └  foreach x ∈ A do t(x) ← t(x) + 1;
6  │  foreach x ∈ M with t(x) = 0 do S ← S ∪ {x};
7  │  while S ≠ ∅ do
8  │  │  let x be an element in S;
9  │  │  S ← S \ {x};     σ(i) ← x;     i ← i + 1;
10 │  │  foreach A ∈ Q with x ∈ A do
11 │  │  │  Q ← Q \ {A};
12 │  │  │  foreach (A, B) ∈ R do
13 │  │  │  │  R ← R \ {(A, B)};
14 │  │  │  │  r(B) ← r(B) − 1;
15 │  │  │  │  if r(B) = 0 then
16 │  │  │  │  │  foreach y ∈ B do
17 │  │  │  │  │  │  t(y) ← t(y) − 1;
18 │  │  │  │  │  └  if t(y) = 0 then S ← S ∪ {y};

19 │  if i = |M| + 1 then return σ;
20 └  else return "No ordering";
```

---

the number of sets $A \in \mathcal{Q}$ with $x \in A$ and $r(A) > 0$. Our algorithm builds the ordering σ from left to right. It is not difficult to see that an element x can be chosen next if and only if $t(x) = 0$. As long as such an element exists, the algorithm chooses one, deletes all tuples (A, B) with $x \in A$ from $\mathcal{R}$ and updates the r- and the t-values. If no such element exists, then the algorithm returns "No ordering".

Theorem 6. *Given a set M, a set $\mathcal{Q} \subseteq \mathcal{P}(M)$ and a relation $\mathcal{R} \subseteq \mathcal{Q} \times \mathcal{Q}$, Algorithm 3 returns a linear ordering σ of M fulfilling the one-before-all property if and only if such an ordering exists. The running time of the algorithm is $\mathcal{O}(|M| + |\mathcal{R}| + \sum_{A \in \mathcal{Q}} |A|)$.*

5 Partial LBFS Orders of Chordal Bipartite Graphs

In [14], Gorzny and Huang showed that the end-vertex problem of LBFS is $\mathcal{NP}$-complete on bipartite graphs but can be solved in polynomial time on AT-free bipartite graphs. In this section we will generalize the latter result in two ways by presenting a polynomial-time algorithm for the partial search order problem on chordal bipartite graphs, a superset of AT-free bipartite graphs.

The following result will be a key ingredient of our approach. It shows that for two vertices x and y in the same layer $N_i(r)$ of a BFS starting in r that have a common neighbor in the succeeding layer $N^{i+1}(r)$, it holds that the neighborhoods of x and y in the preceding layer $N^{i-1}(r)$ are comparable.

Lemma 7. *Let G be a connected chordal bipartite graph and let r be a vertex of G. Let x and y be two vertices in $N^i(r)$. If there is a vertex $z \in N^{i+1}(r)$ which is adjacent to both x and y, then $N(x) \cap N^{i-1}(r) \subseteq N(y)$ or $N(y) \cap N^{i-1}(r) \subseteq N(x)$.*

Algorithm 4 presents the pseudocode of an algorithm for the rooted PSOP of LBFS on chordal bipartite graphs. We assume that the partial order π contains only tuples where both elements are in the same layer of a BFS starting in r. Otherwise, the tuple is trivially fulfilled by any BFS ordering starting in r or no such BFS ordering fulfills the tuple. The algorithm constructs an OBAP-instance with set $\mathcal{Q}_i \subseteq \mathcal{P}(N^i(r))$ and $\mathcal{R}_i \subseteq \mathcal{Q}_i \times \mathcal{Q}_i$ for any layer i of the BFS. First we add the tuple $(\{x\}, \{y\})$ to the set $\mathcal{R}_i$ for every tuple $(x,y) \in \pi$ with $x,y \in N^i(r)$. Now the algorithm iterates through all layers starting in the last one. For any element $(A, B) \in \mathcal{R}_i$ the algorithm inserts a tuple (A'', B') to the relation $\mathcal{R}_{i-1}$. The set A'' contains all neighbors of set A in layer $i-1$ that are not neighbors of set B and whose neighborhood in layer $i-2$ is maximal among all these neighbors. The set B' contains all neighbors of B in the layer $i-1$ that are not neighbors of A. At the end, the algorithm checks whether the OBAP-instance $(N^i(r), \mathcal{Q}_i, \mathcal{R}_i)$ of every layer i can be solved. If this is not the case, then the algorithm rejects. Otherwise, it concatenates the computed OBA-orderings. The resulting ordering ρ is used as tie-breaker for a LBFS$^+$ whose result is returned by the algorithm.

The following lemma is a direct consequence of the construction of the elements of $\mathcal{R}_i$ and Lemma 7.

Lemma 8. *Let $(A, B) \in \mathcal{R}_i$. For any $x \in A$ it holds that $N(A) \cap N^{i-1}(r) \subseteq N(x)$ and if $B \neq \emptyset$ then there is a vertex $y \in B$ with $N(B) \cap N^{i-1}(r) \subseteq N(y)$.*

Using this lemma, we can show the correctness of Algorithm 4.

Theorem 9. *Given a connected chordal bipartite graph G, a partial order π on $V(G)$ and a vertex $r \in V(G)$, Algorithm 4 decides in time $\mathcal{O}(|\pi| \cdot |V(G)|^2)$ whether there is an LBFS ordering of G that starts in r and is a linear extension of π.*

Due to Observation 2, we can solve the end-vertex problem of LBFS on chordal bipartite graphs by solving the rooted PSOP $|V(G)|$ times with a partial order of size $\mathcal{O}(|V(G)|)$. This leads to the following time bound.

Corollary 10. *Given a connected chordal bipartite graph G, we can solve the end-vertex problem of LBFS on G in time $\mathcal{O}(|V(G)|^4)$.*

Similarly, it follows from Proposition 3 that the rooted $\mathcal{F}$-tree recognition problem can be solved in time $\mathcal{O}(|V(G)|^4)$. Different to the general case, we can show that for BFS orderings of bipartite graphs the $\mathcal{L}$-tree recognition problem can also be reduced to the partial search order problem.

Algorithm 4. Rooted PSOP of LBFS on chordal bipartite graphs

Input: Connected chordal bipartite graph G, vertex $r \in V(G)$, partial order π
      on $V(G)$

Output: An LBFS ordering σ of G extending π or "π cannot be linearized"

1   **begin**

2   let k be the maximal distance of a vertex $v \in V(G)$ from r;

3     $\mathcal{Q}_i \leftarrow \{\{x\} \mid x \in N^i(r)\} \quad \forall i \in \{1, \ldots, k\}$;

4     $\mathcal{R}_i \leftarrow \{(\{x\}, \{y\}) \mid x, y \in N^i(r), x \prec_\pi y\} \quad \forall i \in \{1, \ldots, k\}$;

5   **for** $i \leftarrow k$ **downto** _2_ **do**

6   **foreach** $(A, B) \in \mathcal{R}_i$ **do**

7           $A' \leftarrow [N(A) \cap N^{i-1}(r)] \setminus N(B)$;

8           $A'' \leftarrow \{v \in A' \mid N(v) \cap N^{i-2}(r) = N(A') \cap N^{i-2}(r)\}$;

9           $B' \leftarrow [N(B) \cap N^{i-1}(r)] \setminus N(A)$;

10         $\mathcal{Q}_{i-1} \leftarrow \mathcal{Q}_{i-1} \cup \{A'', B'\}$;

11         $\mathcal{R}_{i-1} \leftarrow \mathcal{R}_{i-1} \cup \{(A'', B')\}$;

12   let ρ be an empty vertex ordering;

13   **for** $i \leftarrow k$ **downto** _1_ **do**

14   **if** _there is an OBA-ordering_ σ _for input_ $(N^i(r), \mathcal{Q}_i, \mathcal{R}_i)$ **then**

15           $\rho \leftarrow \sigma \mathbin{+\!\!+} \rho$

16   **else return** "π _cannot be linearized_";

17     $\rho \leftarrow r \mathbin{+\!\!+} \rho$;

18   **return** $LBFS^+(\rho)$ of G;

Proposition 11. _The rooted $\mathcal{L}$-tree recognition problem of any graph search $\mathcal{A}$ that produces BFS orderings can be solved on a bipartite graph G by solving the rooted PSOP of $\mathcal{A}$ on G._

This proposition and the observation above lead to the following time bound for the search tree recognition problems on chordal bipartite graphs.

Corollary 12. _On a chordal bipartite graph G, we can solve the rooted $\mathcal{F}$-tree and the rooted $\mathcal{L}$-tree recognition problem of LBFS in time $\mathcal{O}(|V(G)|^4)$._

6   Partial LBFS and MCS Orders of Split Graphs

Both the end-vertex problem and the $\mathcal{F}$-tree recognition problem of several searches are well studied on split graphs (see [2,4,7]). In this section we will generalize some of these results to the partial search order problem.

Consider a split graph G with a split partition consisting of a clique C and an independent set I. During a computation of an MNS ordering of G, every vertex that has labeled some vertex in I has also labeled every unnumbered vertex contained in C. Therefore, we can choose a vertex of C as the next vertex as long as there are still unnumbered vertices in C. This means that it is not a problem to force a clique vertex to be to the left of an independent vertex in an

MNS ordering. However, forcing a vertex of I to be to the left of a vertex of C is more difficult. We will call a vertex of I that is left to a vertex of C in a vertex ordering σ a *premature vertex of* σ. The neighbors of such a premature vertex must fulfill a strong condition on their positions in σ as the following lemma shows.

Lemma 13 ([1], **Lemma 22**). *Let G be a split graph with a split partition consisting of the clique C and the independent set I. Let σ be an MNS ordering of G. If the vertex $x \in I$ is a premature vertex of σ, then any vertex of C that is to the left of x in σ is a neighbor of x and any non-neighbor of x that is to the right of x in σ is also to the right of any neighbor of x in σ.*

Similar to total orders we will call a vertex $x \in I$ a *premature vertex of partial order* π if there is an element $y \in C$ with $x \prec_\pi y$. To decide whether a partial order π can be extended by an MNS ordering the set of premature vertices of π must fulfill strong properties which we define in the following.

Definition 14. *Let G be a split graph with a split partition consisting of the clique C and the independent set I. Let π be a partial order on $V(G)$ and let A be a subset of I. The tuple (π, A) fulfills the* nested property *if the following conditions hold:*

(N1) If $y \in C$ and $x \prec_\pi y$, then $x \in C \cup A$.
(N2) The neighborhoods of the elements of A can be ordered by inclusion, i.e., there are pairwise disjoint sets $C_1, I_1, C_2, I_2, \ldots, C_k, I_k$ with $\bigcup_{j=1}^{k} I_j = A$ and for any $i \in \{1, \ldots, k\}$ and any $x \in I_i$ it holds that $N(x) = \bigcup_{j=1}^{i} C_j$.
(N3) If $y \in C_i \cup I_i$ and $x \prec_\pi y$, then $x \in C_j \cup I_j$ with $j \leq i$.
(N4) For any $i \in \{1, \ldots, k\}$ there is at most one vertex $x \in I_i$ for which there exists a vertex $y \in C_i$ with $x \prec_\pi y$.

The nested partial order $\pi^N(\pi, A)$ *is defined as the reflexive and transitive closure of the relation containing the following tuples:*

(P1) (x, y) $\forall x, y \in V(G)$ with $x \prec_\pi y$
(P2) (x, y) $\forall x \in I_i \cup C_i, y \in V(G) \setminus \bigcup_{j=1}^{i}(I_j \cup C_j)$
(P3) (x, y) $\forall x \in C, y \in I \setminus A$
(P4) (x, y) $\forall x, y \in I_i$ for which $\exists z \in C_i$ with $x \prec_\pi z$

It is straightforward to check that $\pi^N(\pi, A)$ is a partial order if (π, A) fulfills the nested property. We first show that the set A of the premature vertices of a partial order π must necessarily fulfill the nested property if there is an MNS ordering extending π. Furthermore, any such MNS ordering fulfills a large subset of the constraints given by the nested partial order $\pi^N(\pi, A)$.

Lemma 15. *Let G be a split graph with a split partition consisting of the clique C and the independent set I and let π be a partial order on $V(G)$. Let $A = \{v \in I \mid \exists w \in C$ with $v \prec_\pi w\}$. If there is an MNS ordering σ of G extending π, then (π, A) fulfills the nested property. If $x \prec_\sigma y$ but $(y, x) \in \pi^N(\pi, A)$, then $x \notin A \cup C$.*

The nested property is, in a restricted way, also sufficient for the existence of a MNS ordering extending π. We show that if (π, A) fulfills the nested property, then there is an MNS ordering that fulfills all tuples of π that contain elements of the set A or the clique C. This ordering can be found using an $\mathcal{A}^+$-algorithm.

Lemma 16. *Let G be a split graph with a split partition consisting of the clique C and the independent set I, let π be a partial order on $V(G)$ and A be a subset of I. Assume (π, A) fulfills the nested property and let ρ be a linear extension of $\pi' = \pi^N(\pi, A)$. Then for any graph search $\mathcal{A} \in \{MNS, MCS, LBFS\}$ the ordering $\sigma = \mathcal{A}^+(\rho)$ of G fulfills the following property: If $x \prec_{\pi'} y$, then $x \prec_\sigma y$ or both x and y are not in $A \cup C$.*

After an instance of Algorithm 1 has visited all the clique vertices of a split graph, the labels of the remaining independent vertices do not change anymore. Thus, a vertex x whose label is now smaller than the label of another vertex y will be taken after y. Therefore, it is not enough to consider only the premature vertices of π. Instead, we must also consider all independent vertices x that π forces to be left of another independent vertex y whose label is larger than the label of x if all clique vertices are visited. In the case of MCS this is sufficient to characterize partial orders that are extendable.

Lemma 17. *Let G be a split graph with a split partition consisting of the clique C and the independent set I. Let π be a partial order on $V(G)$. Let $A := \{u \in I \mid \exists v \in V(G) \text{ with } v \in C \text{ or } |N(u)| < |N(v)| \text{ such that } u \prec_\pi v\}$. There is an MCS ordering which is a linear extension of π if and only if (π, A) fulfills the nested property.*

This lemma implies a linear-time algorithm for the PSOP of MCS on split graphs.

Theorem 18. *Given a split graph G and a partial order π on $V(G)$, we can solve the partial search order problem of MCS in time $\mathcal{O}(|V(G)| + |E(G)| + |\pi|)$.*

This is a generalization of the linear-time algorithms for the end-vertex problem [2] and the $\mathcal{F}$-tree recognition problem [3] of MCS on split graphs.

For LBFS there is a characterization of extendable partial orders that is similar to Lemma 17. However, due to the more complex label structure of LBFS, the result is slightly more complicated and uses OBA-orderings.

Lemma 19. *Let G be a split graph with a split partition consisting of the clique C and the independent set I. Let π be a partial order on $V(G)$. Let $A := \{u \in I \mid \exists v \in V(G) \text{ with } v \in C \text{ or } N(u) \subsetneq N(v) \text{ such that } u \prec_\pi v\}$. Let π' be the nested partial order $\pi^N(\pi, A)$ and let $\mathcal{R}$ be the following relation:*

$$\mathcal{R} = \{(X, Y) \mid \exists x, y \in I \setminus A \text{ with } X = N(x) \setminus N(y), \, Y = N(y) \setminus N(x), \, x \prec_\pi y\}$$
$$\cup \{(\{x\}, \{y\}) \mid x, y \in C, \, x \prec_{\pi'} y\}.$$

There is an LBFS ordering extending π if and only if the tuple (π, A) fulfills the nested property and there is an OBA-ordering for $(C, \mathcal{Q}, \mathcal{R})$ where $\mathcal{Q}$ is the ground set of $\mathcal{R}$.

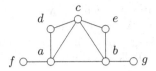

Fig. 1. A split graph consisting of clique $\{a, b, c\}$ and independent set $\{d, e, f, g\}$. Let π be the reflexive and transitive closure of the relation $\{(f, e), (g, d)\}$. There is no MCS ordering extending π since the set A defined in Lemma 17 contains both f and g and, thus, (π, A) does not fulfill the nested property. There is neither an LBFS ordering extending π as there is no OBA ordering for the relation $\mathcal{R} = \{(\{a\}, \{b, c\}), (\{b\}, \{a, c\})\}$ defined in Lemma 19. However, the MNS ordering (f, a, b, c, e, g, d) extends π.

Again, this characterization leads to an efficient algorithm for the PSOP of LBFS on split graphs. However, its running time is not linear.

Theorem 20. *Given a split graph G and a partial order π on $V(G)$, we can solve the partial search order problem of LBFS in time $\mathcal{O}(|V(G)| \cdot |\pi|)$.*

Unfortunately, the ideas of Lemmas 17 and 19 cannot be directly adapted to the PSOP of MNS. A main difficulty of this problem seems to be the identification of independent vertices that have to be premature vertices. To illustrate this, we consider the example given in Fig. 1. The defined partial order π has no premature vertices. Furthermore, the set A defined in Lemma 19 is empty for π. Nevertheless, for any MNS ordering σ extending π, one of the vertices f or g has to be a premature vertex of σ.

7   Further Research

Besides the cases considered in this paper, there are several other combinations of graph classes and searches for which both the end-vertex problem and the $\mathcal{F}$-tree recognition problem can be solved efficiently. Examples are the searches MNS and MCS on chordal graphs [2, 4, 20]. Can all these results be generalized to the PSOP or is there a combination of graph search and graph class where the PSOP is hard but both the end-vertex problem and the $\mathcal{F}$-tree recognition problem can be solved in polynomial time?

As mentioned in the introduction, the graph searches considered in this paper are used to solve several problems on graphs efficiently. This leads to the question whether the construction of a search ordering that extends a special partial order can be used in efficient algorithms for problems besides the end-vertex problem and the search tree recognition problem.

The algorithms given in this paper use the complete partial order as input. Using a Hasse diagram, it is possible to encode a partial order more efficiently. Since there are partial orders of quadratic size where the Hasse diagram has only linear size (e.g. total orders), it could be a good idea to study the running time of the algorithms for instances of the PSOP where the partial order is given as Hasse diagram.

References

1. Beisegel, J., et al.: Recognizing graph search trees. Preprint on arXiv (2018). https://doi.org/10.48550/arXiv.1811.09249
2. Beisegel, J., et al.: On the End-Vertex Problem of Graph Searches. Discrete Math. Theor. Comput. Sci. **21**(1) (2019). https://doi.org/10.23638/DMTCS-21-1-13
3. Beisegel, J., et al.: Recognizing graph search trees. In: Proceedings of Lagos 2019, the tenth Latin and American Algorithms, Graphs and Optimization Symposium. ENTCS, vol. 346, pp. 99–110. Elsevier (2019). https://doi.org/10.1016/j.entcs.2019.08.010
4. Beisegel, J., et al.: The recognition problem of graph search trees. SIAM J. Discrete Math. **35**(2), 1418–1446 (2021). https://doi.org/10.1137/20M1313301
5. Berry, A., Blair, J.R., Heggernes, P., Peyton, B.W.: Maximum cardinality search for computing minimal triangulations of graphs. Algorithmica **39**(4), 287–298 (2004). https://doi.org/10.1007/s00453-004-1084-3
6. Bretscher, A., Corneil, D., Habib, M., Paul, C.: A simple linear time LexBFS cograph recognition algorithm. SIAM J. Discrete Math. **22**(4), 1277–1296 (2008). https://doi.org/10.1137/060664690
7. Charbit, P., Habib, M., Mamcarz, A.: Influence of the tie-break rule on the end-vertex problem. Discrete Math. Theor. Comput. Sci. **16**(2), 57 (2014). https://doi.org/10.46298/dmtcs.2081
8. Chu, F.P.M.: A simple linear time certifying LBFS-based algorithm for recognizing trivially perfect graphs and their complements. Inf. Process. Lett. **107**(1), 7–12 (2008). https://doi.org/10.1016/j.ipl.2007.12.009
9. Corneil, D.G., Dusart, J., Habib, M., Mamcarz, A., De Montgolfier, F.: A tie-break model for graph search. Discrete Appl. Math. **199**, 89–100 (2016). https://doi.org/10.1016/j.dam.2015.06.011
10. Corneil, D.G., Köhler, E., Lanlignel, J.M.: On end-vertices of lexicographic breadth first searches. Discrete Appl. Math. **158**(5), 434–443 (2010). https://doi.org/10.1016/j.dam.2009.10.001
11. Corneil, D.G., Krueger, R.M.: A unified view of graph searching. SIAM J. Discrete Math. **22**(4), 1259–1276 (2008). https://doi.org/10.1137/050623498
12. Corneil, D.G., Olariu, S., Stewart, L.: The LBFS structure and recognition of interval graphs. SIAM J. Discrete Math. **23**(4), 1905–1953 (2009). https://doi.org/10.1137/S0895480100373455
13. Dusart, J., Habib, M.: A new LBFS-based algorithm for cocomparability graph recognition. Discrete Appl. Math. **216**, 149–161 (2017). https://doi.org/10.1016/j.dam.2015.07.016
14. Gorzny, J., Huang, J.: End-vertices of LBFS of (AT-free) bigraphs. Discrete Appl. Math. **225**, 87–94 (2017). https://doi.org/10.1016/j.dam.2017.02.027
15. Hagerup, T.: Biconnected graph assembly and recognition of DFS trees. Technical report A 85/03, Universität des Saarlandes (1985). https://doi.org/10.22028/D291-26437
16. Hagerup, T., Nowak, M.: Recognition of spanning trees defined by graph searches. Technical report A 85/08, Universität des Saarlandes (1985)
17. Hopcroft, J., Tarjan, R.E.: Efficient planarity testing. J. ACM **21**(4), 549–568 (1974). https://doi.org/10.1145/321850.321852
18. Krueger, R., Simonet, G., Berry, A.: A general label search to investigate classical graph search algorithms. Discrete Appl. Math. **159**(2–3), 128–142 (2011). https://doi.org/10.1016/j.dam.2010.02.011

19. Kumar, P.S., Madhavan, C.E.V.: Minimal vertex separators of chordal graphs. Discrete Appl. Math. **89**(1), 155–168 (1998). https://doi.org/10.1016/S0166-218X(98)00123-1
20. Rong, G., Cao, Y., Wang, J., Wang, Z.: Graph searches and their end vertices. Algorithmica (2022). https://doi.org/10.1007/s00453-022-00981-5
21. Rose, D.J., Tarjan, R.E., Lueker, G.S.: Algorithmic aspects of vertex elimination on graphs. SIAM J. Comput. **5**(2), 266–283 (1976). https://doi.org/10.1137/0205021
22. Scheffler, R.: Linearizing partial search orders. Preprint on arXiv (2022). https://doi.org/10.48550/arXiv.2206.14556
23. Tarjan, R.E., Yannakakis, M.: Simple linear-time algorithms to test chordality of graphs, test acyclicity of hypergraphs, and selectively reduce acyclic hypergraphs. SIAM J. Comput. **13**(3), 566–579 (1984). https://doi.org/10.1137/0213035

Minimum Weight Euclidean
$(1 + \varepsilon)$-Spanners

Csaba D. Tóth[1,2]([✉])([iD])

[1] California State University Northridge, Los Angeles, CA, USA
csaba.toth@csun.edu
[2] Tufts University, Medford, MA, USA

Abstract. Given a set S of n points in the plane and a parameter $\varepsilon > 0$, a Euclidean $(1 + \varepsilon)$-spanner is a geometric graph $G = (S, E)$ that contains a path of weight at most $(1 + \varepsilon)\|pq\|_2$ for all $p, q \in S$. We show that the minimum weight of a Euclidean $(1 + \varepsilon)$-spanner for n points in the unit square $[0, 1]^2$ is $O(\varepsilon^{-3/2}\sqrt{n})$, and this bound is the best possible. The upper bound is based on a new spanner algorithm that sparsifies Yao-graphs. It improves upon the baseline $O(\varepsilon^{-2}\sqrt{n})$, obtained by combining a tight bound for the weight of an MST and a tight bound for the lightness of Euclidean $(1 + \varepsilon)$-spanners, which is the ratio of the spanner weight to the weight of the MST. The result generalizes to d-space for all $d \in \mathbb{N}$: The minimum weight of a Euclidean $(1 + \varepsilon)$-spanner for n points in the unit cube $[0, 1]^d$ is $O_d(\varepsilon^{(1-d^2)/d}n^{(d-1)/d})$, and this bound is the best possible. For the $n \times n$ section of the integer lattice, we show that the minimum weight of a Euclidean $(1 + \varepsilon)$-spanner is between $\Omega(\varepsilon^{-3/4}n^2)$ and $O(\varepsilon^{-1}\log(\varepsilon^{-1})n^2)$. These bounds become $\Omega(\varepsilon^{-3/4}\sqrt{n})$ and $O(\varepsilon^{-1}\log(\varepsilon^{-1})\sqrt{n})$ when scaled to a grid of n points in $[0, 1]^2$.

Keywords: Geometric spanner · Yao-graph · Farey sequences

1 Introduction

For a set S of n points in a metric space, a graph $G = (S, E)$ is a *t-spanner* if G contains, between any two points $p, q \in S$, a pq-path of weight at most $t \cdot \|pq\|$, where $t \geq 1$ is the *stretch factor* of the spanner. In other words, a t-spanner approximates the true distances between the $\binom{n}{2}$ pairs of points up to a factor t distortion. Several optimization criteria have been developed for t-spanners for a given parameter $t \geq 1$. Natural parameters are the *size* (number of edges), the *weight* (sum of edge weights), the *maximum degree*, and the *hop-diameter*. Specifically, the *sparsity* of a spanner is the ratio $|E|/|S|$ between the number of edges and vertices; and the *lightness* is the ratio between the weight of a spanner and the weight of an MST on S.

In the geometric setting, S is a set of n points in Euclidean d-space in constant dimension $d \in \mathbb{N}$. For every $\varepsilon > 0$, there exist $(1 + \varepsilon)$-spanners with $O_d(\varepsilon^{1-d})$

Research partially supported by NSF grant DMS-1800734.

sparsity and $O_d(\varepsilon^{-d})$ lightness, and both bounds are the best possible [23]. In particular, the Θ-graphs, Yao-graphs [31], gap-greedy and path-greedy spanners provide $(1+\varepsilon)$-spanners of sparsity $O_d(\varepsilon^{1-d})$. For lightness, Das et al. [9,10,28] were the first to construct $(1+\varepsilon)$-spanners of lightness $\varepsilon^{-O(d)}$. Gottlieb [18] generalized this result to metric spaces with doubling dimension d; see also [6,15]. Recently, Le and Solomon [23] showed that the greedy $(1+\varepsilon)$-spanner in $\mathbb{R}^d$ has lightness $O(\varepsilon^{-d})$; and so it simultaneously achieves the best possible bounds for both lightness and sparsity. The greedy $(1+\varepsilon)$-spanner algorithm [4] generalizes Kruskal's algorithm: It sorts the $\binom{n}{2}$ edges of K_n by nondecreasing weight, and incrementally constructs a spanner H: it adds an edge uv if H does not contain an uv-path of weight at most $(1+\varepsilon)\|uv\|$.

Lightness versus Minimum Weight. Lightness is a convenient optimization parameter, as it is invariant under scaling. It also provides an approximation ratio for the minimum weight $(1+\varepsilon)$-spanner, as the weight of a Euclidean MST (for short, EMST) is a trivial lower bound on the spanner weight. However, minimizing the lightness is not equivalent to minimizing the spanner weight for a given input instance, as the EMST is highly sensitive to the distribution of the points in S. Given that worst-case tight bounds are now available for the lightness, it is time to revisit the problem of approximating the *minimum weight* of a Euclidean $(1+\varepsilon)$-spanner, without using the EMST as an intermediary.

Euclidean Minimum Spanning Trees. For n points in the unit cube $[0,1]^d$, the weight of the EMST is $O_d(n^{1-1/d})$, and this bound is also best possible [14,33]. In particular, a suitably scaled section of the integer lattice attains these bounds up to constant factors. Supowit et al. [34] proved similar bounds for the minimum weight of other popular graphs, such as spanning cycles and perfect matchings on n points in the unit cube $[0,1]^d$.

Extremal Configurations for Euclidean $(1+\varepsilon)$-Spanners. The tight $O_d(\varepsilon^{-d})$ bound on lightness [23] implies that for every set of n points in $[0,1]^d$, there is a Euclidean $(1+\varepsilon)$-spanner of weight $O(\varepsilon^{-d}n^{1-1/d})$. However, the combination of two tight bounds need not be tight; and it is unclear which n-point configurations require the heaviest $(1+\varepsilon)$-spanners. We show that this bound can be improved to $O(\varepsilon^{-3/2}\sqrt{n})$ in the plane. Furthermore, the extremal point configurations are not an integer grid, but an asymmetric grid.

Contributions. We obtain a tight upper bound on the minimum weight of a Euclidean $(1+\varepsilon)$-spanner for n points in $[0,1]^d$.

Theorem 1. *For constant $d \geq 2$, every set of n points in the unit cube $[0,1]^d$ admits a Euclidean $(1+\varepsilon)$-spanner of weight $O_d(\varepsilon^{(1-d^2)/d}n^{(d-1)/d})$, and this bound is the best possible.*

The upper bound is established by a new spanner algorithm, SPARSEYAO, that sparsifies the classical Yao-graph using novel geometric insight (Sect. 3). The weight analysis is based on a charging scheme that charges the weight of the spanner to empty regions (Sect. 4).

The lower bound construction is the scaled lattice with basis vectors of weight $\sqrt{\varepsilon}$ and $\frac{1}{\sqrt{\varepsilon}}$ (Sect. 2); and not the integer lattice $\mathbb{Z}^d$. We analyze the minimum weight of Euclidean $(1 + \varepsilon)$-spanners for the integer grid in the plane.

Theorem 2. *For every $n \in \mathbb{N}$, the minimum weight of a $(1+\varepsilon)$-spanner for the $n \times n$ section of the integer lattice is between $\Omega(\varepsilon^{-3/4} n^2)$ and $O(\varepsilon^{-1} \log(\varepsilon^{-1}) \cdot n^2)$.*

When scaled to n points in $[0, 1]^2$, the upper bound confirms that the integer lattice does not maximize the weight of Euclidean $(1 + \varepsilon)$-spanners.

Corollary 1. *For every $n \in \mathbb{N}$, the minimum weight of a $(1 + \varepsilon)$-spanner for n points in a scaled section of the integer grid in $[0, 1]^2$ is between $\Omega(\varepsilon^{-3/4} \sqrt{n})$ and $O(\varepsilon^{-1} \log(\varepsilon^{-1}) \sqrt{n})$.*

The lower bound is derived from two elementary criteria (the empty ellipse condition and the empty slab condition) for an edge to be present in every $(1+\varepsilon)$-spanner (Sect. 2). The upper bound is based on analyzing the SPARSEYAO algorithm from Sect. 3, combined with results from number theory on Farey sequences (Sect. 5). Closing the gap between the lower and upper bounds in Theorem 2 remains an open problem. Higher dimensional generalizations are also left for future work. In particular, multidimensional variants of Farey sequences are currently not well understood.

Further Related Previous Work. Many algorithms have been developed for constructing $(1+\varepsilon)$-spanners for n points in $\mathbb{R}^d$ [1,8–10,13,19,25,27,29], designed for one or more optimization criteria (lightness, sparsity, hop diameter, maximum degree, and running time). A comprehensive survey up to 2007 is in the book by Narasinham and Smid [28]. We briefly review previous constructions pertaining to the *minimum weight* for n points the unit square (i.e., $d = 2$). As noted above, the recent worst-case tight bound on the lightness [23] implies that the greedy algorithm returns a $(1 + \varepsilon)$-spanner of weight $O(\varepsilon^{-2} \|\text{MST}\|) = O(\varepsilon^{-2} \sqrt{n})$.

A classical method for constructing a $(1 + \varepsilon)$-spanners uses *well-separated pair decompositions (WSPD)* with a hierarchical clustering (e.g., quadtrees); see [20, Chap. 3]. Due to a hierarchy of depth $O(\log n)$, this technique has been adapted broadly to dynamic, kinetic, and reliable spanners [7,8,17,30]. However, the weight of the resulting $(1 + \varepsilon)$-spanner for n points in $[0, 1]^2$ is $O(\varepsilon^{-3} \sqrt{n} \cdot \log n)$ [17]. The $O(\log n)$ factor is due to the depth of the hierarchy; and it cannot be removed for any spanner with hop-diameter $O(\log n)$ [3,11,32].

Yao-graphs and Θ-graphs are geometric proximity graphs, defined as follows. For a constant $k \geq 3$, consider k cones of aperture $2\pi/k$ around each point $p \in S$; in each cone, connect p to the "closest" point $q \in S$. For Yao-graphs, q minimizes the Euclidean distance $\|pq\|$, and for Θ-graphs q is the point that minimizes the length of the orthogonal projection of pq to the angle bisector of the cone. It is known that both Θ- and Yao-graphs are $(1 + \varepsilon)$-spanners for a parameter $k \in \Theta(\varepsilon^{-1})$, and this bound is the best possible [28]. However, if we place $\lfloor n/2 \rfloor$ and $\lceil n/2 \rceil$ equally spaced points on opposite sides of the unit space, then the weight of both graphs with parameter $k = \Theta(\varepsilon^{-1})$ will be $\Theta(\varepsilon^{-1} n)$.

Organization. We start with lower bound constructions in the plane (Sect. 2) as a warm-up exercise. The two elementary geometric criteria build intuition and highlight the significance of $\sqrt{\varepsilon}$ as the ratio between the two axes of an ellipse of all paths of stretch at most $1 + \varepsilon$ between the foci. Section 3 presents Algorithm SparseYao and its stretch analysis in the plane. Its weight analysis for n points in $[0,1]^2$ is in Sect. 4. We analyze the performance of Algorithm SparseYao for the $n \times n$ grid, after a brief review of Feray sequences, in Sect. 5. We conclude with a selection of open problems in Sect. 6. The generalization of Algorithm SparseYao and its analysis are sketched in the full paper [35].

2   Lower Bounds in the Plane

We present lower bounds for the minimum weight of a $(1 + \varepsilon)$-spanner for the $n \times n$ section of the integer lattice (Sect. 2.1); and for n points in a unit square $[0,1]^2$ (Sect. 2.2).

Let $S \subset \mathbb{R}^2$ be a finite point set. We observe two elementary conditions that guarantee that an edge ab is present *in every* $(1+\varepsilon)$-spanner for S. Two points, $a, b \in S$, determine a (closed) line segment $ab = \operatorname{conv}\{a,b\}$; the relative interior of ab is denoted by $\operatorname{int}(ab) = ab \setminus \{a,b\}$. Let $\mathcal{E}_{ab}$ denote the ellipse with foci a and b, and great axis of weight $(1+\varepsilon)\|ab\|$, $\mathcal{L}_{ab}$ be the slab bounded by two lines parallel to ab and tangent lines to $\mathcal{E}_{ab}$; see Fig. 1. Note that the width of $\mathcal{L}_{ab}$ equals the minor axes of $\mathcal{E}_{ab}$, which is $((1+\varepsilon)^2 - 1^2)^{1/2}\|ab\| = (2\varepsilon + \varepsilon^2)^{1/2}\|ab\| > \sqrt{2\varepsilon}\|ab\|$.

- **Empty ellipse condition:** $S \cap \mathcal{E}_{ab} = \{a,b\}$.
- **Empty slab condition:** $S \cap \operatorname{int}(ab) = \emptyset$ and all points in $S \cap \mathcal{L}_{ab}$ are on the line ab.

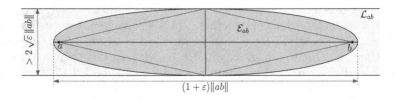

Fig. 1. The ellipse $\mathcal{E}_{ab}$ with foci a and b, and great axis of weight $(1+\varepsilon)\|ab\|$.

Observation 1. *Let $S \subset \mathbb{R}^2$, $G = (S, E)$ a $(1+\varepsilon)$-spanner for S, and $a, b \in S$.*

1. *If ab meets the empty ellipse condition, then $ab \in E$.*
2. *If S is a section of $\mathbb{Z}^2$, $\varepsilon < 1$, and ab meets the empty slab condition, then $ab \in E$.*

Proof. The ellipse $\mathcal{E}_{ab}$ contains all points $p \in \mathbb{R}^2$ satisfying $\|ap\| + \|pb\| \leq (1 + \varepsilon)\|ab\|$. Thus, by the triangle inequality, $\mathcal{E}_{ab}$ contains every ab-path of weight at most $(1 + \varepsilon)\|ab\|$. The empty ellipse condition implies that such a path cannot have interior vertices.

If S is the integer lattice, then $S \cap \text{int}(ab) = \emptyset$ implies that $\vec{ab}$ is a primitive vector (i.e., the x- and y-coordinates of $\vec{ab}$ are relatively prime), hence the distance between any two lattice points along the line ab is at least $\|ab\|$. Given that $\mathcal{E}_{ab} \subset \mathcal{L}_{ab}$, the empty slab condition now implies the empty ellipse condition. $\square$

2.1 Lower Bounds for the Grid

Lemma 1. *For every $n \in \mathbb{N}$ with $n \geq 2\varepsilon^{-1/4}$, the weight of every $(1+\varepsilon)$-spanner for the $n \times n$ section of the integer lattice is $\Omega(\varepsilon^{-3/4}n^2)$.*

Proof. Let $S = \{(s_1, s_2) \in \mathbb{Z}^2 : 0 \leq s_1, s_2 < n\}$ and $A = \{(a_1, a_2) \in \mathbb{Z}^2 : 0 \leq a_1, a_2 < \lceil \varepsilon^{-1/4} \rceil/2\}$. Denote the origin by $o = (0, 0)$. For every grid point $a \in A$, we have $\|oa\| \leq \varepsilon^{-1/4}/\sqrt{2}$. A vector $\vec{oa}$ is *primitive* if $a = (a_1, a_2)$ and $\gcd(a_1, a_2) = 1$. We show that every primitive vector $\vec{oa}$ with $a \in A$ satisfies the empty slab condition. It is clear that $S \cap \text{int}(oa) = \emptyset$. Suppose that $s \in S$ but it is not on the line spanned by oa. By Pick's theorem, $\text{area}(\Delta(oas)) \geq \frac{1}{2}$. Consequently, the distance between s and the line oa is at least $\|oa\|^{-1} \geq \sqrt{2} \cdot \varepsilon^{1/4} \geq 2\varepsilon^{1/2}\|oa\|$; and so $s \notin \mathcal{L}_{oa}$, as claimed.

By elementary number theory, $\vec{oa}$ is primitive for $\Theta(|A|)$ points $a \in A$. Indeed, every $a_1 \in \mathbb{N}$ is relatively prime to $N\varphi(a_1)/a_1$ integers in every interval of length N, where $\varphi(.)$ is Euler totient function, and $\varphi(a_1) = \Theta(a_1)$. Consequently, the total weight of primitive vectors $\vec{oa}$, $a \in A$, is $\Theta(|A| \cdot \varepsilon^{-1/4}) = \Theta(\varepsilon^{-3/4})$.

The primitive edges oa, $a \in A$, form a star centered at the origin. The translates of this star to other points $s \in S$, with $0 \leq s_1, s_2 \leq \frac{n}{2} \leq n - \lceil \varepsilon^{-1/4} \rceil$ are present in every $(1 + \varepsilon)$-spanner for S. As every edge is part of at most two such stars, summation over $\Theta(n^2)$ stars yields a lower bound of $\Omega(\varepsilon^{-3/4}n^2)$. $\square$

Remark 1. The lower bound in Lemma 1 derives from the total weight of primitive vectors $\vec{oa}$ with $\|oa\| \leq O(\varepsilon^{-1/4})$, which satisfy the empty slab condition. There are additional primitive vectors that satisfy the empty ellipse condition (e.g., $\vec{oa}$ with $a = (1, a_2)$ for all $|a_2| < \varepsilon^{-1/3}$). However, it is unclear how to account for all vectors satisfying the empty ellipse condition, and whether their overall weight would improve the lower bound in Lemma 1.

Remark 2. The empty ellipse and empty slab conditions each imply that an edge *must* be present in every $(1 + \varepsilon)$-spanner for S. It is unclear how the total weight of such "must have" edges compare to the the minimum weight of a $(1 + \varepsilon)$-spanner.

2.2 Lower Bounds in the Unit Square

Lemma 2. *For every $n \in \mathbb{N}$ and $\varepsilon \in (0, 1]$, there exists a set S of n points in $[0, 1]$ such that every $(1 + \varepsilon)$-spanner for S has weight $\Omega(\varepsilon^{-3/2}\sqrt{n})$.*

Proof. First let S_0 be a set of $2m$ points, where $m = \lfloor \varepsilon^{-1}/2 \rfloor$, with m equally spaced points on two opposite sides of a unit square. By the empty ellipse property, every $(1+\varepsilon)$-spanner for S_0 contains a complete bipartite graph $K_{m,m}$. The weight of each edge of $K_{m,m}$ is between 1 and $\sqrt{2}$, and so the weight of every $(1 + \varepsilon)$-spanner for S_0 is $\Omega(\varepsilon^{-2})$.

For $n \geq \varepsilon^{-1}$, consider an $\lfloor \sqrt{\varepsilon n} \rfloor \times \lfloor \sqrt{\varepsilon n} \rfloor$ grid of unit squares, and insert a translated copy of S_0 in each unit square. Let S be the union of these $\Theta(\varepsilon n)$ copies of S_0; and note that $|S| = \Theta(n)$. A $(1 + \varepsilon)$-spanner for each copy of S_0 still requires a complete bipartite graph of weight $\Omega(\varepsilon^{-2})$. Overall, the weight of every $(1 + \varepsilon)$-spanner for S is $\Omega(\varepsilon^{-1}n)$.

Finally, scale S down by a factor of $\lfloor \sqrt{\varepsilon n} \rfloor$ so that it fits in a unit square. The weight of every edge scales by the same factor, and the weight of a $(1+\varepsilon)$-spanner for the resulting n points in $[0, 1]^2$ is $\Omega(\varepsilon^{-3/2} \sqrt{n})$, as claimed.   □

Remark 3. The points in the lower bound construction above lie on $O(\sqrt{\varepsilon n})$ axis-parallel lines in $[0, 1]^2$, and so the weight of their MST is $O(\sqrt{\varepsilon n})$. Recall that the lightness of the greedy $(1+\varepsilon)$-spanner is $O(\varepsilon^{-d} \log \varepsilon^{-1})$ [23]. For $d = 2$, it yields a $(1 + \varepsilon)$-spanner of weight $O(\varepsilon^{-2} \log \varepsilon^{-1}) \cdot \|\text{MST}(S)\| = O(\varepsilon^{-3/2} \log(\varepsilon^{-1})\sqrt{n})$.

3   Spanner Algorithm: Sparse Yao-Graphs

Let S be a set of n points in the plane and $\varepsilon \in (0, \frac{1}{9})$. As noted above, the Yao-graph $Y_k(S)$ with $k = \Theta(\varepsilon^{-1})$ cones per vertex is a $(1 + \varepsilon)$-spanner for S. We describe an new algorithm, SPARSEYAO(S, ε), that computes a subgraph of a Yao-graph $Y_k(S)$ (Sect. 3.1); and show that it returns a $(1 + \varepsilon)$-spanner for S (Sect. 3.2). Later, we use this algorithm for n points in the unit square (Sect. 4; and for an $n \times n$ section of the integer lattice (Sect. 5). Our algorithm starts with a Yao-graph that is a $(1 + \frac{\varepsilon}{2})$-spanner, in order to leave room for minor loss in the stretch factor due to sparsification. The basic idea is that instead of cones of aperture $2\pi/k = \Theta(\varepsilon)$, cones of much larger aperture $\Theta(\sqrt{\varepsilon})$ suffice in some cases. (This is idea is flashed out in Sect. 3.2). The angle $\sqrt{\varepsilon}$ then allows us to charge the weight of the resulting spanner to the area of empty regions (specifically, to an empty section of a cone) in Sect. 4.

3.1   Sparse Yao-Graph Algorithm

We present an algorithm that computes a subgraph of a Yao-graph for S. It starts with cones of aperture $\Theta(\sqrt{\varepsilon})$, and refines them to cones of aperture $\Theta(\varepsilon^{-1})$. We connect each point $p \in S$ to the closest points in the larger cones, and use the smaller cones only when "necessary." To specify when exactly the smaller cones are used, we define two geometric regions that will also play crucial roles in the stretch and weight analyses.

Definitions. Let $p, q \in S$ be distinct points; refer to Fig. 2. Let $A(p, q)$ be the line segment of weight $\frac{\sqrt{\varepsilon}}{2} \|pq\|$ on the line pq with one endpoint at p but interior-disjoint from the ray $\overrightarrow{pq}$; and $\widehat{A}(p, q)$ the set of points in $\mathbb{R}^2$ within distance

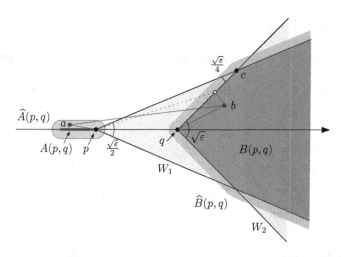

Fig. 2. Wedges W_1 and W_2, line segment $A(p,q)$, and regions $\widehat{A}(p,q)$, $B(p,q)$, and $\widehat{B}(p,q)$ for $p,q \in S$.

$\frac{\varepsilon}{16} \|pq\|$ from $A(p,q)$. Let W_1 be the cone with apex p, aperture $\frac{1}{2} \cdot \sqrt{\varepsilon}$, and symmetry axis $\vec{pq}$; and let W_2 be the cone with apex q, aperture $\sqrt{\varepsilon}$, and symmetry axis $\vec{pq}$. Let $B(p,q) = W_1 \cap W_2$. Finally, let $\widehat{B}(p,q)$ be the set of points in $\mathbb{R}^2$ within distance at most $\frac{\varepsilon}{8} \|pq\|$ from $B(p,q)$.

We show below (cf. Lemma 3) that if we add edge pq to the spanner, then we do not need any of the edges ab with $a \in \widehat{A}(p,q)$ and $b \in \widehat{B}(p,q)$. We can now present our algorithm.

Algorithm SPARSEYAO(S, ε). Input: a set $S \subset \mathbb{R}^2$ of n points, and $\varepsilon \in (0, \frac{1}{9})$.

Preprocessing Phase: Yao-Graphs. Subdivide $\mathbb{R}^2$ into $k := \lceil 16\pi/\sqrt{\varepsilon} \rceil$ congruent cones of aperture $2\pi/k \leq \frac{1}{8} \cdot \sqrt{\varepsilon}$ with apex at the origin, denoted $C_1, \ldots, C_k$. For $i \in \{1, \ldots, k\}$, let $\vec{r}_i$ be the symmetry axis of C_i, directed from the origin towards the interior of C_i. For each $i \in \{1, \ldots, k\}$, subdivide C_i into k congruent cones of aperture $2\pi/k^2 \leq \varepsilon/8$, denoted $C_{i,1}, \ldots, C_{i,k}$; see Fig. 3. For each point $s \in S$, let $C_i(s)$ and $C_{i,j}(s)$, resp., be the translates of cones C_i and $C_{i,j}$ to apex s.

For all $s \in S$ and $i \in \{1, \ldots, k\}$, let $q_i(s)$ be a closest point to s in $C_i(s) \cap (S \setminus \{s\})$; and for all $j \in \{1, \ldots, k\}$, let $q_{i,j}(s)$ be a closest point in $C_{i,j}(s) \cap (S \setminus \{s\})$; if such points exist. For each $i \in \{1, \ldots, k\}$, let L_i be the list of all ordered pairs $(s, q_i(s))$ sorted in decreasing order of the orthogonal projection of s to the directed line $\vec{r}_i$; ties are broken arbitrarily.

Main Phase: Computing a Spanner. Initialize an empty graph $G = (S, E)$ with $E := \emptyset$.

1. For all $i \in \{1, \ldots, k\}$, do:
   - While the list L_i is nonempty, do:
     (a) Let (p, q) be the first ordered pair in L_i.

(b) Add (the unordered edge) pq to E.
(c) For all $i' \in \{i-2, \ldots, i+2\}$ and $j \in \{1, \ldots, k\}$, do:
   If $\|pq_i(p)\| \leq \|pq_{i',j}(p)\|$ and $q_{i',j}(p) \notin B(p,q)$, then add $pq_{i',j}(p)$ to E.
(d) For all $s \in \widehat{A}(p,q)$, including $s = p$, delete the pair $(s, q_i(s))$ from L_i.
2. Return $G = (S, E)$.

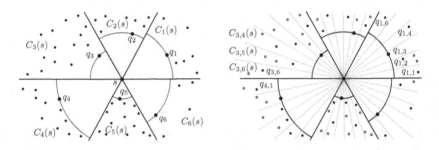

Fig. 3. Cones $C_i(s)$ and $C_{i,j}(s)$ for a point $s \in S$, with $k = 6$.

It is clear that the runtime of Algorithm SPARSEYAO is polynomial in n in the RAM model of computation. In particular, the runtime is dominated preprocessing phase that constructs the Yao-graph with $O(\varepsilon^{-1}n)$ edges: finding the closest points $q_i(s)$ and $q_{i,j}(s)$ is supported by standard range searching data structures [2]. The main phase then computes a subgraph of $Y_{k^2}(S)$ in $O(\varepsilon^{-1}n)$ time. Optimizing the runtime, however, is beyond the scope of this paper.   □

3.2   Stretch Analysis

In this section, we show that $G = \text{SPARSEYAO}(S, \varepsilon)$ is a $(1+\varepsilon)$-spanner for S. In the preprocessing phase, Algorithm SPARSEYAO computes a Yao-graph with $k^2 = \Theta(\varepsilon^{-1})$ cones. The following lemma justifies that we can omit some of the edges $sq_{i,j}$ from G. In the general case, we have $s = a \in \tilde{A}(p,q)$ and $q_{i,j} = b \in \widehat{B}(p,q)$. For technical reasons, we use a slightly larger neighborhood instead of $\widehat{A}(p,q)$. Let $\tilde{A}(p,q)$ be the set of points in $\mathbb{R}^2$ within distance at most $\frac{\varepsilon}{5}$ from $A(p,q)$.

Lemma 3. For all $a \in \tilde{A}(p,q)$ and $b \in \widehat{B}(p,q)$, we have

$$(1+\varepsilon)\|ap\| + \|pq\| + (1+\varepsilon)\|qb\| \leq (1+\varepsilon)\|ab\|. \tag{1}$$

The proof of Lemma 3 is a fairly technical; see the full paper [35]. Next we clarify the relation between $\widehat{A}(p,q)$ and $\tilde{A}(p,q)$.

Lemma 4. Let $p, q \in S$, and assume that $q \in C_{i',j}(p)$ for some $i, j \in \{1, \ldots, k\}$ and $i' \in \{i-1, i, i+1\}$, where $q_{i',j} = q_{i',j}(p)$ is a closest point to p in $C_{i',j}(p)$. Then $\widehat{A}(p, q_i) \subset \tilde{A}(p, q_{i',j})$.

Proof. Since the aperture of $C_i(p)$ is $\frac{1}{8} \cdot \sqrt{\varepsilon}$ and $q_i \in C_i(p)$, then $\angle q_i p q_{i',j} \le \frac{1}{4}\sqrt{\varepsilon}$. Since $\|pq_i\| \le \|pq_{i',j}\|$, then $\|A(p, q_i)\| \le \|A(p, q_{i',j})\|$. Consequently, every point in $A(p, q_i)$ is within distance at most $\|A(p, q_i)\| \sin \angle q_i p q_{i',j} \le \frac{\sqrt{\varepsilon}}{2} \|pq_i\| \cdot \frac{1}{4}\sqrt{\varepsilon} \le \frac{\varepsilon}{8}\|pq_i\|$ from $A(p, q_{i',j})$. By the triangle inequality, the $(\frac{\varepsilon}{16}\|pq_i\|)$-neighborhood of $A(p, q_i)$ is within distance at most $(\frac{\varepsilon}{8} + \frac{\varepsilon}{16})\|pq_i\| < \frac{\varepsilon}{5}\|pq_i\|$ from $A(p, q_{i',j})$. $\square$

The following lemma justifies the role of the regions $\widehat{B}(p, q_i)$. Due to space constraints, its proof is deferred to the full paper [35].

Lemma 5. *Let $p, q \in S$, and assume that $q \in C_{i',j}(p)$ for some $i, j \in \{1, \ldots, k\}$ and $i' \in \{i-1, i, i+1\}$, where $q_{i',j} = q_{i',j}(p)$ is a closest point to p in $C_{i',j}(p)$. If $q \notin B(p, q_i)$ but $q_{i',j} \in B(p, q_i)$, then $q \in \widehat{B}(p, q_i)$.*

Completing the Stretch Analysis. We are now ready to present the stretch analysis for $\text{SPARSEYAO}(S, \varepsilon)$.

Theorem 3. *For every finite point set $S \subset \mathbb{R}^2$ and $\varepsilon \in (0, \frac{1}{9})$, the graph $G = \text{SPARSEYAO}(S, \varepsilon)$ is a $(1+\varepsilon)$-spanner.*

Proof. Let S be a set of n points in the plane. Let L_0 be the list of all $\binom{n}{2}$ edges of the complete graph on S sorted by Euclidean weight (ties broken arbitrarily). For $\ell = 1, \ldots, \binom{n}{2}$, let e_ℓ be the ℓ-th edge in L_0, and let $E(\ell) = \{e_1, \ldots, e_\ell\}$. We show the following claim, by induction, for every $\ell = 1, \ldots, \binom{n}{2}$:

Claim. For every edge $ab \in E(\ell)$, $G = (S, E)$ contains an ab-path of weight at most $(1+\varepsilon)\|ab\|$.

For $\ell = 1$, the claim clearly holds, as the shortest edge pq is necessarily the shortest in some cones $C_i(p)$ and $C_{i'}(q)$, as well, and so the algorithm adds pq to E. Assume that $1 < \ell \le \binom{n}{2}$ and the claim holds for $\ell - 1$. If the algorithm added edge e_ℓ to E, then the claim trivially holds for ℓ.

Suppose that $e_\ell \notin E$. Let $e_\ell = pq$, and $q \in C_{i,j}(p)$ for some $i, j \in \{1, \ldots, k\}$. Recall that $q_i = q_i(p)$ is a closest point to p in the cone C_i; and $q_{i,j} = q_{i,j}(p)$ is a closest point to p in the cone $C_{i,j}(p)$. We distinguish between two cases.

(1) The algorithm added the edge pq_i to E. Note that $\|q_i q\| < \|pq\|$ and $\|q_{i,j} q\| < \|pq\|$. By the induction hypothesis, G contains a $q_i q$-path P_i of weight at most $(1+\varepsilon)\|q_i q\|$ and a $q_{i,j} q$-path $P_{i,j}$ of weight at most $(1+\varepsilon)\|q_{i,j} q\|$. If $q \in \widehat{B}(p, q_i)$, then $pq_i + P_i$ is a pq-path of weight at most $(1+\varepsilon)\|pq\|$ by Lemma 3. Otherwise, $q \notin \widehat{B}(p, q_i)$. In this case, $q_{i,j} \notin B(p, q_i)$ by Lemma 5. This means that the algorithm added the edge $pq_{i,j}$ to E. We have $q \in \widehat{B}(p, q_{i,j})$ by Lemma 5, and so $pq_{i,j} + P_{i,j}$ is a pq-path of weight at most $(1+\varepsilon)\|pq\|$ by Lemma 3.

(2) The algorithm did not add the edge pq_i to E. Then the algorithm deleted (p, q_i) from the list L_i in a step in which it added another edge $p'q_i'$ to E. This means that $p \in \widehat{A}(p', q_i')$, where q_i' is the closest point to p' in the cone $C_i(p')$. As $\texttt{diam}(A(p_i, q_i')) < (\sqrt{\varepsilon} + 2 \cdot \frac{\varepsilon}{16})\|p'q_i'\| < \frac{1}{4}\|p'q_i'\|$ for $\varepsilon \in (0, \frac{1}{9})$, then $p \in \widehat{A}(p', q_i')$ implies $\|pp'\| \le \frac{1}{4}\|p'q_i'\|$. Since L_i is sorted by weight, then $\|p'q_i'\| \le \|pq_i\|$.

Although we have $q \in C_i(p)$, the point q need not be in the cone $C_i(p')$; see Fig. 4. We claim that q lies in the union of three consecutive cones: $q \in C_{i-1}(p') \cup C_i(p') \cup C_{i+1}(p')$. Let $D_i(p')$ be part of the cone $C_i(p')$ outside of the circle of radius $\|p'q_i'\|$ centered at p'. Since $q \in C_i(p)$ and $\|p'q_i'\| < \|pq\|$, then q lies in the translate $D_i(p') + \overrightarrow{p'p}$ of $D_i(p')$. Consider the union of translates:

$$D = D_i(p) + \{\overrightarrow{p'a} : a \in A_{p_i, q_i'}\},$$

and note that $q \in D$. We have $\texttt{diam}(\widehat{A}(p', q_i')) \leq (\frac{\sqrt{\varepsilon}}{2} + 2 \cdot \frac{\varepsilon}{16})\|p'q_i'\| < \frac{1}{4}\|p'q_i'\|$ for $\varepsilon \in (0, \frac{1}{9})$; and recall that the aperture of $C_i(p')$ is $\gamma := 2\pi/k \leq \frac{1}{8} \cdot \sqrt{\varepsilon}$. We can now approximate $\angle qp'q_i'$ as follows; refer to Fig. 4: $\tan \angle qp'q_i' \leq \|p'q_i'\| \tan \gamma / (\|p'q_i'\| - 2\texttt{diam}(A(p', q_i))) \leq 2 \tan \alpha$. Consequently, $\angle qp'q_i' < 2\gamma$. It follows that $q \in \bigcup_{i'=i-2}^{i+2} C_{i'}(p')$. We distinguish between two subcases:

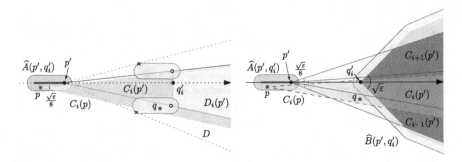

Fig. 4. The relative position of pq and $p'q_i'$. Specifically, $p \in \widehat{A})(p', q_i')$ and $q \in C_i(p)$. Left: the region $D_i(p')$ and translates of $\widehat{A}(p', q_i')$ to two critical points of $D_i(p')$. Right: $q \in C_{i-1}(p') \cup C_i(p') \cup C_{i+1}(p')$ and the region $B(p'q_i')$

(2a) $q \in \widehat{B}(p', q_i')$. By induction, G contains $(1+\varepsilon)$-paths between p and p', and between q and q_i'. By Lemma 3 (with $a = p$ and $b = q$), the concatenation of these paths and the edge $p'q_i'$ is a pq-path of weight at most $(1 + \varepsilon)\|pq\|$.

(2b) $q \notin \widehat{B}(p', q_i')$. Then $q \in C_{i',j'}(p')$ for some $i' \in \{i - 1, i, i + 1\}$ and $j' \in \{1, \ldots, k\}$. By Lemma 5, we have $q_{i',j'} \notin B(p', q_i)$. and so the algorithm added the edge $p'q_{i',j'}$, where $q_{i',j'}$ is the closest point to p' in the cone $C_{i',j'}(p')$. We have $p \in \widehat{A}(p', q_i') \subset \tilde{A}(p', q_{i',j})$ by Lemma 4, and $q \in \widehat{B}(p', q_{i',j})$ by Lemma 5. By induction, G contains $(1+\varepsilon)$-paths between p and p', and between $q_{i',j'}$ and q. The concatenation of these paths and the edge $p'q_{i',j'}$ is a pq-path of weight at most $(1 + \varepsilon)\|pq\|$ by Lemma 3.

4   Spanners in the Unit Square

In this section, we show that for a set $S \subset [0,1]^2$ of n points in the unit square and $\varepsilon \in (0, \frac{1}{9})$, Algorithm SPARSEYAO returns a $(1+\varepsilon)$-spanner of weight $O(\varepsilon^{-3/2}\sqrt{n})$ (cf. Theorem 4).

The spanner $\text{SPARSEYAO}(S, \varepsilon)$ is a subgraph of the Yao-graph with cones of aperture $2\pi/k^2 = O(\varepsilon)$, and so it has $O(\varepsilon^{-1}n)$ edges. Recall that for all $p \in S$ and all $i \in \{1, \ldots, k\}$, there is at most one edge $pq_i(p)$ in G, where $q_i(p)$ is the closest point to p in the cone $C_i(p)$ of aperture $\frac{1}{8}\sqrt{\varepsilon}$. Let

$$F = \{pq_i(p) \in E(G) : p \in S, i \in \{1, \ldots, k\}\}.$$

We first show that the weight of F approximates the weight of the spanner.

Lemma 6. *If Algorithm SPARSEYAO adds $pq_i(p)$ and $pq_{i',j}(p)$ to G in the same iteration, then $\|pq_{i',j}(p)\| < 2\|pq_i(p)\|$.*

Proof. For short, we write $q_i = q_i(p)$ and $q_{i',j} = q_{i',j}(p)$, where $i \in \{i-1, i, i+1\}$. Since SPARSEYAO added $pq_{i',j}$ to G, then $q_{i',j} \notin B(p, q_i)$. Recall (cf. Fig. 2) that $B(p, q_i) = W_1 \cap W_2$, where W_1 and W_2 are cones centered at p and q_i, resp., with apertures $\frac{1}{2}\sqrt{\varepsilon}$ and $\sqrt{\varepsilon}$. Since the aperture of the cone $C_i(p)$ is $\frac{1}{8}\sqrt{\varepsilon}$, then $C_{i-1}(p) \cup C_i(p) \cup C_{i+1}(p) \subset W_1$, hence $(C_{i-1}(p) \cup C_i(p) \cup C_{i+1}(p)) \setminus B(p, q_i) \subset W_1 \setminus W_2$. The line segment pq_i decomposes $W_1 v W_2$ into two isosceles triangles. By the triangle inequality, the diameter of each isosceles triangle is less than $2\|pq_i\|$. This implies $\|pq'\| < 2\|pq_i\|$ for any $q' \in W_1 \setminus W_2$, as claimed.   □

Lemma 7. *For $G = \text{SPARSEYAO}(S, \varepsilon)$, we have $\|G\| = O(\varepsilon^{-1/2}) \cdot \|F\|$.*

Proof. Fix p and $i \in \{1, \ldots, k\}$, let $q_i = q_i(p)$ for short, and suppose $pq_i \in E(G)$. Consider one step of the algorithm that adds the edge pq_i to G, together with up to $3k = \Theta(\varepsilon^{-1/2})$ edges of type $pq_{i',j}$, where $q_{i',j} \notin B(p, q_i)$ and $i' \in \{i-1, i, i+1\}$. By Lemma 6, $\|pq_{i',j}\| < 2\|pq_i\|$. The total weight of all edges $pq_{i',j}$ added to the spanner is

$$\|pq_i(p)\| + \sum_{i'=i-1}^{i+1} \sum_{j=1}^{k} \|pq_{i',j}\| \leq \|pq_i\| + 3k \cdot 2\|pq_i\| \leq O(k\|pq_i\|) \leq O(\varepsilon^{-1/2})\|pq_i\|).$$

Summation over all edges in F yields $\|G\| = O(\varepsilon^{-1/2}) \cdot \|F\|$.   □

It remains to show that $\|F\| \leq O(\varepsilon^{-1}\sqrt{n})$. For $i = 1, \ldots, k$, let $F_i = \{pq_i(p) \in E(G) : p \in S\}$, that is, the set of edges in G between points p and the closest point $q_i(p)$ in cone $C_i(p)$ of aperture $\sqrt{\varepsilon}$. We prove that $\|F_i\| \leq O(\varepsilon^{-1/2}\sqrt{n})$ (in the full version of this paper citefull). Since $k = \Theta(\varepsilon^{-1/2})$ this implies the following.

Theorem 4. *For every set of n points in $[0, 1]^2$ and every $\varepsilon > 0$, Algorithm SPARSEYAO returns a Euclidean $(1 + \varepsilon)$-spanner of weight $O(\varepsilon^{-3/2}\sqrt{n})$.*

Proof. Let $G = \text{SPARSEYAO}(S, \varepsilon)$, and define $F \subset E(G)$ and $F_1, \ldots, F_k$ as above. We prove $\|F\| = \sum_{i=1}^{k} \|F_i\| = O(k\,\varepsilon^{-1/2}\sqrt{n}) = O(\varepsilon^{-1}\sqrt{n})$ in the full paper [35]. Now Lemma 7 yields $\|G\| \leq O(\varepsilon^{-1/2}) \cdot (\|F\| + \sqrt{2}) \leq O(\varepsilon^{-3/2}\sqrt{n})$.   □

5   Spanners for the Integer Grid

Two points in the integer lattice $p, q \in \mathbb{Z}^2$ are *visible* if the line segment pq does not pass through any lattice point. An integer point $(i, j) \in \mathbb{Z}^2$ is visible from the origin $(0,0)$ if i and j are relatively prime, that is, $\gcd(i, j) = 1$. The *slope* of a segment between $(0,0)$ and (i, j) is j/i. For every $n \in \mathbb{N}$, the *Farey set of order* n, $F_n = \left\{ \frac{a}{b} : 0 \le a \le b \le n \right\}$, is the set of slopes of the lines spanned by the origin and lattice points $(b, a) \in [0, n]^2$ with $a \le b$. The *Farey sequence* is the sequence of elements in F_n in increasing order. Note that $F_n \subset [0, 1]$. Farey sets and sequences have fascinating properties, and the distribution of F_n, as $n \to \infty$ is not fully understood [12, 16, 22, 26].

The key result we need is a bound on the average distance to a Farey set F_n. For every $x \in [0, 1]$, let

$$\rho_n(x) = \min_{\frac{p}{q} \in F_n} \left| \frac{p}{q} - x \right|$$

denote the distance between x and the Farey set F_n. Kargaev and Zhigljavsky [21] proved that

$$\int_0^1 \rho_n(x)\, dx = \frac{3}{\pi^2} \frac{\ln n}{n^2} + O\left(\frac{1}{n^2} \right), \qquad \text{as} \qquad n \to \infty. \tag{2}$$

In the full paper [35], we use (2) to prove the following.

Theorem 5. *Let S be the $n \times n$ section of the integer lattice for some positive integer n. Then the graph $G = \text{SPARSEYAO}(S, \varepsilon)$ has weight $O(\varepsilon^{-1} \log(\varepsilon^{-1}) \cdot n^2)$.*

The combination of Lemma 1 and Theorem 5 establishes Theorem 2.

6   Outlook

Our SPARSEYAO algorithm combines features of Yao-graphs and greedy spanners. It remains an open problem whether the celebrated greedy algorithm [4] always returns a $(1 + \varepsilon)$-spanner of weight $O(\varepsilon^{-3/2} \sqrt{n})$ for n points in the unit square (and $O(\varepsilon^{(1-d^2)/d} n^{(d-1)/d})$ for n points in $[0, 1]^d$). The analysis of the greedy algorithm is known to be notoriously difficult [15, 23]. It is also an open problem whether SPARSEYAO or the greedy algorithm achieves an approximation ratio better than the tight lightness bound of $O(\varepsilon^{-d})$ for n points in $\mathbb{R}^d$ (where the approximation ratio compares the weight of the output with the instance-optimal weight of a $(1 + \varepsilon)$-spanner).

All results in this paper pertain to Euclidean spaces. Generalizations to L_p-norms for $p \ge 1$ (or Minkowski norms with respect to a centrally symmetric convex body in $\mathbb{R}^d$) would be of interest. It is unclear whether some or all of the machinery developed here generalizes to other norms. Finally, we note that Steiner points can substantially improve the weight of a $(1 + \varepsilon)$-spanner in Euclidean space [5, 23, 24]. It is left for future work to study the minimum weight of a Euclidean Steiner $(1 + \varepsilon)$-spanner for n points in the unit square $[0, 1]^2$ (or unit cube $[0, 1]^d$); and for an $n \times n$ section of the integer lattice.

References

1. Abu-Affash, A.K., Bar-On, G., Carmi, P.: δ-greedy t-spanner. Comput. Geom. **100**, 101807 (2022). https://doi.org/10.1016/j.comgeo.2021.101807
2. Agarwal, P.K.: Range searching. In: Goodman, J.E., O'Rourke, J., Tóth, C.D. (eds.) Handbook of Discrete and Computational Geometry, chap. 40, 3 edn., pp. 1057–1092. CRC Press, Boca Raton (2017)
3. Agarwal, P.K., Wang, Y., Yin, P.: Lower bound for sparse Euclidean spanners. In: Proceedings of the 16th ACM-SIAM Symposium on Discrete Algorithms (SODA), pp. 670–671 (2005). https://dl.acm.org/citation.cfm?id=1070432.1070525
4. Althöfer, I., Das, G., Dobkin, D., Joseph, D., Soares, J.: On sparse spanners of weighted graphs. Discrete Comput. Geom. **9**(1), 81–100 (1993). https://doi.org/10.1007/BF02189308
5. Bhore, S., Tóth, C.D.: Light euclidean steiner spanners in the plane. In: Proceedings of the 37th Annual Symposium on Computational Geometry (SoCG). LIPIcs, vol. 189, pp. 15:1–15:17. Schloss Dagstuhl (2021). https://doi.org/10.4230/LIPIcs.SoCG.2021.15
6. Borradaile, G., Le, H., Wulff-Nilsen, C.: Greedy spanners are optimal in doubling metrics. In: Proceedings of the 30th ACM-SIAM Symposium on Discrete Algorithms (SODA), pp. 2371–2379 (2019). https://doi.org/10.1137/1.9781611975482.145
7. Buchin, K., Har-Peled, S., Oláh, D.: A spanner for the day after. Discrete Comput. Geom. **64**(4), 1167–1191 (2020). https://doi.org/10.1007/s00454-020-00228-6
8. Chan, T.M., Har-Peled, S., Jones, M.: On locality-sensitive orderings and their applications. SIAM J. Comput. **49**(3), 583–600 (2020). https://doi.org/10.1137/19M1246493
9. Das, G., Heffernan, P.J., Narasimhan, G.: Optimally sparse spanners in 3-dimensional euclidean space. In: Proceedings of the 9th Symposium on Computational Geometry (SoCG), pp. 53–62 (1993). https://doi.org/10.1145/160985.160998
10. Das, G., Narasimhan, G., Salowe, J.S.: A new way to weigh malnourished euclidean graphs. In: Proceedings of the 6th ACM-SIAM Symposium on Discrete Algorithms (SODA), pp. 215–222 (1995). https://dl.acm.org/citation.cfm?id=313651.313697
11. Dinitz, Y., Elkin, M., Solomon, S.: Low-light trees, and tight lower bounds for euclidean spanners. Discrete Comput. Geom. **43**(4), 736–783 (2009). https://doi.org/10.1007/s00454-009-9230-y
12. Dress, F.: Discrépance des suites de farey. J. Théor. Nombres Bordeaux **11**(2), 345–367 (1999)
13. Elkin, M., Solomon, S.: Optimal euclidean spanners: really short, thin, and lanky. J. ACM **62**(5), 1–45 (2015). https://doi.org/10.1145/2819008
14. Few, L.: The shortest path and the shortest road through n points. Mathematika **2**(2), 141–144 (1955). https://doi.org/10.1112/S0025579300000784
15. Filtser, A., Solomon, S.: The greedy spanner is existentially optimal. SIAM J. Comput. **49**(2), 429–447 (2020). https://doi.org/10.1137/18M1210678
16. Franel, J.: Les suites de farey et les problemes des nombres premiers. Gottinger Nachr. **1924**, 198–201 (1924)
17. Gao, J., Guibas, L.J., Nguyen, A.: Deformable spanners and applications. Comput. Geom. **35**(1–2), 2–19 (2006). https://doi.org/10.1016/j.comgeo.2005.10.001
18. Gottlieb, L.: A light metric spanner. In: Proceedings of the 56th IEEE Symposium on Foundations of Computer Science (FOCS), pp. 759–772 (2015). https://doi.org/10.1109/FOCS.2015.52

19. Gudmundsson, J., Levcopoulos, C., Narasimhan, G.: Fast greedy algorithms for constructing sparse geometric spanners. SIAM J. Comput. **31**(5), 1479–1500 (2002). https://doi.org/10.1137/S0097539700382947

20. Har-Peled, S.: Geometric Approximation Algorithms. Mathematics Surveys and Monographs, vol. 173. AMS (2011)

21. Kargaev, P., Zhigljavsky, A.: Approximation of real numbers by rationals: some metric theorems. J. Number Theor. **61**, 209–225 (1996). https://doi.org/10.1006/jnth.1996.0145

22. Landau, E.: Bemerkungen zu der vorstehenden Abhandlung von Herrn Franel. Göttinger Nachr. **8**, 202–206 (1924). Coll. works, (Thales Verlag, Essen)

23. Le, H., Solomon, S.: Truly optimal Euclidean spanners. In: Proceedings of the 60th IEEE Symposium on Foundations of Computer Science (FOCS), pp. 1078–1100. IEEE Computer Society (2019). https://doi.org/10.1109/FOCS.2019.00069

24. Le, H., Solomon, S.: Light euclidean spanners with steiner points. In: Proceedins of the 28th European Symposium on Algorithms (ESA). LIPIcs, vol. 173, pp. 67:1–67:22. Schloss Dagstuhl (2020). https://doi.org/10.4230/LIPIcs.ESA.2020.67

25. Le, H., Solomon, S.: Towards a unified theory of light spanners I: fast (yet optimal) constructions. CoRR abs/2106.15596 (2021). https://arxiv.org/abs/2106.15596

26. Ledoan, A.H.: The discrepancy of farey series. Acta Math. Hungar. **156**(2), 465–480 (2018). https://doi.org/10.1007/s10474-018-0868-x

27. Levcopoulos, C., Narasimhan, G., Smid, M.H.M.: Improved algorithms for constructing fault-tolerant spanners. Algorithmica **32**(1), 144–156 (2002). https://doi.org/10.1007/s00453-001-0075-x

28. Narasimhan, G., Smid, M.H.M.: Geometric Spanner Networks. Cambridge University Press, Cambridge (2007). https://doi.org/10.1017/CBO9780511546884

29. Rao, S., Smith, W.D.: Approximating geometrical graphs via "spanners" and "banyans". In: Proceedings of the 30th Annual ACM Symposium on the Theory of Computing (STOC), pp. 540–550 (1998). https://doi.org/10.1145/276698.276868

30. Roditty, L.: Fully dynamic geometric spanners. Algorithmica **62**(3–4), 1073–1087 (2012). https://doi.org/10.1007/s00453-011-9504-7

31. Ruppert, J., Seidel, R.: Approximating the d-dimensional complete euclidean graph. In: Proceedings of the 3rd Canadian Conference on Computational Geometry (CCCG), pp. 207–210 (1991). https://cccg.ca/proceedings/1991/paper50.pdf

32. Solomon, S., Elkin, M.: Balancing degree, diameter, and weight in euclidean spanners. SIAM J. Discret. Math. **28**(3), 1173–1198 (2014). https://doi.org/10.1137/120901295

33. Steele, J.M., Snyder, T.L.: Worst-case growth rates of some classical problems of combinatorial optimization. SIAM J. Comput. **18**(2), 278–287 (1989). https://doi.org/10.1137/0218019

34. Supowit, K.J., Reingold, E.M., Plaisted, D.A.: The travelling salesman problem and minimum matching in the unit square. SIAM J. Comput. **12**(1), 144–156 (1983). https://doi.org/10.1137/0212009

35. Tóth, C.D.: Minimum weight euclidean $(1 + \varepsilon)$-spanners. CoRR abs/2206.14911 (2022). https://arxiv.org/abs/2206.14911

Author Index

Printed in the United States
by Baker & Taylor Publisher Services